STATISTICAL MECHANICS

STATISTICAL MECHANICS

Peter Riseborough

Temple University, USA

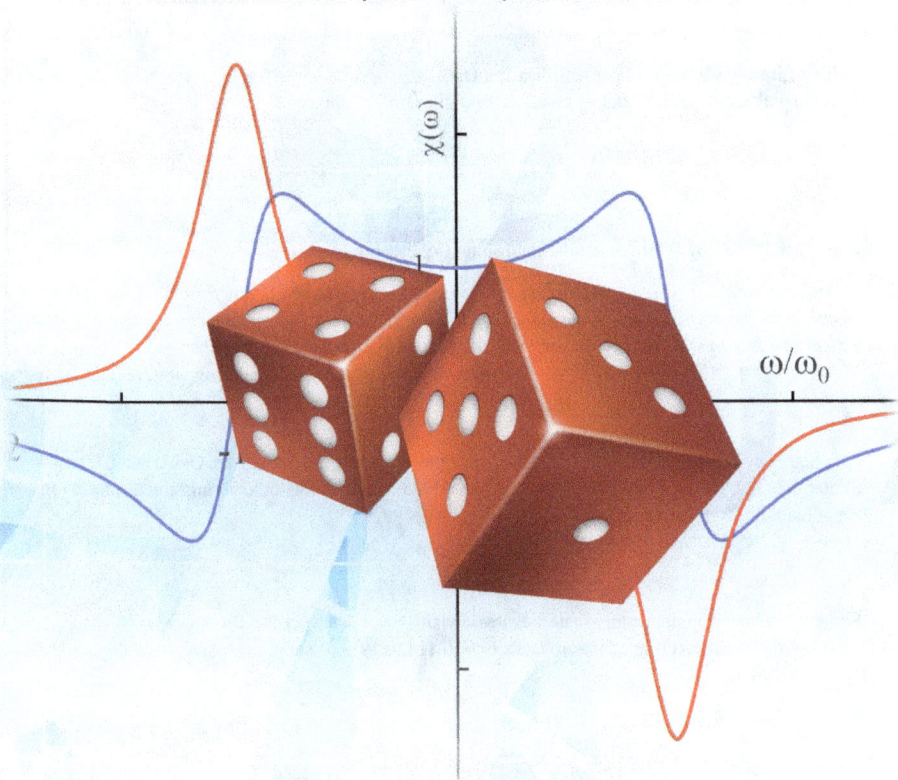

World Scientific

NEW JERSEY · LONDON · SINGAPORE · BEIJING · SHANGHAI · HONG KONG · TAIPEI · CHENNAI · TOKYO

Published by

World Scientific Publishing Co. Pte. Ltd.

5 Toh Tuck Link, Singapore 596224

USA office: 27 Warren Street, Suite 401-402, Hackensack, NJ 07601

UK office: 57 Shelton Street, Covent Garden, London WC2H 9HE

Library of Congress Cataloging-in-Publication Data

Names: Riseborough, Peter, author.

Title: Statistical mechanics / Peter Riseborough, Temple University, USA.

Description: New York : World Scientific, [2021] | Includes bibliographical
references and index.

Identifiers: LCCN 2020026620 | ISBN 9789811223426 (hardcover) |
ISBN 9789811224249 (paperback) | ISBN 9789811223433 (ebook) |
ISBN 9789811223440 (ebook other)

Subjects: LCSH: Statistical mechanics--Textbooks.

Classification: LCC QC174.8 .R57 2021 | DDC 530.13--dc23

LC record available at https://lccn.loc.gov/2020026620

British Library Cataloguing-in-Publication Data

A catalogue record for this book is available from the British Library.

For any available supplementary material, please visit
https://www.worldscientific.com/worldscibooks/10.1142/11913#t=suppl

Dedicated to Vivienne

Preface

Real materials are composed of a huge numbers of particles. For example, one cubic centimeter of copper or one liter of water contains about 10^{23} atoms. The enormity of the number of degrees of freedom prevents one from either determining or storing the initial conditions, let alone from solving the equations of motion. Hence, a detailed microscopic description appears impossible. Nevertheless, the equilibrium states of such materials can be defined by a relative few macroscopic quantities, such as temperature, pressure or volume, etc. These quantities reflect the collective properties of the constituents of the material but still can be measured quite directly by macroscopic means. Likewise, certain non-equilibrium states of the material can also described by a few easily measured quantities such as the voltage drop across or the electrical current flowing through an element of an electrical circuit. Often simple "laws" emerge between the macroscopic quantities that describe the properties of these complex systems. The subject of Thermodynamics is devoted to revealing relations, sometimes expected and sometimes unexpected, between the macroscopic quantities describing materials. Statistical Mechanics provides statistically based methods which bridge the gap between the physics of the individual particles that comprise the materials and the simple thermodynamic "laws" that describe the macroscopic properties of many-particle systems.

This text grew out of lectures given at Brooklyn Polytechnic (NYU Tandem School of Engineering) and Temple University. The text is aimed at providing a graduate level introduction to Statistical Mechanics. In particular, the book is intended to provide students with a working knowledge of the subject and basic skills needed for research in various branches of science. Although Statistical Mechanics has found widespread applications in various disciplines, such as Biology and Commerce, the text focuses on applications found in Physics and Chemistry. Since many applications of Statistical Mechanics provide microscopic bases for Thermodynamics, the first chapter of the book contains a brief summary of the principles of Thermodynamics. The next four chapters describe the concepts and applications of Statistical Mechanics based on Classical Mechanics. The second chapter discusses elements of the foundations of Classical Statistical Mechanics. In particular, it describes thermodynamic measurements in terms of long-time averages of quantities expressed as the microscopic coordinates and momenta, and makes plausible the conceptual jump to Ensemble Averaging. The book does not focus on the mathematical and physical foundations of the subject, since proofs are either incomplete or lack generality. A Quantum Mechanical framework and applications are introduced in the chapters six through eight. Phase Transitions and critical phenomena are treated in chapter nine. The renormalization group technique, which provides a quantitative basis for the description of critical phenomena, is briefly discussed in chapter ten. Non-equilibrium phenomena are discussed in the final chapter. The area of non-equilibrium statistical physics is an active area of research. Unlike the equilibrium state, there is not a unique non-equilibrium state. Therefore, most of what is known about this subject is restricted to systems which are close to equilibrium. The chapter focuses on states in which the deviation from equilibrium can be treated as a perturbation. Since many experimental probes couple weakly to the samples, these types of non-equilibrium phenomena are treated within the framework of linear response theory. The Callen-Welton Theorem, provides an intimate relation between the linear response and time-dependent correlation functions that play central roles in the description of non-equilibrium phenomena. Due to the limitations of size, the book neither describes the quasi-classical Boltzmann equation approach (for which the reader is referred to the excellent text the "Theory of Metals" by Abrikosov), nor does it describe the projection operator approach (for which the reader is referred to the authoritative text "Non-Equilibrium Statistical Mechanics" by Robert Zwanzig).

P.S. Riseborough

Acknowledgments

Statistical Mechanics has numerous applications in many branches of Physics and Chemistry but also finds application in Biology, Commerce and Social Science. These lectures notes are designed to provide a graduate level introduction to the subject of Statistical Mechanics. The notes grew out of courses given at Brooklyn Polytechnic (NYU Tandem School of Engineering) and Temple University. I am indebted to my colleagues and students for their input. I am particularly grateful to Professors K.M. Leung and P. Haenggi who were instrumental in broadening my knowledge of applications in Commerce and Biology. I am also deeply grateful to Professor T.W.B. Kibble who provided me with my first introduction to this subject. It is truly remarkable that, half a century ago, his lectures addressed a number of societal issues that have only received widespread public attention in the past few decades. I have also benefited from the authors of numerous textbooks that have contributed to my understanding of the subject, including H.B. Callen, C. Domb and M. Green, D. Forster, K. Huang, C. Kittel, R. Kubo, Landau and Lifschitz, S-K. Ma, P.C. Martin, R.K. Pathria, D. Ter Haar and many others. Special thanks are due to Professor E. Grinberg for encouraging me to put my lecture notes into printed form.

Contents

Chapter 1

Thermodynamics

Thermodynamics is a branch of science that does not assert new fundamental principles but, instead predicts universal relations between measurable quantities that characterize macroscopic systems. Specifically, thermodynamics involves the study of macroscopic coordinates which can be expressed in terms of an extremely large number of microscopic degrees of freedom, and that describe the macroscopic states of systems.

1.1 The Foundations of Thermodynamics

Macroscopic measurements have the attributes that they involve large numbers of microscopic degrees of freedom (such at the positions and momenta of 10^9 atoms) and, are measured over extremely long time scales compared with the time scales describing the microscopic degrees of freedom (of the order 10^{-7} seconds). In general, for sufficiently large systems and when averaged over sufficiently long time scales, the fluctuations of the macroscopic variables are extremely small and so only the average values need be retained.

Typical macroscopic variables are the internal energy U, the number of particles N and the volume of the system V. The internal energy U is a precisely defined quantity, which in the absence of interactions between the system and its environment, is also a conserved quantity. For systems which contain particles that do not undergo reactions, the number of particles N is also a well-defined and conserved quantity. The volume V of a system can be measured, to within reasonable accuracy. A measurement of the volume usually occurs over time-scales that are long enough so that the long wavelength fluctuations of the atomic positions of the systems boundaries or walls are averaged over.

Thermodynamic measurements often are indirect, and usually involve externally imposed constraints. Systems which are constrained such that they cannot exchange energy, volume or number of particles with their environments are said to be closed. Since one usually only measures changes in the internal energy of a system ΔU, such measurements necessarily involve the system's environment and the assumption that energy is conserved. Such measurements can be performed by doing electrical or mechanical work on the system, and preventing energy in the form of heat from flowing into or out of the system. The absence of heat flow is ensured by utilizing boundaries which are impermeable to heat flow. Boundaries which are impermeable to heat flow are known as adiabatic. Alternatively, one may infer the change in internal energy of a system by putting it in contact with other systems that are monitored and that have been calibrated so that their changes in internal energy can be found. For any process, the increase in the internal energy ΔU can be expressed as the sum of the heat absorbed by the system ΔQ and the work done on the system ΔW.

$$\Delta U = \Delta Q + \Delta W \qquad (1.1)$$

where N is being held constant. This equation is a statement of the conservation of energy. This basic conservation law, in its various forms, forms an important principle of thermodynamics.

1.2 Thermodynamic Equilibrium

Given a macroscopic system, experience shows that this system will evolve to a state in which the macroscopic properties are determined by intrinsic factors and not by any external influences that had been previously exerted on the system. The final states, by definition, are independent of time and are known as equilibrium states.

Postulate I

It is postulated that, in equilibrium, the macroscopic states of a system can be characterized by a set of macroscopic variables. These variables may include variables taken from the set $\{U, V, N\}$ together with any other macroscopic variables that must be added to the set in order to describe the equilibrium state uniquely.

For example, in a ferromagnet this set may be extended by adding the total magnetic moment M of the sample. Thus, for a ferromagnet one might specify the equilibrium state by the macroscopic variables $\{U, N, V, M\}$. Another example is given by the example of a gas containing r different species of atoms, in which case the set of macroscopic variables should be extended to include the number of atoms for each species $\{N_1, N_2, \ldots N_r\}$. Due to the constraint $N = \sum_{i=1}^{r} N_i$, the total number N of atoms should no longer be considered as an independent variable.

The set of variables $\{U, V, N, M, \ldots\}$ are *"Extensive Variables"*, since they scale with the size of the system. This definition can be made more precise as follows: Consider a homogeneous system that is in thermal equilibrium. The value of the variable X for the equilibrated system is denoted by X_0. The variable X is considered as being extensive if, when one considers the system as being composed of λ identical subsystems ($\lambda > 1$), the value of the variable X for each subsystem is equal to $\lambda^{-1} X_0$. This definition assumes that the subsystems are sufficiently large so that the fluctuations δX of X are negligibly small.

The extensive variables $\{U, V, N\}$ that we have introduced, so far, all have mechanical significance. There are extensive variables that only have thermodynamic significance, and these variables can also be used to characterize equilibrium states. One such quantity is the *"Entropy"* S.

Postulate II

The entropy S is only defined for equilibrium states, and takes on a value which is uniquely defined by the state. That is, S is a single-valued function $S(U, V, N)$ of the mechanical extensive variables. The entropy has the property that the entropy of the equilibrium state is maximal with respect to a manifold of constrained equilibrium states formed by varying the constraint.

The constraints must be designed so that, as a constrained is varied, the system varies through a series of constrained equilibrium states. Each of the states in this series should be accessible to the system if the constraint is removed. Thus, if the hypothetical internal constraint characterized by a variable X is imposed on the system, then the entropy of the system depends on the constraint through X and can be denoted by $S(X)$. The maximum value of the entropy of the unconstrained system S is given by the maximum value of $S(X)$ found when X is varied over all possible values.

The function $S(E, V, N)$ for a system is known as the *"Fundamental Relation"*, since all conceivable thermodynamic information on the system can be obtained from it.

Postulate III

The entropy of a system is not only an extensive variable, but also the en-
tropy of a composite system is the sum of the entropies of its components.
The entropy is a continuous, differentiable and a monotonically increasing
function of the internal energy.

Postulate III ensures that when the absolute temperature T is defined
for an equilibrium state, then T will be positive.

Postulate IV

The entropy of a system vanishes in the limit where

$$\left(\frac{\partial S}{\partial U}\right)_{V,N} \to \infty \tag{1.2}$$

The above condition identifies a state for which the absolute temperature
approaches the limiting value $T \to 0$.

Postulate IV is equivalent to Nernst's postulate that the entropy takes
on a universal value when $T \to 0$. The above form of the postulate defines
the universal value of the entropy to be zero.

1.3   The Conditions for Equilibrium

The above postulates allow the conditions for equilibrium to be expressed
in terms of the derivatives of the entropy. Since, entropy is an extensive
quantity, its derivatives with respect to other extensive quantities will be
intensive. That is, the derivatives are independent of the size of the system.
The derivatives will be used to define intensive thermodynamic variables.

First, we shall consider making use of the postulate that entropy is a
single-valued monotonically increasing function of energy. This implies that
the equation for the entropy

$$S = S(U, V, N) \tag{1.3}$$

can be inverted to yield the energy as a function of entropy

$$U = U(S, V, N) \tag{1.4}$$

This inversion may be difficult to do if one is presented with a general
expression for the function, but if the function is presented graphically this

is achieved by simply interchanging the axes. Consider making infinitesimal changes of the independent extensive variables (S, V, N), then the energy $U(S, V, N)$ will change by an amount given by

$$dU = \left(\frac{\partial U}{\partial S}\right)_{V,N} dS + \left(\frac{\partial U}{\partial V}\right)_{S,N} dV + \left(\frac{\partial U}{\partial N}\right)_{S,V} dN \quad (1.5)$$

The three quantities

$$\left(\frac{\partial U}{\partial S}\right)_{V,N}$$

$$\left(\frac{\partial U}{\partial V}\right)_{S,N}$$

$$\left(\frac{\partial U}{\partial N}\right)_{S,V} \quad (1.6)$$

are intensive since, if a system in equilibrium is considered to divided into λ identical subsystems, the values of these parameters for each of the subsystems is the same as for the combined system. These three quantities define the energy intensive parameters. A quantity is "*Intensive*" if its value is independent of the scale of the system. That is, the value of an intensive quantity is independent of the amount of matter used in its measurement. The above intensive quantities can be identified as follows:

By considering a process in which S and N are kept constant and V is allowed to change, one finds

$$dU = \left(\frac{\partial U}{\partial V}\right)_{S,N} dV \quad (1.7)$$

which, when considered in terms of mechanical work ΔW, leads to the identification

$$\left(\frac{\partial U}{\partial V}\right)_{S,N} = -P \quad (1.8)$$

where P is the mechanical pressure. Likewise, when one considers a process in which V and N are kept constant and S is allowed to change, one has

$$dU = \left(\frac{\partial U}{\partial S}\right)_{V,N} dS \quad (1.9)$$

which, when considered in terms of heat flow ΔQ, leads to the identification

$$\left(\frac{\partial U}{\partial S}\right)_{V,N} = T \quad (1.10)$$

where T is the absolute temperature. Finally, on varying N, one finds

$$dU = \left(\frac{\partial U}{\partial N}\right)_{S,V} dN \tag{1.11}$$

which leads to the identification

$$\left(\frac{\partial U}{\partial N}\right)_{S,V} = \mu \tag{1.12}$$

where μ is the chemical potential. Thus, one obtains a relation between the infinitesimal changes of the extensive variables and the intensive parameters

$$dU = T \, dS - P \, dV + \mu \, dN \tag{1.13}$$

This is an expression of the conservation of energy.

Direct consideration of the entropy, leads to the identification of the entropic intensive parameters. Consider making infinitesimal changes of the independent extensive variables (U, V, N), then the entropy $S(U, V, N)$ will change by an amount given by

$$dS = \left(\frac{\partial S}{\partial U}\right)_{V,N} dU + \left(\frac{\partial S}{\partial V}\right)_{U,N} dV + \left(\frac{\partial S}{\partial N}\right)_{U,V} dN \tag{1.14}$$

The values of the coefficients of the infinitesimal quantities are the entropic intensive parameters. By a suitable rearrangement of Eq. (1.13) as

$$dS = \frac{1}{T} \, dU + \frac{P}{T} \, dV - \frac{\mu}{T} \, dN \tag{1.15}$$

one finds that the entropic intensive variables are given by

$$\left(\frac{\partial S}{\partial U}\right)_{V,N} = \frac{1}{T}$$

$$\left(\frac{\partial S}{\partial V}\right)_{U,N} = \frac{P}{T}$$

$$\left(\frac{\partial S}{\partial N}\right)_{U,V} = -\frac{\mu}{T} \tag{1.16}$$

where T is the absolute temperature, P is the pressure and μ is the chemical potential.

The conditions for equilibrium can be obtained from Postulate II and Postulate III, which states that the entropy is maximized in equilibrium and is additive. We shall consider a closed system composed of two systems in contact. System 1 is described by the extensive parameters $\{U_1, V_1, N_1\}$ and system 2 is described by $\{U_2, V_2, N_2\}$. The total energy $U_T = U_1 + U_2$, volume $V_T = V_1 + V_2$ and number of particles $N_T = N_1 + N_2$ are

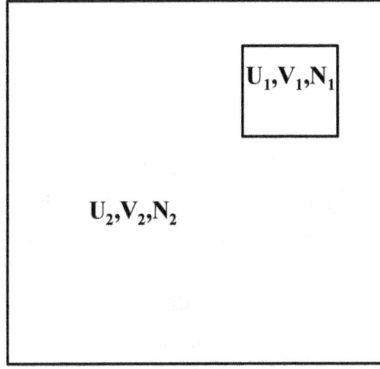

Fig. 1.1   An isolated system composed of two subsystems.

fixed. The total entropy of the combined system S_T is given by

$$S_T = S_1(U_1, V_1, N_1) + S_2(U_2, V_2, N_2) \tag{1.17}$$

and is a function of the variables $\{U_1, V_1, N_1, U_2, V_2, N_2\}$.

1.3.1   *Heat Flow and Temperature*

If one allows energy to be exchanged between the two systems, keeping the total energy $U_T = U_1 + U_2$ constant, then

$$dS_T = \left(\frac{\partial S_1}{\partial U_1}\right)_{V_1, N_1} dU_1 + \left(\frac{\partial S_2}{\partial U_2}\right)_{V_2, N_2} dU_2 \tag{1.18}$$

Since U_T is kept constant, one has $dU_1 = -dU_2$. Therefore, the change in the total entropy is given by

$$dS_T = \left[\left(\frac{\partial S_1}{\partial U_1}\right)_{V_1, N_1} - \left(\frac{\partial S_2}{\partial U_2}\right)_{V_2, N_2}\right] dU_1 \tag{1.19}$$

Furthermore, in equilibrium S_T is maximized with respect to the internal partitioning of the energy, so one has

$$dS_T = 0 \tag{1.20}$$

For this to be true, independent of the value of dU_1, one must satisfy the condition

$$\left(\frac{\partial S_1}{\partial U_1}\right)_{V_1, N_1} = \left(\frac{\partial S_2}{\partial U_2}\right)_{V_2, N_2} \tag{1.21}$$

or, equivalently

$$\frac{1}{T_1} = \frac{1}{T_2} \tag{1.22}$$

Thus, the condition that two systems, which can only exchange internal energy by heat flow, are in thermal equilibrium is simply the condition that the temperatures of the two systems must be equal, $T_1 = T_2$.

Let us consider the same closed system, but one in which the two bodies are initially not in thermal contact with each other. Since the two systems are isolated, they are in a state of equilibrium but may have different temperatures. However, if the two systems are put in thermal contact, the adiabatic constraint is removed and they will no longer be in thermal equilibrium. The system will evolve by exchanging energy, in the form of heat, between the two systems and a new equilibrium state will be established. The new equilibrium state, obtained by removing the internal constraint, will have a larger entropy. Hence, for the two equilibrium states which differ infinitesimally in the partitioning of the energy, $dS_T > 0$ and

$$dS_T = \left[\left(\frac{\partial S_1}{\partial U_1} \right)_{V_1, N_1} - \left(\frac{\partial S_2}{\partial U_2} \right)_{V_2, N_2} \right] dU_1 > 0 \tag{1.23}$$

or

$$\left[\frac{1}{T_1} - \frac{1}{T_2} \right] dU_1 > 0 \tag{1.24}$$

This inequality shows that heat flows from systems with higher temperatures to systems with lower temperatures, in agreement with expectations.

1.3.2 *Work and Pressure*

Consider a system composed of two sub-systems which are in contact, which can exchange energy and also exchange volume. System 1 is described by the extensive parameters $\{U_1, V_1, N_1\}$ and system 2 is described by $\{U_2, V_2, N_2\}$. The total energy is fixed as is the total volume. The energies and volumes of the sub-systems satisfy

$$U_T = U_1 + U_2$$
$$V_T = V_1 + V_2 \tag{1.25}$$

and N_1 and N_2 are kept constant. For an equilibrium state, one can consider constraints that result in different partitionings of the energy and volume. The entropy of the total system is additive

$$S_T = S_1(U_1, V_1, N_1) + S_2(U_2, V_2, N_2) \tag{1.26}$$

The infinitesimal change in the total entropy S_T, found by making infinitesimal changes in U_1 and V_1 is given by

$$dS_T = \left[\left(\frac{\partial S_1}{\partial U_1} \right)_{V_1, N_1} - \left(\frac{\partial S_2}{\partial U_2} \right)_{V_2, N_2} \right] dU_1$$

$$+ \left[\left(\frac{\partial S_1}{\partial V_1} \right)_{U_1, N_1} - \left(\frac{\partial S_2}{\partial V_2} \right)_{U_2, N_2} \right] dV_1 \qquad (1.27)$$

since $dU_1 = -dU_2$ and $dV_1 = -dV_2$. Thus, on using the definitions for the intensive parameters of the sub-systems, one has

$$dS_T = \left[\frac{1}{T_1} - \frac{1}{T_2} \right] dU_1 + \left[\frac{P_1}{T_1} - \frac{P_2}{T_2} \right] dV_1 \qquad (1.28)$$

Since the equilibrium state is that in which S_T is maximized with respect to the variations dU_1 and dV_1, one has $dS_T = 0$ which leads to the conditions

$$\frac{1}{T_1} = \frac{1}{T_2}$$

$$\frac{P_1}{T_1} = \frac{P_2}{T_2} \qquad (1.29)$$

Hence, the pressure and temperature of two the sub-systems are equal in the equilibrium state.

Furthermore, if the systems are initially in their individual equilibrium states but are not in equilibrium with each other, then they will ultimately come into thermodynamic equilibrium with each other by changing their volumes. If the temperatures of the two subsystems are equal but the initial pressures of the two systems are not equal, then the change in entropy that occurs when the volumes are changed by amounts $dV_1 = -dV_2$ is given by

$$dS_T = (P_1 - P_2) \frac{dV_1}{T} > 0 \qquad (1.30)$$

Since $dS_T > 0$, one finds that if $P_1 > P_2$ then $dV_1 > 0$. That is, the system at higher pressure will expand and the system at lower pressure will contract.

1.3.3 *Matter Flow and Chemical Potential*

The above reasoning can be extended to a system with fixed total energy, volume and number of particles, which is decomposed into two sub-systems that exchange energy, volume and number of particles. Since $dU_1 = -dU_2$, $dV_1 = -dV_2$ and $dN_1 = -dN_2$, one finds that an infinitesimal change in the extensive variables yields an infinitesimal change in the total entropy, which is given by

$$dS_T = \left[\left(\frac{\partial S_1}{\partial U_1}\right)_{V_1,N_1} - \left(\frac{\partial S_2}{\partial U_2}\right)_{V_2,N_2}\right] dU_1$$

$$+ \left[\left(\frac{\partial S_1}{\partial V_1}\right)_{U_1,N_1} - \left(\frac{\partial S_2}{\partial V_2}\right)_{U_2,N_2}\right] dV_1$$

$$+ \left[\left(\frac{\partial S_1}{\partial N_1}\right)_{U_1,V_1} - \left(\frac{\partial S_2}{\partial N_2}\right)_{U_2,V_2}\right] dN_1$$

$$= \left[\frac{1}{T_1} - \frac{1}{T_2}\right] dU_1 + \left[\frac{P_1}{T_1} - \frac{P_2}{T_2}\right] dV_1 - \left[\frac{\mu_1}{T_1} - \frac{\mu_2}{T_2}\right] dN_1$$

$$(1.31)$$

Since the total entropy is maximized in equilibrium with respect to the internal constraints, one has $dS_T = 0$ which for equilibrium in the presence of a particle exchange process yields the condition

$$\frac{\mu_1}{T_1} = \frac{\mu_2}{T_2} \tag{1.32}$$

On the other hand, if the systems initially have chemical potentials that differ from each other, then

$$dS_T = (\mu_2 - \mu_1) \frac{dN_1}{T} > 0 \tag{1.33}$$

Hence, if $\mu_2 > \mu_1$ then $dN_1 > 0$. Therefore, particles flow from regions of higher chemical potential to regions of lower chemical potential. Thus, two systems which are allowed to exchange energy, volume and particles have to satisfy the conditions

$$T_1 = T_2$$
$$P_1 = P_2$$
$$\mu_1 = \mu_2 \tag{1.34}$$

if they are in equilibrium.

1.4 The Equations of State

The fundamental relation $S(U, V, N)$ or alternately $U(S, V, N)$, provides a complete thermodynamic description of a system. From the fundamental relation one can derive three equations of state. The expressions for the intensive parameters are equations of state

$$T = T(S, V, N)$$
$$P = P(S, V, N)$$
$$\mu = \mu(S, V, N) \tag{1.35}$$

These particular equations relate the intensive parameters to the independent extensive parameters. If all three equations of state are not known, then one has an incomplete thermodynamic description of the system.

If one knows all three equations of state, one can construct the fundamental relation and, hence, one has a complete thermodynamic description of the system. This can be seen by considering the extensive nature of the fundamental relation and its behavior under a change of scale by s. The fundamental equation is homogeneous and is of first order so

$$U(sS, sV, sN) = s\, U(S, V, N) \tag{1.36}$$

Differentiating the above equation w.r.t s yields

$$U(S, V, N) = \left(\frac{\partial U}{\partial sS}\right)_{sV, sN}\left(\frac{ds S}{ds}\right) + \left(\frac{\partial U}{\partial sV}\right)_{sS, sN}\left(\frac{ds V}{ds}\right)$$
$$+ \left(\frac{\partial U}{\partial sN}\right)_{sS, sV}\left(\frac{ds N}{ds}\right)$$

$$U(S, V, N) = \left(\frac{\partial U}{\partial sS}\right)_{sV, sN} S + \left(\frac{\partial U}{\partial sV}\right)_{sS, sN} V + \left(\frac{\partial U}{\partial sN}\right)_{sS, sV} N$$

which, on setting $s = 1$, results in the Euler Equation

$$\left(\frac{\partial U}{\partial S}\right)_{V, N} S + \left(\frac{\partial U}{\partial V}\right)_{S, N} V + \left(\frac{\partial U}{\partial N}\right)_{S, V} N = U \tag{1.37}$$

When expressed in terms of the intensive parameters, the Euler Equation becomes

$$TS - PV + \mu N = U \tag{1.38}$$

In the entropy representation, one finds the Euler equation in the form

$$\frac{1}{T} U + \frac{P}{T} V - \frac{\mu}{T} N = S \tag{1.39}$$

which has exactly the same content as the Euler equation found from the energy representation. From either of these equations, it follows that knowledge of the three equations of state can be used to find the fundamental relation.

The three intensive parameters cannot be used as a set of independent variables. This can be seen by considering the Euler equation for an infinitesimal variation

$$dU = T\ dS + S\ dT - P\ dV - V\ dP + \mu\ dN + N\ d\mu \qquad (1.40)$$

and comparing it with the form of the first law of thermodynamics

$$dU = T\ dS - P\ dV + \mu\ dN \qquad (1.41)$$

This leads to the discovery that the infinitesimal changes in the intensive parameters are related by the equation

$$0 = S\ dT - V\ dP + N\ d\mu \qquad (1.42)$$

which is known as the Gibbs-Duhem relation. Thus, for a one-component system, there are only two independent intensive parameters, i.e. there are only two thermodynamic degrees of freedom.

1.5   Thermodynamic Processes

Not all processes, that conserve energy, represent real physical processes. Since if the system is initially in a constrained equilibrium state, and an internal constraint is removed, then the final equilibrium state that is established must have a higher entropy.

A quasi-static process is one that proceeds sufficiently slowly that its trajectory in thermodynamic phase space can be approximated by a dense set of equilibrium states. Thus, at each macroscopic equilibrium state one can define an entropy $S_j = S(U_j, V_j, N_j, X_j)$. The quasi-static process is a temporal succession of equilibrium states, connected by non-equilibrium states. Since, for any specific substance, an equilibrium state can be characterized by $\{U, V, N, X\}$, a state can be represented by a point on a hyper-surface $S = S(U, V, N, X)$ in thermodynamic configuration space. The cuts of the hyper-surface at constant U are concave. The quasi-static processes trace out an almost continuous line on the hyper-surface. Since individual quasi-static processes are defined by sequence of equilibrium states connected by non-equilibrium states, the entropy cannot decrease along any part of the sequence if it is to represent a possible process, therefore, $S_{j+1} \geq S_j$. Thus, an allowed quasi-static process must follow a path on the hyper-surface which never has a segment on which S decreases. A reversible process is an allowed quasi-static process in which the entropy difference from beginning to end becomes infinitesimally small. Hence, a reversible process must proceed along a contour of the hyper-surface which has constant entropy. Therefore, reversible process occur on a constant entropy cut of the hyper-surface. The constant entropy cuts of the hyper-surface are convex.

Adiabatic Expansion of Electromagnetic Radiation in a Cavity

Consider a spherical cavity of radius R which contains electromagnetic radiation. The radius of the sphere expands slowly at a rate given by $\frac{dR}{dt}$. The spectral component of wavelength λ contained in the cavity will be changed by the expansion. The change occurs through a change of wavelength $d\lambda$ that occurs at reflection with the moving boundary. Since the velocity $\frac{dR}{dt}$ is much smaller than the speed of light c, one only needs to work to keep terms first-order in the velocity. A single reflection through

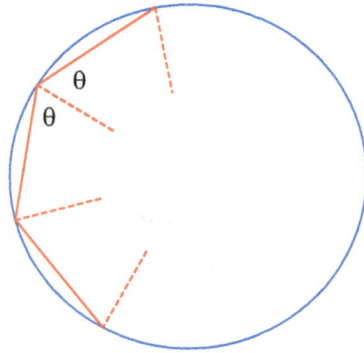

Fig. 1.2 Electromagnetic radiation being reflected through an angle θ from the walls of a slowly expanding spherical electromagnetic cavity.

an angle θ produces a Doppler shift of the radiation by an amount given by

$$d\lambda = 2\,\frac{\lambda}{c}\,\frac{dR}{dt}\,\cos\theta \tag{1.43}$$

The ray travels a distance $2R\cos\theta$ between successive reflections. Hence, the time between successive reflections is given by

$$\frac{c}{2R\cos\theta} \tag{1.44}$$

Thus, the rate at which the wavelength changes is given by

$$\frac{d\lambda}{dt} = \frac{\lambda}{R}\frac{dR}{dt} \tag{1.45}$$

On integrating the above equation, one finds that

$$\frac{\lambda}{R} = \text{Constant} \tag{1.46}$$

Therefore, the wavelength scales with the radius. Quantum mechanically, each state evolves adiabatically so no transitions occur. The wavelength scales with the radius such as to match the boundary condition.

The equation of state for the electromagnetic radiation is

$$P = \frac{1}{3}\frac{U}{V} \tag{1.47}$$

so for adiabatic expansion (with $\mu = 0$ and $dN = 0$), one has

$$dU = -P \, dV \tag{1.48}$$

which leads to

$$\frac{dU}{U} = -\frac{1}{3}\frac{dV}{V} \tag{1.49}$$

Hence, for adiabatic expansion, one has

$$UV^{\frac{1}{3}} = \text{Constant} \tag{1.50}$$

Putting this together with Stefan's law

$$U = \sigma V T^4 \tag{1.51}$$

one finds that, since the entropy is constant,

$$\sigma T^4 V^{\frac{4}{3}} = \text{Constant} \tag{1.52}$$

or $RT = \text{const.}$ Thus, the temperature of the cavity decreases inversely with the radius of the cavity as R increases. Furthermore, since λ scales with R, one finds that the density of each spectral component must scale so that λT is constant.

1.6 Thermodynamic Potentials

The Postulate II is the entropy maximum principle which can be restated as the entropy of a system is maximized in equilibrium (at fixed total energy) with respect to variations of an internal parameter. This is a consequence of the concave geometry of the constant energy cuts of the hyper-surface in thermodynamic configuration space. We have formulated thermodynamics in terms of entropy and the extensive parameters. Thermodynamics has an equivalent formulation in terms of the energy and the extensive parameters. In this alternate formulation, the entropy maximum principle is replaced by the energy minimum principle. The energy minimum principle states that the energy is minimized in an equilibrium state with respect to variations of an internal parameter (for fixed values of the entropy). This is a consequence of the convex nature of the constant entropy cuts of the hyper-surface in thermodynamic configuration space. The statements of the entropy maximum and the energy minimum principles are equivalent as can be seen from the following mathematical proof.

1.6.1 *Equivalence of the Entropy Maximum and Energy Minimum Principles*

The entropy maximum principle can be stated as

$$\left(\frac{\partial S}{\partial X}\right)_U = 0$$

$$\left(\frac{\partial^2 S}{\partial X^2}\right)_U < 0 \tag{1.53}$$

where X is an internal parameter. From the chain rule, it immediately follows that the energy is an extremum in equilibrium since

$$\left(\frac{\partial U}{\partial X}\right)_S = -\left(\frac{\partial U}{\partial S}\right)_X \left(\frac{\partial S}{\partial X}\right)_U$$

$$= -\frac{\left(\frac{\partial S}{\partial X}\right)_U}{\left(\frac{\partial S}{\partial U}\right)_X}$$

$$= -T\left(\frac{\partial S}{\partial X}\right)_U$$

$$= 0 \tag{1.54}$$

The last line vanishes since the entropy is an extremum in equilibrium. Hence, it follows from the entropy maximum principle that, in an equilibrium state the energy satisfies an extremum principle.

The energy extremum principle is an energy minimum principle, this follows by re-writing the second derivative of S as

$$\left(\frac{\partial^2 S}{\partial X^2}\right)_U = \left[\frac{\partial}{\partial X}\left(\frac{\partial S}{\partial X}\right)_U\right]_U \tag{1.55}$$

and designating the internal derivative by A, i.e. let

$$A = \left(\frac{\partial S}{\partial X}\right)_U \tag{1.56}$$

The entropy maximum principle requires that, in equilibrium at constant U,

$$\left(\frac{\partial^2 S}{\partial X^2}\right)_U = \left(\frac{\partial A}{\partial X}\right)_U < 0 \tag{1.57}$$

We need to change the dependent variables from (U, X) to (S, X) if we are to obtain an energy minimum principle. In this case, one can represent U as a function of X and S, i.e. $U = U(X, S)$. Thus, $A(X, S) = A(X, U(X, S))$. Then, the derivative of A w.r.t. X is

$$\left(\frac{\partial A}{\partial X}\right)_S = \left(\frac{\partial A}{\partial X}\right)_U + \left(\frac{\partial A}{\partial U}\right)_X\left(\frac{\partial U}{\partial X}\right)_S \tag{1.58}$$

where the last term vanishes because the energy satisfies an extremum principle. Hence,

$$\left(\frac{\partial A}{\partial X}\right)_S = \left(\frac{\partial A}{\partial X}\right)_U \tag{1.59}$$

Thus, the entropy maximum principle reduces to

$$\left(\frac{\partial^2 S}{\partial X^2}\right)_U = \left(\frac{\partial}{\partial X}\left(\frac{\partial S}{\partial X}\right)_U\right)_S < 0 \tag{1.60}$$

Using the chain rule, the innermost partial derivative can be re-written as

$$\left(\frac{\partial S}{\partial X}\right)_U = -\left(\frac{\partial S}{\partial U}\right)_X\left(\frac{\partial U}{\partial X}\right)_S \tag{1.61}$$

Hence, on substituting this into the entropy maximum principle, one has

$$\left(\frac{\partial^2 S}{\partial X^2}\right)_U = -\left(\frac{\partial}{\partial X}\left[\left(\frac{\partial U}{\partial X}\right)_S\left(\frac{\partial S}{\partial U}\right)_X\right]\right)_S$$

$$= -\left(\frac{\partial^2 U}{\partial X^2}\right)_S\left(\frac{\partial S}{\partial U}\right)_X - \left(\frac{\partial U}{\partial X}\right)_S\left(\frac{\partial}{\partial X}\left(\frac{\partial S}{\partial U}\right)_X\right)_S \tag{1.62}$$

The last term vanishes since, as we have shown, the energy satisfies an extremum principle. Therefore, one has

$$\left(\frac{\partial^2 S}{\partial X^2}\right)_U = -\left(\frac{\partial^2 U}{\partial X^2}\right)_S \left(\frac{\partial S}{\partial U}\right)_X$$

$$= -\frac{1}{T}\left(\frac{\partial^2 U}{\partial X^2}\right)_S$$

$$< 0 \tag{1.63}$$

Thus, since $T > 0$, we have

$$\left(\frac{\partial^2 U}{\partial X^2}\right)_S > 0 \tag{1.64}$$

so we have proved that the energy satisfies a minimum principle, if the entropy satisfies the maximum principle. This proof also shows that the energy minimum principle implies the entropy maximum principle, so the two principles are equivalent.

Sometimes it is more convenient to work with the intensive parameters rather than the extensive parameters. The intensive parameters are defined in terms of the partial derivatives of the fundamental relation $S(U, V, N)$ or equivalently $U(S, V, N)$. Taking partial derivatives usually leads to a loss of information, in the sense that a function can only be re-created from its derivative by integration up to a constant (or more precisely a function) of integration. Therefore, to avoid loss of information, one changes extensive variables to intensive variables by performing Legendre transformations.

1.6.2 *Legendre Transformations*

The Legendre transformation relies on the property of concavity of $S(U, V, N)$ and is introduced so that one can work with a set of more convenient variables, such as T instead of S or P instead of V. This amounts to transforming from an extensive parameter to its conjugate intensive parameter which is introduced as a derivative.

The Legendre transformation is introduced such that the change of variables is easily invertible. Instead of considering the convex function $y = y(x)$[1] being given by the ordered pair (x, y) for each x, one can equally

---

[1]The convexity and concavity of a function implies that the second derivative of the function has a specific sign. All that we shall require is that the second derivative of the function does not go to zero in the interval of x that is under consideration.

describe the curve by an envelope of a family of tangents to the curve. The tangent is a straight line

$$y = px + \psi(p) \tag{1.65}$$

with slope p and has a y-axis intercept denoted by $\psi(p)$. Due to the property of convexity, for each value of p there is a unique tangent to the curve. Hence, we have replaced the sets of pairs (x, y) with a set of pairs (p, ψ). The set of pairs (p, ψ) describes the same curve and has the same information as the set of pairs (x, y).

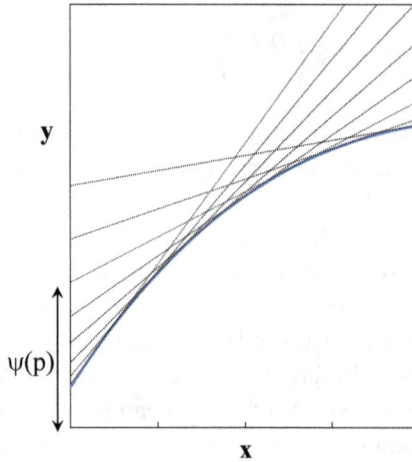

Fig. 1.3   A concave function $y(x)$ is specified by the envelope of a family of tangents with slopes p and y-axis intercepts $\psi(p)$.

Given a curve in the form of $y = y(x)$, one can find $\psi(p)$ by taking the derivative to yield

$$p = \frac{dy}{dx} \tag{1.66}$$

which specifies the slope p of the tangent line at the tangent point x. The above equation can be inverted to yield $x = x(p)$ and, hence, one can obtain $y(p)$ from $y = y(x(p))$. Then, the y-axis intercept of the tangent can be found as a function of p from

$$\psi(p) = y(p) - px(p) \tag{1.67}$$

The function $\psi(p)$ is the Legendre transform of $y(x)$. The quantity $\psi(p)$ contains exactly the same information as $y(x)$ but depends on the variable p instead of the x variable.

The inverse transform can be found by constructing (x, y) from (p, ψ). First the point x at which a tangent with slope p touches the curve is found. Second, after inverting $x(p)$ to yield $p(x)$, one finds $y(x)$ from

$$y = p(x)\, x + \psi(p(x)) \tag{1.68}$$

This formally defines the inverse transformation.

The first step of determining the point of tangency $x(p)$ involves the consideration of a tangent with slope p and a neighboring tangent with slope $p + dp$. The tangent is described by

$$y = px + \psi(p) \tag{1.69}$$

which is valid everywhere on the tangent including the point of tangency which we denote by (x, y). The neighboring tangent which has an infinitesimally different slope $p + dp$ is described by a similar equation, but has a point of tangency $(x + dx, y + dy)$ that differs infinitesimally from (x, y). To first-order in the infinitesimals, one finds that the differences of the coordinates (dx, dy) which describe the separation of the two points of tangency are related by

$$dy = p\, dx + \left(x + \frac{d\psi}{dp} \right) dp \tag{1.70}$$

However, since the two neighboring points of tangency lie on the same curve and because the slope of the tangent is p, one has

$$dy = p\, dx \tag{1.71}$$

so the first terms on the left and right cancel. Thus, we find that the x-coordinate of the point of tangency is determined by the equation

$$x(p) = - \frac{d\psi}{dp} \tag{1.72}$$

The abscissa is given by

$$y(x) = \psi + xp \tag{1.73}$$

in which p has been expressed in terms of x. This is the inverse Legendre transformation.

The inverse Legendre transformation should be compared to the Legendre transformation

$$\psi(p) = y - xp \tag{1.74}$$

in which x has been expressed in terms of p via inversion of

$$p = \frac{dy}{dx} \tag{1.75}$$

Thus, the relation between (x, y) and (p, ψ) is, apart from a minus sign, symmetrical between the Legendre and inverse Legendre transformations.

The Helmholtz Free-Energy F

The Helmholtz Free-Energy is denoted by F is a function of the variables (T, V, N) and is obtained by performing a Legendre transform on the energy $U(S, V, N)$. The process involves defining the temperature T via the derivative

$$T = \left(\frac{\partial U}{\partial S} \right)_{V,N} \tag{1.76}$$

and then defining a quantity F via

$$F = U - TS \tag{1.77}$$

The definition of T is used to express S as a function of T. Then eliminating S from the two terms in the above expression for F, yields the Helmholtz Free-Energy $F(T, V, N)$.

One can show that F does not depend on S by considering an infinitesimal transformation

$$dF = dU - S\, dT - T\, dS \tag{1.78}$$

and then by substituting the expression

$$dU = T\, dS - P\, dV + \mu\, dN \tag{1.79}$$

obtained from $U(S, V, N)$ and by using the definition of the energetic extensive parameters. Substitution of the expression for dU into dF yields

$$dF = -S\, dT - P\, dV + \mu\, dN \tag{1.80}$$

which shows that F only varies as T, V and N are varied. It does not vary as dS is varied. Thus F is a function of the variables (T, V, N). Furthermore, we see that S can be found from F as a derivative

$$S = -\left(\frac{\partial F}{\partial T} \right)_{V,N} \tag{1.81}$$

The Helmholtz Free-Energy has the interpretation that it represents the work done on the system in a process carried out at constant T (and N). This can be seen from the above infinitesimal form of dF since, under the condition that $dT = 0$, one has

$$dF = -P\, dV \tag{1.82}$$

The inverse transform is given found by starting from $F(T, V, N)$ and expressing S as

$$S = -\left(\frac{\partial F}{\partial T}\right)_{V,N} \tag{1.83}$$

This equation is used to express T as a function of S, i.e. $T = T(S)$. The quantity U is formed via

$$U = F + TS \tag{1.84}$$

Elimination of T in favour of S in both terms leads to the energy $U(S, V, N)$.

The Enthalpy $\mathcal{H}$

The enthalpy is denoted by $\mathcal{H}$ and is a function of the variables (S, P, N). It is obtained by a Legendre transform on $U(S, V, N)$ which eliminates the extensive variable V and introduces the intensive variable P. The pressure P is defined by the equation

$$-P = \left(\frac{\partial U}{\partial V}\right)_{S,N} \tag{1.85}$$

and then one forms the quantity $\mathcal{H}$ via

$$\mathcal{H} = U + PV \tag{1.86}$$

which on inverting the equation expressing pressure as a function of V and eliminating V from $\mathcal{H}$, yields the enthalpy $\mathcal{H}(S, P, N)$.

The enthalpy is a function of (S, P, N) as can be seen directly from the infinitesimal variation of $\mathcal{H}$. Since

$$d\mathcal{H} = dU + V\,dP + P\,dV \tag{1.87}$$

and as

$$dU = T\,dS - P\,dV + \mu\,dN \tag{1.88}$$

one finds that

$$d\mathcal{H} = T\,dS + V\,dP + \mu\,dN \tag{1.89}$$

which shows that $\mathcal{H}$ only varies when S, P and N are varied. The above infinitesimal relation also shows that

$$V = \left(\frac{\partial \mathcal{H}}{\partial P}\right)_{S,N} \tag{1.90}$$

The enthalpy has the interpretation that it represents the heat flowing into a system in a process at constant pressure (and constant N). This can be seen from the expression for the infinitesimal change in $\mathcal{H}$ when $dP = 0$

$$d\mathcal{H} = T\, dS \tag{1.91}$$

which is recognized as an expression for the heat flow into the system.

The inverse Legendre transform of $\mathcal{H}(S, P, N)$ is $U(S, V, N)$ and is performed by using the relation

$$V = \left(\frac{\partial \mathcal{H}}{\partial P}\right)_{S,N} \tag{1.92}$$

to express P as a function of V. On forming U via

$$U = \mathcal{H} - PV \tag{1.93}$$

and eliminating P from U, one has the energy $U(S, V, N)$.

The Gibbs Free-Energy G

The Gibbs Free-Energy $G(T, P, N)$ is formed by making two Legendre transformation on $U(S, V, N)$ eliminating the extensive variables S and V and introducing their conjugate intensive parameters T and P. The process starts with $U(S, V, N)$ and defines the two intensive parameters T and $-P$ via

$$T = \left(\frac{\partial U}{\partial S}\right)_{V,N}$$
$$-P = \left(\frac{\partial U}{\partial V}\right)_{S,N} \tag{1.94}$$

The quantity G is formed via

$$G = U - TS + PV \tag{1.95}$$

which on eliminating S and V leads to the Gibbs Free-Energy $G(T, P, N)$.

On performing infinitesimal variations of S, T, V, P and N, one finds the infinitesimal change in G is given by

$$dG = dU - T\, dS - S\, dT + P\, dV + V\, dP \tag{1.96}$$

which on eliminating dU by using the equation

$$dU = T\, dS - P\, dV + \mu\, dN \tag{1.97}$$

leads to

$$dG = -S\, dT + V\, dP + \mu\, dN \tag{1.98}$$

This confirms that the Gibbs Free-Energy is a function of T, P and N, $G(T, P, N)$. It also shows that

$$-S = \left(\frac{\partial G}{\partial T}\right)_{P,N}$$

$$V = \left(\frac{\partial G}{\partial P}\right)_{T,N} \tag{1.99}$$

The inverse (double) Legendre transform of $G(T, P, N)$ yields $U(S, V, N)$ and is performed by expressing the extensive parameters as

$$-S = \left(\frac{\partial G}{\partial T}\right)_{P,N}$$

$$V = \left(\frac{\partial G}{\partial P}\right)_{T,N} \tag{1.100}$$

and using these to express T in terms of S and P in terms of V. The energy is formed via

$$U = G + TS - PV \tag{1.101}$$

and eliminating T and P in favour of S and V, to obtain $S(U, V, N)$.

The Grand-Canonical Potential Ω

The Grand-Canonical Potential $\Omega(T, V, \mu)$ is a function of T, V and μ. It is obtained by making a double Legendre transform on $U(S, V, N)$ which eliminates S and N and replaces them by the intensive parameters T and μ. This thermodynamic potential is frequently used in Statistical Mechanics when working with the Grand-Canonical Ensemble, in which the energy and number of particles are allowed to vary as they are exchanged with a thermal and particle reservoir which has a fixed T and a fixed μ.

The double Legendre transformation involves the two intensive parameters defined by

$$T = \left(\frac{\partial U}{\partial S}\right)_{V,N}$$

$$\mu = \left(\frac{\partial U}{\partial N}\right)_{S,V} \tag{1.102}$$

The quantity Ω is formed as

$$\Omega = U - TS - \mu N \tag{1.103}$$

Elimination of the extensive variables S and N leads to $\Omega(T, V, \mu)$, the Grand-Canonical Potential.

The infinitesimal change in Ω is given by

$$d\Omega = dU - T\ dS - S\ dT - \mu\ dN - N\ d\mu \qquad (1.104)$$

which, on substituting for dU, leads to

$$d\Omega = -S\ dT - P\ dV - N\ d\mu \qquad (1.105)$$

The above equation confirms that Ω only depends on the variables T, V and μ. Furthermore, this relation also shows that

$$-S = \left(\frac{\partial \Omega}{\partial T}\right)_{V,\mu}$$

$$-N = \left(\frac{\partial \Omega}{\partial \mu}\right)_{T,V} \qquad (1.106)$$

The inverse (double) transformation uses the two relations

$$-S = \left(\frac{\partial \Omega}{\partial T}\right)_{V,\mu}$$

$$-N = \left(\frac{\partial \Omega}{\partial \mu}\right)_{T,V} \qquad (1.107)$$

to express T and μ in terms of S and N. The quantity U is formed via

$$U = \Omega + TS + \mu N \qquad (1.108)$$

which on eliminating T and μ leads to the energy as a function of S, V and N, i.e. $U(S, V, N)$.

1.6.3   *Illustrative Processes*

As examples of the use of thermodynamic potentials, we shall consider the processes of Joule Free Expansion and the Joule-Thomson Throttling Process.

Joule Free Expansion

Consider a closed system which is composed of two chambers connected by a valve. Initially, one chamber is filled with gas and the second chamber is evacuated. The valve connecting the two chambers is opened so that the gas can expand into the vacuum.

The expansion process occurs at constant energy, since no heat flows into the system and no work is done in expanding into a vacuum. Hence, the process occurs at constant U.

Due to the expansion the volume of the gas changes by an amount ΔV and, therefore, one might expect that the temperature of the gas may change by an amount ΔT. For a sufficiently small change in volume ΔV, one expects that ΔT and ΔV are related by

$$\Delta T = \left(\frac{\partial T}{\partial V}\right)_{U,N} \Delta V \tag{1.109}$$

where U is being kept constant.

On applying the chain rule, one finds

$$\Delta T = -\left[\left(\frac{\partial U}{\partial V}\right)_{T,N} \Big/ \left(\frac{\partial U}{\partial T}\right)_{V,N}\right] \Delta V \tag{1.110}$$

However, from the expression for the infinitesimal change in U

$$dU = T\,dS - P\,dV \tag{1.111}$$

one finds that the numerator can be expressed as

$$\left(\frac{\partial U}{\partial V}\right)_{T,N} = T\left(\frac{\partial S}{\partial V}\right)_{T,N} - P \tag{1.112}$$

whereas the denominator is identified as

$$\left(\frac{\partial U}{\partial T}\right)_{V,N} = C_V \tag{1.113}$$

which is the specific heat at constant volume.

The quantity proportional to

$$\left(\frac{\partial S}{\partial V}\right)_{T,N} \tag{1.114}$$

is not expressed in terms of directly measurable quantities. It can be expressed as a derivative of pressure by using a Maxwell relation. We note that the quantity should be considered as a function of V, which is being varied and is also a function of T which is being held constant. Processes which are described in terms of the variables V and T can be described by the Helmholtz Free-Energy $F(T, V, N)$, for which

$$dF = -S\,dT - P\,dV + \mu\,dN \tag{1.115}$$

The Helmholtz Free-Energy is an analytic function of T and V, therefore, it satisfies the Cauchy-Riemann condition

$$\left(\frac{\partial^2 F}{\partial V \partial T}\right)_N = \left(\frac{\partial^2 F}{\partial T \partial V}\right)_N \tag{1.116}$$

which on using the infinitesimal form of dF to identify the inner partial differentials of F, yields the Maxwell relation

$$\left(\frac{\partial S}{\partial V}\right)_{T,N} = \left(\frac{\partial P}{\partial T}\right)_{V,N} \qquad (1.117)$$

Hence, the temperature change and volume change that occur in Joule Free Expansion are related via

$$\Delta T = -\left[\frac{T\left(\frac{\partial P}{\partial T}\right)_{V,N} - P}{C_V}\right] \Delta V \qquad (1.118)$$

which can be evaluated with the knowledge of the equation of state.

Since the expansion occurs at constant energy, one finds from

$$dU = T\, dS - P\, dV = 0 \qquad (1.119)$$

that

$$\left(\frac{\partial S}{\partial V}\right)_{U,N} = \frac{P}{T} \qquad (1.120)$$

Thus, the entropy increases on expansion, as it should for an irreversible process.

Joule-Thomson Throttling Process

The Joule-Thomson throttling process involves the constant flow of fluid through a porus plug. The flowing fluid is adiabatically insulated so that heat cannot flow into or out of the fluid. The temperature and pressure of the fluid on either side of the porus plug are uniform but are not equal

$$T_1 \neq T_2$$
$$P_1 \neq P_2 \qquad (1.121)$$

Thus, a pressure drop ΔP defined by

$$\Delta P = P_1 - P_2 \qquad (1.122)$$

and temperature drop ΔT defined by

$$\Delta T = T_1 - T_2 \qquad (1.123)$$

occur across the porus plug.

The Joule-Thomson process is a process for which the enthalpy $\mathcal{H}$ is constant. This can be seen by considering a fixed mass of fluid as it flows

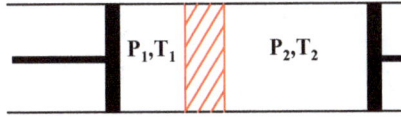

Fig. 1.4 A fluid is confined in a cylindrical tube and two pistons (solid black objects). The pistons force the fluid through the porus plug (orange hatched region). In this process the pressure and temperature on each side of the plug are kept constant but not equal.

through the plug. The pump that generates the pressure difference can, hypothetically, be replaced by two pistons. Consider the volume of fluid contained in the volume V_1 between the piston and the plug, as having internal energy U_1. When this volume of gas has been pushed through the plug, the piston has performed an amount of work $P_1 V_1$. The piston on the other side of the porus plug performs a negative amount of work equal to $-P_2 V_2$ when the gas occupies the volume V_2 between the piston and the plug. The change in internal energy is given by

$$U_2 - U_1 = P_1 V_1 - P_2 V_2 \tag{1.124}$$

This implies that

$$U_1 + P_1 V_1 = U_2 + P_2 V_2 \tag{1.125}$$

or the enthalpy $\mathcal{H}$ of the fluid is constant in the throttling process.

For sufficiently small changes in pressure, the temperature drop is related to the pressure drop by

$$\Delta T = \left(\frac{\partial T}{\partial P} \right)_{\mathcal{H}} \Delta P \tag{1.126}$$

where the enthalpy $\mathcal{H}$ is being kept constant.

On applying the chain rule, one finds

$$\Delta T = -\left[\left(\frac{\partial \mathcal{H}}{\partial P} \right)_{T,N} \bigg/ \left(\frac{\partial \mathcal{H}}{\partial T} \right)_{P,N} \right] \Delta P \tag{1.127}$$

However, from the expression for the infinitesimal change in $\mathcal{H}$

$$d\mathcal{H} = T \, dS + V \, dP \tag{1.128}$$

one finds that the numerator can be expressed as

$$\left(\frac{\partial \mathcal{H}}{\partial P} \right)_{T,N} = T \left(\frac{\partial S}{\partial P} \right)_{T,N} + V \tag{1.129}$$

whereas the denominator is identified as

$$\left(\frac{\partial \mathcal{H}}{\partial T}\right)_{P,N} = C_P \qquad (1.130)$$

which is the specific heat at constant pressure.

The quantity proportional to

$$\left(\frac{\partial S}{\partial P}\right)_{T,N} \qquad (1.131)$$

is not expressed in terms of directly measurable quantities. It can be expressed as a derivative of volume by using a Maxwell relation. We note that the quantity should be considered as a function of P, which is being varied and is also a function of T which is being held constant. Processes which are described in terms of the variables P and T can be described by the Gibbs Free-Energy $G(T, P, N)$, for which

$$dG = -S \ dT + V \ dP + \mu \ dN \qquad (1.132)$$

The Gibbs Free-Energy is an analytic function of T and P, therefore, it satisfies the Cauchy-Riemann condition

$$\left(\frac{\partial^2 G}{\partial P \partial T}\right)_N = \left(\frac{\partial^2 G}{\partial T \partial P}\right)_N \qquad (1.133)$$

which on using the infinitesimal form of dG to identify the inner partial differentials of G, yields the Maxwell relation

$$-\left(\frac{\partial S}{\partial P}\right)_{T,N} = \left(\frac{\partial V}{\partial T}\right)_{P,N} \qquad (1.134)$$

Hence, the pressure change and volume change that occur in the Joule-Thomson process are related via

$$\Delta T = \left[\frac{T\left(\frac{\partial V}{\partial T}\right)_{P,N} - V}{C_P}\right] \Delta P \qquad (1.135)$$

which can be evaluated with the knowledge of the equation of state.

Since the expansion occurs at constant enthalpy, one finds from

$$d\mathcal{H} = T \ dS + V \ dP = 0 \qquad (1.136)$$

that

$$\left(\frac{\partial S}{\partial P}\right)_{\mathcal{H},N} = -\frac{V}{T} \qquad (1.137)$$

Thus, the entropy increases for the irreversible Joule-Thomson process only if the pressure drops across the porus plug.

The description of the above processes used two of the Maxwell's relations. We shall give a fuller description of these relations below.

1.7 Maxwell Relations

Maxwell Relations are statements about the analyticity of the thermodynamic potentials. The Maxwell relations are expressed in the form of an equality between the mixed second derivatives when taken in opposite order. If $B(x, y)$ is a thermodynamic potential which depends on the independent variables x and y, then analyticity implies that

$$\left(\frac{\partial^2 B}{\partial x \partial y} \right) = \left(\frac{\partial^2 B}{\partial y \partial x} \right) \tag{1.138}$$

The Maxwell relations for the four thermodynamic potentials which we have considered are described below:

The Internal Energy $U(S, V, N)$

Since the infinitesimal change in the internal energy is written as

$$dU = T \, dS - P \, dV + \mu \, dN \tag{1.139}$$

one has the three Maxwell relations

$$\left(\frac{\partial T}{\partial V} \right)_{S,N} = -\left(\frac{\partial P}{\partial S} \right)_{V,N}$$

$$\left(\frac{\partial T}{\partial N} \right)_{S,V} = \left(\frac{\partial \mu}{\partial S} \right)_{V,N}$$

$$-\left(\frac{\partial P}{\partial N} \right)_{S,V} = \left(\frac{\partial \mu}{\partial V} \right)_{S,N} \tag{1.140}$$

The Helmholtz Free-Energy $F(T, V, N)$

Since the infinitesimal change in the Helmholtz Free-Energy is written as

$$dF = -S \, dT - P \, dV + \mu \, dN \tag{1.141}$$

one finds the relations

$$-\left(\frac{\partial S}{\partial V}\right)_{T,N} = -\left(\frac{\partial P}{\partial T}\right)_{V,N}$$

$$-\left(\frac{\partial S}{\partial N}\right)_{T,V} = \left(\frac{\partial \mu}{\partial T}\right)_{V,N}$$

$$-\left(\frac{\partial P}{\partial N}\right)_{T,V} = \left(\frac{\partial \mu}{\partial V}\right)_{T,N} \tag{1.142}$$

The Enthalpy $\mathcal{H}(S, P, N)$

Since the infinitesimal change in the enthalpy is written as

$$d\mathcal{H} = T \ dS + V \ dP + \mu \ dN \tag{1.143}$$

one has

$$\left(\frac{\partial T}{\partial P}\right)_{S,N} = \left(\frac{\partial V}{\partial S}\right)_{P,N}$$

$$\left(\frac{\partial T}{\partial N}\right)_{S,P} = \left(\frac{\partial \mu}{\partial S}\right)_{P,N}$$

$$\left(\frac{\partial V}{\partial N}\right)_{S,P} = \left(\frac{\partial \mu}{\partial P}\right)_{S,N} \tag{1.144}$$

The Gibbs Free-Energy $G(T, P, N)$

Since the infinitesimal change in the Gibbs Free-Energy is written as

$$dG = -S \ dT + V \ dP + \mu \ dN \tag{1.145}$$

one has

$$-\left(\frac{\partial S}{\partial P}\right)_{T,N} = \left(\frac{\partial V}{\partial T}\right)_{P,N}$$

$$-\left(\frac{\partial S}{\partial N}\right)_{T,P} = \left(\frac{\partial \mu}{\partial T}\right)_{P,N}$$

$$\left(\frac{\partial V}{\partial N}\right)_{T,P} = \left(\frac{\partial \mu}{\partial P}\right)_{T,N} \tag{1.146}$$

The Grand-Canonical Potential $\Omega(T, V, \mu)$

Since the infinitesimal change in the Grand-Canonical Potential is written as

$$d\Omega = -S\,dT - P\,dV - N\,d\mu \qquad (1.147)$$

one finds the three Maxwell relations

$$-\left(\frac{\partial S}{\partial V}\right)_{T,\mu} = -\left(\frac{\partial P}{\partial T}\right)_{V,\mu}$$

$$-\left(\frac{\partial S}{\partial \mu}\right)_{T,V} = -\left(\frac{\partial N}{\partial T}\right)_{V,\mu}$$

$$-\left(\frac{\partial P}{\partial \mu}\right)_{T,V} = -\left(\frac{\partial N}{\partial V}\right)_{T,\mu} \qquad (1.148)$$

1.8 The Nernst Postulate

The Nernst postulate states that as $T \to 0$, then $S \to 0$. This postulate may not be universally valid. It can be motivated by noting that the specific heat C_V is positive, which implies that the internal energy U is a monotonically increasing function of temperature T. Conversely, if T decreases then U should decrease monotonically. Therefore, U should approach its smallest value as $T \to 0$ and the system should be in a quantum mechanical ground state. The ground state is unique if it is non-degenerate or, in the case where the ground state has a spontaneously broken symmetry, the symmetry broken state adopted by the system may also be unique. In either case, since the entropy is proportional to the logarithm of the degeneracy, one expects the entropy at $T = 0$ to be a minimum. For degeneracies which are not exponential in the size of the system, the entropy is not extensive and, therefore, can be considered as being effectively zero in the thermodynamic limit $N \to 0$. This assumption might not be valid for the case of highly frustrated systems such as ice or spin glasses, since these systems are not equilibrium states and remain highly degenerate as $T \to 0$.

Classically, the entropy can only be defined up to an additive constant. Since classical states are continuous, the "number of microscopic states" depends on the choice of the measure. Because of this, the classical version of Nernst's postulate states that the entropy reaches a universal minimum value in the limit $T \to 0$. Therefore, Walther Nernst's initial 1906 formulation was that the $T = 0$ isotherm is also an isentrope [1]. Max Planck's

1911 restatement of the postulate gave a value of zero to the entropy at $T = 0$. This restatement is frequently attributed to Simon [2].

Nernst Postulate and Specific Heat

Nernst's postulate has a number of consequences. For example, the specific heat vanishes as $T \to 0$. This follows if S approaches zero with a finite derivative, then

$$C_v = T \left(\frac{\partial S}{\partial T} \right)_V \to 0 \qquad \text{as} \quad T \to 0 \qquad (1.149)$$

Likewise,

$$C_P = T \left(\frac{\partial S}{\partial T} \right)_P \to 0 \qquad \text{as} \quad T \to 0 \qquad (1.150)$$

Nernst Postulate and Thermal Expansion

Another consequence of Nernst's postulate is that thermal expansion coefficient also vanishes as $T \to 0$. This can be seen since the Maxwell relation

$$\left(\frac{\partial V}{\partial T} \right)_{P,N} = -\left(\frac{\partial S}{\partial P} \right)_{T,N} \qquad (1.151)$$

shows that

$$\left(\frac{\partial V}{\partial T} \right)_{P,N} \to 0 \qquad \text{as} \quad T \to 0 \qquad (1.152)$$

Hence the coefficient of volume expansion α defined by

$$\alpha = \frac{1}{V} \left(\frac{\partial V}{\partial T} \right)_{P,N} \to 0 \qquad \text{as} \quad T \to 0 \qquad (1.153)$$

vanishes as $T \to 0$.

Nernst's Postulate and the Ideal Gas

Likewise, from the Maxwell relation

$$\left(\frac{\partial P}{\partial T} \right)_{V,N} = \left(\frac{\partial S}{\partial V} \right)_{T,N} \qquad (1.154)$$

one discovers that

$$\left(\frac{\partial P}{\partial T} \right)_{V,N} \to 0 \qquad \text{as} \quad T \to 0 \qquad (1.155)$$

This can be shown to imply that, in the limit $T \to 0$, the difference between the specific heats at constant pressure and constant volume vanish with a higher power of T than the power of T with which the specific heats vanish. Also, from the above formula, one realizes that the classical ideal gas does not satisfy Nernst's postulate. However, quantum mechanical ideal gasses do satisfy Nernst's postulate.

Unattainability of Absolute Zero

Another consequence of the Nernst postulate is that the absolute zero temperature cannot be attained by any means. More precisely, it is impossible by any procedure, no matter how idealized, to reduce the temperature of any system to the absolute zero in a finite number of operations.

To prove this, we shall consider first the final step of a finite process. Cooling a substance below a bath temperature usually requires an adiabatic stage, since otherwise heat would leak from the bath to the system and thereby increase its temperature. Suppose that, by varying a parameter X from X_1 to X_2, one adiabatically cools a system from a finite temperature T_2 to a final temperature T_1. Then the adiabaticity condition requires

$$S(T_1, X_1) = S(T_2, X_2) \tag{1.156}$$

Furthermore, if we reduce the system's final temperature T_2 to zero, the right-hand side vanishes according to Simon's statement of Nernst's principle. Thus, we require

$$S(T_1, X_1) = 0 \tag{1.157}$$

which is impossible for real systems for which S is expected to only approach its minimal value in the limit $T \to 0$. Hence, this argument suggests that the final stages of the process must involve infinitesimal temperature differences.

Such a series of infinitesimal processes is illustrated by a sequence of processes composed of adiabatic expansions between a high pressure P_1 and a low pressure P_2 followed by isothermal contractions between P_2 and P_1. The internal energy and temperature is lowered during the adiabatic expansion stages. The curves of entropy versus temperature for the different pressures must approach each other as $T \to 0$. Hence, both the magnitudes of temperature changes and entropy changes decrease in the successive stages as T approaches zero. Therefore, absolute zero can only be attained for this sequence in the limit of an infinite number of stages.

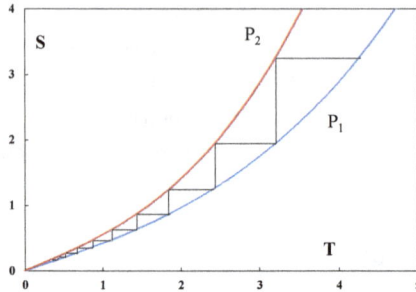

Fig. 1.5   The unattainability of $T = 0$ is illustrated by a substance which undergoes a series of stages composed of an adiabatic expansion followed by an isothermal compression. An infinite number of stages would be required to reach $T = 0$.

For these two example, the unattainability of absolute zero can simply be understood by noting that the adiabat becomes the isotherm as $T \to 0$.

1.9   Extremum Principles for Thermodynamic Potentials

The thermodynamic potentials have been derived from the internal energy $U(S, V, X)$. Here we shall derive the extremum principles for the thermodynamic potentials, in which the extensive parameters are varied and the intensive parameters are held constant, by starting with the energy minimum principle. The Energy Minimum Principle states that the equilibrium value of any unconstrained internal parameter X minimizes $U(S, V, X)$ for a fixed value of S. It can be stated in terms of the first-order and second-order infinitesimal changes

$$dU = 0$$
$$d^2U \geq 0 \tag{1.158}$$

where S is held constant.

Energy-Minimum Principle for a Composite System

The energy-minimum principle can be formulated in terms of a composite system which is composed of a system and a reservoir, for which the total energy and entropy are defined by

$$U_T = U + U_R$$
$$S_T = S + S_R \tag{1.159}$$

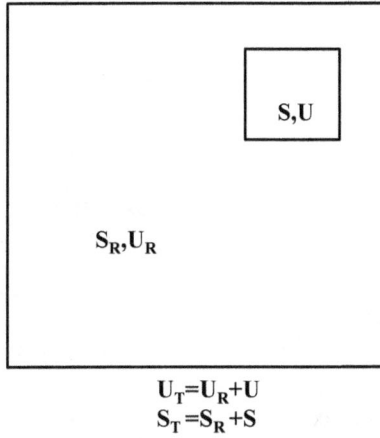

Fig. 1.6   A composite system constructed of a system of interest with energy U and entropy S and a large reservoir with energy U_R and entropy S_R.

The energy minimum principle applied to the combined system becomes

$$dU + dU_R = 0$$
$$d^2(U + U_R) \geq 0 \tag{1.160}$$

where, since S_T is constant, one must have $dS_R = -dS$. As we shall prove below, if the reservoir is sufficiently larger than the system one may set $d^2 U_R = 0$. This allows one to simplify the second inequality to become

$$d^2 U \geq 0 \tag{1.161}$$

Thus, we have shown that the Energy-Minimum Principle applies to a composite system which includes a large reservoir.

For a sufficiently large reservoir, we have asserted that one can set $d^2 U_R \approx 0$. This can be seen by examining the second-order change due to a fluctuation, say of the entropy. For this particular case, where $dS = -dS_R$, the second order change in the reservoir's internal energy $d^2 U_R$ is evaluated as

$$d^2 U_R = \left(\frac{\partial^2 U_R}{\partial S_R^2} \right) (dS_R)^2$$
$$= \left(\frac{\partial^2 U_R}{\partial S_R^2} \right) (dS)^2$$
$$= \frac{T}{C_R} (dS)^2 \tag{1.162}$$

Likewise,

$$d^2U = \left(\frac{\partial^2 U}{\partial S^2}\right)(dS)^2$$

$$= \frac{T}{C}(dS)^2 \tag{1.163}$$

Therefore, if $C_R \gg C$, one has $d^2U \gg d^2U_R$. Applying this type of consideration to the fluctuations of any set of extensive variables leads to the same conclusion.

Extremum Principle for Processes at Constant T

For a system in thermal contact with a reservoir at constant temperature T, the infinitesimal change in internal energy of the reservoir dU_R is given by the heat it absorbs

$$dU_R = T\,dS_R = -T\,dS \tag{1.164}$$

since $dS = -dS_R$. Hence, one has

$$dU + dU_R = 0$$

$$dU - T\,dS = 0 \tag{1.165}$$

which, if T is being held constant, leads to

$$d(U - TS) = 0$$

$$dF = 0 \tag{1.166}$$

where F is defined as

$$F = U - TS \tag{1.167}$$

Since $F = U - TS$ is the Helmholtz Free-Energy $F(T, V, N)$, the Helmholtz Free-Energy F satisfies an extremum principle for processes at constant T.

The extremum principle for F at constant T is a minimum principle since

$$d^2U = d^2(U - TS)$$

$$\geq 0 \tag{1.168}$$

where the first line holds, since T is being held constant and since S is an independent variable so $d^2S \equiv 0$. Therefore, the last term in the second-order change is zero since it can only result in the first-order change $T\,dS$. Thus, one has the condition

$$d^2F \geq 0 \tag{1.169}$$

The Helmholtz Minimum Principle applies to processes at constant T. The Helmholtz Minimum Principle states that, for a system being held at constant temperature T, the equilibrium value of an unconstrained internal parameter minimizes $F(T, V, X)$.

Extremum Principle for Processes at Constant P

For a system in thermal contact with a pressure reservoir of pressure P, the infinitesimal change in internal energy of the reservoir dU_R is equal to the work done on it

$$dU_R = -P \, dV_R = P \, dV \qquad (1.170)$$

since $dV = -dV_R$. Hence, one has

$$dU + dU_R = 0$$
$$dU + P \, dV = 0 \qquad (1.171)$$

which, if P is being held constant, leads to

$$d(U + PV) = 0$$
$$d\mathcal{H} = 0 \qquad (1.172)$$

where $\mathcal{H}$ is defined as

$$\mathcal{H} = U + PV \qquad (1.173)$$

Hence, the quantity $\mathcal{H} = U + PV$ satisfies an extremum principle for process at constant P.

The extremum principle for $\mathcal{H}$ at constant P is a minimum principle since

$$d^2U = d^2(U + PV)$$
$$\geq 0 \qquad (1.174)$$

where the first line holds since P is being held constant and V is an independent variable. Thus, one has the condition

$$d^2\mathcal{H} \geq 0 \qquad (1.175)$$

The Enthalpy Minimum Principle states that, for a system being held at constant pressure P, the equilibrium value of unconstrained internal parameter minimizes $\mathcal{H}(S, P, X)$.

The Gibbs Minimum Principle

For a system in thermal contact with a reservoir at constant temperature T and constant pressure P, the change in energy of the reservoir dU_R can be expressed as

$$dU_R = T\, dS_R - P\, dV_R = -T\, dS + P\, dV \tag{1.176}$$

Hence, the condition of conservation of energy yields

$$dU + dU_R = 0$$
$$dU - T\, dS + P\, dV = 0 \tag{1.177}$$

which, if T and P are being held constant, leads to

$$d(U - TS + PV) = 0$$
$$dG = 0 \tag{1.178}$$

where G is defined as

$$G = U - TS + PV \tag{1.179}$$

Hence, the quantity G satisfies an extremum principle for process at constant T and P.

The extremum principle for G, at constant T and P, is a minimum principle since

$$d^2U = d^2(U - TS + PV)$$
$$\geq 0 \tag{1.180}$$

where the first line holds since T and P are being held constant and since S and V are independent variables. Thus, one has the condition

$$d^2G \geq 0 \tag{1.181}$$

The Gibbs Minimum Principle states that, for a system being held at constant temperature T and pressure P, the equilibrium value of unconstrained internal parameter minimizes $G(T, P, X)$.

Alternate Derivation of Minimum Principles

A perhaps clearer, but less general, derivation of the minimum principle for thermodynamic potentials (when the intensive variables are held constant) can be found directly from the entropy maximum principle. As an example of a minimum principle for a thermodynamic potential, consider a closed system composed of a system and reservoir which are in thermal contact. The entropy of the combined system S_T is given by

$$S_T(U, V, N : U_T, V_T, N_T) = S(U, V, N) + S_R(U_T - U, V_T - V, N_T - N)$$
(1.182)

We shall consider the Taylor expansion of S_T in powers of U, and we shall assume that the reservoir is much bigger than the system so that the terms involving higher-order derivatives are negligibly small

$$S_T(U, V, N : U_T, V_T, N_T)$$
$$= S(U, V, N) + S_R(U_T, V_T - V, N_T - N) - \frac{U}{T_R} + \dots$$
$$= S_R(U_T, V_T - V, N_T - N) + \left(\frac{T_R S(U, V, N) - U}{T_R} \right)$$
(1.183)

where terms of the order N^2/N_R have been neglected and where T_R is the temperature of the thermal reservoir as defined by the partial derivative $\frac{\partial S_R(U_T)}{\partial U_T}$. We note that the term in the round parenthesis is of order N, contains all the U dependence and contains all the information about the subsystem of interest.

In equilibrium, since S_T is an extremum w.r.t. changes of U and since all the U dependence is included in the term enclosed by the large parenthesis, one has

$$\left(\frac{\partial S}{\partial U} \right)_{V,N} = \frac{1}{T_R}$$
(1.184)

One also has

$$\left(\frac{\partial^2 S}{\partial U^2} \right)_{V,N} \leq 0$$
(1.185)

Now consider the convex generalized thermodynamic function $\tilde{F}(U : T_R, V, N)$, previously identified in the expression for S_T, which is defined by

$$\tilde{F}(U : T_R, V, N) = U - T_R \, S(U, V, N)$$
(1.186)

for some constant T_R. The first two derivatives of $\tilde{F}$ w.r.t. U are given by

$$\left(\frac{\partial \tilde{F}}{\partial U} \right)_{V,N} = 1 - T_R \left(\frac{\partial S}{\partial U} \right)_{V,N}$$
(1.187)

and

$$\left(\frac{\partial^2 \tilde{F}}{\partial U^2}\right)_{V,N} = -T_R \left(\frac{\partial^2 S}{\partial U^2}\right)_{V,N} \tag{1.188}$$

which shows that, if the parameter T_R is identified with the temperature T of the system, then $\tilde{F}$ satisfies a minimum principle and that the minimum value of $\tilde{F}$ is given by the Helmholtz Free-Energy $F(T, V, N)$.

1.10   Thermodynamic Stability

The condition of thermodynamic stability imposes the condition of convexity or concavity on the thermodynamic functions characterizing the system. This leads to conditions on certain measurable thermodynamic quantities that must be satisfied if the system is to be thermodynamically stable. We shall first consider the stability conditions obtained from the entropy maximum principle. We shall show the requirement of stability has consequences for physically measurable quantities.

Exchange of Energy

Consider two identical systems in thermal contact. The entropy maximum principle holds for the combined system, of energy $2U$, volume $2V$ and a total number of particles $2N$. For the combined system to be stable against fluctuations of the energy, the entropy function must satisfy the inequality

$$S(2U, 2V, 2N) \geq S(U + \Delta U, V, N) + S(U - \Delta U, V, N) \tag{1.189}$$

for any value of ΔU. Due to the extensive nature of the entropy, this inequality can be re-written as

$$2S(U, V, N) \geq S(U + \Delta U, V, N) + S(U - \Delta U, V, N) \tag{1.190}$$

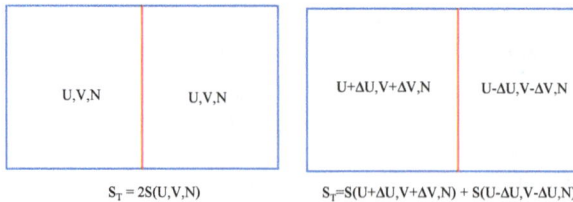

Fig. 1.7   (Left) A system composed of two identical sub-systems. The composite system has with fixed total energy $2U$, volume $2V$ in equilibrium. (Right) A non-equilibrium state in which the sub-systems have exchanged an energy ΔU and a volume ΔV.

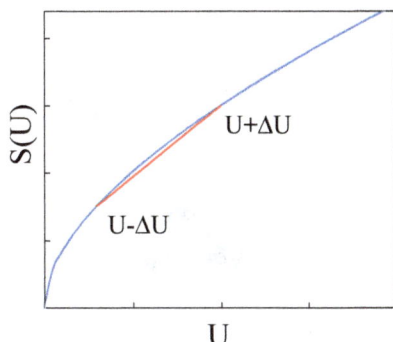

Fig. 1.8   A concave curve representing $S(U)$. Any chord connecting two points on $S(U)$ must lie below the curve.

Geometrically, the inequality expresses the fact that any chord joining two points on the curve $S(U)$ must lie below the curve. Such a curve is known as a concave curve. In the limit $\Delta U \to 0$, one obtains the weaker stability condition

$$0 \geq \left(\frac{\partial^2 S}{\partial U^2}\right)_{V,N} \tag{1.191}$$

This condition must hold if the macroscopic state of the system characterized by U, V, N is an equilibrium state. This condition can be re-stated as

$$
\begin{aligned}
0 &\geq \left(\frac{\partial^2 S}{\partial U^2}\right)_{V,N} \\
&\geq -\frac{1}{T^2}\left(\frac{\partial T}{\partial U}\right)_{V,N} \\
&\geq -\frac{1}{T^2}\frac{1}{C_{V,N}}
\end{aligned}
\tag{1.192}
$$

Thus, we conclude that if a system is to be stable, its heat capacity at constant volume must be positive. This condition implies that the energy is a monotonically increasing function of temperature at constant volume.

Exchange of Energy and Volume

Likewise, if the energy and volume are allowed to fluctuate, the condition for stability becomes

$$2S(U, V, N) \geq S(U + \Delta U, V + \Delta V, N) + S(U - \Delta U, V - \Delta V, N) \tag{1.193}$$

which can be expanded to yield an inequality

$$0 \geq \left(\frac{\partial^2 S}{\partial U^2}\right)_{V,N} \Delta U^2 + 2\left(\frac{\partial^2 S}{\partial U \partial V}\right)_N \Delta U\, \Delta V + \left(\frac{\partial^2 S}{\partial V^2}\right)_{U,N} \Delta V^2 \quad (1.194)$$

This inequality leads to two weak conditions for stability, which are

$$0 \geq \left(\frac{\partial^2 S}{\partial U^2}\right)_{V,N} \quad (1.195)$$

and

$$0 \geq \left(\frac{\partial^2 S}{\partial V^2}\right)_{U,N} \quad (1.196)$$

These are weak conditions, since they only apply if either U is fixed or V is fixed.

However, the right-hand-side of the inequality can be expressed as the sum of two terms

$$0 \geq \frac{1}{\left(\frac{\partial^2 S}{\partial U^2}\right)_{V,N}} \left[\left(\frac{\partial^2 S}{\partial U^2}\right)_{V,N} \Delta U + \left(\frac{\partial^2 S}{\partial U \partial V}\right)_N \Delta V\right]^2$$

$$+ \left[\left(\frac{\partial^2 S}{\partial V^2}\right)_{U,N} - \frac{\left(\frac{\partial^2 S}{\partial U \partial V}\right)_N^2}{\left(\frac{\partial^2 S}{\partial U^2}\right)_{V,N}}\right] \Delta V^2 \quad (1.197)$$

This leads to two stronger conditions for stability, which are

$$0 \geq \left(\frac{\partial^2 S}{\partial U^2}\right)_{V,N} \quad (1.198)$$

and

$$0 \geq \left[\left(\frac{\partial^2 S}{\partial V^2}\right)_{U,N} - \frac{\left(\frac{\partial^2 S}{\partial U \partial V}\right)_N^2}{\left(\frac{\partial^2 S}{\partial U^2}\right)_{V,N}}\right] \quad (1.199)$$

The last condition can be re-stated as

$$\left(\frac{\partial^2 S}{\partial V^2}\right)_{U,N} \left(\frac{\partial^2 S}{\partial U^2}\right)_{V,N} \geq \left(\frac{\partial^2 S}{\partial U \partial V}\right)_N^2 \quad (1.200)$$

which is a condition on the determinant of the matrix of the second-order derivatives. The two by two matrix is a particular example of a Hessian

Matrix which, more generally, is an N by N matrix of the second-order derivatives of a function of N independent variables. The Hessian is the determinant of the Hessian matrix. The Hessian describes the local curvature of the function. Although the above two conditions have been derived for two identical subsystems, they can be applied to any macroscopic part of a homogeneous system since thermodynamic quantities are uniformly distributed throughout the system.

1.10.1 *Stability Conditions for Thermodynamic Potentials*

We have considered the entropy maximum principle. The energy $U(S, V, N)$ satisfies a minimum principle, which is reflected in the behavior of the thermodynamic potentials. Here we shall consider the stability conditions derived from the energy minimum principle and then examine the consequences for physically measurable quantities.

For a system composed of two identical sub-systems each with entropy S, volume V and number of particles N then, according to the energy minimum principle, the condition for equilibrium under interchange of entropy and volume is given by

$$U(S + \Delta S, V + \Delta V, N) + U(S - \Delta S, V - \Delta V, N) > 2U(S, V, N)$$

$$(1.201)$$

Thus, $U(S, V, N)$ satisfies the definition for it to be a convex function of S. For stability against entropy fluctuations, one has

$$\left(\frac{\partial^2 U}{\partial S^2}\right)_{V,N} \geq 0$$

$$\left(\frac{\partial T}{\partial S}\right)_{V,N} \geq 0 \qquad (1.202)$$

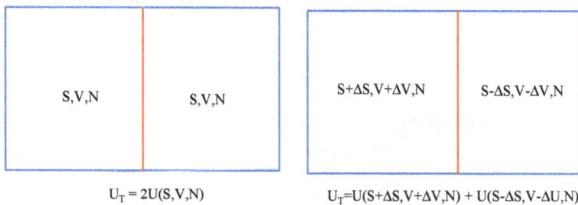

S,V,N   S,V,N   S+ΔS,V+ΔV,N   S-ΔS,V-ΔV,N

$U_T = 2U(S,V,N)$ $U_T = U(S+\Delta S, V+\Delta V, N) + U(S-\Delta S, V-\Delta U, N)$

Fig. 1.9 (Left) A system composed of two identical sub-systems. The composite system has with fixed entropy $2S$, volume $2V$ in equilibrium. (Right) A non-equilibrium state in which the sub-stems have exchanged an entropy ΔS and a volume ΔV.

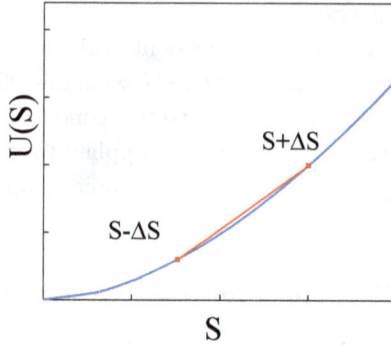

Fig. 1.10   The energy $U(S, V, X)$ is a convex function of the energy U. The energy minimum principle has a geometric interpretation that any chord on the curve $U(X)$ lies above the curve.

which leads to the condition $C_V \geq 0$ i.e. the specific heat at constant volume is always positive. Stability against volume fluctuations leads to

$$\left(\frac{\partial^2 U}{\partial V^2} \right)_{S,N} \geq 0$$

$$-\left(\frac{\partial P}{\partial V} \right)_{S,N} \geq 0 \qquad (1.203)$$

The second line follows immediately from $P = -\left(\frac{\partial U}{\partial V} \right)_{S,N}$. Thus, the entropy is a convex function of the extensive variables and the convexity leads to stability conditions against fluctuations of the extensive variables which always have the same signs.

Stability against fluctuations of both S and V leads to a more complex and more restrictive condition

$$\left(\frac{\partial^2 U}{\partial V^2} \right)_{S,N} \left(\frac{\partial^2 U}{\partial S^2} \right)_{V,N} \geq \left(\frac{\partial^2 U}{\partial S \partial V} \right)_N^2 \qquad (1.204)$$

Later, we shall show that this inequality is equivalent to the condition

$$-\left(\frac{\partial P}{\partial V} \right)_{T,N} \geq 0 \qquad (1.205)$$

i.e. an increase in volume at constant temperature is always accompanied by a decrease in pressure.

Stability Conditions and Legendre Conjugated Variables

We have established that thermodynamic potentials satisfy minimum principles with respect to variations of the extensive variables. The extension of the stability conditions to the thermodynamic potentials with respect to the intensive variables involves some consideration of the properties of the Legendre transform. It will be seen that the thermodynamic potentials are convex functions of their extensive variables but are concave functions of their intensive variables.

Consider a function $y(x)$ which satisfies a minimum condition. The Legendre transform of $y(x)$ is $\psi(p)$. One notes that the Legendre transform and inverse Legendre transform introduces the conjugate variables x and p via

$$p = \left(\frac{\partial y}{\partial x} \right) \tag{1.206}$$

and

$$x = -\left(\frac{\partial \psi}{\partial p} \right) \tag{1.207}$$

These relations lead to

$$\frac{\partial p}{\partial x} = \left(\frac{\partial^2 y}{\partial x^2} \right) \tag{1.208}$$

and

$$\frac{\partial x}{\partial p} = -\left(\frac{\partial^2 \psi}{\partial p^2} \right) \tag{1.209}$$

Thus, on equating the expressions for $\frac{dp}{dx}$, one has

$$\frac{\partial p}{\partial x} = \left(\frac{\partial^2 y}{\partial x^2} \right) = -\frac{1}{\left(\frac{\partial^2 \psi}{\partial p^2} \right)} \tag{1.210}$$

which shows that the sign of the second derivative w.r.t. the conjugate variables changes under the Legendre transform. Therefore, the condition for stability against fluctuations in x (when expressed in terms of the thermodynamic potential y) has the opposite sign to the condition for stability of fluctuations in the conjugate variable p (when expressed in terms of the thermodynamic potential ψ). The stability condition for fluctuations of the other variables (which are not involved in the Legendre transform) have the same sign for both y and ψ.

Stability Conditions and the Helmholtz Free-Energy

The Helmholtz Free-Energy $F(T, V, N)$ is derived from the Legendre transform of $U(S, V, N)$ by eliminating the extensive variable S in favour of the intensive variable T. The condition for stability against temperature fluctuations is expressed in terms of $F(T, V, N)$ as

$$\left(\frac{\partial^2 F}{\partial T^2}\right)_{V,N} \leq 0 \tag{1.211}$$

which has the opposite sign from the stability conditions against entropy fluctuations when expressed in terms of $U(S, V, N)$. Stability against volume fluctuations leads to

$$\left(\frac{\partial^2 F}{\partial V^2}\right)_{T,N} \geq 0 \tag{1.212}$$

which has the same sign as the stability conditions against volume fluctuations when expressed in terms of U.

Stability Conditions and the Enthalpy

The stability condition for the enthalpy $\mathcal{H}(S, P, N)$ against entropy fluctuations is given by

$$\left(\frac{\partial^2 \mathcal{H}}{\partial S^2}\right)_{P,N} \geq 0 \tag{1.213}$$

which has the same sign as the stability conditions against entropy fluctuations when expressed in terms of $U(S, V, N)$. Stability against pressure fluctuations leads to

$$\left(\frac{\partial^2 \mathcal{H}}{\partial P^2}\right)_{S,N} \leq 0 \tag{1.214}$$

which has the opposite sign as the stability conditions against volume fluctuations when expressed in terms of U.

Stability Conditions for the Gibbs Free-Energy

The Gibbs Free-Energy involves a double Legendre transform of U, so both stability conditions have opposite signs. The condition for stability against temperature fluctuations is expressed in terms of $G(T, P, N)$ as

$$\left(\frac{\partial^2 G}{\partial T^2}\right)_{P,N} \leq 0 \qquad (1.215)$$

which has the opposite sign as the stability conditions against entropy fluctuations when expressed in terms of $U(S, V, N)$. Stability against pressure fluctuations leads to the condition

$$\left(\frac{\partial^2 G}{\partial P^2}\right)_{T,N} \leq 0 \qquad (1.216)$$

which has the opposite sign as the stability condition against volume fluctuations when expressed in terms of U. The Gibbs Free-Energy is a concave function of the intensive parameters T and P. The concavity of G does not directly follow from the above two conditions, but also requires that the mixed second-derivative

$$\left(\frac{\partial^2 G}{\partial T \partial P}\right)$$

must satisfy an inequality.

Stability with Respect to Volume Fluctuations at Constant T

The stability against volume fluctuations of a system held at constant temperature is expressed in terms of the second derivative of the Helmholtz Free-Energy as

$$\left(\frac{\partial^2 F}{\partial V^2}\right)_{T,N} \geq 0 \qquad (1.217)$$

We shall show that this inequality is related to the inequality

$$\left(\frac{\partial^2 U}{\partial V^2}\right)_{S,N} \left(\frac{\partial^2 U}{\partial S^2}\right)_{V,N} \geq \left(\frac{\partial^2 U}{\partial S \partial V}\right)_N^2 \qquad (1.218)$$

describing the stability condition obtained the energy minimum principle. We shall prove this by noting that the infinitesimal change in F shows that

$$\left(\frac{\partial F}{\partial V}\right)_{T,N} = -P \qquad (1.219)$$

therefore, one has

$$\left(\frac{\partial^2 F}{\partial V^2}\right)_{T,N} = -\left(\frac{\partial P}{\partial V}\right)_{T,N} \tag{1.220}$$

The derivative of P with respect to V at constant T can be expressed as a Jacobian[2]

$$\left(\frac{\partial^2 F}{\partial V^2}\right)_{T,N} = -\frac{\partial(P,T)}{\partial(V,T)} \tag{1.221}$$

Since we wish to express the inequality in terms of the energy, one should change variables from V and T to S and V. This can be achieved using the properties of the Jacobian

$$\left(\frac{\partial^2 F}{\partial V^2}\right)_{T,N} = -\frac{\partial(P,T)}{\partial(S,V)}\frac{\partial(S,V)}{\partial(V,T)} \tag{1.222}$$

On using the antisymmetric nature of the Jacobian one can recognize that the second factor is a derivative of S with respect to T, with V being held constant

$$\begin{aligned}
\left(\frac{\partial^2 F}{\partial V^2}\right)_{T,N} &= \frac{\partial(P,T)}{\partial(S,V)}\frac{\partial(V,S)}{\partial(V,T)} \\
&= \left(\frac{\partial S}{\partial T}\right)_{V,N}\frac{\partial(P,T)}{\partial(S,V)} \\
&= \left(\frac{\partial S}{\partial T}\right)_{V,N}\left[\left(\frac{\partial P}{\partial S}\right)_V\left(\frac{\partial T}{\partial V}\right)_S - \left(\frac{\partial T}{\partial S}\right)_V\left(\frac{\partial P}{\partial V}\right)_S\right]
\end{aligned} \tag{1.223}$$

where the expression for the Jacobian has been used to obtain the last line. On recognizing that P and T are the energy intensive parameters, one can

---

[2] The Jacobian is defined as

$$\frac{\partial(A,B)}{\partial(x,y)} = \left(\frac{\partial A}{\partial x}\right)_y\left(\frac{\partial B}{\partial y}\right)_x - \left(\frac{\partial A}{\partial y}\right)_x\left(\frac{\partial B}{\partial x}\right)_y$$

The Jacobian is antisymmetric under interchange of A and B. Furthermore, if B is the same as the independent variable y, the Jacobian simply becomes a partial derivative of A w.r.t. x at constant y, i.e.

$$\frac{\partial(A,y)}{\partial(x,y)} = \left(\frac{\partial A}{\partial x}\right)_y$$

write

$$\left(\frac{\partial^2 F}{\partial V^2}\right)_{T,N} = \left(\frac{\partial S}{\partial T}\right)_{V,N}\left[\left(\frac{\partial P}{\partial S}\right)_V\left(\frac{\partial T}{\partial V}\right)_S - \left(\frac{\partial T}{\partial S}\right)_V\left(\frac{\partial P}{\partial V}\right)_S\right]$$

$$= \left(\frac{\partial S}{\partial T}\right)_{V,N}\left[\frac{\partial}{\partial S}\left(\frac{\partial U}{\partial S}\right)_V\frac{\partial}{\partial V}\left(\frac{\partial U}{\partial V}\right)_S\right.$$

$$\left. - \frac{\partial}{\partial S}\left(\frac{\partial U}{\partial V}\right)_S\frac{\partial}{\partial V}\left(\frac{\partial U}{\partial S}\right)_V\right]$$

$$= \left(\frac{\partial S}{\partial T}\right)_{V,N}\left[\left(\frac{\partial^2 U}{\partial S^2}\right)_V\left(\frac{\partial^2 U}{\partial V^2}\right)_S - \left(\frac{\partial^2 U}{\partial S \partial V}\right)^2\right] \qquad (1.224)$$

where the last line has been obtained by using the analyticity of U. Finally, one can write

$$\left(\frac{\partial^2 F}{\partial V^2}\right)_{T,N} = \left(\frac{\partial S}{\partial T}\right)_{V,N}\left[\left(\frac{\partial^2 U}{\partial S^2}\right)_V\left(\frac{\partial^2 U}{\partial V^2}\right)_S - \left(\frac{\partial^2 U}{\partial S \partial V}\right)^2\right]$$

$$= \frac{\left[\left(\frac{\partial^2 U}{\partial S^2}\right)_V\left(\frac{\partial^2 U}{\partial V^2}\right)_S - \left(\frac{\partial^2 U}{\partial S \partial V}\right)^2\right]}{\left(\frac{\partial T}{\partial S}\right)_{V,N}}$$

$$= \frac{\left[\left(\frac{\partial^2 U}{\partial S^2}\right)_V\left(\frac{\partial^2 U}{\partial V^2}\right)_S - \left(\frac{\partial^2 U}{\partial S \partial V}\right)^2\right]}{\left(\frac{\partial^2 U}{\partial S^2}\right)_{V,N}} \qquad (1.225)$$

Thus, we have proved that

$$-\left(\frac{\partial P}{\partial V}\right)_{T,N} = \frac{\left[\left(\frac{\partial^2 U}{\partial S^2}\right)_V\left(\frac{\partial^2 U}{\partial V^2}\right)_S - \left(\frac{\partial^2 U}{\partial S \partial V}\right)^2\right]}{\left(\frac{\partial^2 U}{\partial S^2}\right)_{V,N}} \qquad (1.226)$$

which relates the stability condition against volume fluctuations at constant T to the stability condition for fluctuations in S and V.

1.10.2 *Physical Consequences of Stability*

The convexity of F with respect to V has been shown to lead to the condition

$$\left(\frac{\partial^2 F}{\partial V^2}\right)_{T,N} = -\left(\frac{\partial P}{\partial V}\right)_{T,N} \geq 0 \qquad (1.227)$$

which can be expressed as

$$\left(\frac{\partial^2 F}{\partial V^2}\right)_{T,N} = \frac{1}{V}\frac{1}{\kappa_T} \geq 0 \tag{1.228}$$

Hence, the isothermal compressibility κ_T defined by

$$\kappa_T = -V \left(\frac{\partial P}{\partial V}\right)_T \geq 0 \tag{1.229}$$

must always be positive. Likewise, the concavity of F with respect to T leads to the stability condition

$$C_V \geq 0 \tag{1.230}$$

that the specific heat at constant volume C_V must always be positive.

The Stability of Non-Rigid Systems

Another consequence of thermodynamic stability is that a thermodynamic system which is composed of parts that are free to move with respect to each other, will become unstable if the temperature is negative. The entropy of the α-th component of the system is a function of the internal energy, that is U_α minus the kinetic energy $\frac{p_\alpha^2}{2m_\alpha}$. This follows from Galilean invariance, since S_α describes a property of the system which should be independent of its state of uniform motion. Since the entropies are additive, the total entropy is given by

$$S = \sum_\alpha S_\alpha\left(U_\alpha - \frac{p_\alpha^2}{2m_\alpha}\right) \tag{1.231}$$

but is subject to the constraint that the total momentum is conserved

$$\sum_\alpha \underline{p}_\alpha = \underline{P} \tag{1.232}$$

The entropy has to be maximized subject to the constraint. This can be performed by using Lagrange's method of undetermined multipliers. Thus, Φ is to be maximized with respect to $\underline{p}_\alpha$, where

$$\Phi = \sum_\alpha \left[S_\alpha\left(U_\alpha - \frac{p_\alpha^2}{2m_\alpha}\right) + \underline{\lambda}\cdot\underline{p}_\alpha\right] \tag{1.233}$$

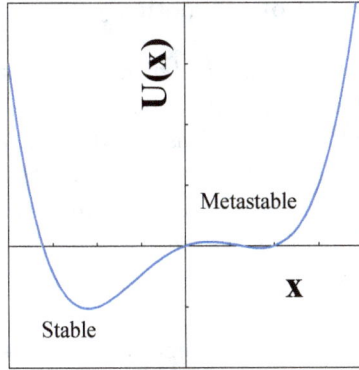

Fig. 1.11 A curve of the internal energy $U(X)$ versus an internal variable X for a system which exhibits a stable and a metastable state.

Maximizing Φ w.r.t. $\underline{p}_\alpha$ leads to the equation

$$
\begin{aligned}
0 &= -\left(\frac{\partial S_\alpha}{\partial U_\alpha}\right)\frac{\underline{p}_\alpha}{m_\alpha} + \underline{\lambda} \\
&= -\frac{1}{T}\frac{\underline{p}_\alpha}{m_\alpha} + \underline{\lambda}
\end{aligned}
\tag{1.234}
$$

which leads to the conclusion that each component must have the same velocity. Thus, no independent internal macroscopic linear motions are allowed in an equilibrium state.

For the stationary state to be stable against the momentum fluctuations of the α-th part, one requires that

$$
-\frac{1}{m_\alpha T} \leq 0
\tag{1.235}
$$

Therefore, stability against break-up of the system requires that $T \geq 0$.

The conditions for stability discussed in this chapter are necessary but not sufficient to establish that the equilibrium is completely stable, since a state may decrease its entropy when there are infinitesimally small fluctuations in its macroscopic parameters, but its entropy may increase if the deviations of the parameters have large values. Such states are known as metastable states. A system which is in a metastable state will remain there until a sufficiently large fluctuation occurs that will take the system into a new state that is more stable.

1.11   Thermodynamics of Magnetic Systems

The axiomatic formulation of thermodynamics postulates that the thermodynamic properties of a system can be obtained from the fundamental relation which expresses the entropy S as an extensive quantity that depends
on the extensive quantities of the system. Since the magnetization $\underline{M}$ is
an extensive quantity, the thermodynamic properties should be obtainable
from the entropy of a magnetic system

$$S = S(U, V, \underline{M}, N) \tag{1.236}$$

which depends on $\underline{M}$. Alternatively, one can use the energy formulation of
thermodynamics which, for a magnetic system, postulates that

$$U = U(S, V, \underline{M}, N) \tag{1.237}$$

One can define an intensive quantity $\underline{H}$ via

$$\underline{H} = \left(\frac{\partial U}{\partial \underline{M}} \right)_{S,V,N} \tag{1.238}$$

Hence, the first law of thermodynamics takes the form

$$dU = T\, dS - P\, dV + \mu\, dN + \underline{H} \cdot d\underline{M} \tag{1.239}$$

The quantity H can be identified as a uniform applied magnetic field. One
may define a thermodynamic potential $\tilde{U}$, by performing a Legendre transformation of the internal energy with respect to M

$$\tilde{U} = U - \underline{H} \cdot \underline{M} \tag{1.240}$$

where $\tilde{U}(S, V, H, N)$ depends on the intensive uniform field $\underline{H}$. The quantity
$\tilde{U}$ includes the interaction energy between the matter in the system and the
applied field while U doesn't. In terms of the Legendre Transformed energy,
the magnetization $\underline{M}$ is determined from

$$\underline{M} = -\left(\frac{\partial \tilde{U}}{\partial \underline{H}} \right)_{S,V,N} \tag{1.241}$$

In the literature, the difference between U and $\tilde{U}$, and all the pairs of
thermodynamic potentials derived from these functions, is often ignored.
However, the identity of the thermodynamic potential being used is immediately revealed by inspection of whether it depends on either the extensive
quantity $\underline{M}$ or the intensive quantity $\underline{H}$. In this text we shall invariably
use Gaussian units.

Identification of $\underline{H}$ with the Applied Magnetic Field

The identification of $\underline{H}$ with the applied magnetic field, can be obtained from consideration of the macroscopic formulation of Maxwell's equations

$$\underline{\nabla} \wedge \underline{E} = -\frac{1}{c} \left(\frac{\partial \underline{B}}{\partial t} \right)$$

$$\underline{\nabla} \cdot \underline{D} = 4\pi \rho$$

$$\underline{\nabla} \wedge \underline{H} = \frac{4\pi}{c} \underline{j} + \frac{1}{c} \left(\frac{\partial \underline{D}}{\partial t} \right)$$

$$\underline{\nabla} \cdot \underline{B} = 0 \tag{1.242}$$

together with the constitutive relations

$$\underline{D} = \underline{E} + 4\pi \ \mathcal{P}$$

$$\underline{B} = \underline{H} + 4\pi \ \mathcal{M} \tag{1.243}$$

where $\mathcal{M}$ and $\mathcal{P}$, respectively, are the magnetization and polarization densities. It should be recognized that the quantities $\underline{E}(\underline{r})$, $\underline{B}(\underline{r})$ are fields that are not necessarily uniform and, therefore, may not be appropriate thermodynamic quantities.

The identification of the dependence of the internal energy on the electromagnetic field can be derived by considering the energy dissipated by a system exposed to an electric field $\underline{E}$. The rate at which heat is generated in the system (i.e. the heat absorbed by the system), P_{EM}, is given by

$$P_{EM} = \int d^3\underline{r} \ \underline{j}(\underline{r}) \cdot \underline{E}(\underline{r}) \tag{1.244}$$

Conservation of energy implies that, in time interval δt, the amount of work done on the system is equal to

$$\delta W_{EM} = -\delta t \int d^3\underline{r} \ \underline{j}(\underline{r}) \cdot \underline{E}(\underline{r}) \tag{1.245}$$

Using Maxwell's equations, one finds that

$$\delta W_{EM} = -\delta t \ \frac{c}{4\pi} \int d^3\underline{r} \left[\underline{\nabla} \wedge \underline{H} - \frac{1}{c} \left(\frac{\partial \underline{D}}{\partial t} \right) \right] \cdot \underline{E} \tag{1.246}$$

Furthermore, on applying a vector identity, the work done on the system is expressed as

$$\delta W_{EM} = -\delta t \ \frac{c}{4\pi} \int d^3\underline{r} \ \underline{\nabla} \cdot [\underline{H} \wedge \underline{E}]$$

$$- \delta t \ \frac{c}{4\pi} \int d^3\underline{r} \left[\underline{H} \cdot (\underline{\nabla} \wedge \underline{E}) - \frac{1}{c} \left(\frac{\partial \underline{D}}{\partial t} \right) \cdot \underline{E} \right] \tag{1.247}$$

Gauss's Theorem allows first term to be represented as a surface integral which, since $\underline{H} \wedge \underline{E}$ is the Poynting vector, is recognized as the energy flow through the surfaces. In the thermodynamic limit, the loss of energy through the surface can be neglected. On using Faraday's law, the work done on the system can be expressed as

$$\delta W_{EM} = \delta t \, \frac{1}{4\pi} \int d^3\underline{r} \left[\underline{H} \cdot \left(\frac{\partial B}{\partial t} \right) + \underline{E} \cdot \left(\frac{\partial D}{\partial t} \right) \right] \qquad (1.248)$$

Hence, the change in the system's internal energy δU in time dt caused by the changes in B and D is given by

$$\delta U_{EM} = \frac{1}{4\pi} \int d^3\underline{r} \left[\underline{H} \cdot \delta \underline{B} + \underline{E} \cdot \delta \underline{D} \right] \qquad (1.249)$$

On substituting the constitutive equations, assuming that the fields are uniform and simply ignoring the contribution of quadratic order in the E and H fields, one finds that the change in the internal energy due to the interaction of the electromagnetic field with matter can be expressed as

$$dU_{EM} = \underline{H} \cdot d\underline{M} + \underline{E} \cdot d\underline{P} \qquad (1.250)$$

where $\underline{M}$ is the uniform magnetization and $\underline{P}$ is the uniform polarization. Thus, one finds that for a magnetic system

$$dU = T \, dS - P \, dV + \mu \, dN + \underline{H} \cdot d\underline{M} \qquad (1.251)$$

where the intensive variable $\underline{H}$ is the uniform applied magnetic field. The neglect of the terms proportional to $\underline{H}^2 \, V$ is done by associating this as the energy in the volume of the vacuum when matter is absent. Whether one treats U and the vacuum energy either together or separately is a matter of choice.

1.12 Problems

Problem 1.1

Prove the reciprocal relation of partial differentiation

$$\left(\frac{\partial x}{\partial y}\right)_z = 1 \bigg/ \left(\frac{\partial y}{\partial x}\right)_z$$

Problem 1.2

Prove Euler's chain rule of partial differentiation

$$\left(\frac{\partial x}{\partial y}\right)_z \left(\frac{\partial y}{\partial z}\right)_x \left(\frac{\partial z}{\partial x}\right)_y = -1$$

Problem 1.3

The mathematical definition of an exact differential dY is that it can be expressed as

$$dY = \sum_{i=1}^{N} \left(\frac{\partial Y}{\partial X_i}\right) dX_i$$

where the X_i are N independent variables.

(i) List three properties of a state function $Y(\{X_i\})$ found by integrating an exact differential.

Consider the first law of thermodynamics of a system, which provides a relationship between δQ, the heat absorbed in a reversible process, and the exact differentials dU and dV which, respectively, denote the increase in internal energy and volume

$$\delta Q = dU + P \, dV$$

where P is known to be a non-trivial function of V and T.

(ii) Show that if U is a state function which depends on V and T, ie. $U = U(V,T)$, then show that δQ is an inexact differential and that Q is not a state function.

(iii) Show that the quantity dS defined by

$$dS = \frac{\delta Q}{T}$$

where T is the absolute temperature, is an exact differential if

$$\left(\frac{\partial U}{\partial V}\right)_T = T \left(\frac{\partial P}{\partial T}\right)_V - P$$

Problem 1.4

(i) Use the extensive nature of the Helmholtz Free-Energy $F(T, V, N)$ to show that

$$F(T, V, N) = V \left(\frac{\partial F}{\partial V} \right)_{T,N} + N \left(\frac{\partial F}{\partial N} \right)_{T,V}$$

Hence, show that the Gibbs Free-Energy G is given by

$$G = \mu N$$

(ii) Extend this result to the case where the system contains n species of particles, so that the Helmholtz Free-Energy F depends on $(T, V, N_1, N_2, \ldots N_n)$.

Problem 1.5

Derive the following thermodynamic relations

$$\left(\frac{\partial E}{\partial V} \right)_T = T \left(\frac{\partial P}{\partial T} \right)_V - P$$

$$\left(\frac{\partial E}{\partial P} \right)_T = -T \left(\frac{\partial V}{\partial T} \right)_V - P \left(\frac{\partial V}{\partial P} \right)_T$$

$$\left(\frac{\partial C_p}{\partial V} \right)_T = T \left(\frac{\partial^2 P}{\partial T^2} \right)_V$$

$$\left(\frac{\partial C_p}{\partial P} \right)_T = -T \left(\frac{\partial^2 V}{\partial T^2} \right)_P$$

Problem 1.6

Starting from the stability condition of a system, with fixed volume, against temperature fluctuations

$$\left(\frac{\partial^2 F}{\partial T^2} \right)_{V,N} \leq 0$$

prove the stability condition

$$\left(\frac{\partial^2 G}{\partial T^2} \right)_P \left(\frac{\partial^2 G}{\partial P^2} \right)_T - \left(\frac{\partial^2 G}{\partial T \partial P} \right)^2 \geq 0$$

The above inequality can be used to complete the proof that G is concave function w.r.t. the intensive parameters.

Problem 1.7

Prove the two equalities

$$C_P - C_V = TV \frac{\alpha^2}{\kappa_T}$$

and

$$\frac{\kappa_S}{\kappa_T} = \frac{C_V}{C_P}$$

Hence, prove that the stability conditions imply the inequalities

$$C_P \geq C_V \geq 0$$

and

$$\kappa_T \geq \kappa_S \geq 0$$

Problem 1.8

The first law of thermodynamics for a magnetic system can be written in the form

$$d\tilde{U} = T \, dS - M \, dH$$

where $\tilde{U}$ is a Legendre Transform of the internal energy which includes the interaction between the magnetic moments and the applied magnetic field. The last term describes the work done on the magnetic moments by the magnetic field.

(i) Derive the Maxwell relations

$$\left(\frac{\partial T}{\partial H}\right)_S = -\left(\frac{\partial M}{\partial S}\right)_H$$

$$\left(\frac{\partial S}{\partial H}\right)_T = \left(\frac{\partial M}{\partial T}\right)_S$$

$$\left(\frac{\partial T}{\partial M}\right)_S = \left(\frac{\partial H}{\partial S}\right)_M$$

$$\left(\frac{\partial S}{\partial M}\right)_T = -\left(\frac{\partial H}{\partial T}\right)_S$$

The specific heats and susceptibilities are defined by

$$C_H = T\left(\frac{\partial S}{\partial T}\right)_H$$

$$C_M = T\left(\frac{\partial S}{\partial T}\right)_M$$

$$\chi_T = \left(\frac{\partial M}{\partial H}\right)_T$$

(ii) Show that

$$(C_H - C_M)\, \chi_T = T\left(\frac{\partial M}{\partial T}\right)_H^2$$

Problem 1.9

The first law of thermodynamics for a magnet can be written as

$$dU = T\, dS + H\, dM$$

where U is the internal energy which does not include the interaction between the applied field the magnetic material. The last term describes the work done on the system and the source of the magnetic field. Thermodynamic stability implies that the Jacobian satisfies the inequality

$$\frac{\partial(S, M)}{\partial(T, H)} \geq 0$$

(i) Show that the inequality implies

$$C_H\, \chi \geq T\left(\frac{\partial M}{\partial T}\right)_M^2$$

Problem 1.10

The tension $\mathcal{T}$ for a rubber band is empirically found to be described by

$$\mathcal{T} = aT\left(\frac{L}{L_0} - \frac{L_0^2}{L^2}\right)$$

where L is its length and a and L_0 are constants. Assume that the specific heat C_L is independent of T.
(i) Find the entropy of the rubber band.
(ii) If the band which is initially held at temperature T_0 is stretched, adiabatically and reversibly, from length L_0 to $2L_0$, what is the final temperature?

Problem 1.11

The entropy of elastic body of length L is found to satisfy

$$\left(\frac{\partial S}{\partial L}\right)_T > 0$$

If the tension is denoted by $\mathcal{T}$, determine the sign of

$$\left(\frac{\partial T}{\partial L}\right)_{\mathcal{T}}$$

Problem 1.12

Consider a rod under tension $\mathcal{T}$. The infinitesimal change in the fundamental equation leads to

$$dU = T \, dS + \mathcal{T} \, dL$$

The isothermal Young's modulus is given by

$$E_T = \frac{L}{A}\left(\frac{\partial \mathcal{T}}{\partial L}\right)_T$$

and the coefficient of linear expansion $\alpha_{\mathcal{T}}$ is defined as

$$\alpha_{\mathcal{T}} = \frac{1}{L}\left(\frac{\partial L}{\partial T}\right)_{\mathcal{T}}$$

(i) Show that when the length of the rod is slowly increased at constant T, the rate at which heat is absorbed is proportional to

$$T\left(\frac{\partial S}{\partial L}\right)_T = T A \, E_T \, \alpha_{\mathcal{T}}$$

Problem 1.13

A system of spins has a finite high temperature limit of the entropy $S(T)$ which is given by

$$\lim_{T \to \infty} S(T) \to N k_B \ln 2$$

It is found that the specific heat can be approximated by

$$C_v = C_0\left(2\,\frac{T}{T_0} - 1\right) \qquad \text{for } T_0 > T > \frac{T_0}{2}$$
$$C_v = 0 \qquad\qquad\qquad \text{otherwise}$$

(i) Determine the value of C_0.

Problem 1.14

Many equations of state for ideal gasses can be written in the form

$$PV = gU$$

where g is a numerical constant.

(i) Show that U must satisfy the thermodynamic relation

$$U = T\left(\frac{\partial U}{\partial T}\right)_V - \frac{V}{g}\left(\frac{\partial U}{\partial V}\right)_T$$

(ii) Show that for this equation to be satisfied, U must have the form

$$U = V^{-g} f(TV^g)$$

where $f(x)$ is an undetermined function.
(iii) If $g = \frac{1}{3}$ and

$$U = V \epsilon(T)$$

determine the temperature dependence of $\epsilon(T)$.

Problem 1.15

The magnetization M of a paramagnetic material is given by $M = \chi H$, where χ is the magnetic susceptibility is a function of T. For a paramagnet at sufficiently high temperatures, the susceptibility may have the form of a Curie law

$$\chi \approx \frac{C}{T}$$

The change in the internal energy $\tilde{U}$ is given by

$$d\tilde{U} = T \, dS - M \, dH$$

where the last term is the work done on the magnetic substance alone.
(i) Show that

$$T \left(\frac{\partial S}{\partial H} \right)_T = T \left(\frac{\partial M}{\partial T} \right)_H$$

(ii) Express the above result in terms of χ. In an isothermal process, does the entropy increase or decrease when H is increased?
(iii) Show that

$$\left(\frac{\partial T}{\partial H} \right)_S = -\frac{TH}{C_H} \left(\frac{\partial \chi}{\partial T} \right)_H$$

(iv) For an adiabatic process, does the temperature increase or decrease when H is increased?

Problem 1.16

A black hole occupies a region of space-time which contains sufficient mass such that nothing, not even a photon, can escape from its interior.

Schwarzschild, using Einstein's theory of general relativity, found that the radius of the black hole R is related to its mass M by the formula

$$R = \left(\frac{2GM}{c^2}\right)$$

Since particles within a radius R of the center cannot escape the black hole, the surface of radius R is known as the event horizon. Typical masses m_p and length scales l_p for gravitation can be found by setting

$$\frac{Gm_p^2}{l_p} = m_p c^2$$

and

$$l_p = \frac{\hbar}{m_p c}$$

which yields the Planck mass m_p and Planck length l_p

$$m_p = \sqrt{\frac{\hbar c}{G}}$$

$$l_p = \sqrt{\frac{\hbar G}{c^3}}$$

Bekenstein [3] showed that the entropy of the black hole can be expressed in term of its surface area A via

$$S = k_B \left(\frac{A}{4l_p^2}\right)$$

and the energy $E = Mc^2$.

(i) When a black hole of mass M_1 collides with a black hole of mass M_2, if radiation can be neglected, what is the change of entropy? In particular, what is the sign of the entropy change?

(ii) Determine the temperature T_H of the surface.

(iii) Show that the specific heat of the black hole is negative.

Result (iii) violates the thermodynamic stability criterion $C_V > 0$ and can be interpreted as indicating that the black hole is ultimately unstable. In particular, it is expected to evaporate. Since the event horizon is held at temperature T, the surface is expected to emit electromagnetic radiation known as Hawking radiation.

(iv) Assume that the surface of a black hole of temperature T_H emits electromagnetic radiation, show that the lifetime τ of the black hole is given by

$$\tau = 5120 \left(\frac{G^2 M^3}{\hbar c^4}\right)$$

Problem 1.17

Consider a system with total energy U. The system is free to rotate about an axis about which it has a moment of inertia I and has a total angular momentum L. The entropy of the system is a function of the internal energy and angular momentum $S(U, L)$. The system can be considered to be composed of N subsystems. Since entropy is an extensive quantity, the system's entropy can be expressed as the sum of the entropies of the subsystems

$$S(U, L) = \sum_{\alpha=1}^{N} S_\alpha(U_\alpha, L_\alpha)$$

where E_α and L_α are the angular momenta of the subsystems. The total energy and angular momenta are also extensive quantities and, therefore, are given by

$$U = \sum_{\alpha=1}^{N} U_\alpha$$

$$L = \sum_{\alpha=1}^{N} L_\alpha$$

The system is rotating about its axis with angular velocity ω. The entropy of the system in the lab frame is a function of the extensive variables

$$S(U, L)$$

while in a rotating frame the entropy is given by

$$S\left(U - \frac{L^2}{2I}, 0\right)$$

The entropy is independent of the coordinate systems as it is a statistical quantity. In fact, it is a measure of the number of accessible microstates of the system.

(i) Show that the corresponding entropic intensive parameters for $S(U, L)$ are

$$\frac{1}{T}$$

$$-\frac{\omega}{T}$$

(ii) Hence show that if the internal subsystems are in equilibrium, they must have the same temperatures and angular momenta.

Chapter 2

Foundations of Statistical Mechanics

Statistical Mechanics provides us with:

(i) A basis for the first-principles calculations of thermodynamic quantities and transport coefficients of matter in terms of the dynamics of its microscopic constituents.

(ii) A physical significance for entropy.

2.1 Phase Space

In general, "*Phase Space*" Γ is the space of a set of ordered numbers which describes the microscopic states of a many-particle system. For a classical system, one can describe the state of a system by a set of continuously varying variables corresponding to the generalized momenta and generalized coordinates of each particle. However, for quantum systems, the Heisenberg uncertainty principle forbids one to know the momentum and position of any single particle precisely. In this case, the quantum states of a particle can be proscribed by specifying the eigenvalues of a mutually commuting set of operators representing physical observables. The eigenvalues can be either continuous or discrete. Thus, the phase space for a quantum system can either consist of a set of discrete numbers or can consist of a set of continuous numbers, as in the classical case.

2.1.1 *Classical Phase Space*

A microscopic state of a classical system of particles can be described by proscribing all the microscopic coordinates and momenta describing the internal degrees of freedom.

For a classical system of N particles moving in a three-dimensional space, the state of one particle, at any instant of time, can be specified by proscribing the values of the three coordinates (q_1, q_2, q_3) and the values of the three canonically conjugate momenta (p_1, p_2, p_3).

The state of a classical many-particle system, at one instant of time, is proscribed by specifying the values of $3N$ coordinates q_i, ($i \in \{1, 2, 3, \ldots 3N\}$) and the values of $3N$ canonically conjugate momenta p_i, ($i \in \{1, 2, 3, \ldots 3N\}$). The space composed of the ordered set of $6N$ components of the coordinates and momenta is the phase space of the N-particle system. This phase space has $6N$ dimensions.

Distinguishable Particles

For *"Distinguishable Particles"*, which are defined as those for which each particle can be given a unique label, each point in phase space represents a unique microscopic state.

Indistinguishable Particles

By contrast, for *"Indistinguishable Particles"* it is not admissible to label particles. The material is invariant under all permutations of the sets of labels assigned to each of the N particles. There are $N!$ such permutations for the N particle system, and each one of these $N!$ permutations can be built by successively permuting the two sets of (six) labels assigned to pairs of particles. To be sure, the permutation of a particle described by the values of the ordered set of variables $\{q_1, q_2, q_3, p_1, p_2, p_3\}$ and a second particle described by the values of the ordered set $\{q_4, q_5, q_6, p_4, p_5, p_6\}$ is achieved by the interchange of the values $\{q_1, q_2, q_3, p_1, p_2, p_3\} \leftrightarrow \{q_4, q_5, q_6, p_4, p_5, p_6\}$. Any of the $N!$ permutations of the sets of labels assigned to the N particles has the action of transforming one point in phase space to a different point. Since it is not permissable to label indistinguishable particles, the resulting $N!$ different points in phase space must represent the same physical state.

2.1.2 *The Number of Microscopic States*

Given the correspondence between points in phase space and microscopic states of the system, it is useful to introduce a measure of the number of microscopic states of a system N_Γ. One such measure is proportional to the *"Volume of Accessible Phase Space"*. Consider an infinitesimal volume element of phase space, defined by the conditions that the generalized momenta p_i lie in the intervals given by

$$\mathcal{P}_i + \Delta p_i > p_i > \mathcal{P}_i \tag{2.1}$$

and the coordinates q_i are restricted to the intervals

$$\mathcal{Q}_i + \Delta q_i > q_i > \mathcal{Q}_i \tag{2.2}$$

for all i. This infinitesimal volume element $\Delta\Gamma$ is given by

$$\Delta\Gamma = \prod_{i=1}^{3N} (\Delta p_i \Delta q_i) \tag{2.3}$$

The infinitesimal volume element of phase space $\Delta\Gamma$ has dimensions of $\Delta p^{3N} \Delta q^{3N}$. To turn this into a dimensionless quantity, one has to divide

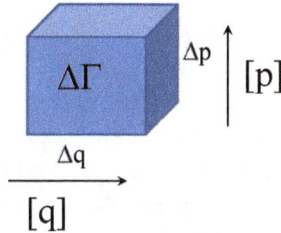

Fig. 2.1 An infinitesimal hyper-cubic volume of phase space $\Delta\Gamma = \Delta p^{3N} \Delta q^{3N}$.

by a quantity with dimensions of $(\text{Action})^{3N}$. Although any quantity with dimensions of the action would do, it is convenient to use $2\pi\hbar$ as the measure for the action. With this particular choice, the dimensionless measure of the volume of phase space is given by

$$\frac{\Delta\Gamma}{(2\pi\hbar)^{3N}} = \prod_{i=1}^{3N} \left(\frac{\Delta p_i \Delta q_i}{2\pi\hbar} \right) \tag{2.4}$$

The identification of $\hbar$ with Planck's constant is convenient since it allows one to make a connection with the number of quantum states, within the quasi-classical limit. The Heisenberg uncertainty principle dictates that the uncertainty in the momentum and position of a single-particle (wave-packet) state cannot be determined to better than $\Delta p_i \Delta q_i > 2\pi\hbar$. Hence, it appears to be reasonable to define the volume of phase space occupied by a single-particle state as $(2\pi\hbar)^3$ and so the dimensionless measure for the number of states for a single-particle system would be given by

$$\prod_{i=1}^{3} \left(\frac{\Delta p_i \Delta q_i}{2\pi\hbar} \right) \tag{2.5}$$

Consequently, the measure of the distinct microscopic states is given by

$$N_\Gamma = \frac{\Delta\Gamma}{(2\pi\hbar)^{3N}} = \prod_{i=1}^{3N}\left(\frac{\Delta p_i \Delta q_i}{2\pi\hbar}\right) \tag{2.6}$$

for a system of N distinguishable particles. If the particles are indistinguishable, the number of distinct microscopic states N_Γ is defined as

$$N_\Gamma = \frac{\Delta\Gamma}{N!(2\pi\hbar)^{3N}} \tag{2.7}$$

where we have divided by $N!$ which is the number of permutations of the N sets of particle labels.

2.2 Trajectories in Phase Space

As time evolves, the system is also expected to evolve with time. For a classical system, the time evolution of the coordinates and momenta are governed by Hamilton's equations of motion, and the initial point in phase space will map out a trajectory in the $6N$-dimensional phase space. For a closed system, where no time-dependent external fields are present, the Hamiltonian is a function of the set of $3N$ generalized momenta and the $3N$ generalized coordinates $H(\{p_i, q_i\})$ and has no explicit time dependence. The rates of change w.r.t. t of the set canonically conjugate coordinates and momenta $\{p_i, q_i\}$, where $i \in \{1, 2, 3, \ldots 3N\}$, are given by the set of Hamilton's equations of motion

$$\frac{dp_i}{dt} = \{p_i, H\}_{PB} = -\frac{\partial H}{\partial q_i}$$

$$\frac{dq_i}{dt} = \{q_i, H\}_{PB} = +\frac{\partial H}{\partial p_i} \tag{2.8}$$

and where P.B. denotes the Poisson Bracket. The Poisson Bracket of two quantities A and B is defined as the antisymmetric quantity

$$\{A, B\}_{PB} = \sum_{i=1}^{3N}\left(\frac{\partial A}{\partial q_i}\frac{\partial B}{\partial p_i} - \frac{\partial B}{\partial q_i}\frac{\partial A}{\partial p_i}\right) \tag{2.9}$$

The trajectory originating form a specific point in phase space will be given by the solution of Hamilton's equations of motion, where the initial conditions correspond to the values of the $6N$ variables at the initial point.

Example: Motion of a Single Particle in One-Dimension

A particle of mass m moving in one dimension in the presence of a potential energy $V(q)$ is described by the Hamiltonian

$$H = \frac{p^2}{2m} + V(q) \tag{2.10}$$

The motion of the particle is described Hamilton's equations of motion which simply reduce to the form

$$\frac{dp}{dt} = -\frac{\partial V}{\partial q}$$

$$\frac{dq}{dt} = \frac{p}{m} \tag{2.11}$$

as is expected.

The time dependence of any physical quantity $A(\{p_i, q_i\} : t)$ can be determined by evaluating A for the set of $\{p_i(t), q_i(t)\}$ which define the system's trajectory in phase space Γ. Hamilton's equation of motion have the consequence that the total derivative of any quantity $A(\{p_i, q_i\} : t)$ can be found from the Poisson Bracket equation of motion

$$\frac{dA}{dt} = \sum_{i=1}^{3N} \left(\frac{dq_i}{dt} \frac{\partial A}{\partial q_i} + \frac{dp_i}{dt} \frac{\partial A}{\partial p_i} \right) + \frac{\partial A}{\partial t}$$

$$= \sum_{i=1}^{3N} \left(\frac{dH}{dp_i} \frac{\partial A}{\partial q_i} - \frac{dH}{dq_i} \frac{\partial A}{\partial p_i} \right) + \frac{\partial A}{\partial t}$$

$$= \{A, H\}_{PB} + \frac{\partial A}{\partial t} \tag{2.12}$$

The first term describes the implicit time dependence of A, due to the dependence of the coordinates and momenta on time, and the second term describes its explicit time dependence.

If a quantity B has no explicit time dependence and the Poisson Bracket of B and H are zero, then B is conserved.

$$\frac{dB}{dt} = \{B, H\}_{PB} + \frac{\partial B}{\partial t}$$

$$= \{B, H\}_{PB}$$

$$= 0 \tag{2.13}$$

where the first two lines follow from our stated assumptions. Since the total derivative governs the change of B as the system flows through phase space, B is conserved. As an example, since our Hamiltonian does not explicitly depend on time, the Poisson Bracket equation of motion shows that the total derivative of the Hamiltonian w.r.t. time is zero. Explicitly, the equation of motion for H is given by

$$\frac{dH}{dt} = \{H, H\}_{PB} + \frac{\partial H}{\partial t}$$
$$= \{H, H\}_{PB}$$
$$= 0 \tag{2.14}$$

where the second line follows from the absence of any explicit time dependence and the last line follows from the antisymmetric nature of the Poisson Bracket. Hence, the energy is a constant of motion for our closed system. That is, the energy is constant over the trajectory traversed in phase space.

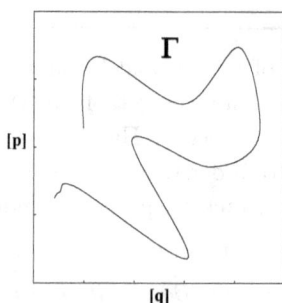

Fig. 2.2 A microscopic state of macroscopic system is described point in phase space and, as it evolves, maps out a very complex trajectory which is governed by Hamilton's equations of motion.

The equations of motion allows us to follow the time evolution of a point in phase space, i.e. the evolution of the microscopic state of the system. The trajectory in phase space may be extremely complicated and rapidly varying. For example in a collision between two neutral molecules, the change in momentum almost exclusively occurs when the separation between the two molecules is of the order of the molecular size. This should be compared with the length scale over which the momentum of each the pair of molecules is constant, which is given by the distance travelled by a molecule between its successive collisions (i.e. the mean free path). The ratio of the mean free path to the molecular size is usually quite large for dilute gasses. If the distances are scaled by the molecular velocities, one concludes that the momentum of the particles changes rapidly at the times when the collisions take place and so the trajectory in phase space changes abruptly. The same distance ratio also implies that the particular form of a trajectory is extremely sensitive to the initial conditions, since a small change in initial conditions determines whether or not a particular collision will occur. The sensitivity to initial conditions and the complexity of the trajectories in phase space prohibit both analytic solution and also numerical solution for realistic materials. Numerical solution is prohibited due to the enormity of the requirements for storing the initial conditions, let alone for implementing the numerical solution of the equations of motion. Despite the complexity of trajectories in phase space, and their sensitivity to initial conditions, the trajectories do have some important common features.

The trajectory of a closed system cannot intersect with itself. This is a consequence of Hamilton's equations of motion which completely specify the future motion of a system, if the set of initial conditions are given and if H has no explicit time dependence. Thus, a trajectory cannot cross itself, since if there was then Hamilton's equations would lead to an indeterminacy at the point of intersection. That is, there would be two possible solutions of Hamilton's equations of motion, if the initial conditions of the system correspond to the crossing point. This is not possible since the information that specifies a trajectory is conserved. However, it is possible that a trajectory closes up on itself and forms a closed orbit.

Secondly, the trajectories only occupy a portion of phase space for which the constants of motion are equal to their initial values.

2.3 Conserved Quantities and Accessible Phase Space

If a system has a set of conserved quantities, then the trajectory followed by the system is restricted to a generalized "surface" in the $6N$-dimensional phase space, on which the conserved quantities take on their initial values. The set of points on the generalized "surface" is known as the "*Accessible Phase Space*" Γ_a.

For a classical system where only the energy is conserved and has the initial value E, the points in the accessible phase space are given by the set of points $\{p_i, q_i\}$ that satisfy the equation

$$H(\{p_i, q_i\}) = E \qquad (2.15)$$

or if the energy is only known to within an uncertainty of ΔE, then the accessible phase space is given by the set of points that satisfy the inequality

$$E + \Delta E > H(\{p_i, q_i\}) > E \qquad (2.16)$$

Example: A One-Dimensional Classical Harmonic Oscillator

The Hamiltonian for a particle of mass m constrained to move in one dimension, subject to a harmonic restoring force, is described by the Hamiltonian

$$H = \frac{p^2}{2m} + \frac{m\omega_0^2}{2} q^2 \qquad (2.17)$$

The phase space Γ of this system corresponds to the entire two-dimensional plane (p, q). If the energy is known to lie in an interval of width ΔE around E, then the accessible phase space Γ_a is determined by

$$E + \Delta E > \frac{p^2}{2m} + \frac{m\omega_0^2}{2} q^2 > E \qquad (2.18)$$

The "surfaces" of constant energy,[1] E, are in the form of ellipses in phase space, with semi-major and semi-minor axes given by the turning points

$$p_{max} = \sqrt{2mE} \qquad (2.19)$$

and

$$q_{max} = \sqrt{\frac{2E}{m\omega_0^2}} \qquad (2.20)$$

The ellipse encloses an area of phase space which is given by

$$\pi \, p_{max} \, q_{max} = 2\pi \frac{E}{\omega_0} \qquad (2.21)$$

Therefore, the accessible phase space Γ_a forms an area enclosed between two ellipses, one ellipse with energy $E + \Delta E$ and another with energy E. Thus, the area of accessible phase space is found as

$$\Gamma_a = 2\pi \frac{\Delta E}{\omega_0} \qquad (2.22)$$

On diving by $2 \pi \hbar$ we can turn Γ_a into a measure of the number of microscopic states accessible to the system N_Γ, we find

$$N_\Gamma = \frac{\Delta E}{\hbar \omega_0} \qquad (2.23)$$

This is a measure of the number of different states accessible to the system, and can be interpreted quantum mechanically as the number of different quantum states which correspond to the energy within the accuracy ΔE that has been specified. The result N_Γ is just the uncertainty in the number of quanta in the system.

---

[1] In this case the "volume" of phase space is an infinite two-dimensional area and, if the energy is specified precisely, the "area" of accessible phase space is a line.

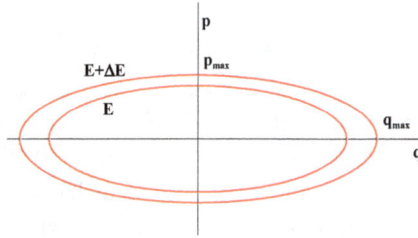

Fig. 2.3   The accessible area of phase space of a one-dimensional Harmonic Oscillator is the area enclosed by the two ellipses.

In the most general case where there are several other conserved quantities $B_j(\{p_i, q_i\})$ (say there are M in number) which have specific values B_j, the accessible phase space will consist of all the points in phase space where the "surfaces" $B_j = B_j(\{p_i, q_i\})$ corresponding to the conserved quantities intersect. That is, the accessible phase space corresponds to all the points which are consistent with the values of the M conserved quantities B_j

$$E = H(\{p_i, q_i\})$$
$$B_j = B_j(\{p_i, q_i\}) \tag{2.24}$$

for all $j \in 1, 2, \ldots M$. In all cases, the physical trajectories of the system are restricted to move within the accessible region of phase space.

2.4   Macroscopic Measurements and Time Averages

The measurement of thermodynamic quantities usually represents a relatively slow process when compared to microscopic time scales. Furthermore, the measurement involves the participation of many of the systems degrees of freedom. This implies that a macroscopic measurement of a quantity A corresponds to a time-average of a quantity $A(\{p_i, q_i\})$ over a trajectory in phase space, over a long period of time. Furthermore, the quantity $A(\{p_i, q_i\})$ must involve many degrees of freedom of the system. For a long period of time T, the macroscopic average is given by

$$\overline{A} = \frac{1}{T} \int_0^T dt\, A(\{p_i(t), q_i(t)\}) \tag{2.25}$$

where $A(\{p_i(t), q_i(t)\})$ varies with time, as the microscopic state changes with time. That is, the set of momenta and coordinates $\{p_i(t), q_i(t)\}$ are considered to be implicit functions of time and are obtained by solving Hamilton's equations using the initial data.

As an example, consider the pressure on a container wall which encloses a dilute gas. The pressure P is defined as the force per unit area. The force F is averaged over a time long compared with the time between molecular collisions with the wall. The force is given by the rate of change of momentum of the molecules impinging on the wall. The force due to a molecular collision occurs over the time-scale which corresponds to the time in which the molecule is in close proximity to the wall. On introducing a short-ranged interaction between the particle and the wall, one finds that the instantaneous force exerted on the wall is given by

$$F_3(t) = \left(\frac{dV(q_3)}{dq_3}\right)\Bigg|_{q_3(t)} \tag{2.26}$$

where $V(q_3)$ is a short-ranged potential due to the interaction of the particle with the wall. Therefore, the instantaneous pressure is given by

$$P(t) = \frac{1}{A}\sum_{i=1}^{N}\left(\frac{dV(q_{3i})}{dq_{3i}}\right)\Bigg|_{q_{3i}(t)} \tag{2.27}$$

where A is the area of the wall. The instantaneous pressure would have the appearance of a sparse sequence of delta-like functions. The thermodynamic pressure is given by the time-average over an interval T in which many collisions occur

$$P = \frac{1}{T}\int_0^T dt\, P(t)$$
$$= \frac{1}{TA}\int_0^T dt \sum_{i=1}^{N}\frac{dV(q_{3i})}{dq_{3i}}\Bigg|_{q_{3i}(t)} \tag{2.28}$$

This result is of the form that we are considering. If the time average is over a long enough time interval, the result should be representative of the equilibrium state in which P does not change with time.

The process of time averaging over long intervals is extremely convenient since it circumvents the question of what microscopic initial conditions should be used. For sufficiently long times, the same average would be obtained for many points on the trajectory. Thus, the long time average is roughly equivalent to an average with a statistical distribution of microscopic initial conditions.

2.5   Ensembles and Averages over Phase Space

The time-average of any quantity over the trajectory in phase space can be replaced by an average over phase space, in which the different volumes are weighted with a distribution function $\rho(\{p_i, q_i\} : t)$. The distribution function may dependent on the point of phase space $\{p_i, q_i\}$, and may also depend on the time t.

Conceptually, the averaging over phase space may be envisaged by introducing an "*Ensemble*" composed of a very large number of identical systems each of which have the same set of values for their measured conserved quantities and all systems must represent the same macroscopic equilibrium state. Although the different systems making up the ensemble correspond to the same macroscopic equilibrium state, the systems may correspond to different microstates. The distribution of the virtual copies of the system in an ensemble is described by a probability distribution for the microscopic states in phase space. The concept of "*Ensemble Averaging*" was first introduced by Maxwell in 1879 and developed more fully by Gibbs in 1909.

There are infinitely many possible choices of ensembles, one trivial example is that each system in the ensemble corresponds to the same initial microstate. Another example corresponds to taking all the different points of a trajectory of one microscopic state as the initial states of the ensemble. Although a mechanical system does evolve in time, this type of ensemble is stationary as it doesn't evolve in time. A frequently used ensemble corresponds to distributing the probability density equally over all points in phase space compatible with the measured quantities of the macroscopic state.

2.5.1   *The Probability Distribution Function*

The classical probability distribution function $\rho(\{p_i, q_i\} : t)$ could, in principle, be measured by measuring the microstates of the systems composing an ensemble at time t and determining the relative number of systems which are found in microstates in the volume $d\Gamma$ of phase space around the point $\{p_i, q_i\}$. In the limit that the number of systems in the ensemble goes to infinity, this ratio reduces to a probability. The probability $dp(t)$ is expected to be proportional to the volume of phase space $d\Gamma$. Therefore, we expect that

$$dp(t) = \rho(\{p_i, q_i\} : t) \, d\Gamma \qquad (2.29)$$

where $\rho(\{p_i, q_i\} : t)$ is the probability distribution function. The probability distribution function is only finite for the accessible volume of phase space. Since probabilities are non-negative, then so is the probability distribution function. Furthermore, since the probabilities are defined to be normalized to unity, the probability distribution function must also be normalized

$$1 = \int dp(t)$$

$$= \int d\Gamma \; \rho(\{p_i, q_i\} : t) \tag{2.30}$$

for all times t. For a macroscopic system, the integration over $d\Gamma$ may be restricted to the volume of accessible phase space with any loss of generality. This is true since, the probability of finding a system in an inaccessible region of phase space is identically zero. The quantum mechanical generalization of the classical probability density is discussed in Chapter 6.

2.5.2 *Ensemble Averages*

Once $\rho(\{p_i, q_i\} : t)$ has been determined, the measured value of any physical quantity $A(\{p_i, q_i\} : t)$ of a system in a macroscopic state at an instant of time, t, can then be represented by an ensemble average. The "*Ensemble Average*" is the average over phase space weighted by the probability distribution function

$$\overline{A}(t) = \int d\Gamma \; A(\{p_i, q_i\} : t) \; \rho(\{p_i, q_i\} : t) \tag{2.31}$$

The ensemble average occurs at an instant of time t and so has a closer resemblance to measurements in quantum mechanics than to measurements of thermodynamic quantities which involve averaging over long time intervals.

If all the different points of a trajectory of one microscopic state is taken to define the initial states of the ensemble, the ensemble averages will coincide with the long-time average for the microscopic state. At the other extreme, if each system in the ensemble corresponds to the same initial microstate, then the ensemble average of a quantity at any time t will simply correspond to the value of the quantity for the microstate at time t.

The fundamental problem of statistical mechanics is to find the probability distribution function for the ensemble that describes measurements on the macroscopic equilibrium states of physical systems most closely. We shall examine the equations that determine the time-dependence of the probability distribution function in the next section.

2.6 Liouville's Theorem

"Liouville's Theorem" concerns how the probability distribution function for finding our N-particle system in some volume element of phase space at time t varies with time.

Since the probability is normalized and since the states of a system evolve on continuous trajectories in phase space, the probability density must satisfy a continuity equation. Consider a volume element $d\Gamma$ of phase space, the number of systems in the ensemble that occupy this volume element is proportional to

$$\rho(\{p_i, q_i\} : t) \, d\Gamma \tag{2.32}$$

and the increase of the number of systems in this volume element that occurs in the infinitesimal time interval dt is proportional to

$$(\rho(\{p_i, q_i\} : t + dt) - \rho(\{p_i, q_i\} : t)) \, d\Gamma \approx \frac{\partial}{\partial t} \rho(\{p_i, q_i\} : t) \, d\Gamma \, dt \tag{2.33}$$

where we have used the Taylor expansion to obtain the right hand side of the equation. Due to the continuous nature of classical trajectories, the increase in the number of trajectories in the volume must be due to system trajectories which cross the surface of our $6N$-dimensional volume. That is, the net increase must be due to an excess of the flow across the bounding surfaces into the volume over the flow out of the volume.

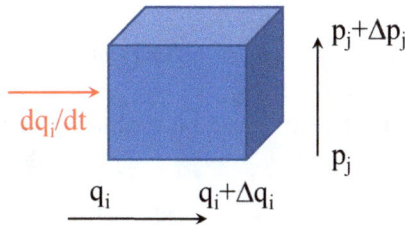

Fig. 2.4 An infinitesimal hyper-cubic element of phase of dimensions $\Delta\Gamma = \prod_{i=1}^{3N} \Delta q_i \Delta p_i$. In time interval dt, the probability density within a distance $\frac{dq_i}{dt}$ perpendicular to the bounding surface at q_i is swept into the volume.

Consider the infinitesimal volume of phase space $d\Gamma$ where the i-th coordinate is restricted to be between q_i and $q_i + \Delta q_i$ and the i-th generalized momentum is restricted to be between p_i and $p_i + \Delta p_i$. The volume element $d\Gamma$ is given by

$$d\Gamma = \prod_{i=1}^{3N} (\Delta q_i \, \Delta p_i) \tag{2.34}$$

The pair of opposite surfaces defined by the coordinates q_i and $q_i + \Delta q_i$ have $6N - 1$ dimensions and have an area given by

$$\prod_{j=1, j\neq i}^{3N} \Delta q_j \prod_{j=1}^{3N} \Delta p_j \tag{2.35}$$

Trajectories which enter or leave the volume element $d\Gamma$ must cross one of its $6N$ boundaries.

Flow In Across a Surface

All the systems of the ensemble in microstates within a distance $\dot{q}_i\, dt$ behind the surface at q_i will enter $d\Gamma$ in time dt. That is, the ensemble systems in the volume $\prod_{j=1, j\neq i}^{3N} \Delta q_j \prod_{j=1}^{3N} \Delta p_j\, \dot{q}_i(\{p_i, q_i\})\, dt$ will enter $d\Gamma$ in the time interval dt. The number of systems in this volume is proportional to

$$\prod_{j=1, j\neq i}^{3N} \Delta q_j \prod_{j=1}^{3N} \Delta p_j\, dt\, \dot{q}_i(\{p_i, q_i\})\, \rho(\{p_i, q_i\} : t) \tag{2.36}$$

Flow Out Across a Surface

All the systems in the ensemble with microstates that are within a distance $\dot{q}_i\, dt$ behind the surface at $q_i + \Delta q_i$ will leave $d\Gamma$ in time dt. The number of systems in this volume is proportional to

$$\prod_{j=1, j\neq i}^{3N} \Delta q_j \prod_{j=1}^{3N} \Delta p_j\, dt\, \dot{q}_i(\{p_i, q_i + \Delta q_i\})\, \rho(\{p_i, q_i + \Delta q_i\} : t) \tag{2.37}$$

where the velocity and density must be evaluated at the position of the second surface.

The Net Flow into the Volume

The net flow into $d\Gamma$ from a pair of coordinate surfaces is given by the difference of the flow crossing the coordinate surface entering the volume and the flow crossing the opposite surface thereby leaving the volume

$$\prod_{j=1, j\neq i}^{3N} \Delta q_j \prod_{j=1}^{3N} \Delta p_j\, dt[\dot{q}_i(\{p_i, q_i\})\, \rho(\{p_i, q_i\} : t)$$

$$- \dot{q}_i(\{p_i, q_i + \Delta q_i\})\, \rho(\{p_i, q_i + \Delta q_i\} : t)]$$

$$\approx -\prod_{j=1}^{3N} \Delta q_j \prod_{j=1}^{3N} \Delta p_j\, dt\, \frac{\partial}{\partial q_i}(\dot{q}_i(\{p_i, q_i\})\, \rho(\{p_i, q_i\} : t)) \tag{2.38}$$

where we have Taylor expanded in powers of Δq_i. Likewise, the net flow into $d\Gamma$ from the pair of momentum surfaces at p_i and $p_i + \Delta p_i$ is given by

$$\prod_{j=1}^{3N} \Delta q_j \prod_{j=1, j \neq i}^{3N} \Delta p_j \, dt[\dot{p}_i(\{p_i, q_i\}) \, \rho(\{p_i, q_i\} : t)$$

$$- \dot{p}_i(\{p_i + \Delta p_i, q_i\}) \, \rho(\{p_i + \Delta p_i, q_i\} : t)]$$

$$\approx -\prod_{j=1}^{3N} \Delta q_j \prod_{j=1}^{3N} \Delta p_j \, dt \, \frac{\partial}{\partial p_i}(\dot{p}_i(\{p_i, q_i\}) \, \rho(\{p_i, q_i\} : t)) \qquad (2.39)$$

On summing over all the $6N$ surfaces, one finds that the net increase of the number of ensemble systems in the volume $d\Gamma$ that occurs in time dt due to their flowing across all its boundaries is proportional to

$$-\prod_{j=1}^{3N} \Delta q_j \prod_{j=1}^{3N} \Delta p_j \, dt \left[\frac{\partial}{\partial q_i}(\dot{q}_i(\{p_i, q_i\}) \, \rho(\{p_i, q_i\} : t)) \right.$$

$$\left. + \frac{\partial}{\partial p_i}(\dot{p}_i(\{p_i, q_i\}) \, \rho(\{p_i, q_i\} : t)) \right] \qquad (2.40)$$

The Continuity Equation

On equating the net increase of the probability in the infinitesimal volume element with the net probability flowing into the volume, one can cancel the factors of dt and $d\Gamma$. Hence, one finds that the probability density satisfies the linear partial differential equation

$$\frac{\partial \rho}{\partial t} + \sum_{i=1}^{3N} \left[\frac{\partial}{\partial q_i}(\dot{q}_i \, \rho) + \frac{\partial}{\partial p_i}(\dot{p}_i \, \rho) \right] = 0 \qquad (2.41)$$

On expanding the derivatives of the products one obtains

$$\frac{\partial \rho}{\partial t} + \sum_{i=1}^{3N} \left[\frac{\partial \dot{q}_i}{\partial q_i} \rho + \dot{q}_i \frac{\partial \rho}{\partial q_i} + \frac{\partial \dot{p}_i}{\partial p_i} \rho + \dot{p}_i \frac{\partial \rho}{\partial p_i} \right] = 0 \qquad (2.42)$$

The above expression simplifies on using Hamilton's equations of motion

$$\dot{q}_i = \frac{\partial H}{\partial p_i}$$

$$\dot{p}_i = -\frac{\partial H}{\partial q_i} \qquad (2.43)$$

so one obtains

$$\frac{\partial \dot{q}_i}{\partial q_i} = \frac{\partial^2 H}{\partial q_i \partial p_i}$$

$$\frac{\partial \dot{p}_i}{\partial p_i} = -\frac{\partial^2 H}{\partial p_i \partial q_i} \tag{2.44}$$

On substituting these two relations in the equation of motion for ρ, the pair of second-order derivatives cancel and one finally obtains Liouville's equation

$$\frac{\partial \rho}{\partial t} + \sum_{i=1}^{3N} \left[\dot{q}_i \frac{\partial \rho}{\partial q_i} + \dot{p}_i \frac{\partial \rho}{\partial p_i} \right] = 0 \tag{2.45}$$

That is, the total derivative of ρ vanishes

$$\frac{d\rho}{dt} = \frac{\partial \rho}{\partial t} + \sum_{i=1}^{3N} \left[\dot{q}_i \frac{\partial \rho}{\partial q_i} + \dot{p}_i \frac{\partial \rho}{\partial p_i} \right] = 0 \tag{2.46}$$

The total derivative is the derivative of ρ evaluated on the trajectory followed by the system. Hence, ρ is constant along the trajectory. Therefore, Liouville's Theorem states that ρ flows like an incompressible fluid.

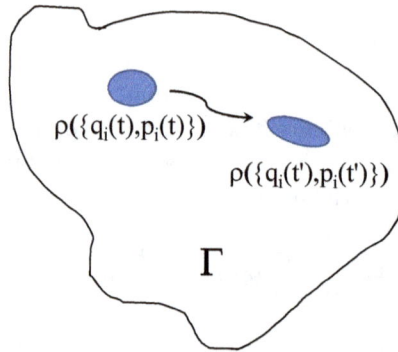

Fig. 2.5   The time evolution of an inhomogeneous probability density $\rho(\{q_i, p_i\})$ satisfies a continuity equation.

On substituting Hamilton's equations for the expressions for $\dot{q}_i$ and $\dot{p}_i$ in Liouville's Theorem, one recovers the Poisson Bracket equation of motion for ρ

$$\frac{d\rho}{dt} = \frac{\partial \rho}{\partial t} + \sum_{i=1}^{3N} \left[\frac{\partial H}{\partial p_i} \frac{\partial \rho}{\partial q_i} - \frac{\partial H}{\partial q_i} \frac{\partial \rho}{\partial p_i} \right]$$

$$= \frac{\partial \rho}{\partial t} + \{\rho, H\}_{PB} = 0 \tag{2.47}$$

which is in a form suitable for Canonical Quantization, in which case ρ and H should be replaced by operators and the Poisson Bracket by a commutator times an imaginary number.

Liouville's Theorem is automatically satisfied for any ρ which has no explicit t-dependence and can be expressed in terms of the constants of motion. Specifically, when ρ is initially uniform over the accessible phase space, Liouville's Theorem ensures that it will remain constant. To be sure, if the distribution satisfies

$$\frac{\partial \rho}{\partial p_i} = 0 \quad \forall i \tag{2.48}$$

and

$$\frac{\partial \rho}{\partial q_i} = 0 \quad \forall i \tag{2.49}$$

for all points $\{p_i, q_i\}$ within the accessible volume of phase space (defined by $H(\{p_i, q_i\}) = E$ and any other relevant conservation laws) then Liouville's Theorem reduces to

$$\frac{\partial \rho}{\partial t} = 0 \tag{2.50}$$

Example: A Particle in a One-Dimensional Box

We shall consider an example that illustrates how a probability density thins and folds as time evolves. The example also shows that for sufficiently large times, the probability distribution is finely divided and distributed over the volume of accessible phase space.

We shall consider an ensemble of systems. Each system is composed of a single particle that is confined in a one-dimensional box of length L. When the particle is not in contact with the walls, the Hamiltonian reduces to

$$H(p, q) = \frac{p^2}{2m} \tag{2.51}$$

The energies of the systems in the ensemble are bounded by

$$E + \Delta E > H(p, q) > E \tag{2.52}$$

which restricts the momenta to the two intervals

$$p_{max} > p > p_{min} \quad \text{and} \quad -p_{min} > p > -p_{max} \tag{2.53}$$

The coordinates are restricted to the interval

$$\frac{L}{2} > q > -\frac{L}{2} \tag{2.54}$$

Thus, the volume of accessible phase space consists of two two-dimensional strips.

The probability distribution $\rho(p, q : t)$ evolves according to Liouville's Theorem

$$\frac{\partial \rho}{\partial t} + \frac{\partial H}{\partial p}\frac{\partial \rho}{\partial q} - \frac{\partial H}{\partial q}\frac{\partial \rho}{\partial p} = 0 \qquad (2.55)$$

which for volumes contained within the spatial boundaries reduces to

$$\frac{\partial \rho}{\partial t} + \frac{p}{m}\frac{\partial \rho}{\partial q} = 0 \qquad (2.56)$$

This equation has the general solution

$$\rho(p, q : t) = A\left(q - \frac{pt}{m}\right)B(p) \qquad (2.57)$$

which is valid everywhere except at the locations of the walls. In the general solution A and B are arbitrary functions which must be fixed by the boundary conditions.

We shall adopt the initial condition that the probability distribution function has the form

$$\rho(p, q : 0) = \delta(q)\,B(p) \qquad (2.58)$$

which initially confines all the particles in the ensemble to the center $q = 0$. The momentum distribution function $B(p)$ is evenly distributed over the allowed range

$$B(p) = \frac{1}{2\,(p_{max} - p_{min})}[\Theta(p - p_{max}) - \Theta(p - p_{min})$$
$$+\, \Theta(p + p_{max}) - \Theta(p + p_{min})] \qquad (2.59)$$

For sufficiently short times, short enough so that the particles in the ensemble have not yet made contact with the walls, the solution is of the form

$$\rho(p, q : t) = \delta\left(q - \frac{p}{m}t\right)B(p) \qquad (2.60)$$

which has the form of two segments of a line. The slope of the line in phase space is given by m/t. For small times the segments are almost vertical, but the slope increases as t increases. The increase in slope is caused by the dispersion of the velocities, and causes the length of the line to increase. The increase in the length of the line does not affect the normalization, which is solely determined by $B(p)$. At a time T_1 some particles in the

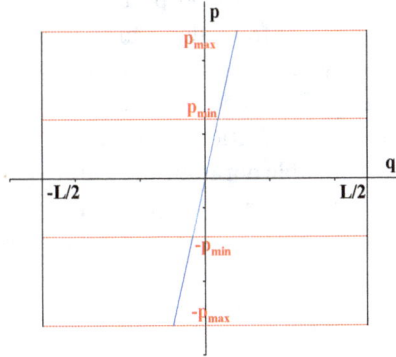

Fig. 2.6   The regions where the probability density for an ensemble of systems composed of a particle in a box is non-zero, at short times, is shown by the solid portion of the blue line. The slope of the line is caused by the dispersion in the velocities. The accessible phase space is enclosed by the red dashed lines between p_{max} and p_{min}, and a similar region in the lower half space.

ensemble will first strike the walls, that is the line segments in available phase space will extend to $q = \pm \frac{L}{2}$. This first happens when

$$T_1 = \frac{Lm}{2p_{max}} \tag{2.61}$$

For times greater than T_1 some of the ensemble's particles will be reflected from the walls. The solution of Liouville's equation can be found by the

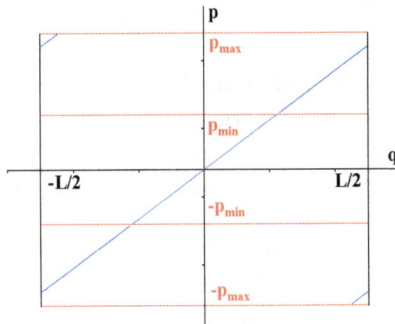

Fig. 2.7   The regions where the probability density for an ensemble of particles in boxes is non-zero, for times slightly greater than the times of the first collision, is shown by the solid portion of the blue line. The two small line segments in the upper left-hand and lower right-hand portion of accessible phase space represents the region of the probability density for systems where the particle has been reflected.

method of images. That is, the reflected portion of the probability density can be thought of as originating from identical systems with identical initial conditions except that they are obtained by spatially reflecting our system at its boundaries $q = \pm \frac{L}{2}$. The reflection requires that $B(p) \to B(-p)$ in the image. The probability distribution emanating from these other systems will enter the volume of available our available phase space at time T_1 which will represent the reflected portion of the probability distribution function. The probability distribution that leaves our system, represents the reflected

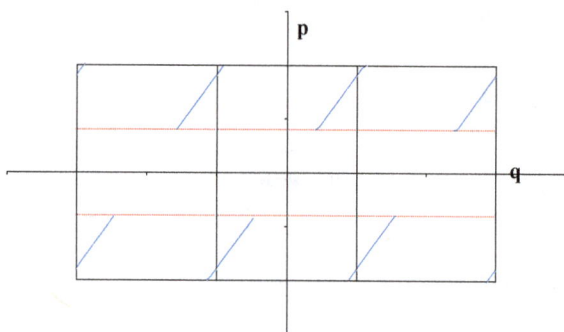

Fig. 2.8 The extended phase space produced by reflecting the central area across its boundaries. In this extended system, the reflected probability density is simply represented by the free evolution of the initial probabilities of the image systems.

portion of the probability distribution for the neighboring systems. Thus, we are mentally extending the region of accessible phase space in the spatial direction. The solution just after the first reflection has occurred, but for times before any system has experienced two reflections is given by

$$\rho(p, q : t) = \sum_{n=-1}^{n=1} \delta\left(q - nL - \frac{p}{m} t\right) B(p) \qquad (2.62)$$

where q is restricted to the interval $\frac{L}{2} > q > -\frac{L}{2}$. The folding of the distribution does not affect its normalization.

For larger times, for which any system in the ensemble has undergone multiple reflections, the set of systems must be periodically continued along the spatial axis. That is, we must consider multiple images of our system. The probability distribution valid at any time obviously has the form

$$\rho(p, q : t) = \sum_{n=-\infty}^{\infty} \delta\left(q - nL - \frac{p}{m} t\right) B(p) \qquad (2.63)$$

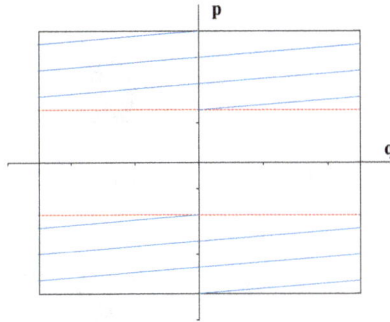

Fig. 2.9   The regions where the probability density for a particle in a box is non-zero, for large times, is shown by the solid blue lines. For large times, particles in the ensemble have experienced different numbers of collisions and is spread over many line segments.

where q is still restricted to the interval of length L. The probability distribution is non-zero on a set of parallel line segments with slope m/t. The line segments are separated by a "distance" $(mL)/t$ along the momentum direction. For sufficiently large times, the slope of the lines will be small and they will be closely spaced. In conclusion, for sufficiently large times, we have shown that the probability distribution will be finely divided and spread throughout the volume of accessible phase space. The phenomenon of thinning and the eventual spreading of the probability distribution throughout all of accessible phase space is known as *"Mixing"* [4].

2.7   The Time Dependence of Averages

Liouville's Theorem shows that the time dependence of any quantity $A(\{p_i, q_i\})$ (with no explicit t dependence) also follows from the Poisson Bracket equations. This can be seen by first multiplying Liouville's equation by $A(\{p_i, q_i\})$ and then integrating over phase space.

$$0 = \int d\Gamma \, A(\{p_i, q_i\}) \, \frac{\partial \rho}{\partial t} + \int d\Gamma \, A(\{p_i, q_i\})\{\rho, H\}_{PB} \qquad (2.64)$$

The derivatives of ρ w.r.t the variables $\{p_i, q_i\}$ that occur in the Poisson Bracket term can be removed by integrating by parts. That is, on noting that ρ vanishes on the boundaries of the integration, then integration by parts yields

$$0 = \int d\Gamma \, A \, \frac{\partial \rho}{\partial t} - \sum_{i=1}^{3N} \int d\Gamma \, \rho \left[\frac{\partial}{\partial q_i}\left(A \, \frac{\partial H}{\partial p_i} \right) - \frac{\partial}{\partial p_i}\left(A \, \frac{\partial H}{\partial q_i} \right) \right] \qquad (2.65)$$

The derivatives of the products can be expanded to yield

$$\int d\Gamma \, A \, \frac{\partial \rho}{\partial t} = \sum_{i=1}^{3N} \int d\Gamma \, \rho \left[\frac{\partial A}{\partial q_i} \frac{\partial H}{\partial p_i} + A \frac{\partial^2 H}{\partial q_i \partial p_i} - \frac{\partial A}{\partial p_i} \frac{\partial H}{\partial q_i} - A \frac{\partial^2 H}{\partial p_i \partial q_i} \right] \tag{2.66}$$

The terms proportional to the second derivative of the Hamiltonian cancel, leading to

$$\int d\Gamma \, A \, \frac{\partial \rho}{\partial t} = \sum_{i=1}^{3N} \int d\Gamma \, \rho \left[\frac{\partial A}{\partial q_i} \frac{\partial H}{\partial p_i} - \frac{\partial A}{\partial p_i} \frac{\partial H}{\partial q_i} \right]$$

$$= \int d\Gamma \, \rho \{ A, H \}_{PB} \tag{2.67}$$

The first term represents the time derivative of the average of A, $\frac{d\overline{A}}{dt}$. This can be seen by considering an ensemble average of A at an instant of time t, given by

$$\overline{A}(t) = \int d\Gamma \, \rho(t) \, A \tag{2.68}$$

so on taking the derivative of the above equation w.r.t. t, one obtains

$$\frac{d\overline{A}}{dt} = \int d\Gamma \, \frac{\partial \rho(t)}{\partial t} \, A \tag{2.69}$$

where the integration over Γ runs over a fixed volume and we have assumed that A has no time dependence. Only ρ is time dependent. Thus, we have shown that

$$\frac{d\overline{A}}{dt} = \int d\Gamma \, \rho \{ A, H \}_{PB} \tag{2.70}$$

Therefore, the time-derivative of the average of A is equated with the average of the Poisson Brackets of A and H. The difference between the two expressions for the rate of change of a quantity A, which has no explicit time dependence, is that in the previous expression the phase space integration volume element is held fixed and $\rho(t)$ is a dynamic quantity, whereas in the last expression the coordinates and momenta are considered as dynamic variables. This is similar to the transformation between the quantum mechanical averages in the Schrödinger and Heisenberg pictures.

The statement that the time-derivative of the average of quantity A which has no explicit time-dependence is given by

$$\frac{d\overline{A}}{dt} = \int d\Gamma \, \frac{\partial \rho}{\partial t} \, A \tag{2.71}$$

has an important consequence for the equilibrium state probability distribution function $\rho(t)$. In equilibrium, all physical quantities should be time-independent, so we require

$$\frac{d\overline{A}}{dt} = \int d\Gamma \, \frac{\partial \rho}{\partial t} \, A = 0 \tag{2.72}$$

for any A. Thus, the requirement that, in an equilibrium state, the average of any time-independent quantity $A(\{p_i, q_i\})$ necessitates that

$$\frac{\partial \rho}{\partial t} = 0 \tag{2.73}$$

since this is independent of any choice for A. Because ρ also satisfies

$$\frac{\partial \rho}{\partial t} = -\{\rho, H\}_{PB} \tag{2.74}$$

one deduces that the Poisson Bracket of ρ and H must vanish for an equilibrium state. This can be achieved if the equilibrium probability distribution function ρ only depends on the Hamiltonian H and any other conserved quantities.

2.8 The Ergodic Hypothesis

In proving Liouville's Theorem, we noted that

$$\frac{\partial \dot{q}_i}{\partial q_i} = \frac{\partial^2 H}{\partial q_i \partial p_i}$$

$$\frac{\partial \dot{p}_i}{\partial p_i} = -\frac{\partial^2 H}{\partial p_i \partial q_i} \tag{2.75}$$

This has the consequence that, if one follows the flow of the systems of the ensemble with microstates contained in a specific volume of phase space $d\Gamma$ at time t, then at time t' the set of microstates will have evolved to occupy a volume of phase space $d\Gamma'$ such that

$$d\Gamma = d\Gamma' \tag{2.76}$$

This can be seen by considering the product of the canonically conjugate pairs of infinitesimal momenta and coordinates at time t

$$dp_i \, dq_i \tag{2.77}$$

At time $t+dt$ the time evolution will have mapped the ends of these intervals onto new intervals such that the lengths of the new intervals are given by

$$dp_i' = dp_i \left(1 + \frac{\partial \dot{p}_i}{\partial p_i} \, dt\right) \tag{2.78}$$

and

$$dq_i' = dq_i \left(1 + \frac{\partial \dot{q}_i}{\partial q_i}\, dt\right) \tag{2.79}$$

Therefore, the product of the new intervals is given by

$$dp_i\, dq_i \left[1 + \left(\frac{\partial \dot{q}_i}{\partial q_i} + \frac{\partial \dot{p}_i}{\partial p_i}\right) dt + O(dt^2)\right] \tag{2.80}$$

which since

$$\frac{\partial \dot{q}_i}{\partial q_i} + \frac{\partial \dot{p}_i}{\partial p_i} = 0 \tag{2.81}$$

leaves the product invariant, to first-order in dt. Hence, since

$$d\Gamma' = \prod_{i=1}^{3N} dp_i' dq_i' \tag{2.82}$$

the size of the volume element occupied by the microstates is invariant, i.e. $d\Gamma = d\Gamma'$. This does not imply that the shape of the volume elements remains unchanged, in fact they will become progressively distorted as time evolves. For most systems for which the trajectories are very sensitive to the initial conditions, the volume elements will be stretched and folded, resulting in the volume being finely divided and distributed over the accessible phase space.

The initial formulation of the *"Ergodic Hypothesis"* was introduced by Boltzmann [5] in 1871. A modified form of the hypothesis asserts that if the volume of accessible phase space is finite, then given a sufficiently long time interval, the trajectories of the microstates initially contained in a volume element $d\Gamma$ will come arbitrarily close to every point in accessible phase space. If this hypothesis is true, then a long-time average of an ensemble containing states initially in $d\Gamma$ will be practically equivalent to an average over the entire volume of accessible phase space with a suitable probability density. That is, the Ergodic Hypothesis leads one to expect that the equation

$$\overline{A} = \frac{1}{T} \int_0^T dt\, A(\{p_i(t), q_i(t)\}) = \int d\Gamma\, A(\{p_i, q_i\})\, \rho(\{p_i, q_i\}) \tag{2.83}$$

holds for some $\rho(\{p_i, q_i\})$ (*"the Ergodic Distribution"*) at sufficiently large times T.

2.8.1 *The Ergodic Theorem*

The *"Ergodic Theorem"* (due to Birkhoff [6] and then improved on by J. von Neumann [7,8]) concerns the time-average of a quantity A along a trajectory that is initially located at almost any point in phase space. The theorem states that, in the limit as the time elapse goes to infinity:

(i) the time-average of the quantity converges to a limit.

(ii) the limit is equal to the weighted average of the quantity over accessible phase space. That is, the trajectory emanating from almost any initial point resembles the whole of the accessible phase space.

The Ergodic Theorem has been proved for collisions of hard spheres and for motion on the geodesics on surfaces with constant negative curvature. Ergodicity can also be demonstrated for systems through computer simulations. The Ergodic Theorem has similar implications to a weaker theorem which is known as the Poincaré Recurrence Theorem.

2.8.2 *Poincaré's Recurrence Theorem*

The *"Poincaré Recurrence Theorem"* [9] states that most systems will, after a sufficiently long time, return to a state very close to their initial states. The Poincaré Recurrence Time T_R is the time interval that has elapsed between the initial time and the time when the system recurs. The theorem was first proved by Henri Poincaré in 1890.

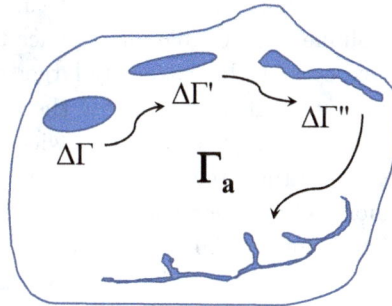

Fig. 2.10 A schematic description of Poincaré's Recurrence Theorem. Under time-evolution, a region $\Delta\Gamma$ of phase space sweeps out a trajectory in phase space. At each instant of time, the region occupies an equal volume of the accessible phase space Γ_a, so that $\Delta\Gamma = \Delta\Gamma' = \Delta\Gamma''$. After a time T_R, the trajectory of the region will come arbitrarily close to the initial region.

The proof is based on the two facts:

(i) The phase-space trajectories of a closed system do not intersect.

(ii) The volume of an infinitesimal region of phase space, whose points evolves with time on their phase-space trajectories, is conserved.

Consider an arbitrarily small neighbourhood around any initial point in accessible phase space and follow the volume's trajectory as the microstates evolve with time. The volume "sweeps out" a tube in phase space as it moves. The tube can never cross the regions that have been already "swept out", since trajectories in phase space do not intersect. Hence, as the accessible phase space is a compact manifold, the total volume available for future motion without recurrence will decrease as the time increases. If the tube has not already returned to the initial neighborhood (in which case recurrence has already occurred), then since the total volume of accessible phase space is finite, in a finite time T_R all the volume of accessible phase space must be exhausted. At that time, the only possibility is that the phase-space tube returns to the neighbourhood of the initial point.

Quod Erat Demonstrandum (QED)

Thus, the trajectory comes arbitrarily close to its initial point at a later time T_R, which is defined as the *"Poincaré Recurrence Time"*. If the trajectories don't repeat and form closed orbits, they must densely fill out all the available phase space. However, if Poincaré recurrence occurs before the entire volume of accessible phase space is swept out, some of it may remain unvisited. Liouville's Theorem implies that the density of trajectories is uniform in the volume of accessible phase space that is visited. The recurrence time T_R is expected to be extremely larger than the time scale for any macroscopic measurement, in which case the Recurrence Theorem cannot be used to justify replacing time averages with ensemble averages.

A simple example, where the Ergodic Theorem holds, is given by the One-dimensional Harmonic Oscillator. The Double-Well Potential exhibits a region where ergodicity applies, but for low energies the motion may become constrained to just one well in which case ergodicity does not apply.

If the Ergodic Theorem holds then, for sufficiently large times T_R, the time-average of any quantity A

$$\overline{A} = \frac{1}{T_R} \int_0^{T_R} dt \ A(\{p_i(t), q_i(t)\}) \tag{2.84}$$

represents the measured value of a macroscopic quantity and the trajectory passes arbitrarily close to every point in phase space. If the system's trajectory dwells in the volume of phase space $\Delta\Gamma$ for time Δt, then the ratio

$$\frac{\Delta t}{T_R} \tag{2.85}$$

has a definite limit which defines the probability that, if the system is observed at some instant of time, it will be found to have a microscopic state in $\Delta\Gamma$.

There are a number of systems which are known not to obey the Ergodic Hypothesis. These include integrable systems, or nearly integrable systems. An "*Integrable System*" has a number of conservation laws B_i equal to half the number of dimensions of phase space. Furthermore, each pair of conserved quantities must be in involution

$$\{B_i, B_j\}_{PB} = 0 \tag{2.86}$$

These sets of conservation laws reduce the trajectory of an integrable system to motion on a $3N$-dimensional surface embedded in the $6N$-dimensional phase space. Furthermore, due to the involution condition, the normals to the surface of constant B_i are orthogonal to the trajectory flows which define a coordinate grid on the surface that does not have a singularity. The singularities of coordinate grids define the topology of the surface. For example, a coordinate grid on the surface of a sphere has two singularities (one at each pole), whereas a coordinate grid on a doughnut does not. The trajectories of an integrable system are confined to surfaces that have the topology of $3N$-dimensional tori. Different initial conditions will lead to different tori that are nested within phase space. The motion on the torus can be separated into $3N$ different types of periodic modes. Since these modes have different frequencies the motion is quasi-periodic. It is the separability of the coordinates that makes integrability a very special property. The very existence of $3N$ conserved quantities implies that most of the conserved quantities are microscopic and their values are not directly measurable by macroscopic means. Hence, they should not be considered as restricting the available phase space for the macroscopic state. Thus, it should be no surprise that integrable systems are generally considered to be non-ergodic.

Example: Circular Billiards

A simple example of an integrable system is given by the circular billiard. In this case, a particle is free to move within a circular area of the plane. The particle is confined to the area of radius R since it is specularly reflected by the perimeter. The spatial paths followed by the particle consists of a succession of cords. Whenever a cord meets the perimeter, the angle α between the cord and the perimeter's tangent is the same for each reflection. The phase space is four-dimensional. However, there are two constant of

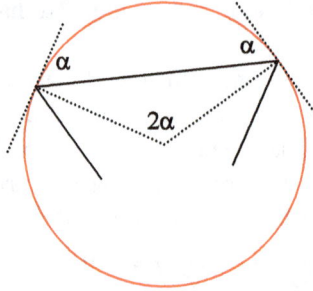

Fig. 2.11 The basic geometry of the scattering for circular billiards.

motion; the energy E and the angular momentum p_φ. Thus, the system is classified as an integrable system. The angle φ satisfies the equation of motion

$$\dot{\varphi} = \frac{p_\varphi}{mr^2} \tag{2.87}$$

The radial motion is described by the pair of equations

$$\dot{r} = \frac{p_r}{m}$$

$$\dot{p}_r = \frac{p_\varphi^2}{mr^3} - V_0\,\delta(r - R) \tag{2.88}$$

where $V_0 \to \infty$. The second equation can be integrated once w.r.t. t by using an integrating factor of p_r on the left and $m\,\dot{r}$ on the right, which introduces a constant of motion which is the (conserved) energy E given by

$$E = \frac{p_r^2}{2m} + \frac{p_\varphi^2}{2mr^2} + V_0\,\Theta(r - R)$$

$$= \frac{p_r^2}{2m} + \frac{p_\varphi^2}{2mr^2} \tag{2.89}$$

where the last line holds for $0 < r < R$. The remaining integration of $\dot{r}$ w.r.t. t leads to the solution for $r(t)$. From this one sees that the radial coordinate $r(t)$ performs anharmonic oscillations between a turning point denoted as a and the radius R (where the particle is reflected). The turning point radius a can be obtained from the principle of conservation of energy and is found as

$$a = \frac{p_\varphi}{\sqrt{2mE}} \tag{2.90}$$

where $0 < a < R$. The ratio of the period for a 2π rotation of φ to the period of the radial motion may not be a rational number. Hence, in general the motion is quasi-periodic. Also since the turning point a is a constant,

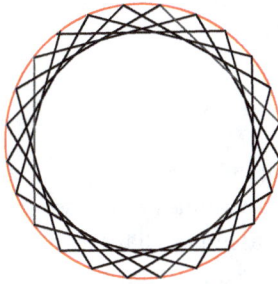

Fig. 2.12   A spatial path traced out by the billiard ball over a long time interval.

the paths in position space are obviously excluded from a circular region of radius a centered on the origin. This area is inaccessible because it does not correspond to the known values of the two-conserved quantities, which are the energy and angular momentum.

When a system does not have $3N$ conserved quantities, the system is "*Non-Integrable*". The trajectories have much fewer restrictions and extend to higher dimensions in phase space. The trajectories are more sensitive to the initial conditions, so the trajectories are inevitably chaotic.

The "*Kolmogorov-Arnold-Moser (KAM) Theorem*" indicates that there is a specific criterion which separates ergodic from non-ergodic behavior.

The trajectories of an integrable system are confined to doughnut-shaped surfaces in phase space, known as invariant tori. If the integrable system is subjected to different initial conditions, its trajectories in phase space will trace out different invariant tori. Inspection of the coordinates of an integrable system shows that the orbits are quasi-periodic. The KAM Theorem specifies the maximum magnitude of a small non-linear perturbation acting an a system (which when non-perturbed is integrable) for which the invariant tori are deformed and survive. The quasi-periodic character of the orbits on these tori is retained. As the strength of the perturbation is increased, a number of the invariant tori are destroyed and their trajectories become chaotic. The KAM Theorem was first outlined by Andrey Kolmogorov [10] in 1954. It was rigorously proved and extended by Vladimir Arnol'd [11] (1963) and by Jürgen Moser [12] (1962). Thus, it has been proved that as the strength of the non-integrable perturbation increases, the number of quasi-periodic orbits is reduced and the number of chaotic orbits increases. There is reason to believe [13] that for perturbed integrable systems, the statistical weight of the quasi-periodic orbits may become negligibly small when $N \to \infty$, so the system may resemble an ergodic system.

2.9   Equal *a priori* Probabilities

The "*Hypothesis of Equal a priori Probabilities*" is a suggestion that assigns equal probabilities to equal volumes of phase space. This hypothesis, first introduced by Boltzmann, assumes that the probability density for ensemble averaging over phase space is uniform. The hypothesis is based on the assumption that dynamics does not preferentially bias some volume elements of available phase space $d\Gamma$ over other elements which have the same volume. This hypothesis is consistent with Liouville's Theorem which ensures that an initially uniform probability distribution will remain uniform at all later times. The Equal a priori Hypothesis is analogous to assuming that the in many consecutive rolling a dice, that each of the six faces of a dice will have an equal probability of appearing.

If the Ergodic Hypothesis holds then, for sufficiently large times T_R, the time-average of any quantity A

$$\overline{A} = \frac{1}{T_R} \int_0^{T_R} dt \; A(\{p_i(t), q_i(t)\}) \tag{2.91}$$

represents the measured value of a macroscopic quantity and, in that time interval, the trajectory passes arbitrarily close to every point in phase space. If the system's trajectory dwells in the volume of phase space $\Delta\Gamma$ for time Δt, then the ratio

$$\frac{\Delta t}{T_R} \tag{2.92}$$

has a definite limit which defines the probability that, if the system is observed at some instant of time, it will be found to be in a microscopic state within $\Delta\Gamma$. However, the Hypothesis of Equal a priori Probabilities assigns the probability density ρ for a system to be found in the volume $\Delta\Gamma$ of accessible phase space to a constant value given by the normalization condition

$$\rho = \frac{1}{\Gamma_a} \tag{2.93}$$

where Γ_a is the entire volume of accessible phase space. The equality of the time-average and ensemble average requires that the two probabilities must be equal

$$\frac{\Delta t}{T_R} = \frac{\Delta\Gamma}{\Gamma_a} \tag{2.94}$$

Hence, the Ergodic Hypothesis implies the Hypothesis of Equal a priori Probabilities, if the trajectory spends equal times in equal volumes of phase space as expected from Liouville's Theorem.

Example: The One-Dimensional Harmonic Oscillator

We shall show that for the one-dimensional harmonic oscillator that the time Δt spent in some volume $\Delta\Gamma = \Delta p\, \Delta q$ of its two-dimensional phase space is proportional to the volume. That is, the trajectory spends equal time in equal volumes.

The Hamiltonian is expressed as

$$H(p, q) = \frac{p^2}{2M} + \frac{M\omega_0^2}{2}\, q^2 \tag{2.95}$$

The equations of motion have the form

$$\frac{dp}{dt} = -M\omega_0^2\, q$$

$$\frac{dq}{dt} = \frac{p}{M} \tag{2.96}$$

The equations of motion for the one-dimensional Harmonic Oscillator can be integrated to yield

$$p(t) = M\omega_0\, A\cos(\omega_0 t + \phi)$$
$$q(t) = A\sin(\omega_0 t + \phi) \tag{2.97}$$

where the amplitude A and initial phase ϕ are constants of integration. The Hamiltonian is a constant of motion, and the accessible phase space is given by

$$E + \Delta E > H(p, q) > E \tag{2.98}$$

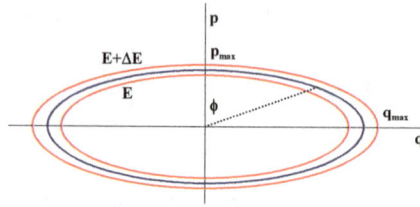

Fig. 2.13　A typical trajectory for the one-dimensional Classical Harmonic Oscillator is shown in blue. The initial phase ϕ is assumed to be unknown. The energy is known to within ΔE so the accessible phase space Γ_a is the area enclosed between the two ellipses.

which leads to the constraint on the amplitude

$$E + \Delta E > \frac{M\omega_0^2}{2} A^2 > E \tag{2.99}$$

From the solution one finds that the orbits are closed and form ellipses in phase space, which pass arbitrary close to every point in accessible phase space. The Poincaré recurrence time T_R is given by

$$T_R = \frac{2\pi}{\omega_0} \tag{2.100}$$

Consider an elemental volume of accessible phase space, $\Delta\Gamma$, bounded by the two constant energy surfaces $E + \Delta E$ and E. One can define generalized phase space vectors $\underline{X}$ with components (p, q) via

$$\underline{X} = (p, q) \tag{2.101}$$

Similarly, one can also define a generalized gradient of H via

$$\underline{\nabla}_X H = \left(\frac{\partial H}{\partial p}, \frac{\partial H}{\partial q} \right) \tag{2.102}$$

which is perpendicular to the surfaces of constant energy. Thus, the infinitesimal vectors $\Delta\underline{X}$ which join the surfaces of constant energy satisfy

$$\Delta E = \underline{\nabla}_X H \cdot \Delta\underline{X} \tag{2.103}$$

This set of vectors contains a unique minimal displacement vector $\Delta\underline{X}_E$, oriented in the direction of $\underline{\nabla}H$ and which has a magnitude given by

$$\Delta X_E = \frac{\Delta E}{|\underline{\nabla}_X H|} \tag{2.104}$$

In time interval Δt, the trajectory undergoes a displacement $\Delta\underline{X}_t$ along the surface of constant energy given by Hamilton's equations of motion

$$\Delta\underline{X}_t = \left(-\frac{\partial H}{\partial q}, \frac{\partial H}{\partial p} \right) \Delta t \tag{2.105}$$

with magnitude

$$\Delta X_t = |\nabla_X H| \, \Delta t \tag{2.106}$$

Hence, the area of the volume element of accessible phase space $\Delta\Gamma$ swept in time Δt is given by the product of the two orthogonal displacements

$$\Delta\Gamma = \Delta X_E \, \Delta X_t$$
$$= \Delta E \, \Delta t \tag{2.107}$$

If we note that all accessible phase space is swept out in time T_R, one may integrate the above expression over an orbit to obtain

$$\Gamma_a = \Delta E \, T_R \tag{2.108}$$

On dividing the above two equations, one finds

$$\frac{\Delta\Gamma}{\Gamma_a} = \left(\frac{\Delta t}{T_R}\right) \tag{2.109}$$

Therefore, we have shown that

$$\frac{\Delta t}{T_R} = \frac{\Delta\Gamma}{\Gamma_a} \tag{2.110}$$

which shows that the trajectory spends equal times in equal volumes of phase space.

The above relation between the volume of accessible phase space Γ_a and T_R is consistent with our previous result for the volume of accessible phase space

$$\Gamma_a = 2\pi \frac{\Delta E}{\omega_0} \tag{2.111}$$

This is simply seen by noting that

$$T_R = \left(\frac{2\pi}{\omega_0}\right) \tag{2.112}$$

leads to

$$\Gamma_a = \Delta E \, T_R \tag{2.113}$$

The above derivation shows that the result does not depend on the detailed shape of the volume of accessible phase space $\Delta\Gamma$, but that the result follows directly from the Hamiltonian description of the dynamics of the closed orbit. This example also illustrates how the average of a property of a system with unknown initial conditions phases (in this case the initial

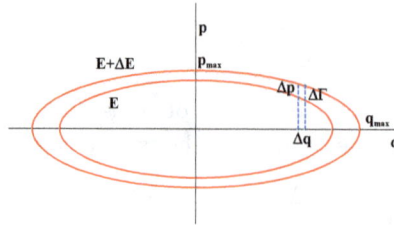

Fig. 2.14 The trajectory crosses an element $\Delta\Gamma$ of accessible phase space with a narrow width Δq in time Δt, the height Δp of the element is determined by the uncertainty in the energy ΔE.

phase ϕ) can be thought of either as a time average or as an ensemble average (i.e. as an average over ϕ).

The hypothesis of equal a priori probabilities does provide a reasonable basis for calculating the equilibrium thermodynamic properties of a large number of physical systems. This anecdotal evidence provides justification for its use. However, one is lead to suspect that ρ is not really uniform but instead is finely dispersed (mixed) throughout the volume of accessible phase space. The phenomenon of mixing may be used to justify a quasi-ergodic hypothesis, in which the system may sample *"typical"* elements of phase space [14] over time scales which are not unreasonably longer than the times over which measurements are made. In our discussion of the Micro-Canonical Ensemble and everything that follows from it, we shall be assuming that the Hypothesis of Equal a priori Probabilities is valid.

2.10　The Physical Significance of Entropy

A system can only make a transition from one macroscopic equilibrium state to another if the external conditions are changed. A change in external conditions, without supplying energy to the system, can be achieved by removing a constraint on the system. The removal of a constraint usually results in an increase in N_Γ the number of microscopic states available to the system. It is convenient to introduce a measure of the number of N_Γ which is extensive, or additive. Since N_Γ is multiplicative, $\ln N_\Gamma$ is additive and represents a measure of the number of microscopic states corresponding to the macroscopic equilibrium state. The removal of a constraint has the effect that $\ln N_\Gamma$ increases, as does the thermodynamic entropy. Therefore, this argument suggest that the entropy may be defined by

$$S = k_B \, \ln N_\Gamma \tag{2.114}$$

Table 2.1   Percentage frequency of occurrence of the letters in English language texts.

a	b	c	d	e
8.17	1.49	2.78	4.25	12.70
f	g	h	i	j
2.23	2.01	6.09	6.97	0.15
k	l	m	n	o
0.77	4.02	2.41	6.75	7.51
p	q	r	s	t
1.93	0.09	5.99	6.33	9.06
u	v	w	x	y
2.76	0.98	2.36	0.15	1.97
z	-	-	-	-
0.07	-	-	-	-

in which case, the entropy is a measure of the dispersivity of the distribution of microscopic states. The factor of k_B (Boltzmann's constant) is required to give the entropy the same dimensions as the thermodynamic entropy.

2.10.1   *Entropy and Information*

Shannon [15] has rigorously proved that the information content S of a probability distribution function of a random process with M possible outcomes is given by

$$S = -\sum_{i=1}^{M} p_i \ln p_i \tag{2.115}$$

where p_i is the probability of the i-th outcome. This is Shannon's entropy.

Consider sending a message consisting of an ordered string of N values of the possible outcomes. The outcomes can be considered as corresponding to the letters of an alphabet with M letters. In this case, the message consists of a word containing N letters. Like languages, the letters don't occur with equal frequency, for example in English language texts the letter e appears most frequently, and those who play the game "scrabble" know that the letters q and z occur very infrequently (See Table 2.1).

The total possible number of messages of length N is just M^N. However, not all messages occur with equal probability, since if the outcome i occurs with a small probability p_i, messages in which the outcome i occurs a significant number of times have very small probabilities of appearing. In the analogy with words of N letters, some allowed words occur so infrequently that they are never listed in a dictionary.

A message of length N could be expected to contain the outcome i an average of $N_i = Np_i$ times. Hence, one can determine the approximate number of times N_i each outcome i will occur in a long "typical message". Since the N_i are fixed, the set of "typical messages" merely differ in the order that these outcomes are listed. The number of these typical messages D_N can be found from the number of different ways of ordering the outcomes

$$D_N = \frac{N!}{\prod_{i=1}^{M} N_i!} \tag{2.116}$$

Hence, the dictionary of typical N-character messages (N-letter words) contains D_N entries. We could index each message in the dictionary by a number. Suppose we wish to transmit a message, instead of transmitting the string of characters of the message we could transmit the index which specifies the place it has in the dictionary. If we were to transmit this index using a binary code, then allowing for all possible messages, one would have to transmit a string of binary digits of length given by

$$\log_2 D_N \approx \log_2 \left(\frac{N!}{\prod_{i=1}^{M} N_i!} \right)$$

$$\approx N \, \log_2 N - \sum_{i=1}^{M} N_i \, \log_2 N_i$$

$$\approx -N \sum_{i=1}^{M} p_i \, \log_2 p_i \tag{2.117}$$

where the second line has been obtained by using Stirling's formula (valid for large N_i) and noting that $N = \sum_{i=1}^{M} N_i$. The last line is found by substituting $N_i = Np_i$. For an uniform probability distribution, this number would be just $N \log_2 M$. The difference in these numbers, divided by N is the information content of the probability distribution function. Shannon's Theorem proves this rigorously.

The Entropy

We shall describe the entropy of a macroscopic state as a phase space average

$$S = -k_B \int d\Gamma \, \rho \, \ln(\rho \, \Gamma_0) \tag{2.118}$$

where the factor of Γ_0 has been introduced to make the argument of the logarithm dimensionless. It is convenient to express Γ_0 for a system of N indistinguishable particles moving in a three-dimensional space as

$$\Gamma_0 = N!(2\pi\hbar)^N \tag{2.119}$$

since the introduction of this factor and the use of the equal a priori hypothesis results in the expression

$$S = k_B \ln N_\Gamma \tag{2.120}$$

if one identifies the number of accessible microstates as

$$N_\Gamma = \frac{\Gamma_a}{\Gamma_0} \tag{2.121}$$

A different choice of Γ_0 will result in the entropy being defined up to an additive constant.

2.10.2 *Entropy and Equilibrium States*

The assumption of equal a priori probabilities is only a simplification of the widely held belief that a system's physical trajectory follows an intricate path which changes rapidly and is finely spread across the volume of accessible phase space. The corresponding physical distribution function will evolve with respect to time, according to Liouville's Theorem. If ρ is expected to describe an equilibrium state S should not evolve.

We shall show that entropy defined by

$$S(t) = -k_B \int d\Gamma \, \rho(\{p_i, q_i\} : t) \, \ln \rho(\{p_i, q_i\} : t) - k_B \ln \Gamma_0 \tag{2.122}$$

is time independent. The last term is an additive constant added to make the argument of the logarithm dimensionless and has no affect on our deliberations. The time derivative of the entropy is given by

$$\frac{dS}{dt} = -k_B \int d\Gamma \left(\frac{\partial \rho}{\partial t} \ln \rho + \frac{\partial \rho}{\partial t} \right)$$

$$= -k_B \int d\Gamma \, \frac{\partial \rho}{\partial t} (\ln \rho + 1) \tag{2.123}$$

Using Liouville's Theorem reduces this to

$$\frac{dS}{dt} = k_B \int d\Gamma \{\rho, H\}(\ln \rho + 1)$$

$$= -k_B \int d\Gamma \sum_{i=1}^{3N} \left[\frac{\partial \rho}{\partial p_i} \frac{\partial H}{\partial q_i} - \frac{\partial \rho}{\partial q_i} \frac{\partial H}{\partial p_i} \right] (\ln \rho + 1) \tag{2.124}$$

The terms linear in the derivatives of ρ can be transformed into factors of ρ, by integrating by parts. This yields

$$\frac{dS}{dt} = k_B \int d\Gamma \sum_{i=1}^{3N} \rho \left[\frac{\partial}{\partial p_i} \left(\frac{\partial H}{\partial q_i} (\ln \rho + 1) \right) - \frac{\partial}{\partial q_i} \left(\frac{\partial H}{\partial p_i} (\ln \rho + 1) \right) \right]$$

(2.125)

since the boundary terms vanish. On expanding the derivatives of the terms in the round parentheses, one finds that some terms cancel

$$\frac{dS}{dt} = k_B \int d\Gamma \left[\rho \frac{\partial^2 H}{\partial p_i \partial q_i} (\ln \rho + 1) + \frac{\partial H}{\partial q_i} \frac{\partial \rho}{\partial p_i} \right.$$

$$\left. - \rho \frac{\partial^2 H}{\partial q_i \partial p_i} (\ln \rho + 1) - \frac{\partial H}{\partial p_i} \frac{\partial \rho}{\partial q_i} \right]$$

$$= k_B \int d\Gamma \left[\frac{\partial H}{\partial q_i} \frac{\partial \rho}{\partial p_i} - \frac{\partial H}{\partial p_i} \frac{\partial \rho}{\partial q_i} \right]$$

(2.126)

which on integrating by parts yields

$$\frac{dS}{dt} = k_B \int d\Gamma \sum_{i=1}^{3N} \left[-\rho \left(\frac{\partial^2 H}{\partial p_i \partial q_i} \right) + \rho \left(\frac{\partial^2 H}{\partial p_i \partial q_i} \right) \right]$$

$$= 0$$

(2.127)

Hence, the entropy of a state with a time-dependent probability density is constant.[2]

From the above discussion, it is clear that the entropy of a system can only change if the Hamiltonian of the system is modified, such as by removing an external constraint. A removal of an external constraint will increase the volume of accessible phase space from Γ_a to Γ'_a. This is not only true for time-dependent ensembles, but is also true for stationary ensembles. In particular, the assumption of equal a priori probabilities implies that in the final equilibrium state the probability density ρ' will be uniformly spread over the increased available phase space. From the normalization condition

$$1 = \int d\Gamma' \, \rho'$$

(2.128)

---

[2]If a relation between the microscopic information carried by a trajectory and entropy is established, this result might follow since, for closed systems described by time-independent Hamiltonians, unitarity ensures the informational content that uniquely specifies a trajectory is conserved.

one finds that in the final state, the probability distribution function is given by

$$\rho' = \frac{1}{\Gamma'_a} \tag{2.129}$$

Since the entropy is given by

$$S = -k_B \int d\Gamma \; \rho \; \ln(\rho \; \Gamma_0) \tag{2.130}$$

where Γ_0 is constant and since $\Gamma'_a > \Gamma_a$, the entropy will have increased by an amount given by

$$\Delta S = k_B \; \ln\left(\frac{\Gamma'_a}{\Gamma_a}\right) \tag{2.131}$$

as expected from thermodynamics.

Example: Joule Free Expansion

We shall consider the Joule Free Expansion of an ideal gas. The gas is initially enclosed by a container of volume V, but the a valve is opened so that the gas can expand into an adjacent chamber which initially contained a vacuum. The volume available to the gas in the final state is V'. Since the adjacent chamber is empty, no work is done in the expansion.

The Hamiltonian for an idea gas can be represented by

$$H = \sum_{i=1}^{3N} \frac{p_i^2}{2m} \tag{2.132}$$

so the sum of the squares of the momenta is restricted by E. The volume of accessible phase space is given by

$$\Gamma_a = \prod_{i=1}^{3N} \left\{ \int dp_i \int dq_i \right\} \delta\left(E - \sum_{i=1}^{3N} \frac{p_i^2}{2m}\right) \tag{2.133}$$

The integrations for the spatial coordinates separates from the integration over the momenta. The integration over the three spatial coordinates for each particle produces a factor of the volume. The integration over the momenta will produce a result which depends on the energy $f(E)$ which is independent of the volume. Hence, the expression for the available phase space has the form

$$\Gamma_a = V^N \; f(E) \tag{2.134}$$

On recognizing that the particles are indistinguishable, one finds that the entropy is given by

$$S = k_B \ln \frac{\Gamma_a}{\Gamma_0}$$

$$= N \, k_B \, \ln V + \, k_B \, \ln f(E) - k_B \, \ln N! - N \, k_B \ln(2\pi\hbar) \tag{2.135}$$

where Γ_0 is the measure of phase space that is used to define a single microscopic state. Thus, the change in entropy is given by

$$\Delta S = N \, k_B \, \ln \left(\frac{V'}{V} \right) \tag{2.136}$$

The same result may be obtained from thermodynamics. Starting from the expression for infinitesimal change of the internal energy

$$dU = T \, dS - P \, dV + \mu \, dN \tag{2.137}$$

and recognizing that Joule Free expansion is a process for which

$$dU = 0$$
$$dN = 0 \tag{2.138}$$

Therefore, one has

$$dS = \frac{P}{T} \, dV \tag{2.139}$$

and on using the equation of state for the ideal gas $PV = N \, k_B T$, one finds

$$dS = N \, k_B \, \frac{dV}{V} \tag{2.140}$$

which integrates to yield

$$\Delta S = N \, k_B \, \ln \left(\frac{V'}{V} \right) \tag{2.141}$$

Hence, the expression for the change in entropy derived by using Statistical Mechanics is in agreement with the expression derived by using Thermodynamics.

2.11 The Central Limit Theorem

The "*Central Limit Theorem*" plays an important role in Statistical Mechanics since a macroscopic system has an extremely large number of microscopic parts. The microscopic components may be approximated as being independent subsystems, and so an extensive quantity X can be considered as the sum of numerous independent microscopic variables x_i. The Central Limit Theorem maintains that, for a system composed of many independent parts, the probability distribution for an extensive variable X can be approximated by a Gaussian distribution. Moreover, the Gaussian distribution is highly peaked in that it has a very narrow width compared to the most probable value $\overline{X}$.

Let x_i be the i-th component of the extensive variable X, so

$$X = \sum_{i=1}^{N} x_i \qquad (2.142)$$

The distribution of X can be expressed in terms of the distributions of x_i. The fluctuations of the extensive variable are defined as $\Delta X = X - \overline{X}$, where $\overline{X}$ denotes the average. The fluctuations ΔX are related to the fluctuations of the independent parts $\Delta x_i = x_i - \overline{x}_i$ via

$$\begin{aligned} \Delta X &= X - \overline{X} \\ &= \sum_{i=1}^{N} (x_i - \overline{x}_i) \\ &= \sum_{i=1}^{N} \Delta x_i \end{aligned} \qquad (2.143)$$

By definition the average of the fluctuation is zero

$$\overline{\Delta x_i} = 0 \qquad (2.144)$$

and since the fluctuations of the parts are independent, one has

$$\begin{aligned} \overline{\Delta X^2} &= \sum_{i=1}^{N} \overline{\Delta x_i^2} \\ &= O(N) \end{aligned} \qquad (2.145)$$

Furthermore, since the variable X is extensive

$$\begin{aligned} \overline{X} &= \sum_{i=1}^{N} \overline{x}_i \\ &= O(N) \end{aligned} \qquad (2.146)$$

the relative squared fluctuation is given by

$$\frac{\overline{\Delta X^2}}{\overline{X}^2} = O(1/N) \tag{2.147}$$

which is negligibly small.

The Central Limit Theorem can be justified by considering the expression for the probability distribution for $\rho(X)$

$$\rho(X) = \prod_{i=1}^{N}\left\{\int_{-\infty}^{\infty} dx_i \; p_i(x_i)\right\} \delta\left(X - \sum_{j=1}^{N} x_j\right) \tag{2.148}$$

where distribution functions for the microscopic variables have been denoted by $p_i(x_i)$. Since the delta function can be expressed as a Fourier Transform, one may express the distribution function for the macroscopic quantity X in terms of a product of the Fourier Transforms of the microscopic distribution functions as

$$\rho(X) = \prod_{i=1}^{N}\left\{\int_{-\infty}^{\infty} dx_i \; p_i(x_i)\right\} \int_{-\infty}^{\infty} \frac{dk}{2\pi} \exp\left[-ik\left(X - \sum_{j=1}^{N} x_j\right)\right]$$

$$= \int_{-\infty}^{\infty} \frac{dk}{2\pi} \exp[-ikX] \prod_{i=1}^{N}\left\{\int_{-\infty}^{\infty} dx_i \; p_i(x_i) \; \exp[ikx_i]\right\} \tag{2.149}$$

Since the microscopic distribution functions are assumed to be independent, there are no cross correlations. One may evaluate the Fourier Transform of the microscopic probabilities $p_i(x_i)$ and express the results in terms of a power series in ik with coefficients involving the $\frac{1}{n!} \overline{x^n}_i$. The series can be exponentiated where the exponent is expressed in quantities such as the average $\overline{x}_i$ and the mean-squared deviations $\overline{\Delta x_i^2} = \overline{x_i^2} - \overline{x}_i^2$ and higher-order cumulants that are frequently used to characterize distributions. Due to the presence of the phase factor which multiplies the normalized microscopic probability distribution function, each individual factor has a magnitude less than one. If the integral over k of the product of N such factors is to yield a contribution to $\rho(X)$ with magnitude of order one, the integral must be dominated by values of k that are of order $O(1/\sqrt{N})$. Therefore, it is reasonable to truncate the series in the exponential. Thus, we approximate the factors by

$$\int_{-\infty}^{\infty} dx_i \; p_i(x_i) \; \exp[ikx_i] \approx \exp[ik\overline{x}_i] \exp\left[-\frac{k^2}{2} \overline{\Delta x_i^2}\right] \tag{2.150}$$

which neglects the higher-order cumulants. Hence, one obtains the approximation

$$\rho(X) \approx \int_{-\infty}^{\infty} \frac{dk}{2\pi} \exp\left[-ik\left(X - \sum_{j=1}^{N} \overline{x}_j\right)\right] \prod_{i=1}^{N} \exp\left[-\frac{k^2}{2} \overline{\Delta x_i^2}\right]$$

$$= \int_{-\infty}^{\infty} \frac{dk}{2\pi} \exp[-ik (X - \overline{X})] \exp\left[-\frac{k^2}{2} \sum_{i=1}^{N} \overline{\Delta x_i^2}\right]$$

$$= \int_{-\infty}^{\infty} \frac{dk}{2\pi} \exp[-ik (X - \overline{X})] \exp\left[-\frac{k^2}{2} \overline{\Delta X^2}\right] \qquad (2.151)$$

where the notations

$$\overline{X} = \sum_{i=1}^{N} \overline{x}_i \qquad (2.152)$$

and

$$\overline{\Delta X^2} = \sum_{i=1}^{N} \overline{\Delta x_i^2} \qquad (2.153)$$

have been introduced. On completing the square and performing the Gaussian integration over k, one finds that the distribution function is Gaussian.

$$\rho(X) = \frac{1}{\sqrt{2\pi \overline{\Delta X^2}}} \exp\left[-\frac{1}{2} \frac{(X - \overline{X})^2}{\overline{\Delta X^2}}\right] \qquad (2.154)$$

Since $\overline{X} = O(N)$ and $\sqrt{\overline{\Delta X^2}} = \sqrt{N}$, the distribution is highly-peaked around $\overline{X}$. This allows one to replace the extensive variable by its average value. Hence, one might expect that most microstates of a system are typical microstates.

A number of examples of the use of the Central Limit Theorem will be seen as this course unfolds.

2.12   Problems

Problem 2.1

In Statistical Mechanics, probabilities are often deduced from geometric considerations by assuming that probability is distributed uniformly throughout space. That is, probabilities are inferred by evaluating "volumes" subject to constraints. The following problem, originally posed by Buffon [16] illustrates the methodology.

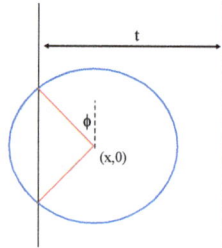

Fig. 2.15   A short Buffon needle of length l crosses the line if $\Theta(\frac{l}{2}\sin\phi - x)$ is non-zero, where the center of mass of the needle is at $(x, 0)$.

Consider a needle of length l which is randomly dropped on a plane ruled with parallel lines that are equally spaced by a distance t. Assume that the center of mass and the angle ϕ of the needle are uniformly distributed. Show that the probability P that a needle intersects a line is given by

$$P = \frac{2l}{\pi t}$$

if the needle is short $t \geq l$, and

$$P = \frac{2}{\pi}\cos^{-1}\left(\frac{t}{l}\right) + \frac{2l}{\pi t}\left[1 - \sqrt{1 - \left(\frac{t}{l}\right)^2}\right]$$

for long needles such that $l \geq t$.

Hence, performing an experiment could yield an approximate value of π which is a rational number. One such rational approximation, good to six decimal places, is

$$\pi \approx \frac{355}{113}$$

Problem 2.2

A simple classical model for a diatomic molecule is given by a rigid rotator consisting of two point masses separated by a fixed distance. In the center of mass frame, the Lagrangian is given by

$$\mathcal{L} = \frac{I}{2}(\dot\theta^2 + \sin^2\theta\ \dot\varphi^2)$$

(i) Find the momenta (p_φ, p_θ) canonically conjugate to (φ, θ) and, hence, find an expression for the Hamiltonian.

The magnitude of the total angular momentum L of the rotator is given by

$$L^2 = p_\theta^2 + \frac{p_\varphi^2}{\sin^2\theta}$$

(ii) Determine the volume of phase space which corresponds to values of the total angular momentum less than L_0.

$$\Gamma(L_0) = \int d\Gamma\ \Theta\left(L_0^2 - p_\theta^2 - \frac{p_\varphi^2}{\sin^2\theta}\right)$$

Show that, in the quasi-classical approximation, this volume corresponds to the number of microstates

$$N_\Gamma = \frac{\Gamma(L_0)}{(2\pi\hbar)^2} = \left(\frac{L_0}{\hbar}\right)^2$$

(iii) Quantum mechanically, $\hat{L}^2$ is quantized where the quantum number is given by $\hbar^2 l(l+1)$ and is $(2l+1)$-fold degenerate, where l has either integer or half-integer values. In either case, use this result to find the number of states available to a quantum rotator with $l < l_0$.

Problem 2.3

Consider the motion of a point particle of mass m constrained to move inside a circle of radius R of a two-dimensional plane. The Hamiltonian is given by

$$H = \frac{p_r^2}{2m} + \frac{p_\varphi^2}{2mr^2} + V_0\ \Theta(r - R)$$

where $V_0 \to \infty$.

(i) Assume that the constants of motion E and p_φ have known values. Find the volume of accessible (p_r, r) phase space.

(ii) Show that the total volume of accessible phase space, $(p_r, r, p_\varphi, \varphi)$, for a precise value of the energy E is

$$\Gamma_a = m\pi^2 R^2$$

(iii) Use the hypothesis of equal a priori probabilities to show that the probability distribution for p_φ is given by

$$P(p_\varphi) = \frac{2}{\pi} \frac{1}{\sqrt{2mER^2}} \sqrt{1 - \frac{p_\varphi^2}{2mER^2}} \, \Theta(2mER^2 - p_\varphi^2)$$

Problem 2.4

Consider a thermalized gas which is trapped in one-dimension. The initial value of the probability distribution for the gas $\rho(q, p; 0)$ is given by

$$\rho(p, q, 0) = \delta(q) \frac{\exp[-\frac{\beta p^2}{2m}]}{\sqrt{2\pi m k_B T}}$$

(i) Use Liouville's Theorem to evaluate $\rho(p, q : t)$.
(ii) Show that

$$\overline{p^2(t)} = m k_B T$$

and

$$\overline{q^2(t)} = \frac{\overline{p^2}}{m^2} t^2$$

Problem 2.5

Consider a gravitational pendulum of mass m and fixed length r. The Lagrangian is given by

$$L = \frac{m}{2} r^2 \dot{\varphi}^2 + mgr \cos \varphi$$

The φ coordinate is defined such that the pendulum is at its lowest height when $\varphi = 0$.
(i) Find the generalized momentum p_φ canonically conjugate to φ.
(ii) Determine an expression for the period of the motion $T(E)$ for $E > mgr$ and for $E < mgr$.
(iii) Find the Hamiltonian $H(p_\varphi, \varphi)$.

(iv) Sketch the possible trajectories in phase space. Determine an expression for the total phase space Γ_T "enclosed" by the trajectories for the cases where $E > mgr$ and $E < mgr$. Indicate the calculated areas on your sketches.

(v) The volume of accessible phase space Γ_a for a system with a spread in energy ΔE is given by

$$\Gamma_a = \left(\frac{\partial \Gamma_T}{\partial E}\right) \Delta E$$

Relate the volume of accessible phase space Γ_a to $T(E)$.

Problem 2.6

Consider two coupled harmonic oscillators described by the Hamiltonian

$$H = \frac{p_1^2}{2m} + \frac{p_2^2}{2m} + \frac{m\omega^2}{2}\left(q_1^2 + q_2^2 - 2q_1 q_2\right)$$

(i) Show that the system is not ergodic.
(ii) Extend this argument to show that any set of coupled harmonic oscillators is not ergodic since energy is never interchanged between the normal modes of oscillation.

Problem 2.7

A diatomic molecule is represented by the classical Hamiltonian

$$H = \frac{p_\theta^2}{2I} + \frac{p_\varphi^2}{2I\sin^2\theta} - Fd\cos\theta$$

where I is the moment of inertia and the last term represents the interaction of the dipole moment with an electric field F.
(i) Show that p_φ is a constant of motion.
(ii) Assume that p_φ is distributed randomly consistent with the value of the total energy E. Determine the volume of accessible phase space.
(iii) By direct integration and using the assumption of equal a priori probabilities, estimate the induced polarization.

Problem 2.8

(A) Consider Weyl's billiard game. In this game, the billiard ball is represented as a point particle which moves freely on the surface of a square

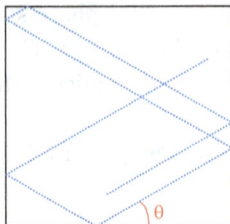

Fig. 2.16   A Weyl billiard on a square executes segments of linear motion followed by elastic reflection at the boundaries of the billiard table.

billiard table with sides of length L. The billiard ball undergoes linear motion but is elastically reflected at the boundaries of the square. Consider the projection of the four-dimensional phase space onto two-dimensional position space.

(i) Show that the projection of the particle's motion onto any side of the square is periodic.

(ii) Show that if the particle's velocity makes an angle θ with respect to one side and if $\tan \theta$ is irrational, then the particle's trajectory comes arbitrarily close to every point in the square.[3]

(B) Consider a two-dimensional harmonic oscillator with the Hamiltonian

$$H = \left(\frac{p_1^2 + p_2^2}{2m} \right) + \frac{m}{2}(\omega_1^2 q_1^2 + \omega_2^2 q_2^2)$$

(iii) Show that the projection of the particle's motion onto any coordinate axis is periodic.

(iv) Show that if the ratio ω_1/ω_2 is irrational, then the motion does not form closed orbits.

The spatial probability distribution function $P(q_1, q_2)$ is defined as a long time average

$$P(q_1, q_2) = \lim_{T \to \infty} \int_0^T \frac{dt}{T} \, \delta(q_1 - q_1(t)) \, \delta(q_2 - q_2(t))$$

where $(q_1(t), q_2(t))$ are the time-dependent coordinates of a trajectory with a general initial condition. The delta functions can be replaced by the expressions

$$\delta(q_i - q_i(t)) = \int_{-\infty}^{\infty} \frac{dk_i}{2\pi} \, \exp[-ik_i(q_i - q_i(t))]$$

---

[3]The probability distribution is also uniform [17].

The integral over time can then be easily evaluated after using the Bessel function generating expansion

$$\exp\left[\frac{x}{2}\left(s - s^{-1}\right)\right] = \sum_{n=-\infty}^{\infty} J_n(x)\, s^n$$

(v) In the case of incommensurate frequencies, show that the particle's spatial probability distribution $P(q_1, q_2)\, dq_1 dq_2$ factorizes and is given by the expression

$$P(q_1, q_2) = \int_{-\infty}^{\infty} \frac{dk_1}{2\pi} \int_{-\infty}^{\infty} \frac{dk_2}{2\pi} \frac{J_0(k_1)\, J_0(k_2)}{A_1 A_2} \exp\left[-i\left(k_1 \frac{q_1}{A_1} + k_2 \frac{q_2}{A_2}\right)\right]$$

where $J_0(x)$ are the Bessel functions of order zero and (A_1, A_2) are the amplitudes of the oscillations along the two coordinate axes.

(vi) Use the integral representation of the Bessel functions,

$$J_0(k_i) = \int_{-\pi}^{\pi} \frac{d\theta}{2\pi} \exp[ik_i \sin\theta]$$

to show that the factors reduce to

$$\frac{1}{A_i} \int_{-\infty}^{\infty} \frac{dk_i}{2\pi} J_0(k_i) \exp\left[-i\frac{k_i q_i}{A_i}\right] = \frac{1}{\pi} \frac{1}{\sqrt{A_i^2 - q_i^2}}$$

Problem 2.9

Consider a particle of mass m which is undergoing linear motion within an ellipse but is reflected specularly from the boundary (Elliptical Billiards). The particle's position can be expressed in elliptical coordinates via

$$x = c \cosh\lambda \cos\theta$$
$$y = c \sinh\lambda \sin\theta$$

(i) Show that the kinetic energy can be expressed as

$$T = \frac{mc^2}{2}(\cosh^2\lambda - \cos^2\theta)(\dot{\lambda}^2 + \dot{\theta}^2)$$

Hence, find the equations of motion.

(ii) Show that the Hamiltonian is given by

$$H = \frac{1}{2mc^2}\left(\frac{(p_\lambda^2 + p_\theta^2)}{\cosh^2\lambda - \cos^2\theta}\right)$$

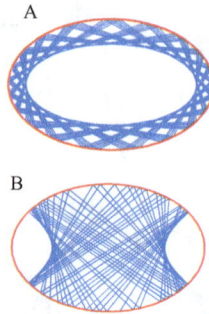

Fig. 2.17 A trajectory reflected from an ellipse is tangential to either a confocal ellipse or hyperbola.

(iii) Show that the product of the angular momentum of the particle as calculated from the foci of the ellipse is

$$\Lambda = \frac{1}{2mc^2} \left(\frac{\sinh^2 \lambda p_\theta^2 - \sin^2 \theta p_\lambda^2}{\cosh^2 \lambda - \cos^2 \theta} \right)$$

and that it is a conserved quantity.

(iv) Hence, show that the trajectories reflected at the ellipse are tangent to the boundary of an inner confocal ellipse for $\Lambda > 0$ (or a confocal hyperbola for $\Lambda < 0$) characterized by

$$\sinh \lambda = \sqrt{\frac{\Lambda}{2mc^2 H}}$$

This example shows that conserved quantities restrict the trajectories to specific regions of space. The motion can also be characterized as being either periodic or irregular.

Problem 2.10

Consider an isotropic two-dimensional harmonic oscillator with Hamiltonian H

$$H = \sum_{n=1}^{2} \left(\frac{p_n^2}{2m} + \frac{m}{2} \omega^2 x_n^2 \right)$$

and total energy E. In quantum mechanics, states can only be specified by the quantum numbers of mutually commuting operators. Hence, the states of our quantum system are determined by only two quantum numbers not the four variables used to description the classical system. For the two-dimensional quantum harmonic oscillator, the normal modes are described by quantized energies E_1 and E_2. The assumption of equal a priori probabilities, appears to amount to an average over E_1 consistent with the total energy $E = E_1 + E_2$ of the combined system. In a classical system, the average is an average over the phases ϕ_1, ϕ_2 and amplitudes squared A_1^2, A_2^2 consistent with conservation of energy

$$\frac{2E}{m\omega^2} = \sum_{n=1}^{2} A_n^2$$

This problem projects the phase space of a our classical system onto two-dimensional position space where the components are given by

$$x_n(t) = A_n \sin(\omega t + \phi_n)$$

and high-lights the roles of averaging over the energies and phases of the classical trajectories.

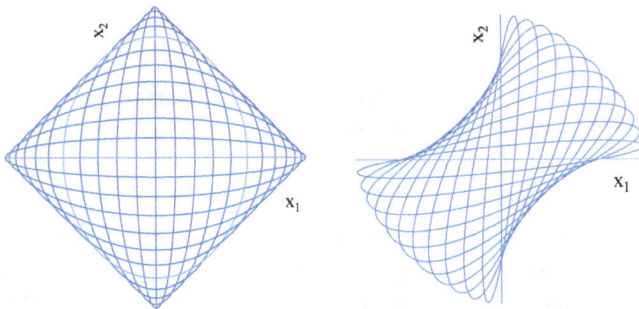

Fig. 2.18 Typical sets of trajectories with various amplitudes (A_1, A_2) (but each trajectory in a set has the same phase difference) consistent with the conservation of energy. The two sets correspond to different values of the phase difference.

(i) Determine the time-averaged (averaged over one period of oscillation) classical probability $P(x_1, x_2)\, dx_1 dx_2$ for finding a particle in an area $dx_1 dx_2$ around the point (x_1, x_2). Show that your result describes a single orbit in two-dimensional position space (x_1, x_2).

It is reasonable to assume that the system is subjected to an extremely small perturbation which, over a long time, causes the system to move between the degenerate states. That is, the constants of motion A_n^2 and the phases ϕ_n should be averaged over, where the averaging respects conservation of energy.

(ii) Average $P(x_1, x_2)$ over the phases ϕ_n. Is the probability density factorizable?

(iii) Perform a further average of $P(x_1, x_2)$ over the square of the amplitudes A_n assuming the total energy is constant. Is the resulting probability distribution uniform over the accessible position space?

Since the energy eigenstates of the quantum system are degenerate and since the values of the mode energies E_1 and E_2 have not been measured, the states could be represented by linear superpositions of products of the single mode energy eigenfunctions $\psi_E(x)$ with complex coefficients. A simple example is given by

$$\Psi_E(x_1, x_2) = C_1 \; \psi_{E_1}(x_1) \; \psi_{E_2}(x_2) + C_2 \; \psi_{E_2}(x_1) \; \psi_{E_1}(x_2)$$

where $E = E_1 + E_2$ and the probabilities are normalized $|C_1|^2 + |C_2|^2 = 1$.

(iv) Discuss whether your results indicate that the assumption of equal a priori probabilities, for a quantum system, includes an assumption of "a priori random phases".

Problem 2.11

Consider an integrable system with a four-dimensional phase space. The integrable system which has two spatial degrees of freedom must have two constants of motion. One constant of motion is the energy H and the other will be designated by B. The problem is to derive the condition that H and B must have a vanishing Poisson Bracket geometrically, without using Hamilton's equations of motion. Furthermore, another aim to show that the motion takes place on a two-dimensional manifold that has the form of a torus.

We define a four-dimensional phase-space vector by

$$\underline{X} = (\underline{p}, \underline{q})$$

and, analogously, define the gradient as

$$\underline{\nabla}_X = (\underline{\nabla}_p, \underline{\nabla}_q)$$

The gradients of H and B define the directions of most rapid variations of E and H. These directions are designated as the "normals" of the surfaces of constant H and B

$$\underline{\nabla}_X H = (\underline{\nabla}_p H, \underline{\nabla}_q H)$$

$$\underline{\nabla}_X B = (\underline{\nabla}_p B, \underline{\nabla}_q B)$$

The motion of the integrable system must be restricted to a two-dimensional manifold $\mathcal{M}$ of constant H and constant B. Thus, the trajectory must be on the $\mathcal{M}$ and have a direction which is orthogonal to the normals. Hence, the motion must be in the plane and have directions defined by the vector fields

$$\underline{X}_H = (-\underline{\nabla}_q H, \underline{\nabla}_p H)$$
$$\underline{X}_B = (-\underline{\nabla}_q B, \underline{\nabla}_p B)$$

(i) Show that the above statement implies that the trajectory

$$\underline{X}(t) = (\underline{p}(t), \underline{q}(t))$$

must satisfy Hamilton's equations of motion up to a multiplicative constant of proportionality.

(ii) Show that, if H and B are to both be constants of motion, their Poisson Bracket must vanish

$$\{B, H\}_{PB} = 0$$

If the volume of accessible phase space is finite, the trajectory falls on a two-dimensional surface and it must be representable by a coordinate system that is devoid of singularities. The "Hairy Ball Theorem" dictates that the manifold $\mathcal{M}$ can be represented by a torus. The torus is called an invariant torus since if the motion is on the torus it will remain on the torus forever.

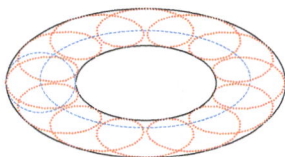

Fig. 2.19   The two-dimensional invariant torus for an asymmetric two-dimensional harmonic oscillator. Trajectories with commensurate frequencies ω_1 and ω_2 form closed orbits. A closed orbit is depicted in red.

Consider the example of a two-dimensional Harmonic Oscillator,

$$H = \sum_{i=1}^{2} \left(\frac{p_i^2}{2m} + \frac{m\omega_i^2}{2} q_i^2 \right)$$

(iii) Show that the system has two independent constants of motion $2\pi \ I_i$ corresponding to the areas of two ellipses traversed by (p_i, q_i) and that the constants of motion can be written as

$$I_i = \frac{1}{2\pi\hbar} \oint p_i \, dq_i$$

$$= \frac{1}{2\pi\hbar} \oint \sqrt{2mE_i - m^2\omega_i^2 q_i^2} \, dq_i$$

such that

$$H = I_1\hbar\omega_1 + I_2\hbar\omega_2$$

(iv) Therefore, show the two-dimensional manifold $\mathcal{M}$ can be depicted by as a two-dimensional torus with cross-sectional areas I_i embedded in three-dimensional space, where a point on the manifold is specified by two time-dependent phases $\phi_i = \omega_i t$. If the frequencies ω_i are commensurate, then the trajectory forms a closed orbit, otherwise it will completely cover the surface of the torus.

The KAM Theorem shows that, when subjected to a weak perturbation, most of the trajectories on an invariant torus will remain on the deformed invariant torus. However, the trajectories that are resonant, or in the neighbourhood of being resonant, will be destroyed. These destroyed trajectories will be chaotically distributed in phase space near the unperturbed torus.

Problem 2.12

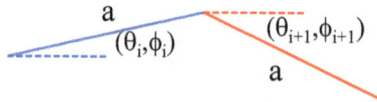

Fig. 2.20   Two successive links of a freely-jointed polymer. The orientation of the i-th monomer of length a is defined by the polar coordinates (θ_i, φ_i).

Consider a static polymer chain composed of N rigid monomers each of length a. The consecutive monomers in the chain are coupled by joints which allow free rotation. That is, if the displacement between one end of the polymer to the other end is used as the polar axis and if the axis of the i-th monomer is denoted by (θ_i, φ_i), then the values of θ_i and ϕ_i may vary over a solid angle 4π. The end to end displacement vector of the polymer has components (L_x, L_y, L_z).

(i) Show that the volume of accessible phase space $\Gamma_a(L_x, L_y, L_z)$ is given by the three-dimensional integral

$$\Gamma_a(L_x, L_y, L_z) = \int_{-\infty}^{+\infty} \frac{d\lambda_x}{2\pi} \exp[i\lambda_x \, L_x] \int_{-\infty}^{+\infty} \frac{d\lambda_y}{2\pi} \exp[i\lambda_y \, L_y]$$

$$\times \int_{-\infty}^{+\infty} \frac{d\lambda_z}{2\pi} \exp[i\lambda_z \, L_z] \, I(\lambda_x, \lambda_y, \lambda_z)^N$$

where

$$I(\lambda_x, \lambda_y, \lambda_z) = \int_0^{2\pi} d\varphi \int_0^{\pi} d\theta \sin\theta$$

$$\times \exp[-ia(\sin\theta(\lambda_x \cos\varphi + \lambda_y \sin\varphi) + \lambda_z \cos\theta)]$$

(ii) Using the hypothesis of equal a priori probabilities, determine an expression for the probability distribution $P(L_x, L_y, L_z)$ for finding a polymer with end to end displacement (L_x, L_y, L_z) in terms of $\Gamma_a(L_x, L_y, L_z)$ and the total volume of phase $\Gamma_0 = (4\pi)^N$.

(iii) Integrate over all possible values of L_x and L_y to find the probability distribution $P(L_z)$

$$P(L_z) = \int_0^\infty \frac{d\lambda_z}{\pi} \, \cos(\lambda_z \, L_z) \left(\frac{\sin\lambda_z \, a}{\lambda_z \, a} \right)^N$$

and show that the resulting probability distribution is properly normalized.

(iv) Evaluate $P(L_z)$ numerically for various values of L_z/a with fixed N, say $N \sim 20$, and plot your results.

(v) On your plot indicate the most probable value of L_z/a.

(vi) Determine the numerical values of the mean-squared fluctuation $\overline{(L_z - \overline{L}_z)^2}/a^2$.

Problem 2.13

Consider a polymer chain composed of N rigid monomers each of length a. The consecutive monomers in the chain are coupled by joints which allow free rotation over solid angle 4π. The vector joining the ends of the polymer is denoted by $\underline{R}$.

(i) Show that the volume of accessible phase space $\Gamma_a(\underline{R})$ can be re-written as

$$\Gamma_a(\underline{R}) = \prod_{n=1}^{N} \left(a^2 \int d^2\Omega_n \right) \delta^3 \left(\underline{R} - \sum_{n=1}^{N} \underline{a}_n \right)$$

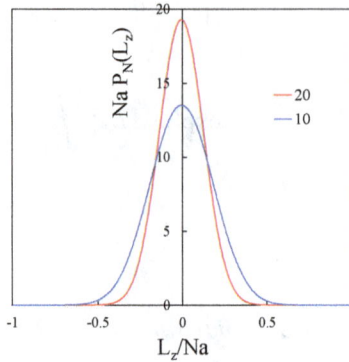

Fig. 2.21 The scaled probability density for finding a z-component of the end to end distance equal to L_z for a flexible polymer consisting of N links.

where the integration $d^2\Omega_n$ is over the direction of the n-th monomer and $\underline{a}_n$ is the vector associated with the n-th monomer.

(ii) Fourier Transform the above expression to obtain

$$\Gamma_a(\underline{Q}) = \left(a^2 \int d^2\Omega \; \exp[i\underline{Q} \cdot \underline{a}]\right)^N$$

$$= \left(4\pi \, a^2 \, \frac{\sin Qa}{Qa}\right)^N$$

(iii) Show that the probability density $P(\underline{R})$ is given by the Inverse Fourier Transform

$$P(\underline{R}) = \frac{1}{(2\pi)^3} \int d^3\underline{Q} \; \exp[-i\underline{Q} \cdot \underline{R}] \left(\frac{\sin Qa}{Qa}\right)^N$$

Evaluate the radial end to end distribution function $4\pi R^2 \, P(R)$ numerically and compare your result with the Gaussian distribution.

Problem 2.14

The method of Problem 2.11 can be extended to a freely jointed polymer in d dimensions.

(i) Evaluate the probability density $P(R)$ for the case where $d = 1$.

(ii) Evaluate $P(R)$ for $d = 1$ directly by using combinatorics. Compare your result with that found for (i).

Problem 2.15

Consider an almost ideal gas that is composed of N indistinguishable hard sphere molecules where each molecule occupies a volume V_0. The system undergoes Joule Free expansion from volume V to $2V$, where $V > N V_0$. What is the change in entropy?

Hint: $\sum_{n=1}^{N} \ln(V - nV_0) \approx \int_1^N dx\ \ln(V - xV_0)$.

Problem 2.16

Consider an insulating box with two compartments. Each compartment contains N atoms of an ideal gas of identical molecules and the compartments are at the same initial temperature T but have different initial pressures P_1 and P_2, $(P_1 \neq P_2)$. A small hole forms in the wall between the compartments which allows atoms to leak from one compartment to the other.

(i) Show that, once equilibrium has been re-established, the temperature retains its initial value.

(ii) Show that the increase in entropy is given by

$$\Delta S = N\ k_B\ \ln\left[\frac{(N_1 + N_2)^2}{4N_1 N_2}\right]$$

where N_1 and N_2 are the number of particles in each compartment in the final equilibrium state, $N_1 + N_2 = 2N$.

Problem 2.17

Consider a classical ideal gas of N particles which are restricted to move in one-dimension on the x-axis between $x = 0$ and $x = L$. The Hamiltonian is given by.

$$H = \sum_{i=1}^{N} \frac{p_{x,i}^2}{2m} + \sum_{i=1}^{N-1} V(x_i, x_{i+1})$$

The particles interact via a repulsive short-ranged potential given by

$$V(x_i, x_{i+1}) = V_0\ \delta(x_i - x_{i+1})$$

The potential has the effect that the particles maintain their spatial order on the line. Hence, one can uniquely label the particles by an index i which denotes their relative positions as described by the inequality

$$x_{i-1} < x_i < x_{i+1}$$

and so are distinguishable.

(i) Determine an integral expression for the volume of accessible phase space. You need not perform the integrals over the momenta. Hence, find an expression for the entropy. Is the positional part of the entropy extensive?

(ii) Assume that the momentum integrals in the expression for the accessible phase space are the same as for non-interacting particles. Compare your result for the entropy with the entropy of a gas of non-interacting indistinguishable particles which can interchange their positions.

Problem 2.18

Use the definition of the delta function to show that

$$\rho(x) = \int dx' \, \rho(x') \, \delta(x - x')$$

Hence, show that the distribution function can be obtained from an inverse Laplace Transform of the moment generating function $M(t)$

$$M(t) = \int_{-\infty}^{\infty} dx \, \rho(x) \, \exp[xt]$$

$$= \int_{-\infty}^{\infty} dx \, \rho(x) \sum_{n=0}^{\infty} \frac{1}{n!} x^n t^n.$$

Problem 2.19

Consider a system made of N identical components. The system is characterized by an extensive variable X which is expressed in terms of the variables x_i of the N independent subsystems

$$X = \sum_{i=1}^{N} x_i$$

The probability distribution $\rho(X)$ for the extensive variable can be expressed in terms of the probability distributions $p(x_i)$ of the components via

$$\rho(X) = \prod_{i=1}^{N} \left\{ \int_{-\infty}^{\infty} dx_i \, p(x_i) \right\} \delta\left(X - \sum_{j=1}^{N} x_j \right)$$

The delta function can be represented by the Fourier Transform

$$\delta\left(X - \sum_{j=1}^{N} x_j \right) = \frac{1}{2\pi} \int_{-\infty}^{\infty} dk \, \exp\left[-ik\left(X - \sum_{j=1}^{N} x_j \right) \right]$$

(i) Evaluate $\rho(X)$ by contour integration if

$$p(x_i) = \exp[-x_i] \quad \text{if } x_i > 0$$
$$= 0 \quad \text{otherwise}$$

(ii) Evaluate $\rho(X)$ if

$$p(x_i) = \frac{1}{\sqrt{\pi}} \exp[-(x_i - \overline{x})^2]$$

(iii) Show that, in the limit $N \to \infty$, both distributions $\rho(X)$ can be approximated by a sharply-peaked Gaussian distribution function. Hence, in the limit $N \to \infty$ the extensive variable is equal to its average value. (The "*Central Limit Theorem*" extends the conclusions of this exercise to a more general class of distributions.)

Chapter 3

The Micro-Canonical Ensemble

The "*Micro-Canonical Ensemble*" is a Statistical Ensemble that represents all the possible microscopic states of a closed system which has a fixed energy E, occupies a fixed volume V and is composed of fixed numbers of particles N.

3.1 Classical Harmonic Oscillators

Consider a set of N classical harmonic oscillators described by the Hamiltonian H

$$H = \sum_{i=1}^{dN} \left[\frac{p_i^2}{2m} + \frac{m\omega_0^2 \, q_i^2}{2} \right] \tag{3.1}$$

The above Hamiltonian may be adopted as a model of the vibrational motion of the atoms in a solid. However, it is being assumed that the vibrations of the atoms are independent and are harmonic and all the oscillators have the same frequency. Furthermore, our treatment will be based on classical mechanics.

We shall consider the system in the Micro-Canonical Ensemble, where the energy is determined to within an uncertainty ΔE

$$E > H > E - \Delta E \tag{3.2}$$

The volume of accessible phase space Γ_a is given by the integral

$$\Gamma_a = \prod_{i=1}^{dN} \left\{ \int_{-\infty}^{\infty} dq_i \int_{-\infty}^{\infty} dp_i \right\} \left[\Theta(E - H) - \Theta(E - \Delta E - H) \right] \tag{3.3}$$

where $\Theta(x)$ is the Heaviside step function. On transforming the coordinates to $\tilde{p}_i = m\omega_0 \, q_i$, the Hamiltonian can be written in the form

$$H = \frac{1}{2m} \sum_{i=1}^{dN} [p_i^2 + \tilde{p}_i^2] \tag{3.4}$$

so the accessible phase space is defined by the inequalities

$$2mE > \sum_{i=1}^{dN} [p_i^2 + \tilde{p}_i^2] > 2m(E - \Delta E) \tag{3.5}$$

so

$$\Gamma_a = \prod_{i=1}^{dN} \left\{ \frac{1}{m\omega_0} \int_{-\infty}^{\infty} d\tilde{p}_i \int_{-\infty}^{\infty} dp_i \right\} [\Theta(E - H) - \Theta(E - \Delta E - H)] \tag{3.6}$$

Thus, the area of accessible phase space is proportional to the volume enclosed between two $2dN$-dimensional hyperspheres of radii $\sqrt{2mE}$ and $\sqrt{2m(E - \Delta E)}$. Therefore, we need to evaluate the volume enclosed by a hypersphere.

3.1.1   The Volume of a d-Dimensional Hypersphere

The volume of a d-dimensional hypersphere of radius R has the form

$$V_d(R) = C \, R^d \tag{3.7}$$

where C is a constant that depends on the dimensionality d. The constant can be determined by comparing two methods of evaluating the integral I_d, given by

$$I_d = \int_{-\infty}^{\infty} dx_1 \int_{-\infty}^{\infty} dx_2 \, \ldots \int_{-\infty}^{\infty} dx_d \, \exp\left[-\sum_{i=1}^{d} x_i^2 \right] \tag{3.8}$$

In a d-dimensional Cartesian coordinate system, the integral can be evaluated as a product of d identical integrals

$$I_d = \prod_{i=1}^{d} \left\{ \int_{-\infty}^{\infty} dx_i \, \exp[-x_i^2] \right\}$$
$$= \{\sqrt{\pi}\}^d$$
$$= \pi^{\frac{d}{2}} \tag{3.9}$$

Alternatively, one may evaluate the integral in hyperspherical polar coordinates as

$$I_d = S_d \int_0^{\infty} dr \, r^{d-1} \, \exp[-r^2] \tag{3.10}$$

where the radial coordinate is defined by

$$r^2 = \sum_{i=1}^{d} x_i^2 \tag{3.11}$$

and S_d is the surface area of a d-dimensional unit sphere. This integral can be re-written in term of the variable $t = r^2$ as

$$I_d = \frac{S_d}{2} \int_0^\infty dt \ t^{\frac{d-2}{2}} \ \exp[-t] \qquad (3.12)$$

The integration is evaluated as

$$I_d = \frac{S_d}{2} \ \Gamma\left(\frac{d}{2}\right) \qquad (3.13)$$

where $\Gamma(n+1) = n!$ is the factorial function. On equating the above two expressions, one obtains the equality

$$\frac{S_d}{2} \ \Gamma\left(\frac{d}{2}\right) = \pi^{\frac{d}{2}} \qquad (3.14)$$

Hence, we find that the surface area of a unit d-dimensional sphere, S_d, is given by

$$S_d = 2 \ \frac{\pi^{\frac{d}{2}}}{\Gamma(\frac{d}{2})} \qquad (3.15)$$

Using this, one finds that the volume of a d-dimensional sphere of radius R is given by

$$V_d(R) = S_d \int_0^R dr \ r^{d-1}$$

$$= S_d \ \frac{1}{d} \ R^d$$

$$= \frac{2}{d} \ \frac{\pi^{\frac{d}{2}}}{\Gamma(\frac{d}{2})} \ R^d$$

$$= \frac{\pi^{\frac{d}{2}}}{\frac{d}{2}\Gamma(\frac{d}{2})} \ R^d$$

$$= \frac{\pi^{\frac{d}{2}}}{\Gamma(\frac{d}{2}+1)} \ R^d \qquad (3.16)$$

which is our final result.

3.1.2   *The Volume of Accessible Phase Space*

The volume of accessible phase space Γ_a is proportional to the volume enclosed by two $2dN$-dimensional hyperspheres of radius $\sqrt{2mE}$ and

$\sqrt{2m(E - \Delta E)}$. Using the above results, one finds

$$\Gamma_a = \frac{\pi^{dN}}{\Gamma(dN+1)} \left(\frac{1}{m\omega_0}\right)^{dN} [(2mE)^{dN} - (2m(E - \Delta E))^{dN}] \qquad (3.17)$$

where the factor of

$$\left(\frac{1}{m\omega_0}\right)^{dN} \qquad (3.18)$$

is the Jacobian for the coordinate transformations. The number of accessible microstates N_Γ is then defined as

$$N_\Gamma = \frac{\Gamma_a}{(2\pi\hbar)^{dN}}$$

$$= \frac{1}{\Gamma(dN+1)} \left(\frac{E}{\hbar\omega_0}\right)^{dN} \left[1 - \left(1 - \frac{\Delta E}{E}\right)^{dN}\right] \qquad (3.19)$$

The second factor in the square brackets is extremely small when compared to unity since the term in the parenthesis is less than unity and the exponent is extremely large. Therefore, it can be neglected

$$N_\Gamma \approx \frac{1}{\Gamma(dN+1)} \left(\frac{E}{\hbar\omega_0}\right)^{dN} \qquad (3.20)$$

This implies that, for sufficiently high dimensions, the volume of the hypersphere is the same as the volume of the hypershell.

3.1.3 *Derivation of Stirling's Approximation*

The Gamma function is defined by the integral

$$\Gamma(n+1) = \int_0^\infty dx \; x^n \; \exp[-x] \qquad (3.21)$$

which, for integer n coincides with $n!$. This can be verified by repeated integration by parts

$$\Gamma(n+1) = \int_0^\infty dx \; x^n \; \exp[-x]$$

$$= -\int_0^\infty dx \; x^n \; \frac{\partial}{\partial x} \exp[-x]$$

$$= -x^n \; \exp[-x]\Big|_0^\infty + n \int_0^\infty dx \; x^{n=-1} \; \exp[-x]$$

$$= n \int_0^\infty dx \; x^{n-1} \; \exp[-x]$$

$$= n \; \Gamma(n) \qquad (3.22)$$

which together with

$$\Gamma(1) = \int_0^\infty dx \ \exp[-x] = 1 \tag{3.23}$$

leads to the evaluation of the integral as

$$\Gamma(n+1) = \int_0^\infty dx \ x^n \ \exp[-x] = n! \tag{3.24}$$

for integer n.

Stirling's approximation to $\ln n!$ can be obtained by evaluating the integral using the method of steepest descents.

$$n! = \int_0^\infty dx \ \exp[-x + n \ \ln x] \tag{3.25}$$

The extremal value of $x = x_c$ is found from equating the derivative of the exponent to zero

$$-1 + \frac{n}{x_c} = 0 \tag{3.26}$$

This yields $x_c = n$. On expanding the integrand to second order in $(x - x_c)$, one has

$$n! \approx \int_0^\infty dx \ \exp[-x_c + n \ \ln x_c] \ \exp\left[-\frac{n}{x_c^2}(x - x_c)^2\right] \tag{3.27}$$

On extending the lower limit of the integration to $-\infty$, one obtains the approximation

$$n! \approx \int_{-\infty}^\infty dx \ \exp[-x_c + n \ \ln x_c] \ \exp\left[-\frac{n}{2x_c^2}(x - x_c)^2\right]$$

$$= \sqrt{\frac{2\pi \ x_c^2}{n}} \ \exp[-x_c + n \ \ln x_c] \tag{3.28}$$

This is expected to be valid for sufficiently large n. On setting $x_c = n$, one has

$$n! \approx \sqrt{2\pi \ n} \ \exp[-n + n \ \ln n] \tag{3.29}$$

Stirling's approximation is obtained by taking the logarithm, which yields

$$\ln n! = n \ \ln n - n + \frac{1}{2} \ \ln(2\pi \ n) \tag{3.30}$$

Stirling's approximation will be used frequently throughout this course.

The Entropy

The entropy S is given by

$$S = k_B \ \ln N_\Gamma$$
$$= dN \ k_B \ \ln \left(\frac{E}{\hbar\omega_0} \right) - k_B \ \ln(dN)! \qquad (3.31)$$

The logarithm of $N!$ can be approximated for large N by Stirling's approximation. This can be quickly re-derived by noting that

$$\ln N! = \ln N + \ln(N - 1) + \ln(N - 2) + \cdots + \ln 2 + \ln 1 \qquad (3.32)$$

For large N, the sum on the right hand side can be approximated by an integral

$$\ln N! \approx \int_0^N dx \ \ln x$$
$$\approx x(\ln x - 1) \Big|_0^N$$
$$\approx N(\ln N - 1) \qquad (3.33)$$

which results in Stirling's approximation

$$\ln N! = N(\ln N - 1) \qquad (3.34)$$

Using Stirling's approximation in the expression for the entropy S, one obtains

$$S(E, N) = dN \ k_B \ \ln \left(\frac{E}{\hbar\omega_0} \right) - k_B \ dN(\ln(dN) - 1)$$
$$= dN \ k_B \ \ln \left(\frac{E}{dN \ \hbar\omega_0} \right) + k_B \ dN \qquad (3.35)$$

which shows that the entropy is an extensive monotonically increasing function of E. This is the fundamental relation. In the Micro-Canonical Ensemble, the energy E is the thermodynamic energy U.

The temperature is defined by the derivative

$$\frac{1}{T} = \left(\frac{\partial S}{\partial U} \right)_N \qquad (3.36)$$

which yields

$$\frac{1}{T} = \frac{dN \ k_B}{U} \qquad (3.37)$$

Hence, we find that the internal energy U is given by

$$U = dN \ k_B T \qquad (3.38)$$

which shows that each degree of freedom carries the thermodynamic energy $k_B T$. The specific heat at constant volume is then found as

$$C_V = dN \, k_B \tag{3.39}$$

which is Dulong and Petit's law [18]. Dulong and Petit's law describes the high-temperature specific heat of solids quite well, but fails at low temperatures where the quantum mechanical nature of the solid manifests itself.

3.2 An Ideal Gas of Indistinguishable Particles

The Hamiltonian for an ideal gas is written as the sum of the kinetic energies

$$H = \sum_{i=1}^{dN} \frac{p_i^2}{2m} \tag{3.40}$$

The gas is contained in a volume V with linear dimensions L, such that

$$V = L^d \tag{3.41}$$

where d is the number of dimensions of space. In the Micro-Canonical Ensemble, the energy is constrained to an interval of width ΔE according to the inequality

$$E > H > E - \Delta E \tag{3.42}$$

The volume of accessible phase space Γ_a is given by the multiple integral

$$\Gamma_a = \prod_{i=1}^{dN} \left\{ \int_0^L dq_i \int_{-\infty}^{\infty} dp_i \right\} [\Theta(E - H) - \Theta(E - \Delta E - H)] \tag{3.43}$$

The integration over the coordinates can be performed, leading to the expression

$$\begin{aligned}
\Gamma_a &= \prod_{i=1}^{dN} \left\{ L \int_{-\infty}^{\infty} dp_i \right\} [\Theta(E - H) - \Theta(E - \Delta E - H)] \\
&= L^{dN} \prod_{i=1}^{dN} \left\{ \int_{-\infty}^{\infty} dp_i \right\} [\Theta(E - H) - \Theta(E - \Delta E - H)] \\
&= V^N \prod_{i=1}^{dN} \left\{ \int_{-\infty}^{\infty} dp_i \right\} [\Theta(E - H) - \Theta(E - \Delta E - H)] \tag{3.44}
\end{aligned}$$

The step functions constrain the momenta such that

$$2mE > \sum_{i=1}^{dN} p_i^2 > 2m \, (E - \Delta E) \tag{3.45}$$

Thus, the integration over the momenta is equal to the volume contained between two dN-dimensional hyperspheres of radii $\sqrt{2mE}$ and $\sqrt{2m(E-\Delta E)}$. Using the expressions for the volume of a dN dimensional hypersphere

$$V_{dN}(R) = \frac{\pi^{\frac{dN}{2}}}{\Gamma\left(\frac{dN}{2}+1\right)} R^{dN} \tag{3.46}$$

which yields

$$\Gamma_a = V^N \frac{\pi^{\frac{dN}{2}}}{\Gamma\left(\frac{dN}{2}+1\right)} (2mE)^{\frac{dN}{2}} \left[1 - \left(1 - \frac{\Delta E}{E}\right)^{\frac{dN}{2}}\right]$$

$$= V^N \frac{\pi^{\frac{dN}{2}}}{\Gamma\left(\frac{dN}{2}+1\right)} (2mE)^{\frac{dN}{2}} \left(1 - \exp\left[-\frac{dN\Delta E}{2E}\right]\right) \tag{3.47}$$

Since

$$1 \gg \exp\left[-\frac{dN\Delta E}{2E}\right] \tag{3.48}$$

the volume of accessible phase space is given by

$$\Gamma_a = V^N \frac{\pi^{\frac{dN}{2}}}{\Gamma\left(\frac{dN}{2}+1\right)} (2mE)^{\frac{dN}{2}} \tag{3.49}$$

However, for an ideal gas of identical particles, we have to take into account that specifying all the momenta p_i and coordinates q_i of the N particles provides us with too much information. Since the N particles are identical, we cannot distinguish between two points of phase space that differ only by the interchange of identical particles. There are $N!$ points corresponding to the different labelings of the particles. These $N!$ points represent the same microstates of the system. To only account for the different microstates, one must divide the volume of accessible phase space by $N!$. Hence, the number of microscopic states N_Γ is given by

$$N_\Gamma = \frac{\Gamma_a}{N!(2\pi\hbar)^{dN}}$$

$$= \frac{V^N}{N!\,\Gamma\left(\frac{dN}{2}+1\right)} \left(\frac{mE}{2\pi\hbar^2}\right)^{\frac{dN}{2}} \tag{3.50}$$

The entropy S is given by the expression

$$S = k_B \ln N_{\Gamma_a}$$

$$= k_B \ln \left[\frac{V^N}{N! \, \Gamma\left(\frac{dN}{2} + 1\right)} \left(\frac{mE}{2\pi\hbar^2}\right)^{\frac{dN}{2}} \right]$$

$$= k_B \ln \left[\frac{V^N}{N! \left(\frac{dN}{2}\right)!} \left(\frac{mE}{2\pi\hbar^2}\right)^{\frac{dN}{2}} \right] \qquad (3.51)$$

On using Stirling's formulae

$$\ln N! = N(\ln N - 1) \qquad (3.52)$$

valid for large N, one finds

$$S = N \, k_B \, \ln \left(\frac{V}{N}\right) + \frac{dN}{2} \, k_B \, \ln \left(\frac{mE}{2\pi\hbar^2 \, \frac{dN}{2}}\right) + N \, k_B + \frac{d}{2} \, N \, k_B \qquad (3.53)$$

or

$$S = N \, k_B \, \ln \left[\left(\frac{V}{N}\right) \left(\frac{mE}{d\pi\hbar^2 \, N}\right)^{\frac{d}{2}} \right] + \left(\frac{d+2}{2}\right) N \, k_B \qquad (3.54)$$

On identifying E with U, the thermodynamic energy, one has the fundamental relation $S(U, V, N)$ from which all thermodynamic quantities can be obtained.

The intensive quantities can be obtained by taking the appropriate derivatives. For example, the temperature is found from

$$\frac{1}{T} = \left(\frac{\partial S}{\partial U}\right)_{V,N} \qquad (3.55)$$

which yields

$$\frac{1}{T} = \frac{dN \, k_B}{2U} \qquad (3.56)$$

Hence, we have recovered the equation of state for an ideal gas

$$U = \frac{dN}{2} \, k_B T \qquad (3.57)$$

Likewise, the pressure is given by

$$\frac{P}{T} = \left(\frac{\partial S}{\partial V}\right)_{U,N} \qquad (3.58)$$

which yields

$$\frac{P}{T} = \frac{N \, k_B}{V} \qquad (3.59)$$

which is the ideal gas law. The chemical potential μ is found from

$$-\frac{\mu}{T} = \left(\frac{\partial S}{\partial N}\right)_{U,V} \tag{3.60}$$

which yields

$$-\frac{\mu}{T} = k_B \ln\left[\left(\frac{V}{N}\right)\left(\frac{mU}{d\pi\hbar^2 \, N}\right)^{\frac{d}{2}}\right] - \left(\frac{d+2}{2}\right) k_B + \left(\frac{d+2}{2}\right) k_B \tag{3.61}$$

Since

$$\frac{U}{dN} = \frac{k_B T}{2} \tag{3.62}$$

one has

$$-\frac{\mu}{T} = k_B \ln\left[\left(\frac{V}{N}\right)\left(\frac{mk_B T}{2\pi\hbar^2}\right)^{\frac{d}{2}}\right] \tag{3.63}$$

This can be re-written as

$$\frac{\mu}{T} = k_B \ln P + f(T) \tag{3.64}$$

where P is the pressure and $f(T)$ is a function of only the temperature T.

On substituting the equation of state

$$U = \frac{dN}{2} k_B T \tag{3.65}$$

into the expression for the entropy, one finds

$$S = N \, k_B \ln\left[\left(\frac{V}{N}\right)\left(\frac{mk_B T}{2\pi\hbar^2}\right)^{\frac{d}{2}}\right] + \left(\frac{d+2}{2}\right) N \, k_B \tag{3.66}$$

This is the Sackur-Tetrode formula for the entropy of an ideal gas.

$$\lambda = \frac{h}{(2\pi m k_B T)^{\frac{1}{2}}} \tag{3.67}$$

The Sackur-Tetrode equation was derived independently by Hugo Martin Tetrode and Otto Sackur, using Maxwell-Boltzmann statistics in 1912.

Note that the factor

$$(2\pi m k_B T)^{\frac{1}{2}} \tag{3.68}$$

has the character of an average thermal momentum of a molecule. We can define λ via

$$\lambda = \frac{h}{(2\pi m k_B T)^{\frac{1}{2}}} \tag{3.69}$$

as a thermal de Broglie wave length associated with the molecule. The entropy can be re-written as

$$S = N \, k_B \, \ln \left[\frac{V}{N \, \lambda^d} \right] + \left(\frac{d+2}{2} \right) N \, k_B \tag{3.70}$$

which shows that the entropy S is essentially determined by the ratio of the volume per particle to the volume λ^d associated with the thermal de Broglie wavelength. The classical description is approximately valid in the limit

$$\frac{V}{N \, \lambda^d} \gg 1 \tag{3.71}$$

where the uncertainties in particle positions are negligible compared with the average separation of the particles. When the above inequality does not apply, quantum effects become important.

3.2.1 The Maxwell-Boltzmann Distribution

The probability that a particle has momentum of magnitude $|p|$ can be obtained using the Micro-Canonical Ensemble. First, we should note that the particle we have selected is being distinguished from the other $N-1$ particles, but the other $(N-1)$ particles remain indistinguishable. Then, the probability that one particle in an ideal gas of N particles has an energy E_p

$$E_p = \sum_{i=1}^{d} \frac{p_i^2}{2m} \tag{3.72}$$

is equal to the probability that the total energy of the other $(N-1)$ particles is $E - E_p$. Therefore, the probability is simply found as the ratio of the number of states in which the $(N-1)$ particles have the total energy $E - E_p$ to the total number of states accessible to the N particle system. Thus, we find the momentum probability distribution function $P(|p|)$ is expressed as

$$P(|\underline{p}|) = \frac{1}{\Gamma_a} \prod_{i=1}^{dN} \left\{ \int_0^L dq_i \right\} \prod_{i=d+1}^{dN} \left\{ \int_{-\infty}^{\infty} dp_i \right\}$$
$$\times \left[\Theta \left(2mE - \sum_{i=1}^{dN} p_i^2 \right) - \Theta \left(2m(E - \Delta E) - \sum_{i=1}^{dN} p_i^2 \right) \right] \tag{3.73}$$

which is evaluated as

$$P(|\underline{p}|) = \frac{\Gamma\left(\frac{dN}{2}+1\right)}{\pi^{\frac{d}{2}}\,\Gamma\left(\frac{d(N-1)}{2}+1\right)}\,\frac{\left(2mE-|\underline{p}|^2\right)^{\frac{d(N-1)}{2}}}{(2mE)^{\frac{dN}{2}}}$$

$$= \frac{\Gamma\left(\frac{dN}{2}+1\right)}{\pi^{\frac{d}{2}}\,\Gamma\left(\frac{d(N-1)}{2}+1\right)}\,\frac{\left(1-\frac{|\underline{p}|^2}{2mE}\right)^{\frac{d(N-1)}{2}}}{(2mE)^{\frac{d}{2}}}$$

$$\approx \left(\frac{dN}{2\pi}\right)^{\frac{d}{2}}\,\frac{\exp\left[-\frac{d(N-1)}{2}\,\frac{|\underline{p}|^2}{2mE}\right]}{(2mE)^{\frac{d}{2}}}$$

$$\approx \left(\frac{dN}{4\pi mE}\right)^{\frac{d}{2}}\,\exp\left[-\frac{d(N-1)}{2}\,\frac{|\underline{p}|^2}{2mE}\right] \tag{3.74}$$

This is the desired result. On using the thermodynamic relation for the energy

$$U = \frac{dN}{2}\,k_B T \tag{3.75}$$

one obtains the Maxwell distribution

$$P(|\underline{p}|) = (2\pi m k_B T)^{-\frac{d}{2}}\,\exp\left[-\frac{|\underline{p}|^2}{2mk_B T}\right] \tag{3.76}$$

which is properly normalized.

3.3   Spin One-Half Particles

A system of spin one-half particles is described by a discrete, not continuous, phase space. The discreteness, is due to the discrete nature of the quantum mechanical eigenvalues.

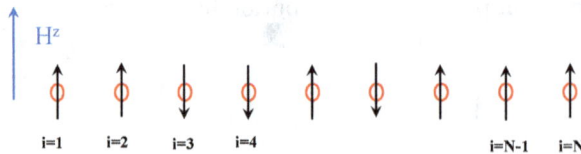

Fig. 3.1   A set of N spins, in the presence of a uniform applied magnetic field H^z directed along the z-axis. The spins are quantized along the z-direction, so their S^z components are given by $\pm\hbar/2$.

Consider a set of N spin one-half particles, in an applied magnetic field. The particles may be either aligned parallel or anti-parallel to the applied magnetic field H^z. The Hamiltonian describing the spins is given by

$$H = -g\,\mu_B \sum_{i=1}^{N} S_i^z H^z \tag{3.77}$$

where $S^z = \pm \frac{1}{2}$. If one defines the total magnetic moment as

$$M^z = g\,\mu_B \sum_{i=1}^{N} S_i^z \tag{3.78}$$

one finds that the energy is determined by the magnetization via

$$H = -M^z H^z \tag{3.79}$$

Therefore, if the energy has a fixed value E, the accessible microstates are determined by the fixed value of the magnetization M^z. We shall introduce the dimensionless magnetization as

$$m = M^z/\mu_B \tag{3.80}$$

Hence, for a fixed energy there are $(N+m)/2$ spin-up particles and $(N-m)/2$ spin-down particles. The number of ways of selecting $(N+m)/2$ particles out of N particles as being spin up is given by

$$\frac{N!}{\left(\frac{N-m}{2}\right)!} \tag{3.81}$$

since there are N ways of selecting the first particle as being spin up, $(N-1)$ ways of selecting the second particle as being spin up, etc. This process continues until the $(N+m)/2$ spin-up particle is chosen which can be selected in $(N+1-(N+m)/2)$ ways. Since the number of choices is multiplicative, the product of the number of choices give the result above. However, not all of these choices lead to independent microstates. Interchanges of the $(N+m)/2$ spin up particles between themselves lead to identical microstates. There are $((N+m)/2)!$ such interchanges. The total number of discrete microstates with magnetization m is found by dividing the above result by $((N+m)/2)!$. The end result is N_Γ given by

$$N_\Gamma = \frac{N!}{\left(\frac{N+m}{2}\right)!\left(\frac{N-m}{2}\right)!} \tag{3.82}$$

The entropy S is found from

$$S = k_B \ln N_\Gamma \tag{3.83}$$

which is evaluated as

$$S = k_B \ln \frac{N!}{\left(\frac{N+m}{2}\right)!\left(\frac{N-m}{2}\right)!}$$

$$= k_B \ln N! - k_B \ln \left(\frac{N+m}{2}\right)! - k_B \ln \left(\frac{N-m}{2}\right)!$$

$$= k_B N \ln N - k_B \left(\frac{N+m}{2}\right) \ln \left(\frac{N+m}{2}\right)$$

$$- k_B \left(\frac{N-m}{2}\right) \ln \left(\frac{N-m}{2}\right) \tag{3.84}$$

where we have used Stirling's approximation in the last line. Hence, the entropy has been expressed in terms of N and m or equivalently in terms of E and N. This is the fundamental relation, from which we may derive all thermodynamic quantities. On identifying the fixed energy with the thermodynamic energy, one may use the definition of temperature

$$\frac{1}{T} = \left(\frac{\partial S}{\partial U}\right)_N \tag{3.85}$$

or

$$\frac{1}{T} = \left(\frac{\partial S}{\partial m}\right)_N \left(\frac{\partial m}{\partial U}\right)$$

$$= -\left(\frac{\partial S}{\partial m}\right)_N \frac{1}{\mu_B H_z} \tag{3.86}$$

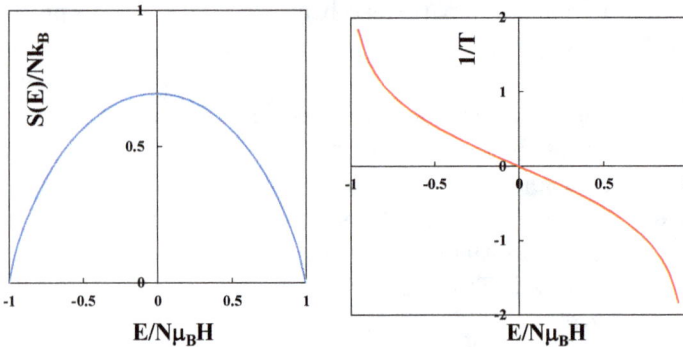

Fig. 3.2 The entropy $S(E)$ as a function of entropy E for a model of a system of spins in a magnetic field is shown in (a). Since the energy is bounded from above, the entropy is not a monotonically increasing function of E. This has the consequence that T can become negative when there is population inversion, as is indicated in (b).

Therefore, one has

$$
\frac{\mu_B H^z}{k_B T} = \frac{1}{2} \ln \left(\frac{N+m}{2} \right) - \frac{1}{2} \ln \left(\frac{N-m}{2} \right)
$$

$$
= \frac{1}{2} \ln \left(\frac{N+m}{N-m} \right) \tag{3.87}
$$

This can be exponentiated to yield

$$
\exp \left[\frac{2\mu_B H^z}{k_B T} \right] = \left(\frac{N+m}{N-m} \right) \tag{3.88}
$$

which can be solved for m as

$$
m = N \left(\frac{\exp \left[\frac{2\mu_B H^z}{k_B T} \right] - 1}{\exp \left[\frac{2\mu_B H^z}{k_B T} \right] + 1} \right)
$$

$$
= N \tanh \left(\frac{\mu_B H^z}{k_B T} \right) \tag{3.89}
$$

Hence, the magnetization is an odd function of H^z and saturates for large fields and low temperatures at $\pm N$. Finally, we obtain the expression for the thermal average of the internal energy

$$
U = -N \mu_B H^z \tanh \left(\frac{\mu_B H^z}{k_B T} \right) \tag{3.90}
$$

which vanishes as the square of the field H^z in the limit of zero applied field. This is expected since the Hamiltonian is linear in H^z and since the magnetization is expected to vanish linearly as H^z vanishes.

Zero Applied Field

We shall now determine the magnetization probability distribution function in the limit of zero applied magnetic field. The spins of the particles may either be aligned parallel or anti-parallel to the axis of quantization. There are a total of 2^N possible microstates. Hence, for zero applied field

$$
N_\Gamma = 2^N \tag{3.91}
$$

Since all microstates are assumed to occur with equal probabilities, the probability of finding a system with magnetization m is given by

$$
P(m) = \frac{1}{N_\Gamma} \frac{N!}{\left(\frac{N+m}{2} \right)! \left(\frac{N-m}{2} \right)!}
$$

$$
= \left(\frac{1}{2} \right)^N \frac{N!}{\left(\frac{N+m}{2} \right)! \left(\frac{N-m}{2} \right)!} \tag{3.92}
$$

which is normalized to unity. On using the more accurate form of Stirling's approximation that we found using the method of steepest descents

$$\ln N! \approx \frac{1}{2}\ln(2\pi N) + N\ln N - N$$
$$\approx \frac{1}{2}\ln(2\pi) + \left(N + \frac{1}{2}\right)\ln N - N \qquad (3.93)$$

in $\ln P(m)$, one obtains

$$\ln P(m) \approx -N\ln 2 - \frac{1}{2}\ln(2\pi) + \left(N + \frac{1}{2}\right)\ln N$$
$$- \frac{N+1}{2}\left(1 + \frac{m}{N+1}\right)\ln\frac{N}{2}\left(1 + \frac{m}{N}\right)$$
$$- \frac{N+1}{2}\left(1 - \frac{m}{N+1}\right)\ln\frac{N}{2}\left(1 - \frac{m}{N}\right) \qquad (3.94)$$

On expanding in powers of m, the expression simplifies to

$$\ln P(m) \approx -\frac{1}{2}\ln N - \frac{1}{2}\ln(2\pi) + \ln 2 - \frac{m^2}{2N} + \dots \qquad (3.95)$$

Hence, one finds that the magnetization probability distribution function $P(m)$ is approximated by a Gaussian distribution

$$P(m) \approx \sqrt{\frac{2}{\pi N}}\,\exp\left[-\frac{m^2}{2N}\right] \qquad (3.96)$$

Therefore, the most probable value of the magnetization is $m = 0$ and the width of the distribution is given by $\sqrt{N}$. This is small compared with the total range of the possible magnetization which is $2N$. Most of the microstates correspond to zero magnetization. This can be seen as total number of available microstates is given by

$$2^N \qquad (3.97)$$

and since the number of states with zero magnetization is given by

$$\frac{N!}{(N/2)!(N/2)!} \sim \sqrt{\frac{2}{\pi N}}\,2^N \qquad (3.98)$$

Thus, this implies that, for $H^z = 0$, the relative size of the fluctuations in the magnetization is small.

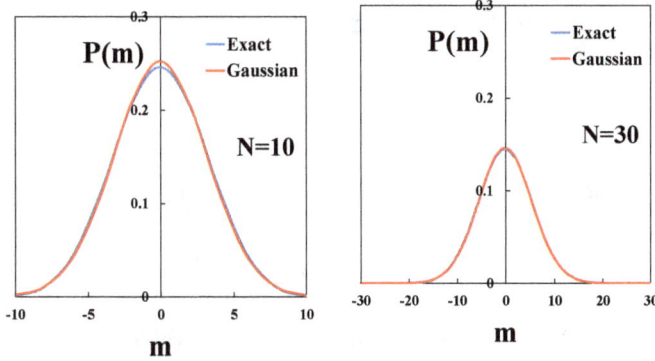

Fig. 3.3 The exact probability distribution $P(m)$ of the magnetic moment m for a system of N spins, and the approximate Gaussian distribution. After scaling m with the size of the system N, the width of the distribution decreases on increasing N.

3.4 The Einstein Model of a Crystalline Solid

The Einstein Model of a crystalline solid [19] considers the normal modes of vibrations of the lattice to be quantized, and it assumes that the frequencies of the all the normal modes are identical and equal to ω_0. It is a reasonable approximation for the optical phonon modes in a solid. For a solid with N unit cells, where there are p atoms per unit cell, one expects there to be $N' = 3(p-1)N$ optic modes. The remaining $3N$ modes are expected to be acoustic modes.

Consider a set of N' quantum mechanical harmonic oscillators in the Micro-Canonical Ensemble. Each oscillator has the same frequency ω_0. The total energy E is given by the sum of the energies of each individual quantum oscillator

$$E = \sum_{i=1}^{N'} \hbar\omega_0 \left(n_i + \frac{1}{2} \right) \tag{3.99}$$

where n_i is the number of quanta in the i-th oscillator. The possible values of n_i are the set of $0, 1, 2, \ldots, \infty$. The last term in the round parenthesis is the zero-point energy of the i-th oscillator.

If we subtract the zero-point energy for each quantum oscillator, then the energy E_{exc} available to distribute amongst the N' quantum mechanical harmonic oscillators is given by

$$E_{exc} = E - N' \left(\frac{\hbar\omega_0}{2} \right) \tag{3.100}$$

The excitation energy E_{exc} is to be distributed amongst the N' quantum oscillators

$$E_{exc} = \sum_{i=1}^{N'} \hbar\omega_0 \, n_i \tag{3.101}$$

The total number of quanta Q available to the entire system is given by

$$\frac{E_{exc}}{\hbar\omega_0} = \sum_{i=1}^{N'} n_i = Q \tag{3.102}$$

which acts as a restriction on the possible sets of values of n_i. Each possible distribution of the Q quanta is described by a set of integer values, $\{n_i\}$, which uniquely describes a microstate of the system. In any allowed microstate the values of $\{n_i\}$ are restricted so that the number of quanta add up to Q.

There are Q quanta which must be distributed between the N' oscillators. We shall count all the possible ways of distributing the Q quanta among the N' oscillators. Let us consider each oscillator as a box and each quanta as a marble. Eventually, the marbles are to be considered as being indistinguishable, so interchanging any number of marbles will lead to the same configuration. We shall temporarily suspend this assumption and instead assume that the marbles could be tagged. Later, we shall restore the assumption of indistinguishability.

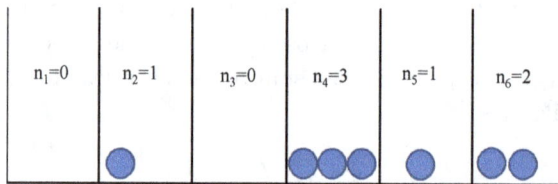

Fig. 3.4 One microscopic state of a system in which Q quanta have been distributed amongst N' oscillators ($Q = 7, N' = 6$).

The Number of Distinguishable Ways

The number of ways of putting Q marbles in N' boxes can be found by arranging the boxes in a row. In this case, a box shares a common wall with its neighboring boxes so there are $N' + 1$ walls for the N' boxes. If one considers both the walls and marbles as being distinguishable objects, then in any distribution of the marbles in the boxes there are $Q + N' + 1$ objects in a row. If there are n_i marbles between two consecutive walls, then that box contains n_i marbles. If there are two consecutive walls in a distribution, then that box is empty. However, the first object and the last object are always walls, so really there are $Q + N' - 1$ objects that can be re-arranged. Therefore, the total number of orderings can be found from the number of ways of arranging $Q + N' - 1$ objects in a row. This can be done in

$$(Q + N' - 1)! \tag{3.103}$$

number of ways. This happens since there are $Q + N' - 1$ ways of selecting the first object. After the first object has been chosen, there are $Q + N' - 2$ objects that remain to be selected, so there are only $Q + N' - 2$ ways of selecting the second object. Likewise, there are $Q + N' - 3$ ways of choosing the third object, and this continues until only the last object is unselected, in which case there is only one possible way of choosing the last object. The number of possible arrangements is given by the product of the number of ways of making each independent choice. Thus, we have found that there are $(Q + N' - 1)!$ possible ways of sequencing or ordering $(Q + N' - 1)$ distinguishable objects.

The Number of Indistinguishable Ways

We do need to consider the walls as being indistinguishable and also the marbles should be considered as indistinguishable. If we permute the indistinguishable walls amongst themselves, the ordering that results is identical to the initial ordering. There are $(N' - 1)!$ ways of permuting the $N' - 1$ walls amongst themselves. Hence, we should divide by $(N' - 1)!$ to only count the number of orderings made by placing the marbles between indistinguishable walls. Likewise, if one permutes the Q indistinguishable marbles, it leads to an identical ordering, and there are $Q!$ such permutations. So we have over-counted the number of orderings by $Q!$, and hence we also need to divide our result by a factor of $Q!$. Therefore, the total number of inequivalent ways N_Γ of distributing Q indistinguishable marbles in N' boxes is given by

$$N_\Gamma = \frac{(N' + Q - 1)!}{(N' - 1)!\, Q!} \tag{3.104}$$

This is equal to the total number of microstates N_Γ consistent with having a total number of quanta Q distributed amongst N' oscillators.

The Entropy

In the Micro-Canonical Ensemble, the entropy S is given by the logarithm of the number of accessible microstates N_Γ

$$S = k_B \ln N_\Gamma \qquad (3.105)$$

On substituting the expression for N_Γ, one obtains

$$S = k_B \ln \left(\frac{(N' + Q - 1)!}{(N' - 1)! \, Q!} \right)$$
$$= k_B \left[\ln(N' + Q - 1)! - \ln(N' - 1)! - \ln Q! \right] \qquad (3.106)$$

On using Stirling's approximation

$$\ln N! \approx N \, (\ln N - 1) \qquad (3.107)$$

valid for large N, for all three terms, after some cancellation one has

$$S \approx k_B \left[(N' + Q - 1) \, \ln(N' + Q - 1) - (N' - 1) \, \ln(N' - 1) - Q \, \ln Q \right] \qquad (3.108)$$

which is valid for large Q and N'. It should be recalled that $Q = (E_{exc}/\hbar\omega_0)$, so S is a function of the total energy E. The above relation between the entropy and the total energy is the same as the relation between the entropy and the thermodynamic energy U. The expression for S in terms of U is the *"Fundamental Relation"* for the thermodynamics of the model.

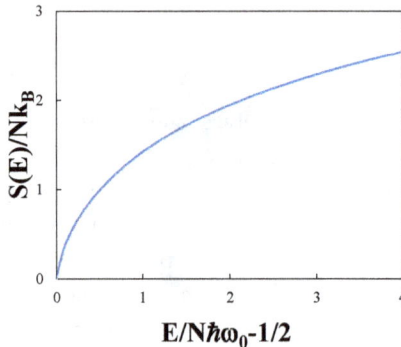

Fig. 3.5   The entropy $S(E)$ versus the dimensionless excitation energy, for the Einstein model for the specific heat a solid.

We shall now consider the system to be in thermal equilibrium with a thermal reservoir held at temperature T. The temperature is defined by

$$\frac{1}{T} = \left(\frac{\partial S}{\partial U} \right)_N \qquad (3.109)$$

which yields

$$\frac{1}{T} = \left(\frac{\partial S}{\partial Q}\right)\left(\frac{\partial Q}{\partial U}\right)$$

$$= k_B \left[\ln(N' + Q - 1) - \ln Q\right] \left(\frac{\partial Q}{\partial U}\right)$$

$$= \frac{k_B}{\hbar\omega_0} \left[\ln(N' + Q - 1) - \ln Q\right]$$

$$= \frac{k_B}{\hbar\omega_0} \ln\left(\frac{N' + Q - 1}{Q}\right)$$

$$= \frac{k_B}{\hbar\omega_0} \ln\left(\frac{\hbar\omega_0\,(N'-1) + U_{exc}}{U_{exc}}\right) \qquad (3.110)$$

where it is now understood that the energy is the thermodynamic value U that is determined by T. On multiplying by $\hbar\omega_0/k_B$ and then exponentiating the equation, one finds

$$\exp\left[\frac{\hbar\omega_0}{k_B T}\right] = \left(\frac{\hbar\omega_0\,(N'-1) + U_{exc}}{U_{exc}}\right) \qquad (3.111)$$

or on multiplying through by U_{exc}

$$U_{exc}\,\exp\left[\frac{\hbar\omega_0}{k_B T}\right] = \hbar\omega_0\,(N'-1) + U_{exc} \qquad (3.112)$$

This equation can be solved to yield U_{exc} as a function of T

$$U_{exc} = \frac{\hbar\omega_0\,(N'-1)}{\exp\left[\frac{\hbar\omega_0}{k_B T}\right] - 1} \qquad (3.113)$$

We can neglect the term 1 compared with N' since, in our derivation we have assumed that N' is very large. Since

$$U_{exc} = \sum_{i=1}^{N'} \hbar\omega_0\,\overline{n}_i \qquad (3.114)$$

we have found that the thermodynamic average number of quanta $\overline{n}$ of energy $\hbar\omega_0$ in each quantum mechanical harmonic oscillator is given by

$$\overline{n} = \frac{1}{\exp\left[\frac{\hbar\omega_0}{k_B T}\right] - 1} \qquad (3.115)$$

If we were to include the zero point energy, then the total thermodynamic energy is given by

$$
\begin{aligned}
U &= \sum_{i=1}^{N'} \hbar\omega_0 \left(\bar{n}_i + \frac{1}{2} \right) \\
&= \sum_{i=1}^{N'} \frac{\hbar\omega_0}{2} \left(\frac{2}{\exp\left[\frac{\hbar\omega_0}{k_B T}\right] - 1} + 1 \right) \\
&= \sum_{i=1}^{N'} \frac{\hbar\omega_0}{2} \left(\frac{\exp\left[\frac{\hbar\omega_0}{k_B T}\right] + 1}{\exp\left[\frac{\hbar\omega_0}{k_B T}\right] - 1} \right) \\
&= N' \frac{\hbar\omega_0}{2} \coth\left(\frac{\hbar\omega_0}{2k_B T} \right)
\end{aligned}
\tag{3.116}
$$

The specific heat of the Einstein model can be found from

$$
C_V = \left(\frac{\partial U}{\partial T} \right)_V
\tag{3.117}
$$

which yields

$$
C_V = N' k_B \left(\frac{\hbar\omega_0}{k_B T} \right)^2 \frac{\exp\left[\frac{\hbar\omega_0}{k_B T}\right]}{\left(\exp\left[\frac{\hbar\omega_0}{k_B T}\right] - 1 \right)^2}
\tag{3.118}
$$

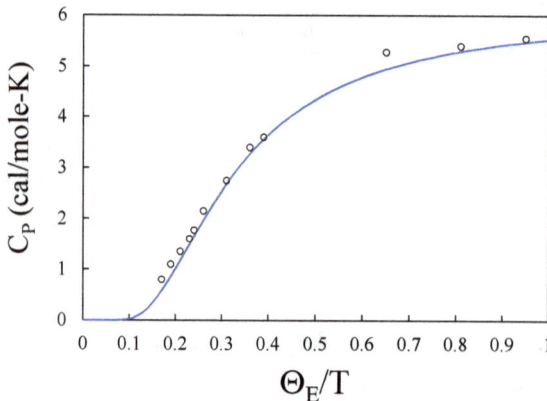

Fig. 3.6 The specific heat of diamond compared with the results of the Einstein Model. The parameter $\Theta_E = \hbar\omega_0/k_B$ is a characteristic temperature that has been assigned the value $\Theta_E = 1320$ K. [After A. Einstein, Ann. Physik **22**, 180-190 (1906).]

The specific heat tends to $N'k_B$ for temperatures $k_B T \gg \hbar\omega_0$, as is expected classically. However, at low temperatures, defined by $k_B T \ll \hbar\omega_0$, the specific heat falls to zero exponentially

$$C_V \approx N'k_B \left(\frac{\hbar\omega_0}{k_B T} \right)^2 \exp\left[-\frac{\hbar\omega_0}{k_B T} \right] \qquad (3.119)$$

Therefore, the specific heat vanishes in the limit $T \to 0$ in accordance with Nernst's law. However, the specific heat of most materials deviate from the prediction of the Einstein model at low temperatures.

3.5 Vacancies in a Crystal

Consider a crystal composed of N identical atoms arranged in a periodic lattice. If an atom is on a proper atomic site, then it has an energy which we shall define to have a constant value denoted by $-\epsilon$. If an atom moves to an interstitial site, it has an energy of zero. This is because it may diffuse to the surface and escape from the crystal. Alternatively, the excitation energy required to unbind an atom from its site thereby creating a vacancy, is given by ϵ. We are considering the number of vacancies to be much smaller than

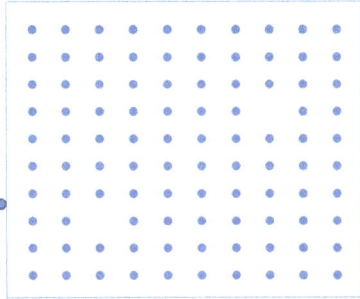

Fig. 3.7 A schematic depiction of a crystalline solid composed of N atoms which contains N_v vacancies.

the number of lattice sites, so that we can neglect the possibility that two vacancies sit on neighboring lattice sites, so we can neglect any effects of their interactions. The number of possible vacancies n_i on the single lattice site i, is restricted to have values zero or one. That is, there either is a vacancy or there is not. Also, we should note that the total number of vacancies is not conserved.

Let us consider a lattice with N_v vacancies. This state has an energy which is greater than the energy of a perfect lattice by the amount $U = N_v \epsilon$. We should note that the vacancies are indistinguishable, since if we permute them, the resulting state is identical. The number of distinct ways of distributing N_v indistinguishable vacancies on N lattice sites is given by

$$N_\Gamma = \left(\frac{N!}{N_v! \, (N - N_v)!} \right) \tag{3.120}$$

This is just the number of ways of choosing the lattice sites for N_v distinguishable vacancies $N!/(N - N_v)!$, divided by the number of permutations of the vacancies $N_v!$. Dividing by the factor $N_v!$ has the effect of counting the vacancies as if they were indistinguishable.

In the Micro-Canonical Ensemble, the entropy is given by

$$S = k_B \ln N_\Gamma \tag{3.121}$$

which on using Stirling's approximation, yields

$$S \approx k_B[N \ln N - N_v \ln N_v - (N - N_v) \ln(N - N_v)] \tag{3.122}$$

which is a function of U since $U = N_v \, \epsilon$.

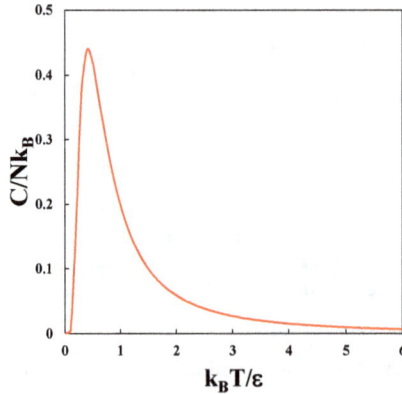

Fig. 3.8   The Schottky specific heat versus temperature of a model of vacancies in a crystalline solid composed of N atoms.

The energy can be expressed in terms of temperature, by using the relation

$$\frac{1}{T} = \left(\frac{\partial S}{\partial U} \right)_N \tag{3.123}$$

since the entropy is a function of energy. This yields

$$\frac{1}{T} = \left(\frac{\partial S}{\partial N_v}\right)\left(\frac{\partial N_v}{\partial U}\right)$$

$$= k_B \left[\ln(N - N_v) - \ln N_v\right] \left(\frac{\partial N_v}{\partial U}\right)$$

$$= \frac{k_B}{\epsilon} \left[\ln(N - N_v) - \ln N_v\right]$$

$$= \frac{k_B}{\epsilon} \ln\left(\frac{N - N_v}{N_v}\right) \tag{3.124}$$

After multiplying by ϵ/k_B and exponentiating, the expression can be inverted to give the number of vacancies

$$N_v = \frac{N}{\exp[\frac{\epsilon}{k_B T}] + 1} \tag{3.125}$$

which shows that the average number of thermally excited vacancies on a site is given by

$$\frac{N_v}{N} = \frac{1}{\exp[\frac{\epsilon}{k_B T}] + 1} \tag{3.126}$$

The thermodynamic energy U at a temperature T is given by the expression

$$U = \frac{N\epsilon}{\exp[\frac{\epsilon}{k_B T}] + 1} \tag{3.127}$$

At low temperatures, $\epsilon \gg k_B T$, this reduces to zero exponentially.

$$U = N\epsilon \, \exp\left[-\frac{\epsilon}{k_B T}\right] \tag{3.128}$$

At high temperatures (where the approximate model is not valid) half the lattice sites would host vacancies.

The specific heat due to the formation of vacancies is given by the expression

$$C = N \, k_B \left(\frac{\epsilon}{2k_B T}\right)^2 \text{sech}^2\left(\frac{\epsilon}{2k_B T}\right) \tag{3.129}$$

which vanishes exponentially at low T as is characteristic of systems with excitation gaps in their excitation spectra. At high temperatures, the specific heat vanishes as the inverse square of T, characteristic of a system with an energy spectrum bounded from above. This form of the specific heat is known as a Schottky anomaly or Schottky peak. The above expression has been derived from the configurational entropy of the vacancies. In real materials, there will also be a vibrational entropy since vacancies will cause local phonon modes to form.

3.6 Problems

Problem 3.1

The Maxwell distribution of the momenta of the molecules of an ideal gas can be derived by using the geometry of a hyper-sphere and the assumption of equal a priori probabilities. The volume of the d-dimensional hyper-sphere of radius R, $V_d(R)$ is defined by

$$V_d(R) = \int_{-\infty}^{\infty} dx_1 \int_{-\infty}^{\infty} dx_2 \ldots \int_{-\infty}^{\infty} dx_d \; \Theta(R - \sqrt{x_1^2 + x_2^2 + \ldots + x_d^2})$$

The surface area of d-dimensional hyper-sphere of radius R, $S_d(R)$ is given by

$$S_d(R) = \frac{\partial V_d(R)}{\partial R}$$

or

$$S_d(R) = \int_{-\infty}^{\infty} dx_1 \int_{-\infty}^{\infty} dx_2 \ldots \int_{-\infty}^{\infty} dx_d \; \delta(R - \sqrt{x_1^2 + x_2^2 + \ldots + x_d^2})$$

(i) Show that the surface area of a d-dimensional hyper-sphere is related to the surface area of a $(d-1)$-dimensional hyper-sphere by

$$S_d(R) = \int_{-R}^{R} dx_d \; \frac{R}{\sqrt{R^2 - x_d^2}} \; S_{d-1}(\sqrt{R^2 - x_d^2})$$

Hint: For those working in a Cartesian coordinate system, it may be convenient to use the identity

$$\int_0^{\infty} dR \; f(R) \; \delta(R - s) = 2 \int_0^{\infty} dR \; R \; f(R) \; \delta(R^2 - s^2)$$

before relating $S_d(R)$ to $S_{d-1}(R')$.

(ii) Hence, show that if the points on the surface area of a d-dimensional hyper-sphere are projected onto any diameter, the density of projected points is given by

$$\rho_d(x_1) = \frac{(d-1) \, V_{d-1}(1)}{d \, V_d(1)} \left(\frac{(\sqrt{R^2 - x_1^2})^{d-3}}{R^{d-2}} \right)$$

where $R > |x_1|$.

Assume that a system is composed of particles with mass m and the system has d translational degrees of freedom. The kinetic energy is given by E_K, hence the sum of the momenta are given by

$$2mE_K = \sum_{n=1}^{d} p_n^2$$

then all the phase space of the momenta are homogeneously distributed over the surface of the d-dimensional hyper-sphere.

(iii) Convert $\rho_d(x_1)$ into a normalized probability density for one component of the momentum $\rho_d(p_1)$, and show that for large d

$$\rho_d(p) = \sqrt{\frac{d}{2\pi(2mE_K)}} \, \exp\left[-\frac{p^2 d}{2mE_K}\right]$$

Hence, the Maxwell distribution holds for either high dimensions, large numbers of particles or both.

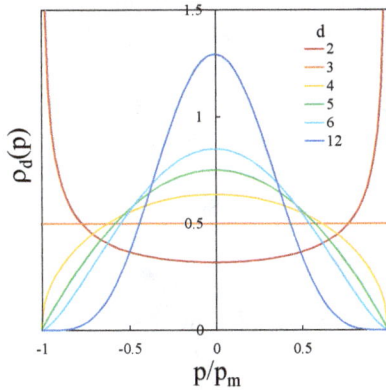

Fig. 3.9   The evolution of the momentum probability distribution function $\rho_d(p)$ with increasing number of translational degrees of freedom of the system.

Problem 3.2

Consider a system of N quantum harmonic oscillators with frequency ω_0, which has a total number of quanta Q. The goal of the problems is to find the probability P_n that a given oscillator has n quanta.

(i) Determine the number of microstates $N_\Gamma(N, Q)$ of the system of N oscillators with a total of Q quanta.

(ii) Determine the number of microstates $N_\Gamma(N-1, Q-n)$ for the system of $N-1$ oscillators with a total of $Q-n$ quanta.

(iii) Argue that P_n is given by

$$P_n = \frac{N_\Gamma(N-1, Q-n)}{N_\Gamma(N, Q)}$$

and, using Stirling's formula, show that for large N and Q, the ratio can be approximated by

$$P_n \approx \frac{N}{N+Q} \left(\frac{Q}{Q+N} \right)^n$$

(iv) Show that P_n is properly normalized.

Problem 3.3

Consider an artificial model of a two-dimensional liquid. Surface tension has the effect of minimizing the length of the surface by keeping it flat. In the $T = 0$ ground state, the liquid fills a two-dimensional area of width L and height h. At finite temperatures, the surface of the liquid ripples but the surface starts and ends at the same height h. The surface height is a one-to-one function of the horizontal distance. The surface of the liquid is discretized on a triangular lattice consisting of horizontal links and links at angles $\pm \frac{\pi}{3}$. Therefore, every link progresses forward either by one unit or by a half unit. The number of horizontal links is denoted by N_0 and the number of $\pm \frac{\pi}{3}$ are denoted by $N_{\pm}$. Since the end points of the surface are of the same height, then

$$N_+ = N_-$$

and since the surface is of length L, one has

$$L = N_0 + (N_+ + N_-)/2$$

The excitation energy of the surface is given by the excess of the total number of links over L

$$E = \sigma(N_0 + N_+ + N_- - L)$$
$$= \sigma(L - N_0)$$

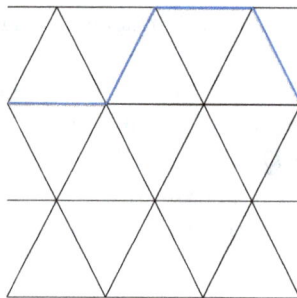

Fig. 3.10   A section of the discretized surface of a two-dimensional liquid.

(i) Show that the number of different surface configurations of energy N_Γ is given by

$$N_\Gamma = \frac{(N_0 + N_+ + N_-)!}{N_0!\, N_+!\, N_-!}$$

$$= \frac{(2L - N_0)!}{N_0!\, (L - N_0)!\, (L - N_0)!}$$

(ii) Determine the entropy $S(E)$ and find an expression for the average value of the squared energy, i.e. $(L - N_0)^2$, in terms of the temperature T.
(ii) Hence, find the temperature-dependence of the total number of links forming the surface.

Problem 3.4

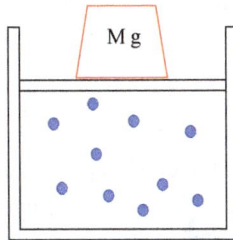

Fig. 3.11 N molecules of an ideal gas are contained inside a frictionless piston of area A. A mass of weight Mg is placed on the piston.

Consider a frictionless piston of area A containing an ideal gas of N molecules with mass m. A mass of weight Mg is placed on the piston. When the piston is at height z, the gas has a volume $V = Az$. Neglect the effect of gravity on the particles of mass m. The energy of the whole system E_T is in the interval between $E + \Delta E$ and E and is related to the kinetic energy of the gas, E_s, by $E_T = E_s + Mgz$.
(i) Determine the entropy $S(E_s, V)$ and, using thermodynamics, determine the temperature T and the pressure P.
(ii) By maximizing the entropy, find the most probable value of z and the value of the pressure at equilibrium.

Problem 3.5

Consider an ideal gas of atoms of mass m held at temperature T. The atoms can emit light with frequency ω_0 which is observed by a detector.

Since the atoms are in motion, the observed spectral line is broadened due to the Doppler effect. From the Lorentz transformation of the photon's momentum four-vector, the Doppler shifted frequency ω observed for light emitted by an atom moving towards the observer with a velocity with a component v subtended to the detector is found to be given by

$$\omega = \omega_0 \sqrt{\frac{1 - \frac{v}{c}}{1 + \frac{v}{c}}}$$

Inverting this relation leads to

$$v = c \left(\frac{\omega_0^2 - \omega^2}{\omega_0^2 + \omega^2} \right)$$

(i) Assuming that the natural line shape is well represented by a delta function centered on ω_0, derive the relation between the frequency distribution $I(\omega)$ and the velocity distribution $P_v(v)$ of the atoms

$$I(\omega)\, d\omega = \frac{c}{\omega} \left(\frac{2\omega\,\omega_0}{\omega_0^2 + \omega^2} \right)^2 P_v \left[c \left(\frac{\omega_0^2 - \omega^2}{\omega_0^2 + \omega^2} \right) \right] d\omega$$

(ii) Therefore, in the non-relativistic approximation, show that the frequency distribution function of the observed light is given by

$$I(\omega) = \left(\frac{mc^2}{2\pi k_B T \omega_0^2} \right)^{\frac{1}{2}} \exp \left[- \beta \frac{mc^2(\omega_0 - \omega)^2}{2\omega_0^2} \right]$$

Problem 3.6

Find the mean squared fluctuations of the speed v of the molecules

$$\overline{\Delta v^2} = \overline{(v - \bar{v})^2}$$

for a gas described by a d-dimensional Maxwell distribution.

Problem 3.7

Consider a column of ideal gas held at temperature T in a constant gravitational field g. Assume the pressure gradient balances the weight of the gas. That is,

$$\frac{dP}{dz} = -mg\,\rho(z)$$

where $\rho(z)$ is the number density.

(i) Use the ideal gas law to show that the pressure varies as

$$P(z) = P(0) \, \exp[-\beta mgz]$$

(ii) Hence, find the number density as a function of z.

Problem 3.8

Consider a Hamiltonian $H(\{p_i, q_i\})$ expressed in terms of generalized coordinates and their canonical momentum. Denote the coordinates q_i and the canonical momenta p_j by x_k. That is, depending on your indexing scheme used for i, x_i could represent either a coordinate or a momentum. Consider the average of the quantity

$$x_i \, \frac{\partial H}{\partial x_j}$$

in the Micro-Canonical Ensemble. This quantity can be used to derive the equipartition of energy. In the Micro-Canonical Ensemble the average of the quantity is over the volume of phase space which has energy between E and $E + \Delta E$

$$\overline{x_i \, \frac{\partial H}{\partial x_j}} = \frac{1}{\Gamma_a} \int d\Gamma \, (\Theta(E + \Delta E - H) - \Theta(E - H)) \, x_i \, \frac{\partial H}{\partial x_j}$$

Define the volume of phase space $\Gamma(E)$ which has energy less than E by

$$\Gamma(E) = \int d\Gamma \, \Theta(E - H)$$

so the volume of accessible phase space Γ_a can be expressed as

$$\Gamma_a = \Gamma(E + \Delta E) - \Gamma(E)$$

(i) Taylor expand the numerator of the expression for $\overline{x_i \, \frac{\partial H}{\partial x_j}}$ and use the fact that E is independent of x_j to obtain

$$\overline{x_i \, \frac{\partial H}{\partial x_j}} = \left[\frac{\Delta E}{\Gamma(E + \Delta E) - \Gamma(E)} \right] \frac{\partial}{\partial E} \left(\int d\Gamma \, \Theta(E - H) \, x_i \, \frac{\partial (H - E)}{\partial x_j} \right)$$

(ii) Show that integrating by parts w.r.t. x_j, the factor in the big round parenthesis by parts w.r.t. x_j is equal to

$$= \delta_{i,j} \int d\Gamma \, \Theta(E - H) \, (E - H)$$

and then that differentiating the numerator w.r.t. E leads to

$$\overline{x_i \, \frac{\partial H}{\partial x_j}} = \left[\frac{\Delta E}{\Gamma(E + \Delta E) - \Gamma(E)} \right] \delta_{i,j} \, \Gamma(E)$$

(iii) Why can one now use the expression for $\Gamma(E)$ as the volume of accessible phase space? Use the expression for the entropy $S(E)$ to argue that

$$\overline{x_i \, \frac{\partial H}{\partial x_j}} \approx \delta_{i,j} \left(\frac{\partial \Gamma(E)}{dE} \right)^{-1} \Gamma(E)$$
$$\approx \delta_{i,j} k_B T$$

where T is the temperature. The above expression is a generalization of the Equipartition Theorem and is valid for classical systems in thermal equilibrium.

Problem 3.9

It has been shown in Chapter Two that the two-dimensional surface area $dp_n \, dq_n$ of a volume of phase space is invariant under time evolution. Extending this result to the thermodynamic average

$$\overline{p_n \, q_n} = \text{Const.}$$

leads to

$$\frac{d}{dt} \overline{p_n \, q_n} = 0$$

Hence, show that for a classical system

$$\overline{\frac{\partial H}{\partial q_n} q_n} = \overline{p_n \frac{\partial H}{\partial p_n}}$$

which, although it does not specify the value of the averages, the equation demonstrates that the validity of the Equipartition Theorem is the same for generalized coordinates as for momenta.

Problem 3.10

Consider an ideal gas of N identical particles of mass m that are confined within a cubic enclosure of volume $V = L^3$. The Hamiltonian of the ideal gas is described by

$$H(\{q_j, p_j\}) = \sum_{i=1}^{3N} H_i(q_i, p_i)$$
$$= \sum_{i=1}^{3N} \left[\frac{p_i^2}{2m} + V_0 \, \Theta(q_i - L) \right]$$

where $V_0 \to \infty$.

(i) Show that, if L is time-dependent and changes very slowly from the initial value L to a new value $L'(t)$, the set of actions, I_i, defined by

$$I_i = \frac{1}{2\pi} \oint dq_i \, p_i$$

are invariant. Hence, argue that the entropy is invariant under an adiabatic expansion.

(ii) Show that, under an adiabatic expansion, the energy of a particle, $\langle E_i \rangle$ averaged over a period, increases at the rate

$$\frac{d}{dt}\langle E_i \rangle \approx -2\langle E_i \rangle \frac{\left(\frac{dL}{dt}\right)}{L}$$

so that if, the volume V is adiabatically increased to V', then

$$\frac{E'}{E} = \left(\frac{V}{V'}\right)^{\frac{2}{3}}$$

Problem 3.11

(A) Consider a system composed of $2N$ identical quantum mechanical Einstein harmonic oscillators of frequency ω_0 with $2Q$ quanta.

(i) Find the number of microscopic states N_Γ of the system with $2N$ particles and $2Q$ quanta. Use Stirling's formula

$$\ln N! = N \left(\ln N - 1 \right) - \frac{1}{2} \ln \left(2N + \frac{1}{3} \right) \pi$$

(ii) Calculate the change in entropy when the system is separated into two subsystems each with N oscillators and Q quanta. Did the entropy increase or decrease?

(B) When the system with $2N$ oscillators and $2Q$ quanta is separated into two subsystems each with N oscillators, it is found that one subsystem has $Q + \Delta Q$ quanta and the second subsystem has $Q - \Delta Q$ quanta.

(iii) Determine the total number of microscopic states $N_\Gamma(\Delta Q)$ for the composite system.

(iv) By considering the change in entropy between A(i) and B(iii), show that the distribution of quanta for a subsystem obeys a Gaussian distribution centered on Q but has a mean-squared fluctuation given by

$$\overline{\Delta Q^2} = \frac{Q(N + Q)}{4N}$$

(v) By forming the dimensionless ratio

$$\frac{\sqrt{\overline{\Delta Q^2}}}{Q}$$

show that the width of the distribution of quanta is extremely narrow for macroscopic systems.

Problem 3.12

Consider a system of vacancies in a crystal. Unlike the Schottky model of vacancies in which the missing atom moves to the surface of the crystal, the Frenkel model considers a lattice with atoms on lattice site in which some atoms may move from the lattice sites and travel to locations that are between the lattice sites (interstitial sites). In the Frenkel model there are N atoms, N lattice sites and the number of interstitial sites is denoted by M. In the model, n defects are created by atoms leaving their lattice positions and moving to interstitial sites. The excitation energy required for an atom to move to an interstitial site is denoted by ϵ. The number of ways N_{Γ_n} of creating n vacancies on the lattice sites and relocating the atoms to interstitial sites is given by

$$N_{\Gamma_n} = \binom{N}{n}\binom{M}{n}$$

(i) Use the Micro-Canonical Ensemble and assume that n is much smaller than N and M, to show that the entropy is given by

$$S(E)/k_B \approx -N\left[\frac{n}{N}\ln\frac{n}{N} + \left(1 - \frac{n}{N}\right)\ln\left(1 - \frac{n}{N}\right)\right]$$
$$-M\left[\frac{n}{M}\ln\frac{n}{M} + \left(1 - \frac{n}{M}\right)\ln\left(1 - \frac{n}{M}\right)\right]$$

(ii) On setting $E = n\epsilon$, show that the temperature is given by

$$\frac{\epsilon}{k_B T} = \ln\left(\frac{(N-n)(M-n)}{n^2}\right)$$

(iii) Show that $n(T)$ is given by the expression

$$n(T) = \left[\frac{\sqrt{(N-M)^2 + 4MN\,\exp[\beta\epsilon]} - (N+M)}{2(\exp[\beta\epsilon] - 1)}\right]$$

and show that for low temperatures

$$n(T) \approx \sqrt{NM}\,\exp\left[-\frac{\beta\epsilon}{2}\right]$$

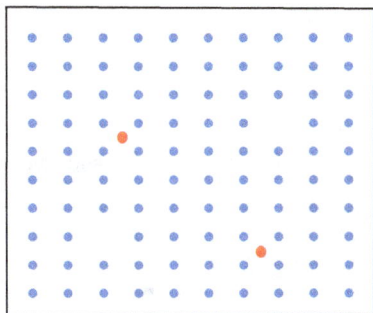

Fig. 3.12 Frenkel defects in a solid. Vacancies have been created on lattice sites when the atoms have moved to interstitial sites.

(iv) Examine the expression for $n(T)$ in the limit when $N = M$ and compare your result with that obtained for Schottky defects.

(v) Find an expression for the entropy S in terms of T, at low temperatures.

(vi) Calculate the specific heat $C(T)$ and compare your result with the result for the specific heat of Schottky defects.

Problem 3.13

Consider the model of Frenkel defects, but now consider thermal equilibrium to have a dynamic character. In particular, consider the principle of detailed balance which implies that the rate of production of Frenkel defects is equal to the rate at which the defects are destroyed. The principle leads to the number of defects being time-independent.

(i) Show that the rate of production of vacancies at regular lattice sites is given by

$$\nu_c \left(\frac{M - n}{N} \right) \exp[-\beta\epsilon] \, (N - n)$$

where $\frac{M-n}{N}$ is the number of vacant interstitial sites per lattice site. This factor represents the number of locations to which the atom at a regular lattice site can be displaced to. The factor $\exp[-\beta\epsilon]$ is the probability that an atom has sufficient energy to escape form the lattice site and ν_c is an attempt frequency for escape. The factor $(N - n)$ is just the number of atoms at regular lattice sites which can be removed to create a vacancy.

(ii) Show that the rate of destruction of interstitial defects is given by

$$\nu_d \left(\frac{n}{M} \right) n$$

where $\frac{n}{M}$ is just the number of vacancies at regular lattice sites, per interstitial site. This factor represents the number of possible final locations of the
interstitial atoms and n represents the number of atoms at the interstitial
sites that can be removed. The factor ν_d is the attempt frequency.

(iii) The principle of detailed balance implies the two rates are equal. From
time-reversal invariance, one expects that $\nu_c = \nu_d$. Use the principle of
detailed balance to show

$$\exp[\beta\epsilon] = \left(\frac{(N-n)(M-n)}{n^2}\right)$$

which is the expression for the equilibrium number of Frenkel defects found
by minimizing the entropy.

Problem 3.14

Consider a system of N spins with $S = 1$ in a tetragonal environment. The
combined effect of spin orbit-coupling and the lattice anisotropy produces
the anisotropic spin Hamiltonian

$$H = D \sum_{i=1}^{N} (S_i^z)^2$$

where $S_i^z = \pm 1, 0$.

(i) If the system has a total energy E, so that $N_e = E/D$ spins are in
excited states, show the number of accessible microstates is given by

$$N_\Gamma = \frac{N!}{(N-N_e)!} \sum_{N_+=0}^{N_e} \frac{1}{N_+!(N_e-N_+)!}$$

$$= 2^{N_e} \frac{N!}{(N-N_e)! \, N_e!}$$

(ii) Find an expression for the entropy $S(E, N)$. Hence, show that the
number of spins in the excited states is given by

$$N_e = N \frac{2 \, \exp[-\beta D]}{1 + 2 \, \exp[-\beta D]}$$

Problem 3.15

Consider a d-dimensional lattice with N sites. At each site there is an
atom which has an internal degree of freedom. Each atom can be in one of
$2d$ internal states, the energy of the first $2m$ internal states of an atom is
denoted by ϵ where $\epsilon > 0$, whereas the remaining $2(d-m)$ internal states
have energy zero.

(i) Show that the number of microscopic states N_Γ corresponding to the macroscopic state with energy $E = Q\epsilon$ is given by

$$N_\Gamma = \binom{N}{Q} (2m)^Q (2(d-m))^{(N-Q)}$$

(ii) Show that the entropy $S(E)$ is given by

$$S(E)/k_B = \ln N_\Gamma$$
$$\approx (N-Q) \ln 2(d-m) + Q \ln 2m$$
$$- (N-Q) \ln \left(1 - \frac{Q}{N}\right) - Q \ln \frac{Q}{N}$$

(iii) Derive the expression for the thermal average value of Q in terms of the temperature T.

$$\overline{Q} = \frac{Nm}{m + (d-m)\ \exp[\beta\epsilon]}$$

(iv) Find an expression for the entropy S as a function of temperature T. Show that the limiting forms of the entropy are given by

$$\lim_{T \to 0} S(T) \to Nk_B \ \ln 2(d-m)$$
$$\lim_{T \to \infty} S(T) \to Nk_B \ \ln 2d$$

Problem 3.16

A quantum violin string vibrates at the fundamental frequency ω_0 and its higher harmonics. The frequency of the p-th harmonic is given by $\omega = p\,\omega_0$. Each harmonic is an independent vibration mode. The excitation energy of the p-th harmonic is given by $E_p = n_p\,p\,\hbar\omega_0$, where the integer n_p is the number of quanta in the mode.

(i) Show that for a system with total excitation energy E where

$$E = N\hbar\omega_0$$

and

$$N = \sum_p p\,n_p$$

that the number of available microstates is given by $P(N)$ where $P(N)$ is the number of integer partitions of N. The first few values are given by $P(0) = 1$, $P(1) = 1$, $P(2) = 2$, $P(3) = 3$, $P(4) = 5$, $P(5) = 7$, $P(6) = 11$, $P(7) = 15$, $P(8) = 22$, $P(9) = 30$, $P(10) = 42$, $P(11) = 56$, $P(12) = 77$, $P(13) = 101$, $P(14) = 135, \ldots, P(100) \approx 1.9 \times 10^6, \ldots,$

$P(1000) \approx 2.4 \times 10^{31}$ etc. The generator function for integer partitioning is given by

$$\sum_{N=0}^{\infty} P(N)\, x^N = \prod_{p=1}^{\infty} \left(\frac{1}{1-x^p} \right)$$

The Hardy-Ramanujan formula gives a useful asymptotic approximation

$$P(N) = \frac{1}{2\pi\sqrt{2}} \exp \left[\pi \sqrt{\frac{2N}{3}} \right]$$

(ii) Determine $S(E)$, hence, find an approximate relation between E and the temperature.

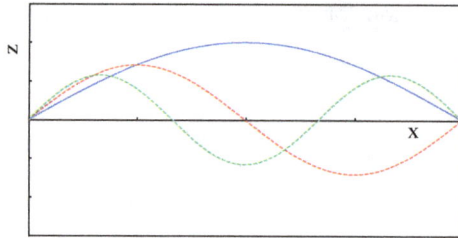

Fig. 3.13   The fundamental $p = 1$ and higher harmonic $p > 1$ vibration modes of a classical violin.

Problem 3.17

The thermodynamics of a rubber band is governed by

$$dF = -S\, dT + \mathcal{T}\, dL$$

where $\mathcal{T}$ is the tension. A simple microscopic model for a rubber band consists of a chain of links, each of length a, with each link equally probable to be directed to the left or right along the chain.

(i) What are the allowed values of L for even N?

(ii) Show that the number of microstates N_Γ corresponding to a chain of $2N$ links of total length L is given by

$$N_\Gamma = \frac{2N!}{\left(N - \frac{L}{2a} \right)! \left(N + \frac{L}{2a} \right)!}$$

(iii) Hence, calculate the entropy S and, by using a Maxwell relation and integrating w.r.t. T, show that the tension $\mathcal{T}$ of the system is given by

$$\mathcal{T} = \frac{k_B T}{(2Na^2)}\, L$$

when $Na \gg L$.

Problem 3.18

(i) Using the expression for the volume accessible phase space $\Gamma(E, V, N)$ of a Classical Ideal Gas with energy E, number of particles N and volume V, calculate the number of accessible microstates $N_\Gamma(E)$.

(ii) Perform the Laplace transform of $N_\Gamma(E)$ wrt energy. Show that if $s = \beta = 1/(k_B T)$, the result is identical to the partition function, $Z_N(\beta)$ given in section (3.2).

(iii) Show that, by analytically continuing β in the complex plane, the inverse Laplace Transform of $Z_N(\beta)$ is $N_\Gamma(E)$.

Problem 3.19

Consider an Ensemble composed of N distinguishable copies of a system such that n_1 systems are in the state 1, n_2 systems are in the state 2, n_3 systems are in state 3, etc.

(i) Show that the number of different ways of distributing the N copies between the states is given by

$$N(\{n_i\}) = \frac{N!}{n_1!\, n_2!\, n_3!\, \cdots}$$

If the probability for finding a system in state i is given by p_i, the probability of this distribution is given by

$$P(\{n_i\}) = \frac{N!}{n_1!\, n_2!\, n_3!\, \cdots} \prod_i \{p_i^{n_i}\}$$

For large N, the probability distribution is very sharply peaked about the average values of $\{n_i\}$ and has a relatively small fluctuations. Therefore, if N is sufficiently large, one may set $n_i \approx N p_i$. The approximate number of ways of placing the systems in the states is then determined by the probabilities

$$N(\{n_i\}) \approx \frac{N!}{(N p_1)!\, (N p_2)!\, (N p_3)!\, \cdots}$$

The entropy of the Ensemble with N copies is given by

$$S = k_B\, \ln N(\{n_i\})$$
$$\approx k_B \left[\ln N! - \sum_i \ln(N p_i)! \right]$$
$$= k_B\, N \left[\ln N - \sum_i p_i \ln N - \sum_i p_i \ln p_i - 1 + \sum_i p_i \right]$$

which, since the probabilities are normalized $\sum_i p_i = 1$, reduces to

$$S = -N\,k_B \sum_i p_i \ln p_i$$

The result per copy coincides with Shannon's entropy formula.

(ii) If all the states have the same energy E and the probability is normalized

$$\sum_{i=1}^{\infty} p_i = 1$$

Use Lagrange's method of undetermined multipliers to show that, if the entropy, S, associated with the probability distribution

$$S(\{p_i\}) = -k_B \sum_i p_i \ln p_i$$

is maximized, all the states have equal probabilities.

(iii) If the probabilities p_i are subject to the constraint that the probabilities are normalized and that the average energy is given by U so

$$U = \sum_{i=1}^{\infty} p_i\,E_i$$

then, using Lagrange's method of undetermined multipliers, maximize $S(\{p_i\})$ wrt the p_i subject to the two constraints.

Chapter 4

The Canonical Ensemble

The "*Canonical Ensemble*" describes a closed system that is divided into two parts, each with a fixed number of particles and fixed volumes. However, the two subsystems can exchange energy with each other. One subsystem is the system which is the focus of our interest. The second subsystem is assumed to be much larger than the system of interest and is known as the environment. The properties of the environment will not be of direct interest and its main role will be to act as a thermal reservoir which absorbs or supplies energy to the system of interest. The distribution function for the subsystem of interest can be derived from the Micro-Canonical Probability Distribution Function for the total system.

4.1 The Boltzmann Distribution Function

The total energy of the complete system E_T is partitioned into the energy of our subsystem E and that of the thermal reservoir E_R

$$E_T = E_R + E \qquad (4.1)$$

where the interaction energy between the system and the environment has been assumed to be negligible. The infinitesimal volume element of total phase space $d\Gamma_T$ is also assumed to be factorizable in terms of the products of the volume elements of the thermal reservoir $d\Gamma_R$ with the volume element of our subsystem $d\Gamma$. This assumes that every degree of freedom of the total system can be uniquely assigned either to the thermal reservoir or to the system of interest. Hence, we assume that

$$d\Gamma_T = d\Gamma_R \, d\Gamma \qquad (4.2)$$

The probability dp_T of finding the total system in the volume element of phase space $d\Gamma$ is described by the constant Micro-Canonical Distribution

Function ρ_{mc}

$$dp_T = \rho_{mc} \ d\Gamma_R \ d\Gamma \tag{4.3}$$

The probability dp for finding the subsystem in the phase space volume element $d\Gamma$ associated with the energy $H = E$ is found by integrating over all the phase space of the reservoir, consistent with the reservoir having the energy $H_R = E_T - E$. Hence,

$$dp = \rho_{mc} \ \Gamma_R(E_T - E) \ d\Gamma \tag{4.4}$$

where $\Gamma_R(E_T - E)$ is the volume of phase space accessible to the reservoir. However, the entropy of the reservoir is related to the volume of its accessible phase space via

$$\Gamma_R(E_T - E) = \Gamma_{R,0} \ \exp[S_R(E_T - E)/k_B] \tag{4.5}$$

where $\Gamma_{R,0}$ is the volume of phase space associated with one microscopic state of the reservoir. Hence, the probability density for the system of interest is given by

$$\left(\frac{dp}{d\Gamma}\right) = \rho_{mc} \ \Gamma_{R,0} \ \exp[S_R(E_T - E)/k_B] \tag{4.6}$$

where ρ_{mc} is just a constant. The energy of our subsystem E is much smaller than the energy of total system, E_T since the energy is extensive and the thermal reservoir is much larger than our subsystem. Hence, it is reasonable to assume that the reservoir's entropy can be Taylor expanded in powers of E and also that the second and higher-order terms can be neglected. That is

$$S_R(E_T - E) = S_R(E_T) - \left.\frac{\partial S_R(E)}{\partial E}\right|_{E_T} E + \dots \tag{4.7}$$

but one recognizes that the derivative of the reservoir's entropy w.r.t. energy is defined as the inverse of the temperature T of the thermal reservoir

$$\left.\frac{\partial S_R(E)}{\partial E}\right|_{E_T} = \frac{1}{T} \tag{4.8}$$

Hence, the probability distribution function ρ_c for finding the system in some region of its phase space, as described in the Canonical Ensemble, depends on the energy of the system E via

$$\left(\frac{dp}{d\Gamma}\right) = \rho_{mc} \ \Gamma_{R,0} \ \exp[S_R(E_T)/k_B] \ \exp[-\beta E]$$

$$= \frac{1}{Z\Gamma_0} \ \exp[-\beta E] \tag{4.9}$$

where Z is a dimensionless constant and Γ_0 is the volume of phase space of the system which is used to define a single microscopic state. The factor $\rho_{mc}\,\Gamma_{R,0}\,\Gamma_0$ is a dimensionless constant which is independent of the specific point of the system's phase space, as is the first exponential factor. It is to be recalled that the region of phase space $d\Gamma$ under consideration corresponds to a specific value of the system's energy E, hence one can express the Canonical Probability Distribution Function as

$$\rho_c(\{p_i, q_i\}) = \frac{1}{Z\Gamma_0}\,\exp[-\beta H(\{p_i, q_i\})] \tag{4.10}$$

The Canonical Distribution Function depends on the point $\{p_i, q_i\}$ of the system's phase space only via the value of the system's Hamiltonian $H(\{p_i, q_i\})$. The dimensionless normalization constant Z is known as the Canonical Partition Function. The normalization condition

$$\begin{aligned}
1 &= \int d\Gamma\,\left(\frac{dp}{d\Gamma}\right) \\
&= \int d\Gamma\,\rho_c(\{p_i, q_i\}) \\
&= \int \frac{d\Gamma}{\Gamma_0}\,\frac{1}{Z}\,\exp[-\beta H(\{p_i, q_i\})]
\end{aligned} \tag{4.11}$$

can be used to express the Canonical Partition Function Z as a weighted integral over the entire phase space of our system

$$Z = \int \frac{d\Gamma}{\Gamma_0}\,\exp[-\beta H(\{p_i, q_i\})] \tag{4.12}$$

where the weighting function depends exponentially on the Hamiltonian H. Hence, in the Canonical Ensemble, the only property of the environment that actually appears in the distribution function is the temperature T. The distribution function $\rho_c(\{p_i, q_i\})$ is known as the Boltzmann Distribution Function. In the Canonical Ensemble, averages of quantities $A(\{p_i, q_i\})$ belonging solely to the system are evaluated as

$$\begin{aligned}
\overline{A} &= \int d\Gamma\,\rho_c(\{p_i, q_i\})\,A(\{p_i, q_i\}) \\
&= \frac{1}{Z}\int \frac{d\Gamma}{\Gamma_0}\,A(\{p_i, q_i\})\,\exp[-\beta H(\{p_i.q_i\})]
\end{aligned} \tag{4.13}$$

where the range of integration runs over all the phase space of our system, irrespective of the energy of the element of phase space. In the Canonical

Distribution Function, the factor that depends exponentially on the Hamiltonian replaces the restriction used in the Micro-Canonical Ensemble where integration only runs over regions of phase space which corresponds to the fixed value of the energy E.

4.1.1   *The Helmholtz Free-Energy*

If the partition function is known, it can be used directly to yield the thermodynamic properties of the system. This follows once the partition function has been related to the Helmholtz Free-Energy $F(T, V, N)$ of our system via

$$Z = \exp[-\beta F] \qquad (4.14)$$

This identification can be made by recalling Eq. (4.9), and cancelling the energy-dependent exponential factors. From this one finds that the partition function Z is related to the Micro-Canonical Distribution Function ρ_{mc} and the entropy of the thermal reservoir with energy E_T via

$$\frac{1}{Z\Gamma_0} = \Gamma_{R,0} \, \rho_{mc} \, \exp[S_R(E_T)/k_B]$$

$$\frac{1}{Z} = \frac{\Gamma_{T,0}}{\Gamma_T(E_T)} \, \exp[S_R(E_T)/k_B] \qquad (4.15)$$

In the second line, the product of the volumes of phase space representing one microscopic state of the reservoir $\Gamma_{R,0}$ and one microscopic state of the subsystem Γ_0 has been assumed to be related to the volume of phase space $\Gamma_{T,0}$ representing one microscopic state of the total system by the equation

$$\Gamma_{T,0} = \Gamma_{R,0} \, \Gamma_0 \qquad (4.16)$$

The ratio in the second line then follows from the relation between the Micro-Canonical Distribution Function of the total system with energy E_T and the volume of accessible phase space for the total system $\Gamma_T(E_T)$. However, for the total system, the ratio is given by

$$\frac{\Gamma_{T,0}}{\Gamma_T(E_T)} = \exp[-S_T(E_T)/k_B]$$

$$= \exp[-(S_R(E_T - U) + S(U))/k_B] \qquad (4.17)$$

where we have used the fact that the thermodynamic value of the entropy is extensive and the thermodynamic entropy of the subsystem is evaluated at the thermodynamic value of its energy U. (One expects from consideration of the maximization of the entropy that the thermodynamic energy U should be equal to the most probable value of the energy. However, as we shall show, the thermodynamic energy also coincides with the average energy $\overline{E}$.) On combining the above two expressions, one finds that

$$\frac{1}{Z} = \exp[-(S_R(E_T - U) - S_R(E_T) + S(U))/k_B] \tag{4.18}$$

which on Taylor expanding the first term in the exponent in powers of the relatively small average energy of the system U yields

$$\frac{1}{Z} = \exp[\beta U - S(U)/k_B] \tag{4.19}$$

where the higher-order terms in the expansion have been assumed to be negligible. Since the Helmholtz Free-Energy of the system is described as a Legendre transformation of the system's energy $U(S, V, N)$

$$F = U - TS(U) \tag{4.20}$$

then F is a function of the variables (T, V, N). Hence one recognizes that the Canonical Partition Function is related to the Helmholtz Free-Energy F of the subsystem of interest via

$$Z = \exp[-\beta F] \tag{4.21}$$

and that it is also a function of the variable (T, V, N). For thermodynamic calculations, it is more convenient to recast the above relation into the form

$$F = -k_B T \ln Z \tag{4.22}$$

4.1.2 Ensemble Averages and Thermodynamics

The above analysis is completed by identifying the thermodynamic energy U with the average energy $\overline{E}$. First we shall note that within the Canonical Ensemble, the average energy $\overline{E}$ is defined as

$$\overline{E} = \frac{1}{Z} \int \frac{d\Gamma}{\Gamma_0} H(\{p_i, q_i\}) \exp[-\beta H(\{p_i.q_i\})] \tag{4.23}$$

which can be re-written as a logarithmic derivative of Z w.r.t. β, since the numerator of the integrand is recognized as the derivative of Z w.r.t. β

$$\overline{E} = -\frac{1}{Z} \frac{\partial}{\partial \beta} \left(\int \frac{d\Gamma}{\Gamma_0} \exp[-\beta H(\{p_i.q_i\})] \right)$$

$$= -\frac{1}{Z} \frac{\partial Z}{\partial \beta}$$

$$= -\frac{\partial \ln Z}{\partial \beta} \tag{4.24}$$

However, $\ln Z$ is also related to the value of the Helmholtz Free-Energy F, so one has

$$
\begin{aligned}
\overline{E} &= \frac{\partial}{\partial \beta}\,(\beta F) \\
&= F + \beta\,\frac{\partial F}{\partial \beta} \\
&= F - T\,\frac{\partial F}{\partial T} \\
&= F + TS
\end{aligned}
\tag{4.25}
$$

where F is the Helmholtz Free-Energy of thermodynamics and the thermodynamic entropy S has been introduced via

$$
S = -\left(\frac{\partial F}{\partial T}\right)_{V,N}
\tag{4.26}
$$

Hence, since the Free-Energy and the thermodynamic energy are related via $F = U - TS$, one finds that

$$
\overline{E} = U
\tag{4.27}
$$

This shows the thermodynamic energy U coincides with the average energy $\overline{E}$ when calculated in the Canonical Ensemble.

4.2   The Equipartition Theorem

The *"Equipartition Theorem"* applied to a quadratic classical Hamiltonian gives the result that each quadratic term in the potential contributes $k_B T/2$ to the total energy and each term in the kinetic energy gives rise to another $k_B T/2$ to the total energy. Hence, for a set of coupled harmonic oscillators the total energy has a contribution of $k_B T$ from each degree of freedom. Thus, if there are n_ν normal modes, the total energy of the classical systems is given by $n_\nu k_B T$.

The result can be generalized to any Hamiltonian $H(\{p_i, q_i\})$. Although we shall use the Hamiltonian formulation, we shall denote the canonical coordinates and momenta by the $2dN$ variables x_k. That is x_k, depending on the indexing system used, could represent either a coordinate or a conjugate momentum. The generalized theorem concerns the Canonical Ensemble average of

$$
x_i\,\frac{\partial H}{\partial x_j}
\tag{4.28}
$$

The average is given by

$$\overline{x_i \frac{\partial H}{\partial x_j}} = \int \frac{d\Gamma}{Z\Gamma_0} x_i \frac{\partial H}{\partial x_j} \exp[-\beta H(\{x_k\})]$$

$$= -k_B T \int \frac{d\Gamma}{Z\Gamma_0} x_i \left(\frac{\partial}{\partial x_j} \exp[-\beta H(\{x_k\})]\right) \qquad (4.29)$$

On integrating by parts w.r.t. x_j, one obtains

$$\overline{x_i \frac{\partial H}{\partial x_j}} = k_B T \int \frac{d\Gamma}{Z\Gamma_0} \frac{\partial x_i}{\partial x_j} \exp[-\beta H(\{x_k\})] \qquad (4.30)$$

where we have assumed that the boundary term vanishes. Since x_i and x_j are either coordinates or canonical momentum, one has

$$\frac{\partial x_i}{\partial x_j} = \delta_{i,j} \qquad (4.31)$$

The remaining integration is the normalization of the probability and yields a factor of unity. Hence, we have shown that

$$\overline{x_i \frac{\partial H}{\partial x_j}} = k_B T \delta_{i,j} \qquad (4.32)$$

On setting $i = j$ and selecting i such that x_i represents a canonically conjugate momentum, then when applied to the classical Hamiltonian with a quadratic kinetic energy

$$p_i \frac{\partial H}{\partial p_i} = \frac{p_i^2}{m} \qquad (4.33)$$

one finds

$$\overline{p_i \frac{\partial H}{\partial p_i}} = \frac{\overline{p_i^2}}{m}$$

$$= k_B T \qquad (4.34)$$

Likewise, when applied to the normal mode coordinates q_i of a Harmonic Hamiltonian

$$q_i \frac{\partial H}{\partial q_i} = m\omega_i^2\, q_i^2 \qquad (4.35)$$

one finds

$$\overline{q_i \frac{\partial H}{\partial q_i}} = m\omega_i^2\, \overline{q_i^2}$$

$$= k_B T \qquad (4.36)$$

There is no coupling of the normal modes when $i \neq j$.

4.3   The Ideal Gas

An ideal gas of N particles moving in a d-dimensional space is described by the Hamiltonian

$$H_N = \sum_{i=1}^{dN} \frac{p_i^2}{2m} \tag{4.37}$$

and the particles are constrained to move within a hypercubic volume with linear dimensions L.

The partition function Z_N is given by

$$Z_N = \frac{1}{N! \, (2\pi\hbar)^{dN}} \prod_{i=1}^{dN} \left\{ \int_{-\infty}^{\infty} dp_i \int_{0}^{L} dq_i \right\} \exp[-\beta H_N] \tag{4.38}$$

Since the Hamiltonian is the sum of independent terms, the expression for Z_N can be expressed as a product of dN terms

$$Z_N = \frac{1}{N! \, (2\pi\hbar)^{dN}} \prod_{i=1}^{dN} \left\{ \int_{-\infty}^{\infty} dp_i \, \exp\left[-\frac{\beta p_i^2}{2m} \right] \int_{0}^{L} dq_i \right\}$$

$$= \frac{1}{N! \, (2\pi\hbar)^{dN}} \prod_{i=1}^{dN} \left\{ \left(\frac{2\pi m}{\beta} \right)^{\frac{1}{2}} L \right\}$$

$$= \frac{1}{N! \, (2\pi\hbar)^{dN}} \left\{ \left(\frac{2\pi m}{\beta} \right)^{\frac{1}{2}} L \right\}^{dN}$$

$$= \frac{V^N}{N!} \left(\frac{mk_BT}{2\pi\hbar^2} \right)^{\frac{dN}{2}} \tag{4.39}$$

On introducing the thermal de Broglie wave length λ, via

$$\lambda = \frac{h}{(2\pi m k_B T)^{\frac{1}{2}}} \tag{4.40}$$

one finds

$$Z_N = \frac{1}{N!} \left(\frac{V}{\lambda^d} \right)^N \tag{4.41}$$

Thermodynamic quantities can be obtained by recalling that the Helmholtz Free-Energy is given by

$$F = -k_B T \, \ln Z_N \tag{4.42}$$

and by using Stirling's approximation

$$\ln N! = N \ln N - N \tag{4.43}$$

which yields

$$F = -N k_B T \ln \left(\frac{eV}{N \lambda^d} \right) \tag{4.44}$$

One can find all other thermodynamic functions from F. Thus, one can obtain the entropy from

$$S = -\left(\frac{\partial F}{\partial T} \right)_{V,N} \tag{4.45}$$

as

$$S = N k_B \ln \left(\frac{Ve}{N \lambda^d} \right) + \frac{d}{2} N k_B \tag{4.46}$$

which is the Sackur-Tetrode formula.

It is quite simple to show that the chemical potential μ is given by

$$\mu = k_B T \ln \frac{N \lambda^d}{V} \tag{4.47}$$

The condition under which the classical description is a expected to be a reasonable approximation is given by

$$\frac{V}{N \lambda^d} \gg 1 \tag{4.48}$$

Hence, we discover that the classical approximation is expected to be valid whenever

$$\exp[-\beta \mu] \gg 1 \tag{4.49}$$

4.4 The Entropy of Mixing

The entropy of mixing is associated with the factor of $N!$ needed to describe the microstates available to a gas of identical particles. This factor is required to make the entropy extensive so that on changing scale by a factor of s we have

$$S(sE, sV, sN) = sS(E, V, N) \tag{4.50}$$

The $N!$ is also needed to make the expression for the chemical potential intensive.

Consider a container partitioned off into two volumes V_1 and V_2. The containers hold N_1 and N_2 gas molecules respectively, and assume that the molecules have the same masses and the gasses are kept at the same temperature (or average energy per particle). Then consider removing the partition. If the gas molecules in the two containers are indistinguishable, then in the Micro-Canonical Ensemble the entropy of the final state is given by

$$S_{indis} = k_B \ln \left[\frac{\Gamma_a}{(N_1 + N_2)! \, (2\pi\hbar)^{d(N_1+N_2)}} \right] \tag{4.51}$$

which corresponds to dividing the enlarged accessible phase space Γ_a by a factor of $(N_1 + N_2)!$ to avoid over-counting the number of microstates. Equivalently, in the Canonical Ensemble the partition function Z is given by

$$Z_{indis} = \frac{\left(\frac{V_1+V_2}{\lambda^d} \right)^{N_1+N_2}}{(N_1 + N_2)!} \tag{4.52}$$

However, if the molecules in the two containers are different, the accessible phase space of the final state is different from that case of indistinguishable particles. Thus, it should be divided by $N_1! \, N_2!$ corresponding to the number of permutations of like molecules. In this case, the final state entropy is given by the expression

$$S_{dis} = k_B \ln \left[\frac{\Gamma_a}{(N_1! \, N_2!)(2\pi\hbar)^{d(N_1+N_2)}} \right] \tag{4.53}$$

or equivalently

$$Z_{dis} = \frac{\left(\frac{V_1+V_2}{\lambda^d} \right)^{N_1+N_2}}{(N_1! \, N_2!)} \tag{4.54}$$

Since in this case the final state consists of a mixture of distinct gasses, the entropy of the mixture must be larger than the entropy of the mixture of identical gasses. That is, it is expected that work would have to be expended to separate the distinct molecules. The entropy of mixing is defined as

$$S_{mix} = S_{indis} - S_{dis} \tag{4.55}$$

and since Γ_a are identical, it is found to be given by

$$
\begin{aligned}
S_{mix} &= k_B \, \ln(N_1 + N_2)! - k_B \, \ln(N_1! \, N_2!) \\
&= (N_1 + N_2) \, k_B \, \ln(N_1 + N_2) \\
&\quad - N_1 \, k_B \ln N_1 - N_2 \, k_B \ln N_2 \\
&= -N_1 \, k_B \ln \frac{N_1}{N_1 + N_2} - N_2 \, k_B \ln \frac{N_2}{N_1 + N_2} \\
&= -(N_1 + N_2) \, k_B \left[\frac{N_1}{N_1 + N_2} \, \ln \frac{N_1}{N_1 + N_2} \right. \\
&\quad \left. + \frac{N_2}{N_1 + N_2} \, \ln \frac{N_2}{N_1 + N_2} \right]
\end{aligned}
\tag{4.56}
$$

which has a form reminiscent of Shannon's entropy.

4.5 The Einstein Model of a Crystalline Solid

We shall revisit the Einstein model of a Crystalline Solid, in the Canonical Ensemble. The Hamiltonian of N' harmonic oscillators with frequency ω_0 takes the form

$$
\hat{H} = \sum_{i=1}^{N'} \hbar \omega_0 \left(\hat{n}_i + \frac{1}{2} \right)
\tag{4.57}
$$

in the number operator representation. The set of possible eigenvalues of $\hat{n}_i$ are the integer values $0, 1, 2, 3, \ldots, \infty$. In this occupation number representation, the partition function $Z_{N'}$ is given by the Trace

$$
\begin{aligned}
Z_{N'} &= \text{Trace} \, \exp[-\beta \hat{H}] \\
&= \text{Trace} \, \exp \left[-\beta \sum_{i=1}^{N'} \hbar \omega_0 \left(n_i + \frac{1}{2} \right) \right]
\end{aligned}
\tag{4.58}
$$

where the Trace is the sum over all the set of quantum numbers n_i for each oscillator. Hence, on recognizing that the resulting expression involves the sum of a geometric series, we have

$$
\begin{aligned}
Z_{N'} &= \prod_{i=1}^{N'} \left\{ \sum_{n_i=0}^{\infty} \exp \left[-\beta \hbar \omega_0 \left(n_i + \frac{1}{2} \right) \right] \right\} \\
&= \prod_{i=1}^{N'} \left\{ \frac{\exp[-\frac{\beta \hbar \omega_0}{2}]}{1 - \exp[-\beta \hbar \omega_0]} \right\} \\
&= \left[2 \, \sinh \frac{\beta \hbar \omega_0}{2} \right]^{-N'}
\end{aligned}
\tag{4.59}
$$

where each normal mode gives rise to an identical factor. The Free-Energy is given by

$$F = N'k_B T \ \ln\left(2\ \sinh\frac{\beta\hbar\omega_0}{2}\right) \tag{4.60}$$

The entropy S is found from

$$S = -\left(\frac{\partial F}{\partial T}\right) \tag{4.61}$$

which yields

$$S = -\frac{F}{T} + N'\ \frac{\hbar\omega_0}{2T}\ \coth\frac{\beta\hbar\omega_0}{2} \tag{4.62}$$

However, since $F = U - TS$, one finds the internal energy U is given by

$$U = N'\ \frac{\hbar\omega_0}{2}\ \coth\frac{\beta\hbar\omega_0}{2} \tag{4.63}$$

This result is the same as that which was previously found using the Micro-Canonical Ensemble.

4.6   Vacancies in a Crystal

The Hamiltonian describing Schottky vacancies in a crystal can be described by

$$H_N = \sum_{i=1}^{N} \epsilon n_i \tag{4.64}$$

where the number of vacancies on site i is defined by n_i. The number of vacancies at site i only n_i has two possible values of unity or zero, since there either is a vacancy or there is not. The partition function Z_N is given by

$$Z_N = \text{Trace}\ \exp[-\beta H_N] \tag{4.65}$$

which is evaluated as

$$Z_N = \text{Trace}\ \exp\left[-\beta \sum_{i=1}^{N} \epsilon n_i\right]$$

$$= \text{Trace} \prod_{i=1}^{N} \exp[-\beta\epsilon n_i] \tag{4.66}$$

The Trace runs over all the set of possible values of the number of vacancies n_i for each atomic site. Thus

$$Z_N = \text{Trace} \prod_{i=1}^{N} \{\exp[-\beta \epsilon n_i]\}$$

$$= \prod_{i=1}^{N} \left\{ \sum_{n_i=0}^{1} \exp[-\beta \epsilon n_i] \right\}$$

$$= \prod_{i=1}^{N} \{1 + \exp[-\beta \epsilon]\}$$

$$= (1 + \exp[-\beta \epsilon])^N \tag{4.67}$$

which leads to the expression for the Free-Energy

$$F = -N k_B T \, \ln(1 + \exp[-\beta \epsilon]) \tag{4.68}$$

Hence, from thermodynamics, one finds that the energy U is given by

$$U = N \epsilon \, \frac{\exp[-\beta \epsilon]}{1 + \exp[-\beta \epsilon]}$$

$$= N \epsilon \, \frac{1}{\exp[\beta \epsilon] + 1} \tag{4.69}$$

which is identical to the expression that was found using the Micro-Canonical Ensemble.

4.7 Quantum Spins in a Magnetic Field

Consider a set of N quantum spins with magnitude S. We shall set $\hbar = 1$ for convenience. The spins interact with a magnetic field H^z through the Zeeman interaction

$$\hat{H}_{int} = - \sum_{i=1}^{n} g\mu_B H^z \, \hat{S}_i^z \tag{4.70}$$

where the $\hat{S}_i^z$ have eigenvalues m_i where $S \geq m \geq -S$.

Since the spins do not interact with themselves, the partition function factorizes as

$$Z = \left\{ \sum_{m=-S}^{S} \exp[\beta g \mu_B H^z m] \right\}^N$$

$$= \left\{ \frac{\exp[\beta g \mu_B H^z (S + \frac{1}{2})] - \exp[-\beta g \mu_B H^z (S + \frac{1}{2})]}{\exp[\beta g \mu_B H^z \frac{1}{2}] - \exp[-\beta g \mu_B H^z \frac{1}{2}]} \right\}^N$$

$$= \left\{ \frac{\sinh[\beta g \mu_B H^z (S + \frac{1}{2})]}{\sinh[\beta g \mu_B H^z \frac{1}{2}]} \right\}^N \tag{4.71}$$

The Free-Energy is given by

$$F = -k_B T \ln Z \tag{4.72}$$

which is evaluated as

$$F = -N k_B T \ln \sinh \left[\beta g \mu_B H^z \left(S + \frac{1}{2} \right) \right]$$

$$+ N k_B T \ln \sinh \left[\beta g \mu_B H^z \frac{1}{2} \right] \tag{4.73}$$

This can be expressed as

$$F = -N k_B T \ln(1 - \exp[-\beta g \mu_B H^z (2S + 1)])$$

$$+ N k_B T \ln(1 - \exp[-\beta g \mu_B H^z]) - N g \mu_B S H^z \tag{4.74}$$

Using thermodynamics, one can obtain the entropy. At high temperatures, the entropy saturates at

$$S \to k_B \ln(2S + 1) \tag{4.75}$$

From the entropy and F, one can find the energy U which is given by

$$U = N g \mu_B H^z \left(\frac{1}{2} \coth \left[\beta g \mu_B H^z \frac{1}{2} \right] - \left(S + \frac{1}{2} \right) \right.$$

$$\times \coth \left[\beta g \mu_B H^z \left(S + \frac{1}{2} \right) \right] \right) \tag{4.76}$$

The internal energy saturates at

$$U = -N g \mu_B S H^z \tag{4.77}$$

in the low temperature limit, $T \to 0$ where the spins are completely aligned with the field. The internal energy vanishes in the high temperature limit, where the different spin orientations have equal probabilities.

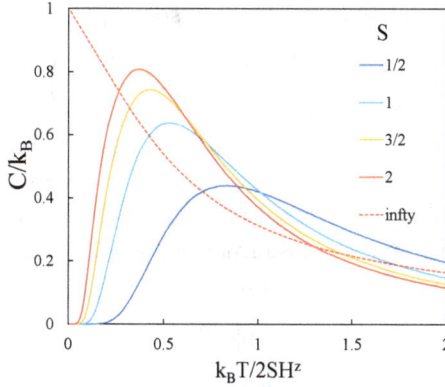

Fig. 4.1 The specific heat for quantum spins with magnitude S. The classical limit is reached when $k_B T \gg \mu_B S H^z$.

4.8 Interacting Ising Spin System

Lenz introduced a one-dimensional model of interacting spins which was intended to be a model that described a phase transition. The model was solved without approximation by Lenz's student Ising. The model became known as the Ising Model [20]. Ising's solution failed to show the expected phase transition. The failure of the one-dimensional model to exhibit a transition at finite temperatures is now understood to be caused by the limited connectivity of the spins by short-ranged interactions on the one-dimensional lattice. In fact, the two-dimensional version of the model is also exactly soluble and does show a phase transition.

Consider a one-dimensional array of spins interacting via the Ising Hamiltonian given by

$$\hat{H} = -\sum_i J S_i^z S_{i+1}^z \tag{4.78}$$

The operator $\hat{S}^z$ has 2 possible eigenvalues which are $-\frac{\hbar}{2}$, $\frac{\hbar}{2}$. The interaction J couples the z-components of $N - 1$ pairs of nearest-neighbor spins. We shall assume that the interaction J has a positive value, so that the lowest energy configuration is ferromagnetic in which all the spins are aligned parallel to each other.

The interaction energy is expressed in terms of the bond variables $S_i^z S_{i+1}^z$. Every microscopic state of the one-dimensional model can be expressed in terms of the set of $(N - 1)$ independent bond variables and the value of any one spin variable, since knowledge of $\frac{2S_i}{\hbar}$ and all the bond variables $S_i^z S_{i+1}^z$ uniquely specifies every spin S_{i+1}^z, since

$$\left(\frac{2S_i^z}{\hbar}\right)\left(\frac{4S_i^z S_{i+1}^z}{\hbar^2}\right) \equiv \left(\frac{2S_{i+1}^z}{\hbar}\right) \tag{4.79}$$

The partition function of the model can be simply evaluated by transforming the Trace over the set of S_i^z to a Trace over the set of bond variables, since the Hamiltonian is expressed in terms of the independent bond variables. We shall perform the transformation by using a projection technique.

The partition function is given by

$$Z = \text{Trace } \exp[-\beta\hat{H}]$$
$$= \text{Trace } \prod_{i=1}^{N-1}\{\exp[\beta J \hat{S}_i^z \hat{S}_{i+1}^z]\} \tag{4.80}$$

which is the product of factors arising from each sequential pair-wise interaction. The factors $\exp[\beta J \hat{S}_i^z \hat{S}_{i+1}^z]$ arising from an interaction can be re-written as

$$\exp[\beta J \hat{S}_i^z \hat{S}_{i+1}^z] = \left(\frac{1}{2} + \frac{2S_i^z S_{i+1}^z}{\hbar^2}\right) \exp\left[+\beta \frac{J\hbar^2}{4}\right]$$
$$+ \left(\frac{1}{2} - \frac{2S_i^z S_{i+1}^z}{\hbar^2}\right) \exp\left[-\beta \frac{J\hbar^2}{4}\right]$$
$$= \left[\cosh\frac{\beta J\hbar^2}{4} + \frac{4}{\hbar^2}\hat{S}_i^z\hat{S}_{i+1}^z \sinh\frac{\beta J\hbar^2}{4}\right] \tag{4.81}$$

since they are to be evaluated on the space where $S_i^z S_{i+1}^z = \pm\frac{\hbar^2}{4}$. Thus

$$Z = \text{Trace } \prod_{i=1}^{N-1}\left\{\cosh\frac{\beta J\hbar^2}{4} + \frac{4}{\hbar^2}\hat{S}_i^z\hat{S}_{i+1}^z \sinh\frac{\beta J\hbar^2}{4}\right\} \tag{4.82}$$

The Trace can be evaluated as a sum over all possible values of the spin eigenvalues

$$\text{Trace} \equiv \prod_{i=1}^{N}\left\{\sum_{S_i^z=-\frac{\hbar}{2}}^{\frac{\hbar}{2}}\right\} \tag{4.83}$$

The Trace runs over all the 2^N possible microstates of the system. The Trace can be evaluated, by noting that the summand in the expression for the partition function only contains one factor which depends on $\hat{S}_1^z$

$$\left[\cosh\frac{\beta J\hbar^2}{4} + \frac{4}{\hbar^2}\hat{S}_1^z\hat{S}_2^z \sinh\frac{\beta J\hbar^2}{4}\right] \tag{4.84}$$

The terms odd in $\hat{S}_1^z$ cancel when taking the Trace. Hence, the Trace over S_1^z contributes a multiplicative factor of

$$2 \, \cosh \frac{\beta J \hbar^2}{4} \tag{4.85}$$

to the partition function, where the factor of two comes from the two spin directions. After the Trace over S_1^z has been performed, only the factor

$$\left[\cosh \frac{\beta J \hbar^2}{4} + \frac{4}{\hbar^2} \, \hat{S}_2^z \hat{S}_3^z \, \sinh \frac{\beta J \hbar^2}{4} \right] \tag{4.86}$$

depends on S_2^z. On taking the Trace over S_2^z, the last term in this factor vanishes and the Trace contributes a second multiplicative factor of $\cosh \frac{\beta J \hbar^2}{4}$ to Z. Each of the $N-1$ interactions contributes a factor of

$$2 \, \cosh \frac{\beta J \hbar^2}{4} \tag{4.87}$$

to the partition function. The Trace over the last spin, produces a multiplicative factor of 2 to Z. Hence, the partition function is given by

$$Z = 2 \left(2 \, \cosh \frac{\beta J \hbar^2}{4} \right)^{N-1} \tag{4.88}$$

The partition reduces to the product of the partition function for any one spin and the products of the $(N-1)$ partition functions for the independent bonds.

The Free-Energy F is given by

$$F = -k_B T \, \ln Z \tag{4.89}$$

which is evaluated as

$$F = -N k_B T \, \ln 2 - (N-1) \, k_B T \, \ln \cosh \frac{\beta J \hbar^2}{4} \tag{4.90}$$

The entropy S is found from

$$S = -\left(\frac{\partial F}{\partial T} \right) \tag{4.91}$$

which yields

$$S = N \, k_B \, \ln 2 + (N-1) \, k_B \, \ln \cosh \frac{\beta J \hbar^2}{4}$$
$$- (N-1) \, k_B \, \frac{\beta J \hbar^2}{4} \, \tanh \frac{\beta J \hbar^2}{4} \tag{4.92}$$

The entropy is seen to reach the value $Nk_B \ln 2$ appropriate to non-interacting spins in the limit $\beta \to 0$ and reaches the value of $k_B \ln 2$ in the limit $T \to 0$. The internal energy U is found from the relation

$$F = U - TS \tag{4.93}$$

as

$$U = -(N-1)\,\frac{J\hbar^2}{4}\,\tanh\frac{\beta J\hbar^2}{4} \tag{4.94}$$

The energy vanishes in the limit $\beta \to 0$ and saturates to the minimal value of $-(N-1)J\frac{\hbar^2}{4}$ appropriate to the $(N-1)$ pair-wise interaction between completely aligned spins in the low temperature limit $T \to 0$. Hence, the ground state is two-fold degenerate and corresponds to minimizing the energy by the spins aligning so that either they are all up or they are all down. The low temperature behavior is to be contrasted with the high temperature results, where the system is dominated by the entropy which is maximized by randomizing the spin directions.

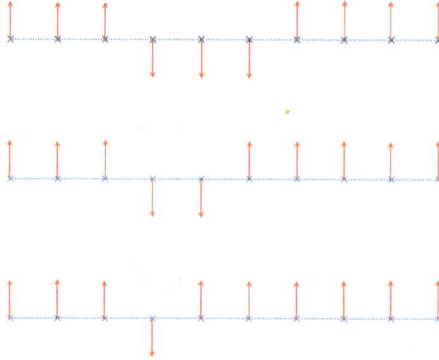

Fig. 4.2 Typical lowest-energy excited states of a one-dimensional Ising Ferromagnet. The energy for reversing a region of consecutive spins is $4\frac{J\hbar^2}{4}$, independent of the size and location of the region.

An infinitesimal magnetic field may be expected to lift the two-fold degeneracy of the $T = 0$ ground state leading to a ferromagnetic state in which all the spins are aligned parallel to each other. Ferromagnetism does not occur at finite temperatures. This can be seen by examining the lowest-energy excitations shown in Fig. 4.2. The excitation energy required to flip one spin is $4\frac{J\hbar^2}{4}$ as two bonds are broken. Likewise, the energy required to flip two, three, or more neighboring spins etc., is also $4\frac{J\hbar^2}{4}$. In fact it costs the same amount of energy, $4\frac{J\hbar^2}{4}$, to flip arbitrarily large regions of spins, and the entropy is proportional to the logarithm of the number of ways of producing the regions of reversed magnetization. Therefore, the presence of these excited states lowers the Free-Energy of the system at finite temperatures and suppresses ferromagnetism.

4.9 Density of States of Elementary Excitations

Consider normal modes of excitation that extend throughout the a hyper-cubic volume $V = L^d$. If the excitations satisfy an isotropic dispersion relation of the form

$$\hbar\omega = \hbar\omega(k) \tag{4.95}$$

where $\omega(k)$ is a monotonically increasing function of k, then this relation can be inverted to yield

$$k = k(\omega) \tag{4.96}$$

Since the normal modes are confined to the system, the normal modes wave functions must vanish on the walls of the system at $x_i = 0$ and $x_i = L$, for $i = 1, 2, \ldots, d$. If the wave functions have the form

$$\underline{\Psi}_\alpha(\underline{r}) = \frac{\epsilon_\alpha(\underline{k})}{\sqrt{V}} \sin \underline{k} \cdot \underline{r} \tag{4.97}$$

for each polarization α, the allowed wave vectors satisfy the d-boundary conditions

$$k_i \, L = n_i \, \pi \tag{4.98}$$

for positive integer values of n_i. Thus, the allowed values of $\underline{k}$ are quantized and can be represented by a vector $\underline{n}$ in n-space

$$\underline{n} = \frac{L\underline{k}}{\pi} \tag{4.99}$$

which has positive integer components $n_1, n_2, n_3, \ldots, n_d$. In n-space, each normal mode with polarization α is represented by a point with positive integer coordinates. Therefore, the normal modes per polarization form a lattice of points arranged on a hyper-cubic lattice with lattice spacing unity. In the segment composed of positive integers, there is one normal mode for each unit volume of the lattice.

Due to the monotonic nature of $\omega(k)$, the number of excitations, per polarization, with energies less than ω, $N(\omega)$, is given by the number of lattice points $\underline{n}$ which satisfy the inequality

$$|\underline{n}| \leq \frac{Lk(\omega)}{\pi} \tag{4.100}$$

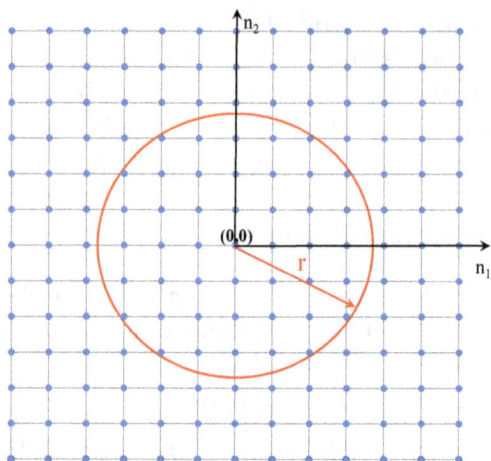

Fig. 4.3   A pictorial representation of n-space in two-dimensions. Each state corresponds to a point (n_1, n_2) for positive integer values of the n's. In the positive quadrant, there is one state per unit area. The states with energy less than E are located in an area of the positive quadrant enclosed by a circular arc of radius r given by $r = Lk(E)/\pi$.

or more explicitly

$$\sqrt{\sum_{i=1}^{d} n_i^2} \leq \frac{Lk(\omega)}{\pi} \tag{4.101}$$

Since the segment of n-space with positive integers,

$$n_1 \geq 0$$
$$n_2 \geq 0$$
$$\ldots$$
$$n_{d-1} \geq 0$$
$$n_d \geq 0 \tag{4.102}$$

is a fraction of $\frac{1}{2^d}$ of the entire volume of n-space, the number of normal modes with energy less than ω is given by $\frac{1}{2^d}$ of the volume enclosed by a radius

$$r = \frac{Lk(\omega)}{\pi} \tag{4.103}$$

where we have recalled that there is one normal mode for each unit cell in n-space and that each cell has a volume 1^d. Hence, on dividing the expression for the volume of a hypersphere of radius r by 2^d, one finds

$$N(\omega) = \frac{1}{2^d} \frac{S_d}{d} \left(\frac{Lk(\omega)}{\pi} \right)^d$$

$$= \frac{S_d}{d} \left(\frac{Lk(\omega)}{2\pi} \right)^d \qquad (4.104)$$

This assumes that no points lie on the bounding surface of the hypersphere, or if they do their numbers are negligible. The surface area of a unit dimensional hypersphere is given by

$$S_d = \frac{2\pi^{\frac{d}{2}}}{\Gamma(\frac{d}{2})} \qquad (4.105)$$

so

$$N(\omega) = \frac{2\pi^{\frac{d}{2}}}{d\Gamma(\frac{d}{2})} V \left(\frac{k(\omega)}{2\pi} \right)^d \qquad (4.106)$$

The number of excitations, per polarization, with energy less that $\hbar\omega$ can be expressed as an integral of the density of states $\rho(\omega)$, per polarization, defined by

$$\rho(\omega) = \sum_{\underline{k}} \delta(\omega - \omega(k)) \qquad (4.107)$$

as

$$N(\omega) = \int_{-\infty}^{\omega} d\omega' \, \rho(\omega')$$

$$= \int_{-\infty}^{\omega} d\omega' \sum_{\underline{k}} \delta(\omega' - \omega(k))$$

$$= \sum_{\underline{k}} \Theta(\omega - \omega(k)) \qquad (4.108)$$

where Θ is the Heaviside step function. The step function restricts the summation to the number of normal modes with frequencies less than ω, which are counted with weight unity. Thus, the density of states per polarization can be found from $N(\omega)$ by taking the derivative w.r.t. ω

$$\rho(\omega) = \frac{d}{d\omega} N(\omega) \qquad (4.109)$$

Hence, we find that the density of states can be represented by

$$\rho(\omega) = \frac{V\pi^{\frac{d}{2}}}{\pi\Gamma(\frac{d}{2})} \left(\frac{k(\omega)}{2\pi} \right)^{d-1} \frac{dk(\omega)}{d\omega} \qquad (4.110)$$

The total density of states is given by the sum of the density of states for each polarization.

4.10   The Debye Model of a Crystalline Solid

Consider a crystalline solid consisting of N atoms on a d-dimensional lattice. The model [21] considers the lattice vibrations as being isotropic sound waves. The sound waves are characterized by their wave vectors $\underline{k}$ and by their polarizations. The vibrational modes consist of N longitudinal modes and $(d-1)N$ transverse modes. The dispersion relations for the modes will be denoted by $\omega_\alpha(\underline{k})$. The Hamiltonian is given by

$$\hat{H} = \sum_{\underline{k},\alpha} \hbar\omega_\alpha(\underline{k}) \left(n_{\underline{k},\alpha} + \frac{1}{2} \right) \tag{4.111}$$

where $n_{\underline{k},\alpha}$ is a integer quantum number.

The partition function Z is given by

$$
\begin{aligned}
Z &= \prod_{\underline{k},\alpha} \left\{ \sum_{n_{\underline{k},\alpha}=0}^{\infty} \right\} \exp\left[-\beta \sum_{\underline{k},\alpha} \hbar\omega_\alpha(\underline{k}) \left(n_{\underline{k},\alpha} + \frac{1}{2} \right) \right] \\
&= \prod_{\underline{k},\alpha} \left\{ \sum_{n_{\underline{k},\alpha}=0}^{\infty} \exp\left[-\beta\hbar\omega_\alpha(\underline{k}) \left(n_{\underline{k},\alpha} + \frac{1}{2} \right) \right] \right\} \\
&= \prod_{\underline{k},\alpha} \left\{ \frac{\exp[-\frac{1}{2}\beta\hbar\omega_\alpha(\underline{k})]}{1 - \exp[-\beta\hbar\omega_\alpha(\underline{k})]} \right\}
\end{aligned} \tag{4.112}
$$

where we have performed the sum over a geometric series. The Free-Energy F is given by

$$
\begin{aligned}
F &= -k_B T \ln Z \\
&= k_B T \sum_{\underline{k},\alpha} \ln\left[\exp\left[+\frac{1}{2}\beta\hbar\omega_\alpha(\underline{k}) \right] - \exp\left[-\frac{1}{2}\beta\hbar\omega_\alpha(\underline{k}) \right] \right] \\
&= k_B T \int_{-\infty}^{\infty} d\omega\, \rho(\omega) \ln\left[\exp\left[+\frac{1}{2}\beta\hbar\omega \right] - \exp\left[-\frac{1}{2}\beta\hbar\omega \right] \right] \\
&= \int_{-\infty}^{\infty} d\omega\, \rho(\omega) \left[\frac{\hbar\omega}{2} + k_B T \ln(1 - \exp[-\beta\hbar\omega]) \right]
\end{aligned} \tag{4.113}
$$

where we have introduced the density of states $\rho(\omega)$ via

$$\rho(\omega) = \sum_{\underline{k},\alpha} \delta(\omega - \omega_{\underline{k},\alpha}) \tag{4.114}$$

Since the density of states from the different polarizations is additive, one has

$$\rho(\omega) = \frac{V S_d}{(2\pi)^d} \left(\frac{1}{c_L^d} + \frac{(d-1)}{c_T^d} \right) \omega^{d-1} \tag{4.115}$$

where the dispersion relation for the longitudinal modes is given by $\omega = c_L\, k$ and the dispersion relation for the $(d-1)$ transverse modes is given by $\omega = c_T\, k$.

Since the lattice vibrations are only defined by the motion of point particles arranged on a lattice, there is an upper limit on the wave vectors $\underline{k}$ and, hence, a maximum frequency. The maximum frequency ω_D is determined from the condition that the total number of normal modes is dN. Thus,

$$\int_0^{\omega_D} \rho(\omega) = dN \tag{4.116}$$

which yields

$$\frac{VS_d}{d(2\pi)^d}\left(\frac{1}{c_L^d} + \frac{(d-1)}{c_T^d}\right)\omega_D^d = dN \tag{4.117}$$

Hence, we may write the density of states as

$$\rho(\omega) = d^2 N\,\frac{\omega^{d-1}}{\omega_D^d} \tag{4.118}$$

for $\omega_D \geq \omega \geq 0$, add is zero otherwise.

The Free-Energy is given in terms of the density of states as

$$F = \int_{-\infty}^{\infty} d\omega\, \rho(\omega)\left[\frac{\hbar\omega}{2} + k_B T\,\ln(1 - \exp[-\beta\hbar\omega])\right] \tag{4.119}$$

and the entropy S is found from

$$S = -\left(\frac{\partial F}{\partial T}\right)_V \tag{4.120}$$

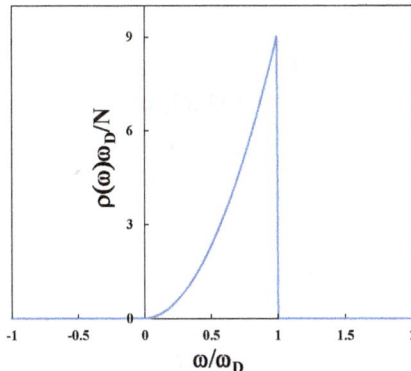

Fig. 4.4 The density of states for the Debye model of a three-dimensional solid containing N atoms, with an upper cut-off frequency ω_D.

The internal energy is found from F and S as

$$U = \int_0^{\omega_D} d\omega \, \rho(\omega) \, \hbar\omega \left(\frac{1}{2} + \frac{1}{\exp[\beta\hbar\omega] - 1} \right) \tag{4.121}$$

The specific heat at constant volume is found from

$$C_V = \left(\frac{\partial E}{\partial T} \right)_V \tag{4.122}$$

which yields

$$C_V = k_B \beta^2 \int_0^{\omega_D} d\omega \, \rho(\omega) \, \hbar^2 \omega^2 \frac{\exp[\beta\hbar\omega]}{(\exp[\beta\hbar\omega] - 1)^2} \tag{4.123}$$

or

$$C_V = \frac{\hbar^2}{k_B T^2} \int_0^{\omega_D} d\omega \, \rho(\omega) \, \omega^2 \frac{\exp[\beta\hbar\omega]}{(\exp[\beta\hbar\omega] - 1)^2} \tag{4.124}$$

On substituting for the density of states, one finds

$$C_V = \frac{\hbar^2}{k_B T^2} \frac{d^2 N}{\omega_D^d} \int_0^{\omega_D} d\omega \, \omega^{d+1} \frac{\exp[\beta\hbar\omega]}{(\exp[\beta\hbar\omega] - 1)^2} \tag{4.125}$$

The specific heat can be evaluated in two limits. In the high temperature limit, $k_B T \gg \hbar\omega_D$, then $k_B T \gg \hbar\omega$. In this limit, one can expand the integrand in powers of $\beta\hbar\omega$, which leads to

$$C_V = k_B \frac{d^2 N}{\omega_D^d} \int_0^{\omega_D} d\omega \, \omega^{d-1}$$

$$\approx dN k_B \tag{4.126}$$

Thus, at high temperatures, the Debye model of a solid reproduces Dulong and Petit's law.

At low temperatures, $k_B T \ll \hbar\omega_D$, one can introduce a dimensionless variable

$$x = \beta\hbar\omega \tag{4.127}$$

The maximum frequency ω_D corresponds to a maximum value x_D

$$x_D \gg 1 \tag{4.128}$$

The specific heat can be expressed as

$$C_V = k_B \frac{d^2 N}{x_D^d} \int_0^{x_D} dx \, x^{d+1} \frac{\exp[x]}{(\exp[x] - 1)^2} \tag{4.129}$$

or on extending the upper limit to infinity

$$C_V = k_B \, (k_B T)^d \frac{d^2 N}{(\hbar\omega_D)^d} \int_0^\infty dx \, x^{d+1} \frac{\exp[x]}{(\exp[x] - 1)^2}$$

$$= k_B \, (k_B T)^d \frac{d^2 N}{(\hbar\omega_D)^d} (d+1) \int_0^\infty dx \, \frac{x^d}{(\exp[x] - 1)} \tag{4.130}$$

where the last line has been found through integration by parts. The integration produces a constant $\Gamma(d+1)\,\xi(d+1)$, so the end result is

$$C_V = Nk_B \left(\frac{k_BT}{\hbar\omega_D}\right)^d d^2\,\Gamma(d+2)\,\xi(d+1) \tag{4.131}$$

Hence, the specific heat at low temperatures is proportional to T^d, which is in accord with experimental observation.

For $d = 3$, one finds

$$C_V = Nk_B \left(\frac{k_BT}{\hbar\omega_D}\right)^3 216\,\frac{\pi^4}{90} \tag{4.132}$$

4.11 Electromagnetic Cavities

Electromagnetic waves can flow through a vacuum, therefore an empty cavity can support electromagnetic modes. The normal modes can be represented by their wave vectors $\underline{k}$ and by their two possible polarizations. However, unlike the sound waves in a solid, there is no upper cut off for the wave vector of an electromagnetic wave. The Hamiltonian is given by

$$\hat{H} = \sum_{\underline{k},\alpha} \hbar\omega_\alpha(\underline{k})\left(n_{\underline{k},\alpha} + \frac{1}{2}\right) \tag{4.133}$$

where $n_{\underline{k},\alpha}$ is a quantum number (the number of photons) which has the allowed values of $0, 1, 2, 3, \ldots, \infty$.

The partition function Z is given by

$$
\begin{aligned}
Z &= \prod_{\underline{k},\alpha}\left\{\sum_{n_{\underline{k},\alpha}=0}^{\infty}\right\}\exp\left[-\beta\sum_{\underline{k},\alpha}\hbar\omega_\alpha(\underline{k})\left(n_{\underline{k},\alpha}+\frac{1}{2}\right)\right] \\
&= \prod_{\underline{k},\alpha}\left\{\sum_{n_{\underline{k},\alpha}=0}^{\infty}\exp\left[-\beta\hbar\omega_\alpha(\underline{k})\left(n_{\underline{k},\alpha}+\frac{1}{2}\right)\right]\right\} \\
&= \prod_{\underline{k},\alpha}\left\{\frac{\exp[-\frac{1}{2}\beta\hbar\omega_\alpha(\underline{k})]}{1-\exp[-\beta\hbar\omega_\alpha(\underline{k})]}\right\}
\end{aligned} \tag{4.134}
$$

and the Free-Energy is given by

$$
\begin{aligned}
F &= -k_BT\,\ln Z \\
&= k_BT\sum_{\underline{k},\alpha}\ln\left[\exp\left[+\frac{1}{2}\beta\hbar\omega_\alpha(\underline{k})\right]-\exp\left[-\frac{1}{2}\beta\hbar\omega_\alpha(\underline{k})\right]\right] \\
&= k_BT\int_{-\infty}^{\infty}d\omega\,\rho(\omega)\,\ln\left[\exp\left[+\frac{1}{2}\beta\hbar\omega\right]-\exp\left[-\frac{1}{2}\beta\hbar\omega\right]\right] \\
&= \int_{-\infty}^{\infty}d\omega\,\rho(\omega)\left[\frac{\hbar\omega}{2}+k_BT\,\ln(1-\exp[-\beta\hbar\omega])\right]
\end{aligned} \tag{4.135}
$$

where the density of states $\rho(\omega)$ is given by

$$\rho(\omega) = \sum_{\underline{k},\alpha} \delta(\omega - \omega_{\underline{k},\alpha}) \tag{4.136}$$

The first term in the Free-Energy represents the (infinite) zero-point energy of the electromagnetic modes. It is divergent, because the electromagnetic cavity can support modes of arbitrarily high frequency. Divergences due to the presence of modes with arbitrarily large frequencies are known as ultra-violet divergences. Since only excitation energies are measured, the zero-point energy can usually be ignored. However, if the boundaries of the cavity are changed, there may be a measurable change in the zero-point energy of the cavity such as found in the Casimir effect [22]. That is, although it may be reasonable to speculate that the divergence in the zero-point energy may merely reflect our ignorance of the true physics at ultra-short distances, the zero-point energy cannot be dismissed since it does have some physical reality.

The density of states for the $(d-1)$ transverse electromagnetic modes can be described by

$$\rho(\omega) = (d-1) \frac{S_d V \omega^{d-1}}{(2\pi)^d \, c^d} \tag{4.137}$$

Hence, the Free-Energy is given by the integral

$$F = (d-1) \frac{S_d V}{(2\pi)^d \, c^d} \int_0^\infty d\omega \, \omega^{d-1}$$

$$\times \left[\frac{\hbar\omega}{2} + k_B T \, \ln(1 - \exp[-\beta\hbar\omega]) \right] \tag{4.138}$$

The internal energy is given by

$$U = F + TS$$

$$= F - T \left(\frac{\partial F}{\partial T} \right) \tag{4.139}$$

which leads to

$$U = (d-1) \frac{S_d V}{(2\pi c)^d} \int_0^\infty d\omega \, \omega^{d-1} \left[\frac{\hbar\omega}{2} + \frac{\hbar\omega}{\exp[\beta\hbar\omega] - 1} \right] \tag{4.140}$$

The first term is divergent and represents the zero-point energy. The second term, ΔU, represents the energy of thermally excited photons. The second term can be evaluated by changing variable to x defined by

$$x = \beta\hbar\omega \tag{4.141}$$

Thus

$$\Delta U = \frac{4(d-1)\,\pi^{\frac{d+2}{2}}}{\Gamma(\frac{d}{2})}\, V\hbar c \left(\frac{k_B T}{2\pi\hbar c}\right)^{d+1} \int_0^\infty dx \, \frac{x^d}{\exp[x]-1} \tag{4.142}$$

which leads to

$$\Delta U = \frac{4(d-1)\,\pi^{\frac{d+2}{2}}}{\Gamma(\frac{d}{2})}\, V\hbar c \left(\frac{k_B T}{2\pi\hbar c}\right)^{d+1} \Gamma(d+1)\,\xi(d+1) \tag{4.143}$$

In three-dimensions, the thermal energy of the cavity is given by

$$\Delta U = \frac{\pi^2 k_B^4 T^4 V}{15\,\hbar^3 c^3} \tag{4.144}$$

since $3!\,\xi(4) = \frac{\pi^4}{15}$. This result is closely related to the Stefan-Boltzmann law which describes the energy flux radiating from an electromagnetic cavity held at temperature T.

The energy flux escaping through a wall of the cavity is given by the energy that passes through a unit area, per unit time. In a unit time, a photon travels a distance $c\cos\theta$ perpendicular to the wall, where θ denotes the angle subtended by the photon's velocity to the normal to the wall. Thus, the number of photons incident on a unit area of the wall, in a unit time, is given by the integral

$$\mathcal{F}_N = \int \frac{d\Omega}{S_d}\, c\cos\theta \int_0^\infty d\omega\, \frac{N_\omega}{V} \tag{4.145}$$

where the integration over the angle θ runs from 0 to $\frac{\pi}{2}$ as the flux only includes light that is traveling towards the wall. In this expression $\int d\omega\, N_\omega/V$ is the photon density. Therefore, the energy flux is given by

$$\begin{aligned}
\mathcal{F}_E &= \int \frac{d\Omega}{S_d}\, c\cos\theta \int_0^\infty d\omega\, \hbar\omega\, \frac{N_\omega}{V} \\
&= \frac{S_{d-1}}{S_d} \int_0^{\frac{\pi}{2}} d\theta\, \sin^{d-2}\theta\, c\cos\theta \int_0^\infty d\omega\, \hbar\omega\, \frac{N_\omega}{V} \\
&= \frac{c S_{d-1}}{(d-1)\, S_d} \int_0^\infty d\omega\, \hbar\omega\, \frac{N_\omega}{V}
\end{aligned} \tag{4.146}$$

The density of photons in a frequency interval $d\omega$ is given by

$$\frac{N_\omega}{V}\, d\omega = \frac{(d-1)\, S_d}{(2\pi c)^d}\, \frac{\omega^{d-1}}{\exp[\beta\hbar\omega]-1}\, d\omega \tag{4.147}$$

Hence, the energy flux is given by

$$\begin{aligned}
\mathcal{F}_E &= \frac{c S_{d-1}}{(2\pi c)^d} \int_0^\infty d\omega\, \frac{\hbar\omega^d}{\exp[\beta\hbar\omega]-1} \\
&= 2\pi\hbar c^2\, S_{d-1}\, \Gamma(d+1)\,\xi(d+1) \left(\frac{k_B T}{2\pi\hbar c}\right)^{d+1}
\end{aligned} \tag{4.148}$$

For three dimensions, this reduces to

$$\mathcal{F}_E = \frac{\pi^2 k_B^4 T^4}{60\ \hbar^3 c^2} \tag{4.149}$$

which is the Stefan-Boltzmann law inferred from experiment [23] and then deduced theoretically [24]

$$\mathcal{F}_E = \sigma T^4 \tag{4.150}$$

where Stefan's constant is given by

$$\sigma = \frac{\pi^2 k_B^4}{60\ \hbar^3 c^2} \tag{4.151}$$

Mathematical Interlude: Evaluation of the Riemann zeta function

Consider the function

$$f(k) = \int_0^\infty dx\ \frac{\sin(kx)}{\exp[x] - 1} \tag{4.152}$$

where the integrand is seen to be finite at $x = 0$. Its Taylor expansion can be written as

$$f(k) = \sum_{n=0}^\infty (-1)^n\ \frac{k^{2n+1}}{(2n+1)!} \int_0^\infty dx\ \frac{x^{2n+1}}{\exp[x] - 1}$$

$$= \sum_{n=0}^\infty (-1)^n\ k^{2n+1}\ \xi(2n+2) \tag{4.153}$$

so $f(k)$ can be regarded as the generating function for the Riemann zeta functions. The value of the coefficient of k^{2n+1} is simply related to the Riemann zeta function $\xi(2n + 2)$. As $\sin(kx)$ is the imaginary part of $\exp[ikx]$, one can re-write the integral as the imaginary part of a related complex function

$$f(k) = \lim_{\eta \to 0} \Im m \int_\eta^\infty dx\ \frac{\exp[ikx]}{\exp[x] - 1} \tag{4.154}$$

where, due to the finite value of η, the integration avoids the pole at the origin. The real function $f(k)$ can be evaluated by considering an integral of the related complex function over a contour C

$$\oint_C dz\ \frac{\exp[ikz]}{\exp[z] - 1} \tag{4.155}$$

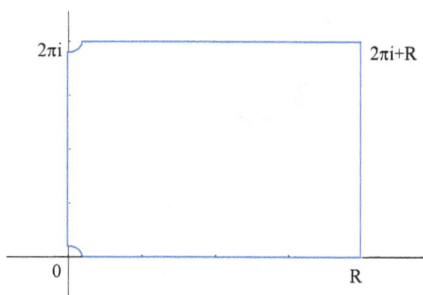

Fig. 4.5   The contour of integration which avoids the poles on the imaginary axis.

The integrand has simple poles at the points $2\pi n i$ with integer n on the imaginary axis, and has residues $\exp[-2\pi k n]$ at these points. The contour C runs from η to R along the real axis, then to $R + 2\pi i$ along the imaginary axis, then the contour runs back parallel to the real axis to $\eta + 2\pi i$. Then, to avoid the pole at $2\pi i$, the contour follows a clockwise circle of radius η centered on $2\pi i$ from $\eta + 2\pi i$ to the point $-i\eta + 2\pi i$. The contour then runs down the imaginary axis from $-i\eta + 2\pi i$ to $i\eta$, and finally returns to η, by following a quarter circle of radius η centered on zero, thereby avoiding the pole at zero. The integral will be evaluated in the limit where $R \to \infty$ and $\eta \to 0$.

Since there are no poles enclosed by the integration contour, Cauchy's Theorem yields

$$\oint_C dz \, \frac{\exp[ikz]}{\exp[z] - 1} = 0 \tag{4.156}$$

In the limit $R \to \infty$, the contribution from the segment from R to $R + 2\pi i$ tends to zero, as the integrand vanishes due to the denominator. The integrations over the segments parallel to the real axis from η to R and from $R + 2\pi i$ to $\eta + 2\pi i$ can be combined to yield

$$[1 - \exp[-2\pi k]] \int_\eta^\infty dx \, \frac{\exp[ikx]}{\exp[x] - 1} \tag{4.157}$$

which has an imaginary part that is related to $f(k)$. The integrations over the quarter circles about the simple poles are both clockwise and are given by $-i\frac{\pi}{2}$ times the residues at the poles, and can be combined to yield

$$- i \, \frac{\pi}{2} \, [1 + \exp[-2\pi k]] \tag{4.158}$$

The remaining contribution runs from $-i\eta + 2\pi i$ to $i\eta$

$$i \int_{-\eta+2\pi}^{\eta} dy \, \frac{\exp[-ky]}{\exp[iy] - 1}$$

$$= -\int_{\eta}^{-\eta+2\pi} dy \, \frac{\exp[-(2k+i)\frac{y}{2}]}{2 \sin \frac{y}{2}}$$

$$= -\frac{1}{2} \int_{\eta}^{-\eta+2\pi} dy \, \exp[-ky] \left(\cot \frac{y}{2} - i \right) \tag{4.159}$$

which, in the limit $\eta \to 0$, has an imaginary part that is given by

$$\frac{1}{2} \int_{0}^{2\pi} dy \, \exp[-ky] = \frac{1}{2} \left(\frac{\exp[-2\pi k] - 1}{-k} \right) \tag{4.160}$$

If one now takes the imaginary part of the entire integral of Eq. (4.156) and takes the limit $\eta \to 0$, one obtains

$$0 = [1 - \exp[-2\pi k]] \, f(k) - \frac{\pi}{2} \, [1 + \exp[-2\pi k]]$$

$$+ \frac{1}{2k} \, [1 - \exp[-2\pi k]] \tag{4.161}$$

On rearranging the equation, $f(k)$ is found to be given by

$$f(k) = \frac{\pi}{2} \, \coth \pi k - \frac{1}{2k} \tag{4.162}$$

Since the series expansion of $\coth(\pi k)$ is given by

$$\coth \pi k = \frac{1}{\pi k} + \frac{\pi k}{3} - \frac{1}{45} \, (\pi k)^3 + \frac{2}{945} \, (\pi k)^5 + \dots \tag{4.163}$$

then

$$f(k) = \frac{\pi^2}{6} \, k - \frac{\pi^4}{90} \, k^3 + \frac{\pi^6}{945} \, k^5 + \dots \tag{4.164}$$

so the values of the Riemann zeta function are given by

$$\xi(2) = \frac{\pi^2}{6}$$

$$\xi(4) = \frac{\pi^4}{90}$$

$$\xi(6) = \frac{\pi^6}{945} \tag{4.165}$$

etc.

4.12 Energy Fluctuations

The Canonical Distribution Function can be used to calculate the entire distributions of most physical quantities of the system. However, for most applications it is sufficient to consider the average values $\overline{A}$ and the moments of the fluctuation $\overline{\Delta A^n}$ where the fluctuation is defined as

$$\Delta A = A - \overline{A} \tag{4.166}$$

The average fluctuation $\overline{\Delta A}$ is identically zero since

$$\overline{\Delta A} = \overline{A} - \overline{A} = 0 \tag{4.167}$$

However, the mean squared fluctuation is given by

$$\overline{\Delta A^2} = \overline{(\overline{A} - \overline{A})^2}$$
$$= \overline{A^2} - \overline{A}^2 \tag{4.168}$$

which can be non-zero.

In the Canonical Ensemble, the energy is no longer fixed. However, the average value of the energy $\overline{E}$ was found to be equal to the thermodynamic value U, as

$$\overline{E} = \frac{1}{Z} \left(\int \frac{d\Gamma}{\Gamma_0} H \, \exp[-\beta H] \right)$$
$$= -\frac{1}{Z} \frac{\partial}{\partial \beta} \left(\int \frac{d\Gamma}{\Gamma_0} \exp[-\beta H] \right)$$
$$= -\frac{\partial \ln Z}{\partial \beta}$$
$$= \frac{\partial}{\partial \beta}(\beta F)$$
$$= U \tag{4.169}$$

The mean squared fluctuation in the energy can be expressed as

$$\overline{\Delta E^2} = \frac{1}{Z} \left(\int \frac{d\Gamma}{\Gamma_0} H^2 \, \exp[-\beta H] \right)$$
$$- \frac{1}{Z^2} \left(\int \frac{d\Gamma}{\Gamma_0} H \, \exp[-\beta H] \right)^2$$
$$= \frac{1}{Z} \frac{\partial^2 Z}{\partial \beta^2} - \frac{1}{Z^2} \left(\frac{\partial Z}{\partial \beta} \right)^2$$
$$= \frac{\partial}{\partial \beta} \left(\frac{1}{Z} \frac{\partial Z}{\partial \beta} \right)$$
$$= \frac{\partial^2 \ln Z}{\partial \beta^2} \tag{4.170}$$

It should be noted that the mean squared energy fluctuation can also be expressed as a derivative of the average energy w.r.t. β

$$\overline{\Delta E^2} = -\frac{\partial}{\partial \beta}(\overline{E}) \tag{4.171}$$

Therefore, on expressing this as a derivative w.r.t T, we find that

$$\overline{\Delta E^2} = k_B T^2 \left(\frac{\partial \overline{E}}{\partial T}\right)_{V,N} \tag{4.172}$$

Hence, the mean squared energy fluctuation can be expressed in terms of the specific heat at constant volume

$$\overline{\Delta E^2} = k_B T^2 \, C_{V,N} \tag{4.173}$$

From this we deduce that the relative magnitude of the energy fluctuations given by the dimensionless quantity

$$\frac{\overline{\Delta E^2}}{\overline{E}^2} \tag{4.174}$$

is of the order of $1/N$ since

$$\frac{\overline{\Delta E^2}}{\overline{E}^2} = \frac{k_B T^2 \, C_{V,N}}{\overline{E}^2} \sim \frac{1}{N} \tag{4.175}$$

where the similarity follows since C_V is extensive and proportional to N as is $\overline{E}$. Thus, on taking the square root, one sees that the relative magnitude of the root mean squared (rms) fluctuation in the energy vanishes in the thermodynamic limit, since

$$\frac{\Delta E_{rms}}{\overline{E}} = \frac{\sqrt{k_B T^2 \, C_{V,N}}}{\overline{E}} \sim \frac{1}{\sqrt{N}} \tag{4.176}$$

Therefore, the relative fluctuations of the energy are negligible in the thermodynamic limit. This suggest the reason why quantities calculated with the Canonical Distribution Function agree with those found with the Micro-Canonical Distribution Function.

The underlying reason for the Canonical and Micro-Canonical Ensemble yielding identical results is that the energy probability distribution function is sharply peaked. The probability density that the system has energy E is given by $P(E)$, where

$$\begin{aligned} P(E) &= \int d\Gamma \; \delta(E - H(\{p_i, q_i\})) \; \rho_c(\{p_i, q_i\}) \\ &= \frac{1}{Z} \int \frac{d\Gamma}{\Gamma_0} \; \delta(E - H(\{p_i, q_i\})) \; \exp[-\beta H(\{p_i, q_i\})] \\ &= \frac{1}{Z} \frac{\Gamma(E)}{\Gamma_0} \; \exp[-\beta E] \end{aligned} \tag{4.177}$$

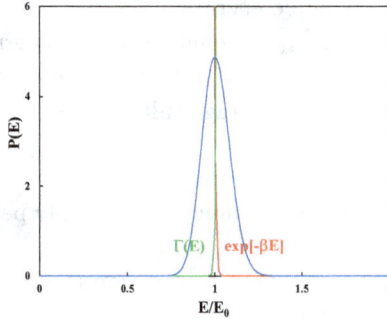

Fig. 4.6 The energy distribution function $P(E)$ in the Canonical Ensemble (shown in blue). The distribution is sharply peaked since it is the product of an exponentially decreasing factor $\exp[-\beta E]$ (red) and a rapidly increasing function $\Gamma(E)$ (green).

which is the product of an exponentially decreasing function of energy $\exp[-\beta E]$ and $\Gamma(E)$ the volume of accessible phase space for a system in the Micro-Canonical Ensemble with energy E. We recognize that $\Gamma(E)$ is an extremely rapidly increasing function of E, since for a typical system $\Gamma(E) \sim E^{\alpha N}$ where α is a number of the order of unity. The most probable value of energy E_{max} can be determined from the condition for the maximum of the energy distribution function

$$\frac{dP(E)}{dE}\bigg|_{E_{max}} = 0 \tag{4.178}$$

which leads to

$$\frac{d}{dE}(\Gamma(E)\ \exp[-\beta E])|_{E_{max}} = 0 \tag{4.179}$$

On representing $\Gamma(E)$ in terms of the entropy $S(E)$ one finds that the most probable value of the energy is given by the solution for E_{max} of the equation

$$\frac{d}{dE}\ (\exp[-\beta E + S(E)/k_B])|_{E_{max}} = 0 \tag{4.180}$$

or after some simplification

$$-\frac{1}{T} + \frac{\partial S(E)}{\partial E}\bigg|_{E_{max}} = 0 \tag{4.181}$$

This equation is satisfied if the temperature T of the thermal reservoir is equal to the temperature of the system. This condition certainly holds true in thermal equilibrium, in which case

$$E_{max} = U \tag{4.182}$$

where U is the thermodynamic energy. Thus, we find that the most probable value of the energy E_{max} is equal to U, the thermodynamic energy. From our previous consideration, we infer that the most probable value of the energy is also equal to the average value of the energy $\overline{E}$,

$$E_{max} = \overline{E} \tag{4.183}$$

Thus, the probability distribution function is sharply peaked at the average energy.

The energy probability distribution function $P(E)$ can be approximated by a Gaussian expression, centered on U. This follows by Taylor expanding the exponent of $P(E)$ in powers of $(E - U)$

$$P(E) = \frac{1}{Z} \exp\left[-\beta F + \frac{1}{2k_B} \left(\frac{d^2 S}{dE^2}\right)\bigg|_U (E - U)^2 + \ldots \right] \tag{4.184}$$

or on cancelling the factor of Z with $\exp[-\beta F]$, one finds

$$P(E) = \exp\left[\frac{1}{2k_B} \left(\frac{d^2 S}{dE^2}\right)\bigg|_U (E - U)^2 + \ldots \right] \tag{4.185}$$

The energy width of the approximate Gaussian distribution is governed by the quantity

$$-\frac{1}{k_B} \left(\frac{d^2 S}{dE^2}\right)\bigg|_U = -\frac{1}{k_B} \left(\frac{\partial \frac{1}{T}}{\partial U}\right)_{V,N}$$

$$= \frac{1}{k_B T^2} \left(\frac{\partial T}{\partial U}\right)_{V,N}$$

$$= \frac{1}{k_B T^2 \, C_V} \tag{4.186}$$

Hence, the mean square fluctuations in the energy are given by

$$\overline{\Delta E^2} = k_B T^2 \, C_V \tag{4.187}$$

in accordance with our previous calculation. We note that in the thermodynamic limit $N \to \infty$, the energy distribution is so sharply peaked that the energy fluctuations usually can be ignored.

4.12.1 *Characteristics of Probability Distributions*

A probability distribution function $P(A)$ can be characterized by the set of its moments M_n which are defined by

$$M_n = \int dA \, A^n \, P(A) \tag{4.188}$$

for $n = 0, 1, 2, \ldots \infty$. The zeroth moment is the normalization condition. The probability distribution function can be characterized by the moment generating function $M(t)$

$$
\begin{aligned}
M(t) &= \int dA \ P(A) \ \exp[At] \\
&= \int dA \ P(A) \sum_{n=0}^{\infty} \frac{1}{n!} A^n \ t^n \\
&= \sum_{n=0}^{\infty} M_n \frac{t^n}{n!}
\end{aligned}
\tag{4.189}
$$

The moment generating function is a power series in t with coefficients given by the moments. The n-th order moment can be simply found from the n-th order derivative of the moment generating function evaluated at $t = 0$

$$
M_n = \frac{\partial^n M(t)}{\partial t^n} \bigg|_{t=0}
\tag{4.190}
$$

It is often more convenient to describe the distribution function in terms of cumulants, since cumulants provide a more intuitive description of unimodular probability distribution functions. The cumulant generating function $K(t)$ is defined as the logarithm of the moment generating function

$$
K(t) = \ln M(t)
\tag{4.191}
$$

The cumulants are obtained as a power series

$$
K(t) = \sum_{n=0}^{\infty} \kappa_n \frac{t^n}{n!}
\tag{4.192}
$$

The cumulants are related to the moments through the relation

$$
\sum_n \kappa_n \frac{t^n}{n!} = \ln \left(\sum_m M_m \frac{t^m}{m!} \right)
\tag{4.193}
$$

simply by taking the n-th order derivatives of each side and setting $t = 0$. The zeroth-order moment is $\kappa_0 = 0$ since $M_0 = 1$. The cumulants and the moments are related by the recursion formula

$$
\kappa_n = M_n - \sum_{m=1}^{n-1} \binom{n-1}{m-1} \kappa_m \ M_{n-m}
\tag{4.194}
$$

The first few cumulants are given by

$$\kappa_0 = 0$$
$$\kappa_1 = M_1$$
$$\kappa_2 = M_2 - M_1^2$$
$$\kappa_3 = M_3 - 3M_2M_1 + 2M_1^3$$
$$\kappa_4 = M_4 - 4M_3M_1 - 3M_2^2 + 12M_2M_1^2 - 6M_1^4 \qquad (4.195)$$

The lowest non-zero cumulant is the average or mean

$$\kappa_1 = \int dA \; P(A) \; A = \overline{A} \qquad (4.196)$$

The next is just the mean-squared fluctuation

$$\kappa_2 = \int dA \; P(A)(A - \overline{A})^2 \qquad (4.197)$$

The third cumulant is known as the skew and is given by

$$\kappa_3 = \int dA \; P(A)(A - \overline{A})^3 \qquad (4.198)$$

The fourth cumulant is the kurtosis (or peakedness) and isn't simply given by the fourth-order fluctuation. However, cumulants take on an intuitive meaning that becomes apparent when one compares the distribution $P(A)$ with a Gaussian distribution with the same mean and mean-squared fluctuation. Cumulants are frequently found in physics since they are linked to connectedness and, therefore, express correlations. The energy fluctuations in the Canonical Ensemble are trivially related to the connection with the heat bath. Likewise, as we shall see later, the number fluctuations in the Grand-Canonical Ensemble are also related to the connection with the heat bath.

Example: Moment Generators for the Energy

In the Canonical Ensemble, the probability $P(E)$ of finding a state with energy E is given by

$$P(E) = \frac{1}{Z(\beta)} \int d\Gamma \exp[-\beta H] \; \delta(E - H) \qquad (4.199)$$

Therefore, the energy moment generating function is calculated as

$$M_E(t) = \int dE \; P(E) \; \exp[tE]$$

$$= \frac{1}{Z(\beta)} \int dE \int d\Gamma \exp[-\beta H] \; \delta(E - H) \; \exp[tE]$$

$$= \frac{1}{Z(\beta)} \int d\Gamma \exp[-(\beta - t)H]$$

$$= \frac{Z(\beta - t)}{Z(\beta)} \tag{4.200}$$

where the partition function $Z(T, V, N)$ is considered to be a function of β. Use of the moment generating function yields the expression for the moments as

$$\overline{E^n} = \frac{(-1)^n}{Z(\beta)} \left(\frac{\partial^n Z(\beta)}{\partial \beta^n} \right) \tag{4.201}$$

Likewise, the energy cumulants are generated by the function

$$K_E(t) = \ln \left(\frac{Z(\beta - t)}{Z(\beta)} \right) \tag{4.202}$$

which characterizes the energy distribution.

4.13 Correlations and Fluctuations in the $d = 1$ Ising Model

The one-dimensional Ising model has the Hamiltonian

$$\hat{H} = -J \sum_{i=1}^{N-1} S_i^z S_{i+1}^z \tag{4.203}$$

where S_i^z is the value of the spin on the i-th site, where in units of $\hbar = 2$, one finds that $S_i^z = \pm 1$. The partition function Z is evaluated from

$$Z_N = \text{Trace} \prod_{i=1}^{N-1} \exp[\beta J S_i^z S_{i+1}^z]$$

$$= \text{Trace} \prod_{i=1}^{N-1} (\cosh \beta J + S_i^z S_{i+1}^z \; \sinh \beta J) \tag{4.204}$$

The Trace over the spin S_1^z can be performed. Since only the first factor of the product depends on S_1^z, one has

$$\sum_{S_1^z = -1}^{S_1^z = 1} (\cosh \beta J + S_1^z S_2^z \; \sinh \beta J) = 2 \; \cosh \beta J \tag{4.205}$$

since the term which is odd in S_1^z vanishes. Hence, one has

$$Z_N = 2 \ \cosh \beta J \ \mathrm{Trace}_{i=2,\ldots,N} \ \prod_{i=2}^{N-1} (\cosh \beta J + S_i^z S_{i+1}^z \ \sinh \beta J)$$

$$= 2 \ \cosh \beta J Z_{N-1} \tag{4.206}$$

In the first line, the Trace over S_1^z has been performed, so the remaining Trace only involves $(N-1)$ spins and $(N-2)$ interactions. Thus, the remaining factor is identified as the partition function for $(N-1)$ interacting spins, Z_{N-1}. This process can be iterated a total of $N-1$ times, giving $(N-1)$ factors of $2 \cosh \beta J$ times the partition function for one spin Z_1. Thus,

$$Z_N = (2 \ \cosh \beta J)^{N-1} Z_1$$
$$= 2(2 \ \cosh \beta J)^{N-1}$$
$$= 2^N (\cosh \beta J)^{N-1} \tag{4.207}$$

The second line occurs since the partition function for one atom is given by $Z_1 = 2$. That is

$$Z_1 = \sum_{S_N^z = -1}^{1} 1 = 2 \tag{4.208}$$

is independent of J since an interaction must involve a pair of spins.

One can also evaluate thermally averaged quantities, such as the correlation function

$$\langle S_i^z S_j^z \rangle \tag{4.209}$$

In the Canonical Ensemble, the above average is defined as

$$\langle S_i^z S_j^z \rangle = \mathrm{Trace} \ \frac{1}{Z} \ S_i^z S_j^z \ \exp \left[\beta J \sum_{i=1}^{N-1} S_k^z S_{k+1}^z \right] \tag{4.210}$$

The exponential factor can be expressed in terms of a product

$$\prod_{k=1}^{N-1} \exp[\beta J S_k^z S_{k+1}^z] = \prod_{k=1}^{N-1} (\cosh \beta J + S_k^z S_{k+1}^z \ \sinh \beta J) \tag{4.211}$$

as before. On inserting the expression for Z, one finds that the average can be expressed as

$$\langle S_i^z S_j^z \rangle = \frac{1}{2^N} \ \mathrm{Trace} \ S_i^z S_j^z \ \prod_{k=1}^{N-1} (1 + S_k^z S_{k+1}^z \ \tanh \beta J) \tag{4.212}$$

Our approach to evaluating the Trace is simply noting that the only terms that are non-zero, after the Trace is taken, must have had even powers (including zero) of every spin variable S_k^z. Therefore, we shall arrange the expression so that terms of even order in S_i^z and S_j^z can easily be identified.

We shall assume that $j > i$, and re-arrange the expression so that the position of a spin variable S_k^z in the product is ordered according to its spatial position index k. This results in the Trace being taken over three factors. One factor coming from $k < i$, a second factor is associated with the range $i < k < j$ and the last factor corresponds to $j < k$. Thus, the correlation function is written as

$$\langle S_i^z S_j^z \rangle = \frac{1}{2^N} \text{ Trace} \prod_{k=1}^{i-1} (1 + S_k^z S_{k+1}^z \tanh \beta J)$$

$$\times S_i^z \prod_{k=i}^{j-1} (1 + S_k^z S_{k+1}^z \tanh \beta J)$$

$$\times S_j^z \prod_{k=j}^{N-1} (1 + S_k^z S_{k+1}^z \tanh \beta J) \tag{4.213}$$

The Trace over the first $(i-1)$ spins yields a factor of 2^{i-1} since the terms involving an odd power of any of these spins vanish. Thus,

$$\langle S_i^z S_j^z \rangle = \frac{2^{i-1}}{2^N} \text{ Trace}_{i,\dots,N} S_i^z \prod_{k=i}^{j-1} (1 + S_k^z S_{k+1}^z \tanh \beta J)$$

$$\times S_j^z \prod_{k=j}^{N-1} (1 + S_k^z S_{k+1}^z \tanh \beta J) \tag{4.214}$$

The sum over S_i^z involves the product of S_i^z and one of the factors which originated from the exponential Boltzmann distribution factor

$$\sum_{S_i^z=-1}^{S_i^z=1} S_i^z (1 + S_i^z S_{i+1}^z \tanh \beta J) \tag{4.215}$$

Furthermore, since the terms odd in S_i^z vanish and since $(S_i^z)^2 = 1$, the Trace over S_i^z is evaluated as

$$\sum_{S_i^z=-1}^{S_i^z=1} S_i^z (1 + S_i^z S_{i+1}^z \tanh \beta J) = 2 \tanh \beta J S_{i+1}^z \tag{4.216}$$

Only the second factor in the parenthesis survived the Trace and the factor of 2 is simply the Trace over $(S_i^z)^2 = 1$. The next sum in the Trace is a sum over S_{i+1}^z and involves the factor

$$2 \tanh \beta J \sum_{S_{i+1}^z=-1}^{S_{i+1}^z=1} S_{i+1}^z (1 + S_{i+1}^z S_{i+2}^z \tanh \beta J) = (2 \tanh \beta J)^2 S_{i+2}^z$$

(4.217)

where again, only the term quadratic in S_{i+1}^z survives. By inductive reasoning, one recognizes that the Trace over all the spins with labels k between i and $j-1$ results in a factor of

$$\text{Trace}_{i,\ldots,j-1} \; S_i^z \prod_{k=i}^{j-1} (1 + S_k^z S_{k+1}^z \tanh \beta J) = (2 \tanh \beta J)^{j-i} S_j^z \quad (4.218)$$

Hence, the average value of the two-spin correlation function is given by

$$\langle S_i^z S_j^z \rangle = \frac{2^{j-1}}{2^N} (\tanh \beta J)^{j-i}$$

$$\times \text{Trace}_{j,\ldots,N} \; (S_j^z)^2 \prod_{k=j}^{N-1} (1 + S_k^z S_{k+1}^z \tanh \beta J)$$

$$= \frac{2^N}{2^N} (\tanh \beta J)^{j-i} \tag{4.219}$$

since $(S_j^z)^2 = 1$ and since

$$\text{Trace}_{j,\ldots,N} \prod_{k=j}^{N-1} (1 + S_k^z S_{k+1}^z \tanh \beta J) = 2^{N-j+1} \tag{4.220}$$

In this expression we have assumed that $j > i$. More generally, one has

$$\langle S_i^z S_j^z \rangle = (\tanh \beta J)^{|j-i|} \tag{4.221}$$

Since the tanh factor is less than unity, one sees that the average value falls of exponentially with the distance between site i and site j. Therefore, we find

$$\langle S_i^z S_j^z \rangle = \exp\left[-\frac{|j-i|}{\xi} \right] \tag{4.222}$$

where ξ is defined to be the correlation length and

$$\xi = -\frac{1}{\ln \tanh(\beta J)} \tag{4.223}$$

The correlation length increases as the temperature decreases. As $T \to 0$ the correlation length diverges $\xi \to \infty$.

Having identified the thermal averaged value of $\langle S_i^z S_j^z \rangle$, one can insert the value of the nearest-neighbor correlation function into the result for the average value of the energy

$$\langle H \rangle = -J \sum_{i=1}^{N-1} \langle S_i^z S_{i+1}^z \rangle$$

$$= -J(N-1) \tanh \beta J \qquad (4.224)$$

and recognize that this is identical to the value of the thermodynamic energy U obtained from the partition function. Therefore, one may conclude that the specific heat, which we have previously shown to be a measure of energy-fluctuations, is also related to a nearest-neighbor spin-spin correlation function.

The magnetic susceptibility χ can also be obtained as a derivative of the magnetization M^z with respect to the magnetic field.

$$M^z = \sum_i \langle S_i^z \rangle \qquad (4.225)$$

where the magnetic field is contained in the Canonical Ensemble's probability density

$$P = \frac{1}{Z} \exp[-\beta H] \qquad (4.226)$$

via the Zeeman interaction energy

$$H_{int} = -B^z \sum_i S_i^z \qquad (4.227)$$

On taking the derivative of the magnetization with respect to the applied field and then taking the limit $B^z \to 0$, one finds

$$\chi = \left(\frac{\partial M^z}{\partial B^z} \right) \Bigg|_{B_z=0}$$

$$= \beta \sum_{i,j} (\langle S_i^z S_j^z \rangle - \langle S_i^z \rangle \langle S_j^z \rangle)$$

$$= \beta \sum_{i,j} (\tanh \beta J)^{|j-i|}$$

$$= \beta N \left(1 + \frac{2 \tanh \beta J}{1 - \tanh \beta J} \right)$$

$$= \beta N \left(\frac{1 + \tanh \beta J}{1 - \tanh \beta J} \right) \qquad (4.228)$$

where we have noted that $\langle S_i^z \rangle = 0$ in the limit of zero field and have ignored finite size or end effects. The second line is important, it shows that the susceptibility is actually a measure of the fluctuations in the spin directions

$$\Delta S_i^z = S_i^z - \langle S_i^z \rangle \tag{4.229}$$

The magnetic susceptibility simply has the form of the thermal average of the fluctuations in the spin directions

$$\chi = \beta \sum_{i,j} \langle \Delta S_i^z \, \Delta S_j^z \rangle \tag{4.230}$$

Thus, the susceptibility is a measure of the spin distribution function. The above analysis indicates that there is a relation between fluctuations and the measured response of the system to an externally applied field. This is the content of the *"Fluctuation-Dissipation Theorem"*.

One should note that the denominator of the susceptibility diverges when $T \to 0$, since

$$\tanh \beta J \to 1 \tag{4.231}$$

The divergence of the magnetic susceptibility signals the onset of a finite magnetization due to the presence of an infinitesimal field. That is, there appears to be a phase transition at the point $T = 0$. The onset of the $T = 0$ magnetization is accompanied by the onset of long-ranged order since, no matter how distant i is from j

$$\lim_{T \to 0} \langle S_i^z S_j^z \rangle \to 1 \tag{4.232}$$

It is important to note that the phase transition breaks the symmetry between the two-fold degenerate lowest-energy states. In the absence of the infinitesimal field, the degenerate states differ only in that one has all the spins pointing up and the other has all the spins pointing down. The infinitesimal field that we considered was important in making one of this pair of states more favorable than the other. If the field had not been considered, the magnetization would have been calculated to be identically zero since it would have been evaluated as a Trace over two sets of states. In one set of states the spins have specific directions and the states in the second set are obtained from the states of the first set by merely by reversing the direction of each spin. Therefore, for each state in the first set there is a degenerate partner state in the second set which has the opposite value for the magnetization. The contributions of these two sets of states to the magnetization are equal and opposite and, thus, must cancel. The effect of the imperceptible tiny field has caused the system to adopt a single state that does not have the full symmetry of the Hamiltonian. This is the phenomenon of spontaneous symmetry breaking.

4.14 The Boltzmann Distribution from Entropy Maximization

The general expression for entropy in terms of the probability distribution function $\rho_c(\{p_i, q_i\})$ is given by the integral over phase space

$$S = -k_B \int d\Gamma \left[\rho_c(\{p_i, q_i\}) \, \ln \rho_c(\{p_i, q_i\})\Gamma_0\right] \tag{4.233}$$

This is trivially true in the Micro-Canonical Ensemble and is also true in the Canonical Ensemble where

$$\Gamma_0 \, \rho_c(\{p_i, q_i\}) = \frac{1}{Z} \, \exp[-\beta H(\{p_i, q_i\})] \tag{4.234}$$

This can be seen by substituting the equation for $\rho_c(\{p_i, q_i\})$ in the expression for S, which leads to

$$S = k_B \int d\Gamma \, \rho_c(\{p_i, q_i\})[\beta H(\{p_i, q_i\}) + \ln Z] \tag{4.235}$$

However, we know that

$$\ln Z = -\beta F \tag{4.236}$$

and the distribution function is normalized

$$\int d\Gamma \, \rho_c(\{p_i, q_i\}) = 1 \tag{4.237}$$

so on multiplying by T we find

$$TS = \int d\Gamma \left[\rho_c(\{p_i, q_i\}) \, H(\{p_i, q_i\})\right] - F \tag{4.238}$$

Since the average energy $\overline{E}$ is defined as

$$\overline{E} = \int d\Gamma \left[\rho_c(\{p_i, q_i\}) \, H(\{p_i, q_i\})\right] \tag{4.239}$$

then we end up with an expression for the Helmholtz Free-Energy

$$F = \overline{E} - TS \tag{4.240}$$

Finally, since $\overline{E} = U$ (the thermodynamic energy), we have shown that

$$S = -k_B \int d\Gamma \left[\rho_c(\{p_i, q_i\}) \, \ln \rho_c(\{p_i, q_i\})\Gamma_0\right] \tag{4.241}$$

which we shall regard as the fundamental form of S for any distribution function.

Derivation

Given the above form of S one can derive the Canonical Distribution Function as the distribution function ρ which maximizes the functional $S[\rho]$, subject to the requirements that the average energy is U and that the distribution function is normalized. That is ρ_c must maximize the functional $S[\rho]$ subject to the constraints that

$$1 = \int d\Gamma \; \rho(\{p_i, q_i\})$$

$$U = \int d\Gamma \; \rho(\{p_i, q_i\}) \; H(\{p_i, q_i\}) \qquad (4.242)$$

The maximization of S subject to the constraints is performed by using Lagrange's method of undetermined multipliers. In this method, one forms the functional $\Phi[\rho]$ defined by

$$\Phi[\rho] = -k_B \int d\Gamma \; [\rho(\{p_i, q_i\}) \; \ln \rho(\{p_i, q_i\})\Gamma_0]$$

$$+ \alpha \left(1 - \int d\Gamma \; \rho(\{p_i, q_i\}) \right)$$

$$+ \gamma \left(U - \int d\Gamma \; \rho(\{p_i, q_i\}) \; H(\{p_i, q_i\}) \right) \qquad (4.243)$$

where α and γ are undetermined numbers. If ρ satisfies the two constraints then $\Phi[\rho]$ is equal to $S[\rho]$, and then maximizing Φ is equivalent to maximizing S.

If $\Phi[\rho]$ is to be maximized by $\rho_c(\{p_i, q_i\})$ then a small change in $\rho(\{p_i, q_i\})$ by $\delta\rho(\{p_i, q_i\})$ should not change Φ. That is, if we set

$$\rho(\{p_i, q_i\}) = \rho_c(\{p_i, q_i\}) + \lambda \; \delta\rho(\{p_i, q_i\}) \qquad (4.244)$$

where $\delta\rho$ is an arbitrary deviation, then the first-order change in Φ, $\delta\Phi^{(1)}$ defined by

$$\Phi[\rho_c + \lambda\delta\rho] - \Phi[\rho_c] = \lambda \; \delta\Phi^{(1)} + O(\lambda)^2 \qquad (4.245)$$

must vanish

$$\delta\Phi^{(1)} = 0 \qquad (4.246)$$

If this condition was not satisfied, then a specific choice for the sign of λ would cause Φ to increase further. Thus, the requirement that $\Phi[\rho]$ is maximized by ρ_c, leads to the condition

$$\int d\Gamma \; \delta\rho(\{p_i, q_i\})[-k_B \; \ln \rho_c(\{p_i, q_i\})\Gamma_0 - k_B - \alpha - \gamma H(\{p_i, q_i\})] = 0$$

$$(4.247)$$

This integral must vanish for any choice of $\delta\rho$. This can be achieved by requiring that the quantity inside the square brackets vanishes at every point in phase space. That is

$$k_B \ln \rho_c(\{p_i, q_i\})\Gamma_0 = -k_B - \alpha - \gamma H(\{p_i, q_i\}) \tag{4.248}$$

where α and γ are undetermined constants. Hence, on exponentiating, we have

$$\rho_c(\{p_i, q_i\})\Gamma_0 = \exp[-1 - \alpha/k_B] \ \exp[-\gamma H(\{p_i, q_i\})/k_B] \tag{4.249}$$

The constants α and γ are determined by ensuring that the two constraints are satisfied. The conditions are satisfied by requiring that

$$1 = \exp[-1 - \alpha/k_B] \int \frac{d\Gamma}{\Gamma_0} \ \exp[-\gamma H(\{p_i, q_i\})/k_B]$$

$$U = \exp[-1 - \alpha/k_B] \int \frac{d\Gamma}{\Gamma_0} \ H(\{p_i, q_i\}) \ \exp[-\gamma H(\{p_i, q_i\})/k_B]$$

$$\tag{4.250}$$

which then has the effect that ρ_c maximized S. The two constraints suggests that one should rewrite the parameters as

$$\gamma = \frac{1}{T} \tag{4.251}$$

and

$$Z = \exp[1 + \alpha/k_B] \tag{4.252}$$

In fact, if the form of ρ_c is substituted back into S and one constraint is used to express S in terms of U and the second constraint to produce a constant term (independent of U), then if one demands that

$$\left(\frac{\partial S}{\partial U}\right)_V = \frac{1}{T} \tag{4.253}$$

then one finds $\gamma = \frac{1}{T}$. Thus, the distribution that maximizes $S[\rho]$ is recognized as being the Boltzmann Distribution Function

$$\rho_c(\{p_i, q_i\})\Gamma_0 = \frac{1}{Z} \ \exp[-\beta H(\{p_i, q_i\})] \tag{4.254}$$

In summary, we have shown that the Boltzmann Distribution Function maximizes $S[\rho]$ subject to the two constraints

$$1 = \int d\Gamma \ \rho_c(\{p_i, q_i\})$$

$$U = \int d\Gamma \ \rho_c(\{p_i, q_i\}) \ H(\{p_i, q_i\}) \tag{4.255}$$

4.15   The Gibbs Ensemble

The "*Gibbs Ensemble*" corresponds to the situation where a closed system is partitioned into two parts. A smaller part of which is our system of interest and the other part comprises its environment. The system and its environment are allowed to exchange energy, and also a partition that separates the system from its environment is free to move. Therefore, the volume of the subsystem can be interchanged with the environment. The total energy E_T is partitioned as

$$E_T = E + E_R \tag{4.256}$$

and the volume is partitioned as

$$V_T = V + V_R \tag{4.257}$$

The probability that the partition, considered by itself, will be found such that the volume of the system is in a range dV around V is assumed to be given by the ratio dV/V_T. The probability dp that the closed system (including the partition) is in the joint volume element $d\Gamma_T$ and dV,

$$dp = \frac{1}{V_T} \, \rho_{mc} \, d\Gamma_T \, dV \tag{4.258}$$

On factorizing the infinitesimal volume element of total phase space $d\Gamma_T$ into contributions from the reservoir $d\Gamma_R$ and the system $d\Gamma$, one has

$$dp = \frac{1}{V_T} \, \rho_{mc} \, d\Gamma_R \, d\Gamma \, dV \tag{4.259}$$

We are assuming that the phase space $d\Gamma$ is consistent with the position of the partition defining the volume V and also that the system has energy E. The probability dp that the system is in the volume element $d\Gamma$ irrespective of the microstates of the reservoir is obtained by integrating over all of the reservoir's accessible phase space, consistent with the energy $E_T - E$ and volume $V_T - V$. The result is

$$dp = \frac{1}{V_T} \, \rho_{mc} \, \Gamma_R(E_T - E, V_T - V) \, d\Gamma \, dV \tag{4.260}$$

The Gibbs Probability Distribution Function ρ_G is defined via

$$dp = \left(\frac{dp}{d\Gamma \, dV} \right) d\Gamma \, dV$$
$$= \rho_G \, d\Gamma \, dV \tag{4.261}$$

and is found as

$$\rho_G = \frac{1}{V_T} \, \rho_{mc} \, \Gamma_{R,0} \, \exp[S_R(E_T - E, V_T - V)/k_B] \tag{4.262}$$

However, one also has

$$
\begin{aligned}
\rho_{mc} &= \frac{1}{\Gamma_{T,0}} \exp[-S_T(E_T, V_T)/k_B] \\
&= \frac{1}{\Gamma_{T,0}} \exp[-(S_R(E_T - U, V_T - \overline{V}) + S(U, \overline{V}))/k_B] \quad (4.263)
\end{aligned}
$$

The phase space volumes representing single microscopic states of the total system, reservoir and subsystem are assumed to satisfy the relation

$$
\Gamma_{T,0} = \Gamma_{R,0} \, \Gamma_0 \quad (4.264)
$$

Hence, we can express the Gibbs Probability Distribution Function as

$$
\begin{aligned}
\rho_G \, \Gamma_0 = \frac{1}{V_T} \exp[(S_R(E_T - E, V_T - V) \\
- S_R(E_T - U, V_T - \overline{V}) - S(U, \overline{V}))/k_B] \quad (4.265)
\end{aligned}
$$

The first term in the exponent can be expanded in powers of the energy and volume fluctuations

$$
\begin{aligned}
S_R(E_T &- U, V_T - V) \\
&= S_R(E_T - U, V_T - \overline{V}) \\
&+ (U - E) \left(\frac{\partial S_R(E_R, V_T - \overline{V})}{\partial E_R} \right) \bigg|_{E_T - U} \\
&+ (\overline{V} - V) \left(\frac{\partial S_R(E_T - U, V_R)}{\partial V_R} \right) \bigg|_{V_T - \overline{V}} + \cdots \\
&= S_R(E_T - U, V_T - \overline{V}) + (U - E) \frac{1}{T} + (\overline{V} - V) \frac{P}{T} + \cdots \quad (4.266)
\end{aligned}
$$

On substituting this in the expression for ρ_G, one finds that

$$
\rho_G \, \Gamma_0 = \frac{1}{V_T} \exp[\beta G] \exp[-\beta(E + PV)] \quad (4.267)
$$

where G is the Gibbs Free-Energy $G(T, P, N)$ of the system

$$
G = U - TS + P\overline{V} \quad (4.268)
$$

The only quantities in the Gibbs Distribution Function pertaining to the reservoir is its temperature and pressure. On introducing the Gibbs Partition Function Y via

$$
Y = \exp[-\beta G] \quad (4.269)
$$

one can express the Gibbs Probability Distribution Function as

$$
\rho_G \, \Gamma_0 = \frac{1}{V_T} \exp[-\beta PV] \frac{1}{Y} \exp[-\beta H] \quad (4.270)
$$

4.15.1   *The Gibbs Partition Function*

The normalization condition for the probability distribution function is given by

$$
\begin{aligned}
1 &= \int_0^{V_T} dV \int d\Gamma \, \rho_G \\
&= \int_0^{V_T} dV \, \frac{1}{V_T} \, \exp[-\beta PV] \, \frac{1}{Y} \int \frac{d\Gamma}{\Gamma_0} \, \exp[-\beta H] \\
&= \int_0^{V_T} dV \, \frac{1}{V_T} \, \exp[-\beta PV] \, \frac{1}{Y} \, Z(V)
\end{aligned}
\tag{4.271}
$$

where $Z(V)$ is the Canonical Partition Function. Hence, one finds that the Gibbs Partition Function Y is determined from

$$
Y = \int_0^{V_T} dV \, \frac{1}{V_T} \, \exp[-\beta PV] \, Z(V)
\tag{4.272}
$$

which only involves quantities describing the system. Since the Canonical Partition Function is a function of the variable (T, V, N), the Gibbs partition function is a function of (T, P, N). Once Y has been determined from the above equation, thermodynamic quantities can be evaluated from the Gibbs Free-Energy $G(T, P, N)$ which is expressed in terms of Y as

$$
G = -k_B T \, \ln Y
\tag{4.273}
$$

4.15.2   *The Ideal Gas*

The Gibbs Partition Function for the ideal gas is given by

$$
\begin{aligned}
Y(T, P, N) = \frac{1}{N!} \int_0^{V_T} \frac{dV}{V_T} \, \exp[-\beta PV] \\
\times \prod_{i=1}^{3N} \left\{ \iint \frac{dp_i dq_i}{2\pi\hbar} \right\} \exp\left[-\beta \sum_{i=1}^{3N} \frac{p_i^2}{2m} \right]
\end{aligned}
\tag{4.274}
$$

which on integrating over the particle's coordinates and momenta becomes

$$
Y(T, P, N) = \int_0^{V_T} \frac{dV}{V_T} \, \exp[-\beta PV] \, \frac{V^N}{N!} \left(\frac{\sqrt{2\pi m k_B T}}{2\pi\hbar} \right)^{3N}
\tag{4.275}
$$

The integral over V is easily evaluated by changing variable to

$$
x = \beta PV
\tag{4.276}
$$

and yields

$$Y(T, P, N) = \frac{1}{\beta P V_T} (\beta P)^{-N} \left(\frac{\sqrt{2\pi m k_B T}}{2\pi \hbar} \right)^{3N} \qquad (4.277)$$

where the factor of $N!$ has cancelled.

On ignoring the non-extensive contributions, the Gibbs Free-Energy is given by

$$\begin{aligned} G &= -k_B T \ \ln Y \\ &= N k_B T \left[\ln P - \frac{5}{2} \ \ln(k_B T) + \frac{3}{2} \ \ln \left(\frac{2\pi \hbar^2}{m} \right) \right] \qquad (4.278) \end{aligned}$$

Since the infinitesimal variation in G is given by

$$dG = -S \ dT + \overline{V} \ dP + \mu \ dN \qquad (4.279)$$

one finds that the average volume is given by

$$\begin{aligned} \overline{V} &= \left(\frac{\partial G}{\partial P} \right)_T \\ &= \frac{N k_B T}{P} \qquad (4.280) \end{aligned}$$

which is the ideal gas law.

The value of the enthalpy $\mathcal{H} = \overline{E} + P\overline{V}$ can be calculated directly from its average in the Gibbs Distribution

$$\begin{aligned} \mathcal{H} &= \overline{E} + P\overline{V} \\ &= \frac{1}{Y} \int_0^{V_T} \frac{dV}{V_T} \int \frac{d\Gamma}{\Gamma_0} \ (H(\{p_i, q_i\}) + PV) \\ & \quad \times \exp[-\beta(H(\{p_i, q_i\}) + PV)] \qquad (4.281) \end{aligned}$$

which can simply be expressed as a derivative w.r.t. β

$$\begin{aligned} \mathcal{H} &= -\frac{1}{Y} \left(\frac{\partial Y}{\partial \beta} \right)_P \\ &= -\frac{\partial \ln Y}{\partial \beta} \qquad (4.282) \end{aligned}$$

Thus, we find that the enthalpy is given by

$$\mathcal{H} = \frac{5}{2} \ N k_B T \qquad (4.283)$$

from which the specific heat at constant pressure C_P is found as

$$C_P = \frac{5}{2} \ N k_B \qquad (4.284)$$

as is expected for the ideal gas.

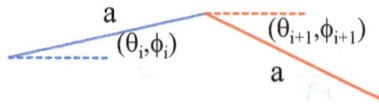

Fig. 4.7 Two successive links of a polymer. The orientation of the i-th monomer is defined by the polar coordinates (θ_i, φ_i) in which the polar axis is defined by displacement vector of the whole polymer which runs from one end to the other.

4.15.3 A Flexible Polymer

Consider a polymer made of a large number N monomers of length a. The length of the polymer is variable since, although the monomers are joined end to end, the joints are assumed to be freely flexible. That is, the polymers are joined in way that allows free rotation at the ends. The length L of the polymer is defined by its end to end distance, and this definition of the length also defines a preferred (polar) axis which has the direction of the vector joining the ends. The orientational degrees of freedom of the i-th monomer is given by the polar coordinates (θ_i, φ_i). Hence, the length of the polymer is given by

$$L = \sum_{i=1}^{N} a \cos \theta_i \qquad (4.285)$$

Although there is only one polymer, the fact that it is composed of a very large number of monomers allows one to consider it as being in the thermodynamic limit and to use statistical mechanics effectively. The Hamiltonian

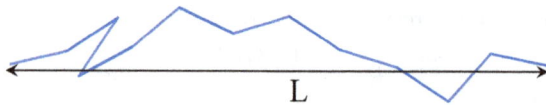

Fig. 4.8 A polymer chain consisting of N links and length L. The polar axis is defined by the orientation of the displacement vector defining the length of the polymer.

is set equal to zero, since we are assuming that the monomers are freely jointed and have negligible masses. The partition function $Z(L)$ for the polymer of length L is given by

$$Z(L) = \prod_{i=1}^{N} \left\{ \int_0^{2\pi} d\varphi_i \int_0^{\pi} \sin \theta_i \, d\theta_i \right\} \qquad (4.286)$$

where the integrations are restricted by the constraint

$$L = \sum_{i=1}^{N} a \cos \theta_i \tag{4.287}$$

In this case Z coincides with Γ since $H = 0$.

If a tension $\mathcal{T}$ is applied to the polymer it is described by a Gibbs Distribution Function $Y(T, \mathcal{T})$, defined by

$$Y = \int_0^{Na} dL \; \exp[\beta \mathcal{T} L] \prod_{i=1}^{N} \left\{ \int_0^{2\pi} d\varphi_i \int_0^{\pi} \sin \theta_i \; d\theta_i \right\}$$
$$\times \delta \left(L - \sum_{i=1}^{N} a \cos \theta_i \right) \tag{4.288}$$

where the delta function has absorbed the factor of the normalization of the length probability density. The use of the Gibbs Distribution is justified since $-\mathcal{T}$ and L are analogous to P and V. On performing the integral over L, one obtains

$$Y = \prod_{i=1}^{N} \left\{ \int_0^{2\pi} d\varphi_i \int_0^{\pi} \sin \theta_i \; d\theta_i \right\} \exp \left[\beta \mathcal{T} a \sum_{i=1}^{N} \cos \theta_i \right] \tag{4.289}$$

which is no longer subject to a constraint. The constraint on the length has been replaced by a non-uniform weighting function. The Gibbs Partition Function, Y can be evaluated as

$$Y = \prod_{i=1}^{N} \left\{ \int_0^{2\pi} d\varphi_i \int_0^{\pi} \sin \theta_i \; d\theta_i \; \exp[\beta \mathcal{T} a \cos \theta_i] \right\}$$
$$= \prod_{i=1}^{N} \left\{ \int_0^{2\pi} d\varphi_i \int_{-1}^{1} d\cos \theta_i \; \exp[\beta \mathcal{T} a \cos \theta_i] \right\}$$
$$= \prod_{i=1}^{N} \left\{ 2\pi \int_{-1}^{1} d\cos \theta_i \; \exp[\beta \mathcal{T} a \cos \theta_i] \right\}$$
$$= \prod_{i=1}^{N} \left\{ 2\pi \left(\frac{\exp[\beta \mathcal{T} a] - \exp[-\beta \mathcal{T} a]}{\beta \mathcal{T} a} \right) \right\}$$
$$= \left[2\pi \left(\frac{\exp[\beta \mathcal{T} a] - \exp[-\beta \mathcal{T} a]}{\beta \mathcal{T} a} \right) \right]^N$$
$$= \left[4\pi \left(\frac{\sinh[\beta \mathcal{T} a]}{\beta \mathcal{T} a} \right) \right]^N \tag{4.290}$$

This has the form expected N for non-interacting monomers, where the only quantity which couples the monomers is the tension across the polymer. The reason for the form is recognized most clearly in the limit $\beta \, \mathcal{T} \, a \to 0$, where the monomers are expected to be distributed uniformly over the unit solid angle 4π. Since, in this limit,

$$\lim_{\beta \mathcal{T} \to 0} \left(\frac{\sinh[\beta \mathcal{T} a]}{\beta \mathcal{T} a} \right) \to 1 \qquad (4.291)$$

it is seen that Y reduces to the products of the unit solid angles for each monomer.

The Gibbs Free-Energy G is given by

$$Y = \exp[-\beta G] \qquad (4.292)$$

where in this case G is given by

$$G = U - TS - \mathcal{T}\bar{L} \qquad (4.293)$$

in which $\bar{L}$ is the average length. Since the change in the (thermodynamic) internal energy is given by

$$dU = T \, dS + \mathcal{T} \, d\bar{L} \qquad (4.294)$$

then G is a function of T and $\mathcal{T}$

$$dG = -S \, dT - \bar{L} \, d\mathcal{T} \qquad (4.295)$$

Furthermore, the average length can be determined from thermodynamics, as a partial derivative of G with respect to the tension $\mathcal{T}$ at constant T

$$\bar{L} = -\left(\frac{\partial G}{\partial \mathcal{T}} \right)_T \qquad (4.296)$$

or

$$\bar{L} = k_B T \left(\frac{\partial \ln Y}{\partial \mathcal{T}} \right)_T \qquad (4.297)$$

which leads to the expression for the average length as a function of tension

$$\bar{L} = k_B T N \frac{\frac{\partial}{\partial \mathcal{T}} \left(\frac{\sinh[\beta \mathcal{T} a]}{\beta \mathcal{T} a} \right)}{\left(\frac{\sinh[\beta \mathcal{T} a]}{\beta \mathcal{T} a} \right)} \qquad (4.298)$$

This length is evaluated as

$$\bar{L} = N a \left(\coth[\beta \mathcal{T} a] - \frac{1}{\beta \mathcal{T} a} \right) \qquad (4.299)$$

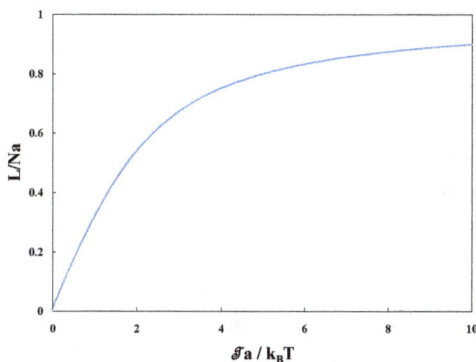

Fig. 4.9   The length-tension relation for a polymer.

It is seen that the effect of tension is that of extending the length of the polymer, whereas the temperature acts to contract the polymer. For small values of the ratio of the tension to temperature, $\beta \mathcal{T} a \ll 1$, the relationship becomes approximately linear

$$\overline{L} \approx Na \left(\frac{1}{3} (\beta \mathcal{T} a) - \frac{1}{45} (\beta \mathcal{T} a)^3 + \dots \right) \tag{4.300}$$

Therefore, for small tensions the polymer acts like a rubber band. The polymer will contract as the temperature is increased. However, for large values of the tension $\beta \mathcal{T} a \gg 1$ the average length saturates at the value Na as the length is given by

$$L \approx Na \left(1 - \frac{1}{\beta \mathcal{T} a} + \dots \right) \tag{4.301}$$

with exponentially small corrections. This occurs since, for large ratios of the tension to the temperature, all the segments are aligned parallel.

The Helmholtz Free-Energy F is found from

$$F = G + \mathcal{T}\overline{L} \tag{4.302}$$

as

$$F = -Nk_B T \, \ln \left(4\pi \, \frac{\sinh[\beta \mathcal{T} a]}{\beta \mathcal{T} a} \right)$$
$$- Nk_B T + Na\mathcal{T} \coth \beta \mathcal{T} a \tag{4.303}$$

where $\mathcal{T}$ should be expressed in terms of $\overline{L}$.

4.16   Problems

Problem 4.1

Consider a gas of N non-interacting particles moving freely in a d-dimensional volume $V = L^d$. The particles have the relativistic dispersion relation

$$\epsilon = c \, |\underline{p}|$$

and are in thermal equilibrium at a temperature T.
(i) Find the partition function Z_N.
(ii) Find the pressure P, energy U and the specific heat C_V as functions of temperature T.

Problem 4.2

(i) Use the Equipartition Theorem to find the contribution to the energy originating from the kinetic and potential energy terms of the anharmonic Hamiltonian

$$H = \sum_i \left(\frac{p_i^2}{2m} + k \, q_i^{2n} \right)$$

where n is an arbitrary positive integer.
(ii) Use the Equipartition Theorem to find the contribution to the energy originating from the kinetic and potential energy terms of the Hamiltonian

$$H = \sum_i \left(\frac{p_i^2}{2m} - k \, q_i^{-n} \right)$$

where n is an arbitrary positive integer.

Problem 4.3

Apply the Equipartition Theorem to find the contribution to the energy of a gas of ultra-relativistic particles moving in d-dimensional space

$$H = \sum_{n=1}^{N} c \, \sqrt{\underline{p}_n^2}$$

where c is the speed of light.
(i) Show that the kinetic energy per particle is given by dk_BT, which is twice the contribution from the kinetic energy of non-relativistic particles.

Problem 4.4

Consider a column of a classical ideal gas of N particles of mass m, in thermal equilibrium and in a uniform gravitational field g.
(i) Find the partition function Z_N.
(ii) Find the average energy of a particle.
(iii) The heat capacity of the particles.

Problem 4.5

Consider a classical diatomic gas. Each atom has a magnetic dipole moment μ oriented along the molecule's axis and the gas is subjected to an external magnetic field B. The rotational part of the molecular Hamiltonian H_r is given by

$$H_r = \frac{p_\theta^2}{2I} + \frac{p_\varphi^2}{2I \, \sin^2 \theta} - \mu B \cos \theta$$

(i) Evaluate the rotational factor, Z_r

$$Z_r = \int_{-\infty}^{\infty} \int_0^\pi \left(\frac{dp_\theta d\theta}{2\pi\hbar} \right) \int_{-\infty}^{\infty} \int_0^{2\pi} \left(\frac{dp_\varphi d\varphi}{2\pi\hbar} \right) \exp[-\beta H_r]$$

of the partition function of a single molecule.
(ii) Evaluate the magnetic moment M defined by

$$M = -\left(\frac{\partial F_r}{\partial B} \right)$$

Problem 4.6

A model a polyatomic molecule consists of a classical rigid object, with principle moments of inertia I_1, I_2 and I_3. The Hamiltonian can be expressed in terms of the Euler angles (θ, ϕ, ψ), and the associated generalized momenta via

$$H = \frac{1}{2I_1 \, \sin^2 \theta} \left[(p_\phi - \cos \theta p_\psi) \cos \psi - \sin \theta \sin \psi \, p_\theta \right]^2$$
$$+ \frac{1}{2I_2 \, \sin^2 \theta} \left[(p_\phi - \cos \theta p_\psi) \sin \psi + \sin\theta \cos \psi \, p_\theta \right]^2 + \frac{p_\psi^2}{2I_3}$$

Derive the rotational part of the partition function Z_r

$$Z_r = \sqrt{\pi} \left(\frac{2I_1 \, k_B T}{\hbar^2} \right)^{\frac{1}{2}} \left(\frac{2I_2 \, k_B T}{\hbar^2} \right)^{\frac{1}{2}} \left(\frac{2I_3 \, k_B T}{\hbar^2} \right)^{\frac{1}{2}}$$

Hint: The range of integration over the Euler angle θ is between 0 and π, whereas the integrals over ϕ and ψ have a range of 2π. It may be of interest to note that, if the rotational Hamiltonian for a classical rigid body is quantized, the system could have half-integer values of the angular momentum. Historically, this could have been used to justify Uhlenbeck and Goudsmidt's 1925 postulate for the existence of spin.

Problem 4.7

Consider a gas consisting of three coexisting gasses made of neutral Hydrogen atoms, ionized Hydrogen atoms and free electrons. The combined gas is electrically neutral and so contains equal total numbers of electrons and protons which are denoted by N. The three gasses are held at constant temperature T and occupy the same volume V. Neglect the effect of the Coulomb interaction and treat the gasses as being classical ideal gasses. The partition function of the combined system is given by

$$Z = Z_e \, Z_p \, Z_H$$

so the Free-Energy F is given by

$$F = F_e + F_p + F_H$$

(i) Using the relations between the number of un-ionized Hydrogen atoms N_H, ionized hydrogen atoms N_p and the number of free electrons N_e and N that express electrical neutrality and conservation of the total number of protons

$$N_e = N_p$$
$$N = N_H + N_p$$

show that the partition function depends on one independent variable particle number which can be chosen as N_e.

(ii) Determine the value of N_e that extremalizes F, hence, show that in equilibrium

$$\frac{N_e^2}{N - N_e} = \frac{V}{\lambda^3} \, \exp[-\beta R]$$

where λ is the thermal de Broglie wavelength for the electrons and R is the Rydberg constant. This equation is known as the Saha equation.

Problem 4.8

Consider an ideal gas of N identical classical particles of mass m confined in a cylinder of radius R and height h. The gas is in equilibrium with the cylinder which is rotating about its axis with constant angular velocity $\underline{\omega}$. An underlying assumption is that the interactions between the rotating gas particles and the walls of the rotating cylinder cause them to co-rotate.
(i) Show that the Lagrangian for the gas molecules can be expressed as

$$L = \frac{m}{2} \sum_{i=1}^{N} (\dot{z}_i^2 + \dot{\rho}_i^2 + \rho_i^2(\dot{\varphi}_i + \omega)^2)$$

when written in terms of cylindrical coordinates rotating with the cylinder.
(ii) Hence, show that in the rotating frame, the Hamiltonian is expressed as

$$H' = \sum_{i=1}^{N} \left(\frac{p_{z,i}^2}{2m} + \frac{p_{\rho,i}^2}{2m} + \frac{p_{\varphi,i}^2}{2m\rho_i^2} - p_{\varphi,i}\,\omega \right)$$

where the last term can be identified with the scalar product of the total angular momentum $\underline{L}$ with the angular velocity ω, that is $-\underline{L} \cdot \underline{\omega}$. The Hamiltonian can be re-written as

$$H' = \sum_{i=1}^{N} \left(\frac{p_{z,i}^2}{2m} + \frac{p_{\rho,i}^2}{2m} + \frac{(p_{\varphi,i} - m\rho_i^2\omega)^2}{2m\rho_i^2} - \frac{m\rho_i^2\omega^2}{2} \right)$$

The last term is the "fictitious" centrifugal potential. The expression shows that the kinetic energy associated with $p_{\varphi,i}$ is minimized when

$$p_{\varphi,i} = m\rho_i^2\,\omega$$

or $\dot{\varphi}_i = 0$. This reflects the underlying assumption that, in the rotating frame, angular momentum is conserved by the collisions between the particles and the cylinder's walls.
(iii) Write an expression for the partition function $Z_N(\omega)$ in the rotating frame.
(iv) Find the probability density $\rho(r)$

$$\rho(r) = \langle \delta(r - \rho_i) \rangle \tag{4.304}$$

for finding a particle at the radial distance $\rho_i = r$ from the center of the cylinder.

(v) Evaluate the Helmholtz Free-Energy $F'(\omega)$ and the internal energy $U'(\omega)$ in the rotating frame.

(vi) Determine the total angular momentum of the gas $\underline{L}$

$$\underline{L} = -\left(\frac{\partial F'}{\partial \underline{\omega}}\right)$$

(vii) The energy in the lab frame is given by

$$U = U' + \underline{L} \cdot \underline{\omega}$$

Hence, determine the energy U of the rotating gas.

Problem 4.9

Consider a two-dimensional ideal gas of N particles confined to move inside a circle of radius R. The state of the system is described by two conserved quantities; the energy E and the angular momentum L.

(i) Use Lagranges's method of undetermined multipliers to determine the probability densities P_α that maximize the entropy subject to the constraints. Hence, show that P_α has the form

$$P_\alpha = \frac{1}{Z_N} \exp[-\beta(H - \nu L)]$$

The Hamiltonian is given by

$$H = \sum_{i=1}^{N} \left(\frac{p_{x,i}^2 + p_{y,i}^2}{2m}\right)$$

and the angular momentum is given by

$$L = \sum_{i=1}^{N} (x_i\, p_{y,i} - y_i\, p_{x,i})$$

(ii) Show that the partition function Z_N is given by the expression

$$Z_N = \frac{1}{N!} \left(\frac{k_B T}{\hbar \nu}\right)^{2N} \left[\exp\left[\frac{m\nu^2 R^2}{2k_B T}\right] - 1\right]^N$$

(iii) Find a relation between the average angular momentum, $\overline{L}$, and ν.

(iv) Hence, determine the density of particles as a function of the radial distance.

Problem 4.10

An isotropic two-dimensional harmonic oscillator has the Lagrangian

$$L = \frac{m}{2} (\dot{r}^2 + r^2 \dot{\varphi}^2) - \frac{m\omega^2}{2} r^2$$

(i) Find an expression for the Hamiltonian and determine the constants of motion.

Consider an equilibrium Ensemble of N such systems, which has an average energy U and an average angular momentum L.

(ii) Show that the partition function can be written as

$$Z_N = \left(\int_0^\infty dr \int_0^\infty dp_r \int_0^{2\pi} d\varphi \int_{-\infty}^\infty dp_\varphi \; \frac{\exp[-\beta(H - \nu p_\varphi)]}{(2\pi\hbar)^2} \right)^N$$

What is the physically acceptable range of ν?

(iii) Hence, determine the internal energy, U, the entropy S and the average value of the angular momentum $\overline{L}$.

(iv) What is the probability of finding a particle in an interval dr at the radial distance r from the origin?

Problem 4.11

Consider an ideal gas of indistinguishable particles trapped in d-dimensional spatial region by a confining potential $V(\underline{r})$. The density of particles $\rho(\underline{r})$ is defined by

$$\rho(\underline{r}) = \sum_{i=1}^N \delta^d(\underline{r} - \underline{r}_i)$$

(i) Show that the density of particles at $\underline{r}$ is generally given by

$$\rho(\underline{r}) = N \frac{\exp[-\beta V(\underline{r})]}{\int d^d\underline{r}' \; \exp[-\beta V(\underline{r}')]}$$

(ii) Hence, show that if the confining potential is defined so that $V(0) = 0$, the entropy can be expressed as

$$S = N k_B \left[\left(\frac{d+2}{2} \right) - \ln(\rho(0) \, \lambda^d) - \frac{T}{\rho(0)} \left(\frac{\partial \rho(0)}{\partial T} \right) \right]$$

where λ is the thermal de Broglie wavelength and $\rho(0)$ is the density at $\underline{r} = 0$.

Problem 4.12

A quantum violin string vibrates at the fundamental frequency ω_0 and its higher harmonics. The frequency of the p-th harmonic is given by $\omega = \omega_0$. Each harmonic is an independent vibration mode. The excitation energy of the p-th harmonic is given by $E_p = n_p p \hbar \omega_0$, where the integer n_p is the number of quanta in the mode.

(i) Show that the partition function for the string can be written as

$$Z = \prod_{p=1}^{\infty} \left(\frac{1}{1 - \exp[-\beta p \hbar \omega_0]} \right)$$

(ii) Show that average excitation energy $\overline{E}$ is given by the expression

$$\overline{E} = \sum_{p=1}^{\infty} \frac{p \hbar \omega_0}{\exp[\beta p \hbar \omega_0] - 1}$$

while the Free-Energy is given by

$$F = k_B T \sum_{p=1}^{\infty} \ln(1 - \exp[-\beta p \hbar \omega_0])$$

(iii) Hence, find an expression for the entropy.

(iv) Evaluate the sum over p to show that, at high temperatures, the Free-Energy is given by

$$F = -\zeta(2) \left(\frac{k_B^2 T^2}{\hbar \omega_0} \right)$$

Problem 4.13

Consider the Frenkel Defect Model using the Canonical Ensemble. Denote the total number of atoms by N, the number of lattice sites by N, the number of interstitial sites by M and the number of interstitial defects by n. The energy of an atom on an interstitial site is denoted by ϵ.

(i) Show that the partition function Z_N is approximately given by

$$Z_N = \sum_{n=0}^{\min(N,M)} \binom{N}{n} \binom{M}{n} \exp[-\beta \epsilon n]$$

This summation cannot be easily evaluated in closed form. The above approximate expression for $Z_N(T)$ is compatible with the expression for $S(T)$

found from the Micro-Canonical Ensemble as can be seen by considering the inequalities

$$Z_N > \max_n \binom{N}{n}\binom{M}{n} \exp[-\beta \epsilon n]$$

$$Z_N < (N+1) \max_n \binom{N}{n}\binom{M}{n} \exp[-\beta \epsilon n]$$

in which it is recognized that summation is larger than its largest term and that the summation is smaller than $N+1$ times its largest term.

(ii) Using the above two inequalities together with the equality

$$\max_n \binom{N}{n}\binom{M}{n} \exp[-\beta \epsilon n] = \exp\left[-\frac{(U-TS)}{k_B T}\right]$$

and using the extensive character of the Helmholtz Free-Energy F

$$F = U - TS$$

show that the results of the Canonical and Micro-Canonical Ensembles are compatible.

Problem 4.14

Consider a set of quantum spins of magnitude S in a magnetic field.

(i) Determine the magnetization M^z defined by

$$M^z = -\left(\frac{\partial F}{\partial H^z}\right)$$

and the susceptibility $\chi^{z,z}$ which is defined as

$$\chi^{z,z} = \left(\frac{\partial M^z}{\partial H^z}\right)$$

(ii) Find the zero field limit of the susceptibility.

Problem 4.15

Consider a system of N magnetic moments held at temperature T. In the polar representation, a classical magnetic moment μ makes an angle θ with respect to the polar axis. In the presence of a magnetic field, B, aligned with the polar axis, the energy is given by

$$E = -\underline{\mu} \cdot \underline{B}$$

(i) Show that the average magnetization is given by

$$\overline{M}^z = N\mu\mathcal{L}(\beta\mu B)$$

where the Langevin function $\mathcal{L}(x)$ is given by

$$\mathcal{L}(x) = \coth x - \frac{1}{x}$$

(ii) Show that the quantum mechanical expression for the average magnetization for a system of spins with magnitude j and moment $\mu = g\mu_B j$ is given by

$$\overline{M}^z = N\mu\mathcal{B}_j(\beta\mu B)$$

where the Brillouin function $\mathcal{B}_j(x)$ is given by

$$\mathcal{B}_j(x) = \left[\left(\frac{2j+1}{2j}\right)\coth\left(\frac{2j+1}{2j}x\right) - \frac{1}{2j}\coth\left(\frac{x}{2j}\right)\right]$$

(iii) Show that the quantum mechanical magnetization reduces to the classical expression when $j \gg 1$.

Problem 4.16

(i) Show that the partition function Z_N of a one-dimensional Ising model with N spins and periodic boundary conditions, ie $S_{N+1}^z = S_1^z$, is given by

$$Z_N = 2^N \left(\cosh^N\left(\frac{\beta J\hbar^2}{4}\right) + \sinh^N\left(\frac{\beta J\hbar^2}{4}\right)\right)$$

(ii) Determine the energy E and the entropy S.
(iii) Evaluate the expressions for S and E in the limit $T \to 0$ when J is greater than zero.
(iv) Evaluate the $T \to 0$ limit of the expressions for S and E when N is odd and $J < 0$.

Problem 4.17

Random systems are often modeled by the Ising Hamiltonian in which the interactions J between the pairs of atoms are randomly given values $+J$ and $-J$. These competing interactions may result in the lowest-energy state being highly degenerate. Consider the one-dimensional Ising model with N spins and periodic boundary conditions, i.e. $S_{N+1}^z = S_1^z$.

(i) Show that if all the interactions have the value $+J$ and one interaction has the value of $-J$, then

$$Z_N = 2^N \left(\cosh^N \left(\frac{\beta J \hbar^2}{4} \right) - \sinh^N \left(\frac{\beta J \hbar^2}{4} \right) \right)$$

(ii) Show that in the $T \to 0$ limit, the lowest energy state is highly degenerate.

This phenomenon where competing interactions lead to the lowest-energy state having a large degeneracy is known as frustration. Frustration also occurs in higher-dimensions where the Free-Energy can be written as a sum of contributions from closed paths. The phenomenon of frustration can occur when either J has random or non-random values.

Problem 4.18

Find the density of states for particles moving in a three-dimensional space obeying the dispersion relation

$$\omega = ck^n \qquad \text{for } n > 0$$

Problem 4.19

(i) Find the degeneracy of the energy levels of three-dimensional harmonic potential with frequency ω_0 and energy E

$$E = \hbar\omega_0 \left(n_x + n_y + n_z + \frac{3}{2} \right)$$

where n_x, n_y and n_z are non-negative integers.
(ii) Derive the expression for the density of state $\rho(E)$

$$\rho(E) = \frac{1}{2\hbar\omega_0} \left[\left(\frac{E}{\hbar\omega_0} \right)^2 - \frac{1}{4} \right]$$

valid for $E \geq \hbar\omega_0$. Note that the above results is exact and has been obtained by partitioning an integer Q into three parts. The result can be seen to be exact since as $\rho(\frac{3}{2}\hbar\omega_0) = 1$ the above result does accurately include the ground state.

Problem 4.20

Consider the Debye Model of the specific heat, in which the Debye cut-off in the density of states is denoted by ω_D.

(i) Show that for $\hbar\omega_D \gg k_BT$

$$-\beta F/N = -\frac{9}{8}\left(\frac{\hbar\omega_D}{k_BT}\right) + \frac{\pi^4}{5}\left(\frac{k_BT}{\hbar\omega_D}\right)^3 + \dots$$

$$\beta U/N = \frac{9}{8}\left(\frac{\hbar\omega_D}{k_BT}\right) + \frac{3\pi^4}{5}\left(\frac{k_BT}{\hbar\omega_D}\right)^3 + \dots$$

$$S/(Nk_B) = \frac{4\pi^4}{5}\left(\frac{k_BT}{\hbar\omega_D}\right)^3 + \dots$$

(ii) Show that for $k_BT \gg \hbar\omega_D$

$$-\beta F/N = -\frac{9}{8}\left(\frac{\hbar\omega_D}{k_BT}\right) - 3\,\ln\left(\frac{k_BT}{\hbar\omega_D}\right) + 1\ \dots$$

$$\beta U/N = \frac{9}{8}\left(\frac{\hbar\omega_D}{k_BT}\right) + 3 + \dots$$

$$S/(Nk_B) = -3\,\ln\left(\frac{\hbar\omega_D}{k_BT}\right) + 4 + \dots$$

Problem 4.21

(i) Show that the thermal energy, per unit volume, of electromagnetic radiation with frequency ω in the range $d\omega$ is given by

$$\frac{U_\omega}{V}\,d\omega = \frac{\hbar\omega^3}{\pi c^3}\,(\exp[\beta\hbar\omega] - 1)^{-1}\,d\omega$$

The spectrum of emitted radiation from a perfect emitter is a universal function of temperature which was first devised by Planck [25].

(ii) Show that at high temperatures ($k_BT \gg \hbar\omega$) it reduces to the Rayleigh-Jeans Law [26], [27]

$$\frac{U_\omega}{V}\,d\omega \approx \frac{\omega^2}{\pi c^3}\,k_BT\,d\omega$$

in which each mode has an energy k_BT.

Problem 4.22

(i) Show that the average number of photons in an electromagnetic cavity of volume V with walls held at temperature T is given by

$$N = 16\pi\,\zeta(3)V\left(\frac{k_BT}{hc}\right)^3$$

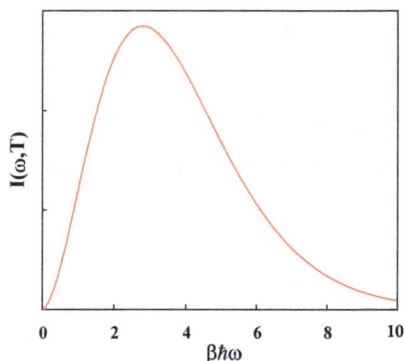

Fig. 4.10 The spectrum of electromagnetic radiation emitted from a cavity held at a temperature T.

where

$$\zeta(3) = \frac{1}{(n-1)!} \int_0^\infty dx \, \frac{x^{n-1}}{\exp[x] - 1}$$

(ii) Show that the energy U and the entropy S per photon is given by

$$\frac{U}{N} = \left(\frac{\pi^4}{30 \, \zeta(3)}\right) k_B T$$

$$\frac{S}{N} = \left(\frac{2\pi^4}{45 \, \zeta(3)}\right) k_B$$

The entropy per photon is a constant and is independent of temperature.

Problem 4.23

Since the Universe is expanding it is not in thermal equilibrium. The density of matter is so small that the microwave background can be considered to be decoupled from the matter. Therefore, the universe can be considered as an electromagnetic cavity filled with microwave radiation that obeys the Planck distribution with a temperature of approximately 2.725 K [28]. Because of the expansion of the universe, the photons are Doppler shifted. The Doppler shift reduces the photons' energies and squeezes the frequencies ω to new frequencies ω'.

(i) If the linear-dimension of the universe is changed from L to L', how does the frequency of a photon change?

(ii) Show that the spectral density of thermally activated photons $N_\omega \, d\omega$ retains its form after the expansion but that the temperature is changed from T to T'. What is the final temperature T' in terms of the initial temperature T and L'/L?

(iii) Assume that expansion of the universe is adiabatic, and then use thermodynamics to determine the change in temperature of the background microwave radiation when the linear-dimension L of the universe is increased to L'.

Problem 4.24

Consider a Hamiltonian describing a one-dimensional array of $N+1$ coupled quantum mechanical harmonic oscillators. The positions of the particles are denoted by n where $-\frac{N}{2} < n < \frac{N}{2}$ and their transverse displacements are denoted by q_n. The particle at the site $n = 0$ has mass M and all the other particles have mass m. The Hamiltonian is given by

$$\hat{H} = \frac{\hat{p}_0^2}{2M} + \sum_{n \neq 0} \frac{\hat{p}_n^2}{2m} + \sum_n \frac{m\omega_0^2}{2} \left(\hat{q}_n - \hat{q}_{n+1} \right)^2$$

(i) Use Bloch's Theorem to describe the amplitudes of the oscillations with wave vector k, in which the atom of mass M acts as an impurity that produces forward scattered and back scattered waves.

(ii) Show that the continuous part of the spectrum is described by the dispersion relation

$$\omega^2 = 4 \, \omega_0^2 \, \sin^2 \frac{ka}{2}$$

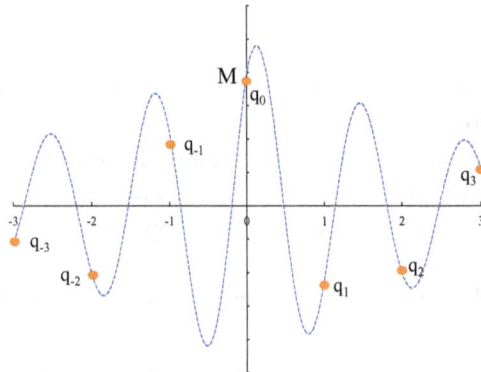

Fig. 4.11 A one-dimensional array of coupled harmonic oscillators. The particles all have mass m, except for the central particle which has mass M. The transverse displacements of the particles are labeled by q_n.

(iii) By considering the state at the origin, show that the ratio of the amplitude of the transmitted wave to incident wave is given by

$$T(k) = \frac{1}{1 + i(1 - \frac{M}{m})\tan\frac{ka}{2}}$$

(iv) Assume that the system is subjected to periodic boundary conditions at $n = \pm\frac{N}{2}$ and show that the values of k must satisfy the condition

$$Nka + \delta(k) = m\pi$$

where m is an integer and $\delta(k)$ is the phase shift.

(v) By inverting $\omega = \omega(k)$ show that the change in the density of states $\Delta\rho(\omega)$ induced by the impurity can be expressed in terms of the phase shifts $\delta(k)$. via

$$\Delta\rho(\omega) = \frac{1}{\pi}\frac{\partial\delta(k(\omega))}{\partial\omega}$$

(vi) Determine an expression for the change in the Free-Energy ΔF due to the impurity which comes from the continuous spectrum.

Problem 4.25

Consider a Hamiltonian describing a one-dimensional array of $N+1$ coupled harmonic oscillators. The positions of the particles are denoted by the index n where $-\frac{N}{2} < n < \frac{N}{2}$. The particle at the site $n = 0$ has mass M and all the other particles have mass m. The Hamiltonian is given by

$$\hat{H} = \frac{\hat{p}_0^2}{2M} + \sum_{n\neq0}\frac{\hat{p}_n^2}{2m} + \sum_n\frac{m\omega_0^2}{2}(\hat{q}_n - \hat{q}_{n+1})^2$$

(i) By examining the expression for the amplitudes of the transmitted and reflected waves and performing an analytic continuation of k from real to imaginary values, find a criterion (which involves M/m) for the spectrum to contain bound states.

(ii) By considering the change in the density of states of the continuous spectrum show that, when a bound state exist, the continuum spectrum has lost one state.

(iii) By analytic continuation of k to imaginary values and by assuming $\omega > 2\omega_0$, show that the amplitude of the local mode on the n-th site C_n is given in terms of the amplitude C_0 on the central site by

$$C_n = \exp[-\kappa|n|]\,C_0$$

where

$$\exp[-\kappa|n|] = \left[1 - \frac{\omega^2}{2\omega_0^2} \pm \sqrt{\frac{\omega^2(\omega^2 - 4\omega_0^2)}{4\omega_0^4}}\right]^{|n|}$$

Which value of the sign does one need to choose in order to describe vibrations localized around the impurity?

(iv) Show that the equation of motion for the oscillator at $n = 0$ can be expressed as

$$\left(2 - \frac{\omega^2}{\omega_0^2} \frac{M}{m}\right) C_0 = (C_1 + C_{-1})$$

(v) Hence, show that the bound state frequency is given by

$$\omega = \frac{2m}{\sqrt{M(2m - M)}} \omega_0$$

(vi) Determine the bound state contribution to the Free-Energy.

Problem 4.26

(i) Show that for an ideal gas, the energy fluctuations in the Canonical Ensemble are such that

$$\frac{\sqrt{\overline{\Delta E^2}}}{E} = \sqrt{\frac{2}{3N}}$$

(ii) Hence, show that the relative energy fluctuations vanish in the thermodynamic limit $N \to \infty$.

Problem 4.27

(i) Calculate $\overline{\Delta E^3}$ for a system in the Canonical Ensemble.
(ii) Hence, show that

$$\overline{\Delta E^3} = k_B^2 T^2 \left[T^2 \frac{\partial C_v}{\partial T} + 2TC_V\right]$$

and evaluate this for an ideal gas.

Problem 4.28

(i) Prove that

$$\overline{\Delta E^n} = (-1)^n \frac{\partial^n \ln Z}{\partial \beta^n}$$

holds for the Canonical Ensemble for $n = 2$ and $n = 3$. Show that the result for $n = 4$, involves $(\overline{\Delta E^2})^2$.

(ii) Hence, deduce that the higher-order moments of the energy fluctuations are all proportional to N.

Problem 4.29

Consider a classical system with discrete variables with a Hamiltonian H that is given by

$$H = H_0 + H_{int}$$

The partition function Z_H of the perturbed system is given by

$$Z_H = \exp[-\beta F_H]$$
$$= \text{Trace} \left[\exp[-\beta H]\right]$$

and the partition function Z_0 for the unperturbed system is

$$Z_0 = \exp[-\beta F_0]$$
$$= \text{Trace} \left[\exp[-\beta H_0]\right]$$

The change in Free-Energy ΔF_{int} can be calculated as

$$\exp[-\beta \Delta F_{int}] = \frac{Z_H}{Z_0}$$
$$= \text{Trace} \, \frac{\exp[-\beta H_{int}] \, \exp[-\beta H_0]}{Z_0}$$

(i) Show that the first few terms in the perturbation expansion for F_{int} are given by

$$-\beta \, \Delta F_{int} = \sum_n \frac{(-1)^n}{n!} \, A_n \, \beta^n$$

where

$$A_1 = \langle H_{int} \rangle_0$$
$$A_2 = \langle H_{int}^2 \rangle_0 - \langle H_{int} \rangle_0^2$$
$$A_3 = \langle H_{int}^3 \rangle_0 - 3\langle H_{int}^2 \rangle_0 \langle H_{int} \rangle_0 + 2\langle H_{int} \rangle_0^3$$

(ii) Relate the coefficients A_n to the cumulants of the interaction.

(iii) Will this analysis hold up for a quantum system?

Problem 4.30

Consider the Gaussian distribution which has the probability density $P(x)$

$$P(x) = \frac{1}{\sqrt{2\pi\sigma^2}} \exp\left[-\frac{(x-\bar{x})^2}{2\sigma^2}\right]$$

(i) Determine the moment and cumulant generating functions.
(ii) Show that the Gaussian distribution is characterized by only two non-zero cumulants.
(iii) Show that

$$\overline{x^{2n}} = (2n-1)!!\,\sigma^{2n}$$

Problem 4.31

Consider the sum of N identically distributed independent variables x_i which have zero mean. The normalized sum of the variables is given by X where

$$X = \frac{1}{\sqrt{N}} \sum_{i=1}^{N} x_i$$

(i) Show that the moment generating function $M_X(z)$ is given by in terms of the moment generating function of the x_i, $M_x(z)$ by

$$M_X(z) = M_x^N(z/\sqrt{N})$$

Hence, determine the cumulant generating function $K(z)$.
(ii) Show that the first non-zero cumulant of the distributions of X and x are identical. Furthermore, show that in the limit $N \to \infty$ the higher-order cumulants vanish.

Problem 4.32

Consider the two-dimensional Ising model on a square lattice with N lattice sites

$$H = \sum_{i,j} S_i^z S_j^z$$

where the sum is over pairs of nearest-neighbor spins. The partition function can be written as

$$Z_N(\beta J) = \text{Trace} \prod_{i,j} [\cosh \beta J(1 + S_i^z S_j^z \tanh \beta J)]$$

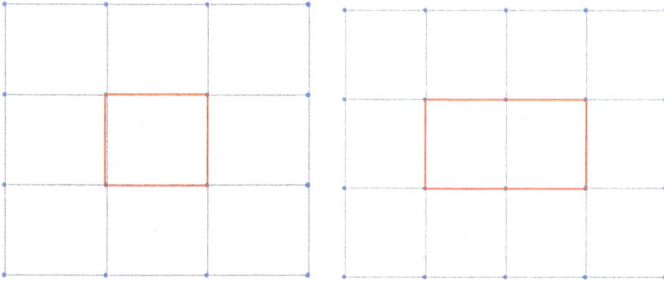

Fig. 4.12   The first two non-trivial terms in the high-temperature expansion of the 2d Ising Model. The expansion proceeds by categorizing non-overlapping closed configurations according to the number of bonds. The smallest loop corresponds to a square with four bonds and has degeneracy of unity, and the next excitation corresponds to a rectangle with six bonds and has a degeneracy of two.

The product can be expanded in powers of $\tanh \beta J$. The only terms that survive the Trace contain even powers of each spin and, therefore, must corresponds to closed polygons of bonds.

(i) Perform the Trace to show that the first few terms of the expansion of Z_N are given by

$$Z_N(\beta J) = 2^N \, (\cosh \beta J)^{2N} \left[1 + N \, \tanh^4 \beta J + 2N \, \tanh^6 \beta J \right.$$
$$\left. + \frac{1}{2} \, N(N+9) \, \tanh^8 \beta J + \ldots \right]$$

The factor of 2^N corresponds to the Trace over the N spins and the factor $\cosh^{2N} \beta J$ corresponds to the four interactions associated with every spin. It should be noted that every bond is shared between two spins. The factor $N \tanh^4 \beta J$ occurs from the bonds that form square plaquettes, there is one plaquette per lattice site, etc. This expansion may be expected to converge at sufficiently high temperatures.

(ii) Show that the terms of order N^2 can be decomposed into contributions from connected and disconnected loops. Using the fact that the Free-Energy is extensive, provide an argument which shows that the expansion of the Free-Energy can only involves the contribution from connected loops.

Consideration of the low temperature ferromagnetic state, shows that the maximal contribution to the trace occurs from the two states where every spin is aligned parallel to each other yielding a contribution to the low-temperature partition function of

$$2 \, \exp[2N\beta J]$$

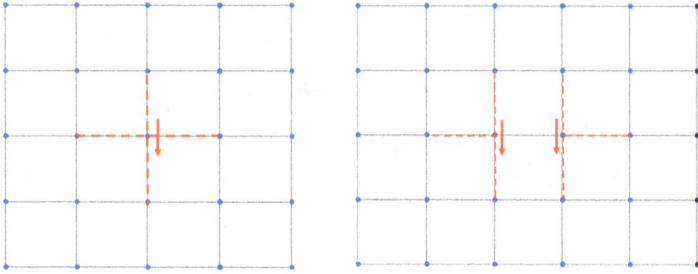

Fig. 4.13   The first two non-trivial terms in the low temperature expansion of the 2d Ising Model. The expansion proceeds by enumerating the low-energy excitations of the ferromagnetic state. The lowest-energy excitation corresponding to the reversal of one spin and has degeneracy of unity, and the next lowest-energy excitation corresponds to reversing two neighboring spins and has a degeneracy of two.

The other contributions to Z require one or more spins to be flipped. Flipping a single spin increases the energy by a factor of $8J$, since the signs of the four interactions on neighboring sites have changed from $-J$ to $+J$, and the spin which is flipped can be chosen to be on any one of the N lattice sites. The next-lowest energy excited state occurs when two nearest-neighbor spins have been flipped which has an excitation energy of $12\,J$ due to the interaction of the two flipped spins with the rest of their nearest-neighbors. The two nearest neighboring spins that are flipped can be chosen in $2N$ ways.

(iii) Continuing this expansion, valid at sufficiently low temperatures, show that

$$Z_N(\beta J) = 2\,\exp[2N\beta J]\left[1 + N\,\exp[-8\beta J] + 2N\,\exp[-12\beta J]\right.$$
$$\left. + \frac{1}{2}\,N(N+9)\,\exp[-16\beta J] + \dots\right]$$

It has been shown that the coefficients of each term in the low temperature series expansion must agree with the corresponding coefficient in the high temperature series expansion. Moreover, the terms in the square brackets of the high and low temperature expansion are equal under the transformation

$$\tanh\beta J \to \exp[-2\beta J]$$

Since the factors in Z_N outside the big square parenthesis are analytic, any non-analyticity must arise from the terms in the square brackets. The equality of the possibly non-analytic terms establishes a relation between the partition function at low temperature β^* with the partition function at a high temperature β. The relation between the temperatures is given by

$$\exp[-2\beta^* J] = \tanh \beta J$$

(iv) Show that the relation between the temperatures can be re-written in a more symmetric way as

$$\sinh 2\beta^* J = \frac{1}{\sinh 2\beta J}$$

This is known as the Kramers-Wannier duality.
(v) Use this duality to find the equation that relates $Z_N(\beta J)$ to $Z_N(\beta^* J)$.

Problem 4.33

Consider a Ising model on a hypercubic d-dimensional lattice with N lattice sites

$$H = \sum_{i,j} S_i^z S_j^z$$

where the sum is over pairs of nearest-neighbor spins. The partition function can be written as

$$Z_N(\beta J) = \text{Trace} \prod_{i,j} [\cosh \beta J (1 + S_i^z S_j^z \tanh \beta J)]$$

The product can be expanded in powers of $\tanh \beta J$. The only terms that survive the Trace contain even powers of each spin and, therefore, must corresponds to closed loops (polygons) of bonds.
(i) On performing the Trace show that the first few terms of the series expansion of Z_N are given by

$$Z_N(\beta J) = 2^N (\cosh \beta J)^{2N} \left[1 + N \frac{d(d-1)}{2!} \tanh^4 \beta J \right.$$

$$\left. + N \left(12 \frac{d(d-1)(d-2)}{3!} + 2 \frac{d(d-1)}{2!} \right) \tanh^6 \beta J + \dots \right.$$

and evaluate the term of order $\tanh^8 \beta J$.
(ii) Expand the partition function $Z_n(\beta J)$ for the d-dimensional Ising model about the low-temperature ferromagnetic state in powers of $\exp[-4\beta J]$. Evaluate the first few terms, and by examining them discuss whether it likely that the d-dimensional model exhibits a duality between the high and low temperature phases.

(iii) Are there any fundamental differences between the nearest-neighbor ferromagnetic and antiferromagnetic Ising models on either a simple hypercubic lattice or a body-centered hypercubic lattice? Are your conclusions valid for a face-centered hypercubic lattice?

(iv) Discuss, by using arguments concerning convergence, whether any non-analytic behavior of the Free-Energy is related to the distribution of long-length scale closed paths.

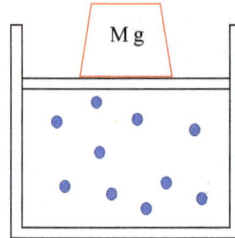

Fig. 4.14 N molecules of an ideal gas are contained inside a frictionless piston of area A. A mass of weight Mg is placed on the piston.

Problem 4.34

Consider an ideal gas composed of N indistinguishable molecules of mass m enclosed within a frictionless piston of area A. The piston has weight Mg placed on it which exerts a pressure P on the gas. The Hamiltonian of the combined system is given by

$$H = \sum_i \frac{p_i^2}{2m} + \left(\frac{P_z^2}{2M} + Mgz \right)$$

(i) Show that the partition function of the combined system is given by

$$Y_N(T, P) = \left(\frac{Ak_BT}{Mg\lambda^3} \right)^N \sqrt{\frac{k_B^3 T^3}{2\pi\hbar^2 Mg^2}}$$

(ii) Hence, find the Gibbs Free-Energy and, by using thermodynamic arguments, determine the volume of the gas.

Problem 4.35

A system has quantum states with discrete energy levels E_n and volumes V_n. The system and has an average energy U and average volume V.

(i) Show that the probability p_n that a state n is occupied is given by

$$p_n = \frac{1}{Y} \exp[-\beta(E_n + P\,V_n)]$$

where Y is a normalization constant.

(ii) On assuming that Y is the Gibbs Partition function, show that the parameters β and P correspond to the inverse temperature and pressure.

Problem 4.36

Using the Gibbs Ensemble, prove that the average square fluctuations in volume is given by

$$\overline{\Delta V^2} = k_B T \kappa_T \overline{V}$$

where κ_T is the isothermal compressibility

$$\kappa_T = -\frac{1}{\overline{V}}\left(\frac{\partial \overline{V}}{\partial P}\right)_T$$

and show that the relative volume fluctuation vanishes in the thermodynamic limit.

Problem 4.37

A linear polymer is made of N monomers arranged in a chain. Each monomer has an internal degree of freedom corresponding to two states. In state α, the energy of the monomer is E_α and the monomer has length a. In state β, the energy of the monomer is given by E_β and its length by b. The length of the polymer is denoted by L and a tension $\mathcal{T}$ is exerted on the polymer.

(i) Show that the Gibbs partition function Y can be written as

$$Y = \sum_{N_\alpha} \left(\frac{N!}{N_\alpha!(N - N_\alpha)!}\right) \exp[-\beta((E_\alpha - \mathcal{T}a)N_\alpha + (E_\beta - \mathcal{T}a)N_\beta)]$$

$$= (\exp[\beta(\mathcal{T}a - E_\alpha)] + \exp[\beta(\mathcal{T}b - E_\beta)])^N$$

(ii) Using the thermodynamic relation

$$\overline{L} = k_B T \left(\frac{\partial \ln Y}{\partial \mathcal{T}}\right)_{T,N}$$

calculate the average length $\overline{L}$ in terms of the tension $\mathcal{T}$ of the polymer.

Problem 4.38

Consider a static string composed of N segments that are equally spaced along the x-axis, and which is confined in a horizontal plane. The string is governed by the Hamiltonian

$$H = \sum_{i=1}^{N} \frac{\kappa}{2} (y_i - y_{i+1})^2$$

(i) Find the partition function if the end-points are free.

(ii) Determine the partition function if the end-end length to be fixed at L. That is, $y_1 = 0$, $y_{N+1} = L$.

4.17 Appendix: The Transfer Matrix Method: $d = 1$ Ising Model

We shall re-scale the interaction J, the magnetic field H^z and the spin variables S^z, so that the spin values are restricted to $S^z = \pm 1$. That is, we shall write $S^z \to \frac{\hbar}{2} S^z$, introduce a new variable $B = g\mu_B \frac{\hbar}{2} H^z$ and transform J such that $J\frac{\hbar^2}{4} \to J$. We shall also apply periodic boundary conditions to a one-dimensional chain of length N, so that $S^z_{N+1} = S^z_1$, so the system of spins becomes equivalent to a ring of spins with N links. The partition function can be written as the Trace over N factors each of which depends on the values of two neighboring spins

$$Z = \text{Trace} \left[\prod_{i=1}^{N} \{T(S^z_i, S^z_{i+1})\} \right] \tag{4.305}$$

where the factors T have been symmetrized wrt to the pairs of spins

$$T(S^z_i, S^z_{i+1}) = \exp \left[\beta J S^z_i S^z_{i+1} + \beta \frac{B}{2} (S^z_i + S^z_{i+1}) \right] \tag{4.306}$$

The factors $T(S^z_i, S^z_{i+1})$ can be regarded as elements of a matrix $\hat{T}$ (the transfer matrix)

$$T(S^z_i, S^z_{i+1}) = \langle S^z_i | \hat{T} | S^z_{i+1} \rangle \tag{4.307}$$

where $|S^z_i\rangle$ is a column vector

$$|S^z_{i+1}\rangle = \begin{pmatrix} \left(\frac{1+S^z_{i+1}}{2} \right) \\ \left(\frac{1-S^z_{i+1}}{2} \right) \end{pmatrix} \tag{4.308}$$

and $\langle S^z_i |$ is a row vector

$$\langle S^z_i | = \left(\left(\frac{1+S^z_i}{2} \right) \quad \left(\frac{1-S^z_i}{2} \right) \right) \tag{4.309}$$

Thus, the matrix $\hat{T}$ is given by

$$\hat{T} = \begin{pmatrix} \exp[+\beta(J + B)] & \exp[-\beta J] \\ \exp[-\beta J] & \exp[+\beta(J - B)] \end{pmatrix} \tag{4.310}$$

Since

$$\sum_{S^z_{i+1}} |S^z_{i+1}\rangle \langle S^z_{i+1}| = \hat{I} \tag{4.311}$$

where $\hat{I}$ is the unit 2×2 matrix

$$\hat{I} = \begin{pmatrix} 1 & 0 \\ 0 & 1 \end{pmatrix} \tag{4.312}$$

the summation over S^z_{i+1} in the expression can be performed as

$$\sum_{S^z_{i+1}} \langle S^z_i|\hat{T}|S^z_{i+1}\rangle\langle S^z_{i+1}|\hat{T}|S^z_{i+2}\rangle = \langle S^z_i|\hat{T}\hat{T}|S^z_{i+2}\rangle$$

$$= \langle S^z_i|\hat{T}^2|S^z_{i+2}\rangle \qquad (4.313)$$

Using the completeness property iteratively, the successive Traces over the variables S^z_i for $i = 2,\ldots,N$, in the expression for Z can be replaced by successive multiplication of the matrices $\hat{T}$. Thus

$$Z = \sum_{S^z_1=\pm 1} \langle S^z_1|\hat{T}^N|S^z_1\rangle \qquad (4.314)$$

The partition function can be expressed in terms of the eigenvalues λ_i of $\hat{T}$ defined by the eigenvalue equation

$$\hat{T}|\chi_i\rangle = \lambda_i|\chi_i\rangle \qquad (4.315)$$

for $i = 1$ or $i = 2$. We shall assume that $\lambda_1 > \lambda_2$. On defining a 2×2 matrix $\hat{S}$ as a row vector of the two column vectors $|\chi_1\rangle$ and $|\chi_2\rangle$

$$\hat{S} = \left(\,|\chi_1\rangle\ \ |\chi_2\rangle\,\right) \qquad (4.316)$$

then, as the similarity transform based on $\hat{S}$ diagonalizes $\hat{T}$, the inverse transformation is given by

$$\hat{T} = \hat{S}\begin{pmatrix} \lambda_1 & 0 \\ 0 & \lambda_2 \end{pmatrix}\hat{S}^{-1} \qquad (4.317)$$

Then, the partition function can be evaluated as

$$Z = \sum_{S^z_1=\pm 1} \langle S^z_1|\hat{S}\begin{pmatrix} \lambda^N_1 & 0 \\ 0 & \lambda^N_2 \end{pmatrix}\hat{S}^{-1}|S^z_1\rangle \qquad (4.318)$$

and on utilizing the cyclic invariance of the Trace, one finds the result

$$Z = \sum_{S^z_1=\pm 1} \langle S^z_1|\begin{pmatrix} \lambda^N_1 & 0 \\ 0 & \lambda^N_2 \end{pmatrix}|S^z_1\rangle$$

$$= \lambda^N_1 + \lambda^N_2 \qquad (4.319)$$

The Free-Energy F is given by

$$F = -k_BT\ \ln(\lambda^N_1 + \lambda^N_2)$$

$$= -k_BTN\ \ln\lambda_1 - k_BT\ \ln\left(1 + \frac{\lambda^N_2}{\lambda^N_1}\right) \qquad (4.320)$$

The second term can be neglected since it is of order unity. Thus,

$$F \approx -k_BTN\ \ln\lambda_1 \qquad (4.321)$$

The eigenvalues are determined from the secular equation

$$\begin{vmatrix} \exp[+\beta(J+B)] - \lambda & \exp[-\beta J] \\ \exp[-\beta J] & \exp[+\beta(J-B)] - \lambda \end{vmatrix} = 0 \qquad (4.322)$$

which can be expressed as the quadratic equation

$$\lambda^2 - 2\,\lambda\,\exp[\beta J]\,\cosh\beta B + 4\,\cosh\beta J\,\sinh\beta J = 0 \qquad (4.323)$$

The quadratic equation has the two solutions

$$\lambda = \exp[\beta J]\,\cosh\beta B \pm \sqrt{\exp[+2\beta J]\,\sinh^2\beta B + \exp[-2\beta J]} \qquad (4.324)$$

Thus, the final result for the Free-Energy is given by

$$F = -N\,k_B T\,\ln\left[\exp[\beta J]\,\cosh\beta B \right.$$
$$\left. + \sqrt{\exp[+2\beta J]\,\sinh^2\beta B + \exp[-2\beta J]}\,\right] \qquad (4.325)$$

The magnetization is defined as

$$M = -\left(\frac{\partial F}{\partial B}\right) \qquad (4.326)$$

which yields

$$M = N\,\frac{\exp[\beta J]\,\sinh\beta B}{\sqrt{\exp[+2\beta J]\,\sinh^2\beta B + \exp[-2\beta J]}} \qquad (4.327)$$

Since the Free-Energy is an analytic function of B and T, and as $M(B,T)$ is also an analytic function of B for all T, the system does not exhibit a phase transition at any finite T.

Chapter 5

The Grand-Canonical Ensemble

The "*Grand-Canonical Ensemble*" allows one to thermally average over a system which is able to exchange both energy and particles with its environment. The probability distribution function for the Grand-Canonical Ensemble can be derived by considering the closed system consisting of a subsystem and its larger environment. The probability distribution for the total closed system is calculated in the Micro-Canonical Ensemble. It is assumed that the number of particles and the energy can be uniquely partitioned into contributions from either the subsystem or its environment

$$E_T = E + E_R$$
$$N_T = N + N_R \tag{5.1}$$

Likewise, one assumes that for a given value of N, the infinitesimal phase space volume element can also be uniquely partitioned into factors representing the system and its environment. In this case, the probability dp for finding the entire closed system in a volume element $d\Gamma_T$ of its phase space

$$dp = \rho_{mc} \, d\Gamma_T \tag{5.2}$$

can be expressed as

$$dp = \rho_{mc} \, d\Gamma_N \, d\Gamma_{R,N_T-N} \tag{5.3}$$

where the subsystem's phase space element is composed of the contributions from N particles and has energy E, while the reservoir has $N_T - N$ particles and has energy $E_T - E$. Since we are only interested in the probability distribution for finding the subsystem in the a volume element $d\Gamma_N$ corresponding to having N particles and energy E and are not interested in the environment, we shall integrate over the phase space available to the environment. This results in the probability for finding the subsystem in a state with N particles and in a volume of phase space $d\Gamma_N$ with energy E being given by

$$dp = \rho_{mc} \, \Gamma_{R,N_T-N}(E_T - E) \, d\Gamma_N \tag{5.4}$$

where $\Gamma_{R,N_T-N}(E_T - E)$ is the volume of accessible phase space for the reservoir which has $N_T - N$ particles and energy $E_T - E$. The Micro-Canonical Distribution Function ρ_{mc} can be expressed as

$$\rho_{mc} = \frac{1}{\Gamma_{T,N_T}(E_T)} \tag{5.5}$$

where $\Gamma_{T,N_T}(E_T)$ is the entire volume of phase space accessible to the closed system with energy E_T. Since $\Gamma_{T,N_T}(E_T)$ can be expressed in terms of the total entropy of the closed system $S_T(E_T)$, the Micro-Canonical Distribution Function can be expressed as

$$\rho_{mc} = \frac{1}{\Gamma_{N_T,0}} \, \exp[-S_T(E_T, N_T)/k_B] \tag{5.6}$$

where $\Gamma_{N_T,0}$ is the volume of phase space which represents one microscopic state of the system with N_T particles. The volume of accessible phase space for the reservoir can also be written in terms of its entropy

$$\Gamma_{R,N_T-N}(E_T - E) = \Gamma_{N_R,0} \, \exp[S_R(E_T - E, N_T - N)/k_B] \tag{5.7}$$

where $\Gamma_{N_R,0}$ is the volume of phase space which represents a single microscopic state of the reservoir which contains N_R particles. Hence, the probability dp for finding the N particle system in an infinitesimal volume of phase space $d\Gamma_N$ with energy E is given by

$$dp_{N,E} = \left(\frac{dp}{d\Gamma_N}\right) d\Gamma_N$$

$$= \exp[(S_R(E_T - E, N_T - N) - S_T(E_T, N_T))/k_B] \, \frac{d\Gamma_N}{\Gamma_{N,0}} \tag{5.8}$$

where we have assumed that

$$\Gamma_{N_T,0} = \Gamma_{N_R,0} \, \Gamma_{N,0} \tag{5.9}$$

Since the environment has been assumed to be much larger than the subsystem both E and N are small compared to E_T and N_T. Therefore, it is reasonable to assume that the entropy of the reservoir can be Taylor expanded in powers of the fluctuations of E from the thermodynamic value U and the fluctuations of N from its thermodynamic value $\overline{N}$.

$$S_R(E_T - E, N_T - N) = S_R(E_T - U, N_T - \overline{N})$$

$$+ (U - E) \left(\frac{\partial S_R(E_R, N_T - \overline{N})}{\partial E_R} \right) \Bigg|_{E_T - U}$$

$$+ (\overline{N} - N) \left(\frac{\partial S_R(E_T - U, N_R)}{\partial N_R} \right) \Bigg|_{N_T - \overline{N}} + \ldots$$

$$(5.10)$$

On using the definitions of the thermal reservoir's temperature

$$\left(\frac{\partial S_R(U_R, \overline{N}_R)}{\partial U_R} \right)_{\overline{N}_R} = \frac{1}{T} \qquad (5.11)$$

and its chemical potential

$$\left(\frac{\partial S_R(U_R, \overline{N}_R)}{\partial \overline{N}_R} \right)_{U_R} = -\frac{\mu}{T} \qquad (5.12)$$

the expansion becomes

$$S_R(E_T - E, N_T - N) = S_R(E_T - U, N_T - \overline{N})$$

$$+ \frac{(U - E)}{T} - \frac{\mu(\overline{N} - N)}{T} + \ldots \qquad (5.13)$$

where μ and T are the chemical potential and temperature of the reservoir. The total entropy $S_T(E_T, N_T)$ is extensive and can be decomposed as

$$S_T(E_T, N_T) = S_R(E_T - U, N_T - \overline{N}) + S(U, \overline{N}) \qquad (5.14)$$

Thus, the Grand-Canonical Distribution Function can be written as

$$\left(\frac{dp}{d\Gamma} \right)_{N,E} \Gamma_{N,0} = \exp[-\beta(E - \mu N)] \, \exp[\beta(U - \mu \overline{N}) - S(U, \overline{N})/k_B] \qquad (5.15)$$

or

$$\left(\frac{dp}{d\Gamma} \right)_{N,E} \Gamma_{N,0} = \exp[-\beta(E - \mu N)] \, \exp[\beta\Omega] \qquad (5.16)$$

where Ω is the Grand-Canonical Potential $\Omega(T, V, \mu)$

$$\Omega = U - TS - \mu \overline{N} \qquad (5.17)$$

which describes the thermodynamics of the subsystem. Once again, we note that the quantities E, N and Ω in the probability distribution function are properties of the system and that the only quantities which describe the environment are the temperature T and the chemical potential μ. The Grand-Canonical Partition Function Ξ is defined by

$$\Xi = \exp[-\beta\Omega] \tag{5.18}$$

so the Grand-Canonical Distribution Function can be written as

$$\left(\frac{dp}{d\Gamma}\right)_N \Gamma_{N,0} = \frac{1}{\Xi} \exp[-\beta(H_N - \mu N)] \tag{5.19}$$

where H_N is the Hamiltonian for the N particle system. The exponential factor containing the Hamiltonian automatically provides different weights for the regions of N-particle phase space. The quantity

$$dp_N = \left(\frac{dp}{d\Gamma}\right)_N d\Gamma_N \tag{5.20}$$

is the probability for finding the subsystem to have N particles and be in the volume of phase space $d\Gamma_N$. Hence, the Grand-Canonical Probability Distribution Function can be used in determining the average of any function defined on the N-particle phase space A_N via

$$\overline{A} = \sum_{N=0}^{\infty} \int d\Gamma_N \left(\frac{dp}{d\Gamma}\right)_N A_N \tag{5.21}$$

or, more explicitly

$$\overline{A} = \frac{1}{\Xi} \sum_{N=0}^{\infty} \int \frac{d\Gamma_N}{\Gamma_{N,0}} \exp[-\beta(H_N - \mu N)] \, A_N \tag{5.22}$$

and involves an integration over the N-particle phase space and a summation over all possible particle numbers.

5.1   The Grand-Canonical Partition Function

The quantity

$$dp_N = \left(\frac{dp}{d\Gamma}\right)_N d\Gamma_N \tag{5.23}$$

is the probability for finding the system to have N particles in the volume of phase space $d\Gamma_N$. The probability p_N for finding the system as having N particles anywhere in its phase space is found by integrating over all $d\Gamma_N$

$$p_N = \frac{1}{\Xi} \int \frac{d\Gamma_N}{\Gamma_{N,0}} \exp[-\beta(H_N - \mu N)]$$

$$= \exp[\beta\mu N] \int \frac{d\Gamma_N}{\Gamma_{N,0}} \exp[-\beta H_N]$$

$$= \frac{1}{\Xi} \exp[\beta\mu N] Z_N \tag{5.24}$$

Since the probability p_N must be normalized, one requires that

$$\sum_{N=0}^{\infty} p_N = 1 \tag{5.25}$$

since a measurement of the number of particles in the system will give a result which is contained in the set $0, 1, 2, \ldots, \infty$. This normalization condition determines Ξ as being given by

$$\Xi = \sum_{N=0}^{\infty} \int \frac{d\Gamma_N}{\Gamma_{N,0}} \exp[-\beta(H_N - \mu N)]$$

$$= \sum_{N=0}^{\infty} \exp[\beta\mu N] \int \frac{d\Gamma_N}{\Gamma_{N,0}} \exp[-\beta H_N]$$

$$= \sum_{N=0}^{\infty} \exp[\beta\mu N] Z_N \tag{5.26}$$

which relates the Grand-Canonical Partition Function Ξ to a sum involving the Canonical Partition Functions for the N particle systems Z_N.

5.2 Ensemble Averages and Ω

The above normalization condition can be used to evaluate Ξ and, hence, the Grand-Canonical Potential Ω. One can then evaluate thermodynamic averages directly from $\Omega(T, V, \mu)$, via

$$\Omega = -k_B T \, \ln \Xi \tag{5.27}$$

Thus for example, knowing Ω one can find the average number of particles $\overline{N}$ via the thermodynamic relation

$$\overline{N} = -\left(\frac{\partial \Omega}{\partial \mu}\right)_T \tag{5.28}$$

which can be expressed as

$$\overline{N} = k_B T \left(\frac{\partial \ln \Xi}{\partial \mu} \right)_T$$

$$= k_B T \frac{1}{\Xi} \left(\frac{\partial \Xi}{\partial \mu} \right)_T$$

$$= k_B T \frac{1}{\Xi} \sum_{N=0}^{\infty} \left(\frac{\partial \exp[\beta \mu N] \, Z_N}{\partial \mu} \right)_T$$

$$= \frac{1}{\Xi} \sum_{N=0}^{\infty} N \, \exp[\beta \mu N] \, Z_N \tag{5.29}$$

Hence, p_N defined by

$$p_N = \frac{1}{\Xi} \, \exp[\beta \mu N] \, Z_N \tag{5.30}$$

appears to be the probability for the system to have N particles, as

$$\overline{N} = \sum_{N=0}^{\infty} N p_N \tag{5.31}$$

Since Z_N is given by

$$Z_N = \int \frac{d\Gamma_N}{\Gamma_{N,0}} \, \exp[-\beta H_N] \tag{5.32}$$

on substituting in the expression for p_N and combining the exponentials, we can express p_N as

$$p_N = \frac{1}{\Xi} \int \frac{d\Gamma_N}{\Gamma_{N,0}} \, \exp[-\beta(H_N - \mu N)] \tag{5.33}$$

which is in accord with our previous identification.

More generally, given the average of a quantity $\overline{A}$ defined by

$$\overline{A} = \frac{1}{\Xi} \sum_{N=0}^{\infty} \int \frac{d\Gamma_N}{\Gamma_{N,0}} \, \exp[-\beta(H_N - \mu N)] \, A_N \tag{5.34}$$

and on defining the thermodynamic average of A in the Canonical Ensemble with N particles as

$$\overline{A}_N = \frac{1}{Z_N} \int \frac{d\Gamma_N}{\Gamma_{N,0}} \, \exp[-\beta H_N] \, A_N \tag{5.35}$$

one finds that

$$\overline{A} = \frac{1}{\Xi} \sum_{N=0}^{\infty} \exp[\beta\mu N] \, Z_N \, \overline{A}_N$$

$$= \sum_{N=0}^{\infty} p_N \, \overline{A}_N \tag{5.36}$$

as is expected.

5.3 The Ideal Gas

The Grand-Canonical Partition Function Ξ is given by

$$\Xi = \sum_{N=0}^{\infty} \frac{\exp[\beta\mu N]}{N!(2\pi\hbar)^{dN}} \int d\Gamma_N \, \exp[-\beta H_N]$$

$$= \sum_{N=0}^{\infty} \exp[\beta\mu N] \, Z_N \tag{5.37}$$

However, for an ideal gas the Canonical Partition Function Z_N is given by

$$Z_N = \frac{1}{N!} \left(\frac{V}{\lambda^d}\right)^N \tag{5.38}$$

Therefore, one has

$$\Xi = \sum_{N=0}^{\infty} \frac{1}{N!} \left(\frac{\exp[\beta\mu] \, V}{\lambda^d}\right)^N$$

$$= \exp\left[\frac{\exp[\beta\mu] \, V}{\lambda^d}\right] \tag{5.39}$$

This leads to the expression for the Grand-Canonical Potential Ω

$$\Omega = -k_B T \, \ln \Xi$$

$$= -k_B T \, \exp[\beta\mu] \, \frac{V}{\lambda^d} \tag{5.40}$$

The average number of particles is given by

$$\overline{N} = -\left(\frac{\partial \Omega}{\partial \mu}\right)_{V,T}$$

$$= \exp[\beta\mu] \, \frac{V}{\lambda^d} \tag{5.41}$$

Thus, the chemical potential μ is given by the equation

$$\mu = k_B T \, \ln \left(\frac{\overline{N} \lambda^d}{V} \right) \tag{5.42}$$

which is identical to the result found by using the Canonical Ensemble. Furthermore, on using the thermodynamic relation

$$P = -\left(\frac{\partial \Omega}{\partial V} \right)_{\mu, T}$$

$$= \exp[\beta\mu] \, \frac{k_B T}{\lambda^d} \tag{5.43}$$

and on combining with the expression for $\overline{N}$, one recovers the ideal gas law

$$P = \overline{N} \, \frac{k_B T}{V} \tag{5.44}$$

5.4 Equilibria of Chemical Reactions

Consider several species of molecules which are in equilibrium. The molecules may undergo chemical reactions such as

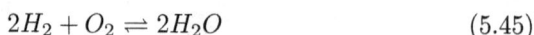

$$2H_2 + O_2 \rightleftharpoons 2H_2O \tag{5.45}$$

If the type of n-th molecule in the reaction is denoted by the symbol A_n and its stoichiometric coefficient is denoted by ν_n, the reaction can be expressed as

$$\sum_{\nu} \nu_n A_n = 0 \tag{5.46}$$

where the coefficients of initial substances have positive signs and the end products have negative signs. In this example, $A_1 = H_2$, $A_2 = O_2$ and $A_3 = H_2O$, with $\nu_1 = 2$, $\nu_2 = 1$ and $\nu_3 = -2$. In general there may be a number of different chemical reactions that could occur in the system, which would be represented by a number of such equations. We shall determine the concentration of the different molecular species in chemical equilibrium.

For a process at constant temperature and pressure, the Gibbs Free-Energy is extremal and satisfies

$$dG = \sum_{n} \mu_n \, dN_n = 0 \tag{5.47}$$

In the reaction the change in the number of molecules of n-th species, $dN_n = \nu_n \, dx$, is proportional to dx, therefore dx can be factorized out.

Hence, the equilibrium condition reduces to

$$\sum_n \mu_n \, \nu_n = 0 \qquad (5.48)$$

The above equation relates the chemical potentials of the molecules for a system in chemical equilibrium. Given the Grand-Canonical partition functions Ξ_n for the component gasses, one may calculate the number of molecules N_n of each species as a function of μ_n. If this relation can be inverted, then one may express μ_n in terms of a function of N_n, or preferably the concentration c_n. When the relation between μ_n and c_n is substituted into Eq. (5.48), one obtains the law of mass action.

Since the gasses occupy the same volume and are held at the same temperature, the Grand-Canonical partition for the gas mixture Ξ can be expressed as a product over the molecular species

$$\Xi = \prod_n (\Xi_n) \qquad (5.49)$$

For a mixture of ideal gasses, the Grand-Canonical partition function can be written as

$$\Xi = \exp\left[\sum_n Z_n^{(1)} \, \exp[\beta \mu_n] \right] \qquad (5.50)$$

where $Z_n^{(1)}$ is the partition function for a single molecule of species n which includes its internal degrees of freedom. The single molecule partition function has the form

$$Z_n^{(1)} = \frac{V}{\lambda_n^d} \, Z_n^{int} \qquad (5.51)$$

where λ_n is the thermal de Broglie wavelength for species n. The average number of molecules of the n-th species, $\overline{N}_n$, is given by

$$\overline{N}_n = Z_n^{(1)} \, \exp[\beta \mu_n] \qquad (5.52)$$

The above equation can be solved for μ_n, merely by taking the logarithm, yielding

$$\mu_n = k_B T \, \ln\left(\frac{\overline{N}_n}{Z_n^{(1)}} \right) \qquad (5.53)$$

It should be noted that the arguments of the logarithms are intensive quantities and can be expressed in terms of intensive quantities such as P, T. On substituting the above expressions for μ_n and $Z_n^{(1)}$ into the equation

$$\sum_n \mu_n \, \nu_n = 0 \qquad (5.54)$$

and cancelling a common factor of $k_B T$, one obtains

$$\sum_n \nu_n \ln \left(\frac{\overline{N}_n \lambda_n^d}{V Z_n^{int}} \right) = 0 \tag{5.55}$$

Using Dalton's law of partial pressures

$$PV = \sum_n \overline{N}_n k_B T$$

$$= N k_B T \tag{5.56}$$

to eliminate V, and then on expressing the concentrations c_n of the different chemical species as

$$c_n = \frac{\overline{N}_n}{N} \tag{5.57}$$

one obtains the relation

$$\sum_n \nu_n \ln c_n = -\sum_n \nu_n \ln \left[\frac{P \lambda_n^d}{k_B T Z_n^{int}} \right] \tag{5.58}$$

On exponentiating the above equation, one finds a product involving the concentrations of the different species

$$\prod_n c_n^{\nu_n} = \left(\frac{k_B T}{P} \right)^{\sum_n \nu_n} \prod_n \left[\frac{Z_n^{int}}{\lambda_n^d} \right]^{\nu_n}$$

$$= K(P, T) \tag{5.59}$$

which is the law of mass action. The function $K(P, T)$ is the equilibrium constant of the chemical reaction. It depends on the properties of the reacting molecules. It should be noted that the pressure dependence of $K(P, T)$ is governed by the sum $\sum_n \nu_n$. Typically, the pressure dependence is determined by the least abundant species, whereas if the sum is zero the equilibrium concentrations are independent of pressure.

5.5 The Einstein Coefficients

Since the number of photons is not conserved, in equilibrium the number of photons should be calculated by extremalizing the Free-Energy w.r.t. the number of photons. Hence, the chemical potential μ is set to zero. In processes involving the emission and absorption of photons, it is self-evident that the number of photons does fluctuate. Dirac calculated the quantum mechanical transition rates for emission and absorption of photons by atoms, using a fully quantized theory of the electromagnetic field [29].

Dirac's calculation was carried out in Fock space using eigenstates of the photon mode number operators. Dirac proved that the rate of emission of electromagnetic radiation has two components, one component represents spontaneous emission in which the electron in an atom makes a transition from a higher-energy to a lower-energy state. The other contribution occurs due to stimulated emission of radiation in which pre-existing electromagnetic radiation of the same frequency induces the atomic electron to make a transition from a higher-energy to a lower-energy state. This latter process can be considered to be the time-reversal partner of absorption of electromagnetic radiation, which obviously cannot proceed without the presence of electromagnetic radiation.

Previously, Einstein [30] had consider the emission and absorption processes and had reached a similar conclusion using thermodynamic arguments. Einstein was able to determine two constants, independent of frequency or temperature, that describe the probabilities of spontaneous emission and stimulated emission. Einstein's argument was based on the *"Principle of Detailed Balance"*, which assumes that equilibrium is a dynamic process in which transitions do occur, however the transitions do not lead to a noticeable change in the macroscopic equilibrium state. That is, in the macroscopic equilibrium state, the probabilities that the electrons are in the various atomic states are time-independent. This *"Dynamic Equilibrium"* requires that the number of transitions, per unit time, in which electrons lower their energies are balanced by the number of transitions, per unit time, in which the electrons increase their energies.

Consider an atom with two levels 1 and 2 with energies E_1 and E_2. The energy of state 2, E_2, is higher than the energy of state 1, E_1 so, $E_2 > E_1$. The population of state 1 is denoted by p_1 and the population of state 2 is denoted by p_2. In equilibrium, one would expect that the ratio p_2/p_1 is governed by the Boltzmann distribution. Hence,

$$\frac{p_2}{p_1} = \frac{D_2}{D_1} \exp[-\beta(E_2 - E_1)] \tag{5.60}$$

where D_2 and D_1 are the degeneracies of the respective states. Energy conservation for the emission and absorption of photons with frequency ω leads to the ratio being given by

$$\frac{p_2}{p_1} = \frac{D_2}{D_1} \exp[-\beta\hbar\omega] \tag{5.61}$$

Emission Processes

In the emission process the atom makes a transition from state 2 to state 1 by emitting a photon of energy $\hbar\,\omega$, given by

$$E_2 - E_1 = \hbar\omega \tag{5.62}$$

The transition can occur spontaneously by emitting a photon. The change in population p_2, per unit time, due to spontaneous emission is given by

$$-p_2\,A_{2\to1} \tag{5.63}$$

where A is the Einstein A coefficient. The transition can also occur by stimulated emission, where the presence of photons with frequency ω induces the transition to occur. The change in the population p_2 due to stimulated emission is given by

$$-p_2\,B_{2\to1}\,u(\omega) \tag{5.64}$$

where $u(\omega)$ is the energy density of the electromagnetic radiation with frequency ω and $B_{2\to1}$ is the Einstein B coefficient. Both processes lead to the emission of a photon and produce an increase of one in the total number of photons.

Absorption Process

In the absorption process an atom makes a transition from state 1 to state 2 by the absorption of a photon. The population of the lower energy state decreases but the population of the energy level 2 increases by an amount

$$p_1\,B_{1\to2}\,u(\omega) \tag{5.65}$$

where $B_{1\to2}$ is the Einstein B coefficient. Time-reversal symmetry would suggest that $B_{1\to2} = B_{2\to1}$.

5.5.1 *Detailed Balance*

The Einstein coefficients A and B coefficients are transition probabilities per unit time. The coefficients have definite values which depend on the atom and the properties of the two atomic states. The relation between the coefficients can be found by using the principle of detailed balance. The principle of detailed balance asserts that thermodynamic equilibrium is a dynamic equilibrium, in the sense that transitions do occur between the two atomic states, but that there is no net change in the populations of the states as is required for the system to be in equilibrium. Therefore, the number of transitions, per unit time, which decrease the thermal population of level 2 must be equal to the number of transitions, per unit time, that increase the population of level 2. Hence, on equating the rates one must have

$$p_1 \, B_{1 \to 2} \, u(\omega) = p_2 \, (B_{2 \to 1} \, u(\omega) + A_{2 \to 1}) \tag{5.66}$$

Substitution of the ratio of the thermal populations p_2/p_1 in the above equation leads to

$$B_{1 \to 2} \, u(\omega) = \frac{D_2}{D_1} \, \exp[-\beta \hbar \omega] \, (B_{2 \to 1} \, u(\omega) + A_{2 \to 1}) \tag{5.67}$$

This equation must be true for every temperature. The electromagnetic energy density $u(\omega)$ is given by

$$u(\omega) = \left(\frac{\hbar \omega^3}{\pi^2 c^3} \right) \frac{1}{\exp[\beta \hbar \omega] - 1} \tag{5.68}$$

Substituting the energy density in the detailed balance equation and multiplying the equation by the factor

$$\exp[\beta \hbar \omega] - 1 \tag{5.69}$$

leads to

$$\left(\frac{\hbar \omega^3}{\pi^2 c^3} \right) \frac{D_1}{D_2} \, B_{1 \to 2} \, \exp[\beta \hbar \omega] = (\exp[\beta \hbar \omega] - 1) \, A_{2 \to 1}$$
$$+ \left(\frac{\hbar \omega^3}{\pi^2 c^3} \right) B_{2 \to 1} \tag{5.70}$$

Since this equation must be satisfied for all T, the exponential terms must balance

$$\frac{D_1}{D_2} \, B_{1 \to 2} = A_{2 \to 1} \tag{5.71}$$

and the constant term must also balance

$$A_{2 \to 1} = \left(\frac{\hbar \omega^3}{\pi^2 c^3} \right) B_{2 \to 1} \tag{5.72}$$

The above equations express the relations between the Einstein coefficients.

The above relations can be substituted into the expression for the transition rates for the combined (spontaneous and stimulated) emission process. This leads to the emission rate, per state, being given by

$$B_{2 \to 1} \, u(\omega) + A_{2 \to 1} = A_{2 \to 1} \left(\frac{1}{\exp[\beta \hbar \omega] - 1} + 1 \right) \tag{5.73}$$

On recognizing that the thermal average number of photons of frequency ω, $N(\omega)$, is given by

$$N(\omega) = \frac{1}{\exp[\beta\hbar\omega] - 1} \tag{5.74}$$

then the combined emission rate can be expressed as

$$A_{2\to 1}(N(\omega) + 1) \tag{5.75}$$

The absorption rate, per state, can then be expressed as

$$A_{2\to 1}\, N(\omega) \tag{5.76}$$

which is proportional to the average number of photons of frequency ω.

Dirac's theory shows that the expressions remain valid when the electromagnetic cavity is not in thermal equilibrium, if the thermally averaged number of photons $N(\omega)$ is replaced the initial (non-equilibrium) number. That is, the Einstein relations must hold in general, even when the electromagnetic radiation is not in thermal equilibrium with the atomic system. Furthermore, Dirac's theory shows that the emitted radiation is, in all respects, identical to the stimulating radiation. Dirac's calculation which was carried out in the basis of photon number states, remains silent on the phase of the emitted photons. This is expected due to the number-phase uncertainty relation. However, if one uses a coherent state basis for the photons [31, 32], it could be argued that since absorption does not destroy the phase coherence of the electromagnetic radiation then stimulated radiation must also be in phase with the stimulating radiation. When an electromagnetic mode has a high intensity, stimulated emission can dominate over spontaneous emission. Stimulated emission from excited atoms can be used in an electromagnetic cavity to create a high-intensity beam of radiation. The coherence of stimulated emission plays an essential role in the operation of LASERs.

5.6   Fluctuations in the Number of Particles

In the Grand-Canonical Ensemble the probability of finding the system in a state with N particles in a volume element $d\Gamma_N$ of phase space is given by

$$
\begin{aligned}
dp_{N,E} &= \left(\frac{dp}{d\Gamma}\right)_{N,E} d\Gamma_N \\
&= \frac{1}{\Xi}\, \exp[\beta\mu N]\, \exp[-\beta H_N(\{p_i, q_i\})]\, \frac{d\Gamma_N}{\Gamma_{N,0}} \tag{5.77}
\end{aligned}
$$

where the Grand-Canonical Partition Function is given by

$$\Xi = \sum_{N=0}^{\infty} \exp[\beta\mu N] \int \frac{d\Gamma_N}{\Gamma_{N,0}} \exp[-\beta H_N(\{p_i, q_i\})] \tag{5.78}$$

The average number of particles $\overline{N}$ is defined by the expression

$$\overline{N} = \frac{1}{\Xi} \sum_{N=0}^{\infty} N \exp[\beta\mu N] \int \frac{d\Gamma_N}{\Gamma_{N,0}} \exp[-\beta H_N(\{p_i, q_i\})] \tag{5.79}$$

where, since we are not interested in the position of the N-particle system in its phase space, we have integrated over $d\Gamma_N$. The above expression can be re-written in terms of a logarithmic derivative of Ξ w.r.t. the product $\beta\mu$. Alternatively, on defining the *"Fugacity"* z as

$$z = \exp[\beta\mu] \tag{5.80}$$

one may express $\overline{N}$ as

$$\overline{N} = \frac{1}{\Xi} \sum_{N=0}^{\infty} N z^N \int \frac{d\Gamma_N}{\Gamma_{N,0}} \exp[-\beta H_N(\{p_i, q_i\})] \tag{5.81}$$

which can be expressed as the derivative w.r.t. z

$$\overline{N} = \frac{1}{\Xi} \left(z \frac{\partial \Xi}{\partial z} \right)$$

$$= z \frac{\partial \ln \Xi}{\partial z} \tag{5.82}$$

The average value $\overline{N^2}$ can also be written as a second derivative w.r.t. $\beta\mu$ or

$$\overline{N^2} = \frac{1}{\Xi} \left(z \frac{\partial}{\partial z} + z^2 \frac{\partial^2}{\partial z^2} \right) \Xi$$

$$= \frac{1}{\Xi} \left(z \frac{\partial}{\partial z} \right)^2 \Xi \tag{5.83}$$

Likewise, the average squared fluctuations of N is given by

$$\overline{\Delta N^2} = \overline{N^2} - \overline{N}^2$$

$$= \frac{1}{\Xi} \left(z \frac{\partial}{\partial z} \right)^2 \Xi - \frac{1}{\Xi^2} \left(z \frac{\partial \Xi}{\partial z} \right)^2$$

$$= \left(z \frac{\partial}{\partial z} \right)^2 \ln \Xi$$

$$= \left(z \frac{\partial}{\partial z} \right) \overline{N}$$

$$= k_B T \frac{\partial \overline{N}}{\partial \mu} \tag{5.84}$$

The relative fluctuation of the particle number is given by

$$\frac{\overline{\Delta N^2}}{\overline{N}^2} = \frac{k_B T}{\overline{N}^2} \left(\frac{\partial \overline{N}}{\partial \mu} \right)_{V,T} \tag{5.85}$$

which is of order $1/\overline{N}$ and vanishes in the thermodynamic limit.

The above expression for the relative fluctuations is not expressed in terms of quantities that are not easily measurable. However, the factor $(\frac{\partial \overline{N}}{\partial \mu})_T$ can be re-written in terms of $(\frac{\partial V}{\partial P})_T$ which is quite easily measurable. On combining the expression for the infinitesimal change in the Grand-Canonical Potential

$$d\Omega = -S \, dT - P \, dV - \overline{N} \, d\mu \tag{5.86}$$

with

$$\Omega = -PV \tag{5.87}$$

one finds the relation

$$-V \, dP = -S \, dT - \overline{N} \, d\mu \tag{5.88}$$

For a process at constant T this reduces to a relation between P and μ

$$V \, dP = \overline{N} \, d\mu \tag{5.89}$$

which on dividing by dV becomes

$$V \left(\frac{\partial P}{\partial V} \right)_T = \overline{N} \left(\frac{\partial \mu}{\partial V} \right)_T \tag{5.90}$$

This second term in this relation can be expressed in terms of the derivative of the volume per particle

$$v = \frac{V}{\overline{N}} \tag{5.91}$$

as

$$V \left(\frac{\partial P}{\partial V} \right)_T = \left(\frac{\partial \mu}{\partial v} \right)_T \tag{5.92}$$

However, the second term can be reinterpreted as the derivative of μ w.r.t. $\overline{N}$ at constant V. Therefore, one obtains

$$V \left(\frac{\partial P}{\partial V} \right)_T = \left(\frac{\partial \overline{N}}{\partial v} \right)_V \left(\frac{\partial \mu}{\partial \overline{N}} \right)_T \tag{5.93}$$

but since $\overline{N} = V/v$, one has

$$V \left(\frac{\partial P}{\partial V} \right)_T = -\frac{V}{v^2} \left(\frac{\partial \mu}{\partial \overline{N}} \right)_T$$

$$= -\frac{\overline{N}^2}{V} \left(\frac{\partial \mu}{\partial \overline{N}} \right)_T \tag{5.94}$$

On inverting this relation and substituting this in the expression for the relative fluctuations in the number of particles, one finds that

$$\frac{\overline{\Delta N^2}}{\overline{N}^2} = -\frac{k_B T}{V^2} \left(\frac{\partial V}{\partial P} \right)_T \tag{5.95}$$

Again, one notices that the relative fluctuations of the particle number is inversely proportional to the volume and thus vanishes in the thermodynamic limit.

5.7 Energy Fluctuations in the Grand-Canonical Ensemble

The average energy $\overline{E}$ in the Grand-Canonical Ensemble can be represented as a derivative of the logarithm of the Grand-Canonical Partition Function w.r.t. β as

$$\begin{aligned}
\overline{E} &= -\frac{1}{\Xi} \left(\frac{\partial \Xi}{\partial \beta} \right)_z \\
&= -\left(\frac{\partial \ln \Xi}{\partial \beta} \right)_{z,V}
\end{aligned} \tag{5.96}$$

where the fugacity z is held constant. Likewise, the mean squared energy is given by

$$\overline{E^2} = \frac{1}{\Xi} \left(\frac{\partial^2 \Xi}{\partial \beta^2} \right)_{z,V} \tag{5.97}$$

Hence, the mean squared fluctuation of the energy is given by

$$\begin{aligned}
\overline{\Delta E^2} &= \overline{E^2} - \overline{E}^2 \\
&= \left(\frac{\partial^2 \ln \Xi}{\partial \beta^2} \right)_{z,V} \\
&= -\left(\frac{\partial \overline{E}}{\partial \beta} \right)_{z,V} \\
&= k_B T^2 \left(\frac{\partial \overline{E}}{\partial T} \right)_{z,V}
\end{aligned} \tag{5.98}$$

The above relations are similar to the relations found for the Canonical Ensemble, but are different because the derivatives are evaluated at constant N for the Canonical Ensemble and at constant fugacity for the Grand-Canonical Ensemble. Hence, the energy fluctuations are different in the Canonical and the Grand-Canonical Ensembles.

The cause for the difference between the fluctuations in the Grand-Canonical and the Canonical Ensembles is not easy to discern from the above expression since the fugacity is difficult to measure. The difference can be made explicit by using thermodynamics, in which case we identify the average energy $\overline{E}$ in the Grand-Canonical Ensemble with U. That is, since on holding V fixed

$$\overline{N} = \overline{N}(T, z) \tag{5.99}$$

one has

$$U = U(T, \overline{N}(T, z)) \tag{5.100}$$

so the infinitesimal variation in U can be expressed as

$$
\begin{aligned}
dU &= \left(\frac{\partial U}{\partial T}\right)_{\overline{N}} dT + \left(\frac{\partial U}{\partial \overline{N}}\right)_T d\overline{N} \\
&= \left(\frac{\partial U}{\partial T}\right)_{\overline{N}} dT + \left(\frac{\partial U}{\partial \overline{N}}\right)_T \left[\left(\frac{\partial \overline{N}}{\partial T}\right)_z dT + \left(\frac{\partial \overline{N}}{\partial z}\right)_T dz\right]
\end{aligned}
\tag{5.101}
$$

Hence, the derivative of U w.r.t. T with z kept constant is given by

$$
\begin{aligned}
\left(\frac{\partial U}{\partial T}\right)_{z,V} &= \left(\frac{\partial U}{\partial T}\right)_{\overline{N},V} + \left(\frac{\partial U}{\partial \overline{N}}\right)_{T,V} \left(\frac{\partial \overline{N}}{\partial T}\right)_{z,V} \\
&= C_{N,V} + \left(\frac{\partial U}{\partial \overline{N}}\right)_{T,V} \left(\frac{\partial \overline{N}}{\partial T}\right)_{z,V}
\end{aligned}
\tag{5.102}
$$

Therefore, part of the energy fluctuation in the Grand-Canonical Ensemble is the same as the energy fluctuations in the Canonical Ensemble where the number of particles and the volume are fixed and the other contribution originates from the temperature dependence of the number of particles.

One can obtain more insight into the origin of the energy fluctuations in the Grand-Canonical Ensemble by using a specific thermodynamic relation involving the factor

$$\left(\frac{\partial U}{\partial \overline{N}}\right)_{T,V} \tag{5.103}$$

and another relation for

$$\left(\frac{\partial \overline{N}}{\partial T}\right)_{z,V} \tag{5.104}$$

On considering the infinitesimal change

$$dU = T\, dS - P\, dV + \mu\, d\overline{N} \tag{5.105}$$

with constant T and V, then on dividing by $d\overline{N}$ one obtains

$$\left(\frac{\partial U}{\partial \overline{N}}\right)_{T,V} = T\left(\frac{\partial S}{\partial \overline{N}}\right)_{T,V} + \mu \tag{5.106}$$

Substitution of the Maxwell relation

$$\left(\frac{\partial S}{\partial \overline{N}}\right)_{T,V} = -\left(\frac{\partial \mu}{\partial T}\right)_{\overline{N},V} \tag{5.107}$$

obtained from the analyticity of the Helmholtz Free-Energy $F(T,V,\overline{N})$, yields the first thermodynamic relation

$$\left(\frac{\partial U}{\partial \overline{N}}\right)_{T,V} = \mu - T\left(\frac{\partial \mu}{\partial T}\right)_{\overline{N},V} \tag{5.108}$$

The thermodynamic relation for the factor

$$\left(\frac{\partial \overline{N}}{\partial T}\right)_{z,V} \tag{5.109}$$

is obtained from $\overline{N}(T,V,\mu)$ by expressing $\mu = \mu(T,z)$ so

$$\left(\frac{\partial \overline{N}}{\partial T}\right)_{z,V} = \left(\frac{\partial \overline{N}}{\partial T}\right)_{\mu,V} + \left(\frac{\partial \overline{N}}{\partial \mu}\right)_{T,V}\left(\frac{\partial \mu}{\partial T}\right)_{z,V} \tag{5.110}$$

The first term on the right-hand side can be re-written, yielding

$$\left(\frac{\partial \overline{N}}{\partial T}\right)_{z,V} = -\left(\frac{\partial \overline{N}}{\partial \mu}\right)_{T,V}\left(\frac{\partial \mu}{\partial T}\right)_{\overline{N},V} + \left(\frac{\partial \overline{N}}{\partial \mu}\right)_{T,V}\left(\frac{\partial \mu}{\partial T}\right)_{z,V} \tag{5.111}$$

The factor of the derivative of μ at constant z in the last term is found by recognizing that μ is related to the fugacity by

$$\mu = k_B T \ln z \tag{5.112}$$

which leads to

$$\left(\frac{\partial \mu}{\partial T}\right)_z = \frac{\mu}{T} \tag{5.113}$$

Hence, we have

$$\left(\frac{\partial \overline{N}}{\partial T}\right)_{z,V} = -\left(\frac{\partial \overline{N}}{\partial \mu}\right)_{T,V}\left(\frac{\partial \mu}{\partial T}\right)_{\overline{N},V} + \frac{\mu}{T}\left(\frac{\partial \overline{N}}{\partial \mu}\right)_{T,V} \tag{5.114}$$

$$= \frac{1}{T}\left(\frac{\partial \overline{N}}{\partial \mu}\right)_{T,V}\left[\mu - T\left(\frac{\partial \mu}{\partial T}\right)_{\overline{N},V}\right] \tag{5.115}$$

On substituting the relation (5.108) in the above expression, one finds

$$\left(\frac{\partial \overline{N}}{\partial T}\right)_{z,V} = \frac{1}{T} \left(\frac{\partial \overline{N}}{\partial \mu}\right)_{T,V} \left(\frac{\partial U}{\partial \overline{N}}\right)_{T,V} \tag{5.116}$$

This analysis yields the two equivalent expressions for the energy fluctuations in the Grand-Canonical Ensemble

$$\overline{\Delta E^2} = k_B T^2 \, C_{N,V} + k_B T \left(\frac{\partial \overline{N}}{\partial \mu}\right)_{T,V} \left[\mu - T \left(\frac{\partial \mu}{\partial T}\right)_{\overline{N},V}\right]^2 \tag{5.117}$$

and

$$\overline{\Delta E^2} = k_B T^2 \, C_{N,V} + k_B T \left(\frac{\partial \overline{N}}{\partial \mu}\right)_{T,V} \left(\frac{\partial U}{\partial \overline{N}}\right)_{T,V}^2$$

$$= k_B T^2 \, C_{N,V} + \overline{\Delta N^2} \left(\frac{\partial U}{\partial \overline{N}}\right)_{T,V}^2 \tag{5.118}$$

where, in the last line, we have used the equality

$$\overline{\Delta N^2} = k_B T \left(\frac{\partial \overline{N}}{\partial \mu}\right)_{T,V} \tag{5.119}$$

The second expression for $\overline{\Delta E^2}$ shows that the mean-squared energy fluctuations have two contributions, one originating from the mean squared energy fluctuation with a fixed number of particles and the second contribution comes from the mean squared fluctuations in the particle number where each particle that is exchanged with the reservoir carries with it the energy $(\frac{\partial E}{\partial \overline{N}})_T$.

5.8   Problems

Problem 5.1

Derive the probability distribution function $\rho_{gc}(N, \{p_i, q_i\}_N)$ for the Grand-Canonical Ensemble by maximizing the entropy subject to the three constraints

$$\sum_{N=0}^{\infty} \int d\Gamma_N \, \rho_{gc} = 1$$

$$\sum_{N=0}^{\infty} \int d\Gamma_N \, H_N \, \rho_{gc} = U$$

$$\sum_{N=0}^{\infty} N \int d\Gamma_N \, \rho_{gc} = \overline{N}$$

Problem 5.2

The probability distribution function for the Grand-Canonical Ensemble is given by

$$\rho_{gc} = \frac{1}{\Xi \, \Gamma_{N,0}} \, \exp[-\beta(H_N - \mu N)]$$

Show that Shannon's entropy formula

$$S = -k_B \sum_{N=0}^{\infty} \int d\Gamma_N \, \rho_{gc} \, \ln[\rho_{gc} \, \Gamma_{N,0}]$$

leads to the thermodynamic expression for the entropy in terms of the Grand-Canonical Potential Ω.

Problem 5.3

Consider a classical ideal gas in a uniform gravitational field g. Assume that the gas is in thermal and chemical equilibrium.
(i) Determine the chemical potential and show that the particle density is given by

$$\rho(z) = \rho_0 \, \exp[-\beta mgz]$$

Problem 5.4

Consider an ideal gas of atoms represented by the Grand-Canonical Ensemble.
(i) Show that the probability P_N of finding a subsystem with N atoms is given by

$$P_N = \frac{1}{N!} \overline{N}^N \exp[-\overline{N}]$$

where $\overline{N}$ is the average number of atoms.
(ii) Calculate the moment and cumulant generating functions, $M(t)$ and $K(t)$, for the probability distribution P_N.

Problem 5.5

Consider an ideal gas.
(i) Show that the mean squared number fluctuations are given by

$$\overline{\Delta N^2} = \overline{N}$$

(ii) Calculate the next highest order cumulant that characterizes the number distribution.

Problem 5.6

The problem is to re-derive the Poisson distribution for $P(N_0)$ for finding N_0 particles of a classical ideal gas in a volume V_0 where the gas has an average density $\rho = N/V$ directly by using probability theory. Consider a large volume V which contains N non-interacting particles and encloses a smaller volume element V_0.
(i) Express the probability p for a given particle being found in the volume V_0 in terms of the density ρ and N.
(ii) Show that the probability for finding N_0 particles in the volume V_0 is given by the binomial expression

$$P(n) = \binom{N}{N_0} p^{N_0} (1-p)^{(N-N_0)}$$

(iii) Use Stirling's approximation to show that $N! \approx N^{N_0} (N - N_0)!$ when $N \gg N_0$.
(iv) Assume that N and V both tend to infinity but ρ is kept constant. Show that, in this limit, $P(N_0)$ tends to the Poisson distribution where average number of particles N_0 is $\overline{N}_0 = \rho V_0$.

Problem 5.7

Show that the specific heat at constant N is related to the specific heat at constant μ via the relation

$$C_{V,N} = C_{V,\mu} - T \frac{\left(\frac{\partial \bar{N}}{\partial T}\right)_{\mu,V}^2}{\left(\frac{\partial \bar{N}}{\partial \mu}\right)_{T,V}}$$

Problem 5.8

Using the Gibbs-Duhem relation, derive the relation

$$\overline{\Delta N^2} = k_B T \, \kappa_T \, \frac{\bar{N}^2}{V}$$

between the fluctuations in the particle number $\overline{\Delta N^2}$ and the isothermal compressibility κ_T defined by

$$\kappa_T = -\frac{1}{V} \left(\frac{\partial V}{\partial P}\right)_{N,T}$$

Problem 5.9

Consider a system in the Grand-Canonical Ensemble with chemical potential μ, and average energy U, show that the correlation between the energy E and the number of particles N is given by

$$\overline{EN} - \bar{E}\bar{N} = k_B T \left(\frac{\partial U}{\partial \mu}\right)_{T,V}$$

Problem 5.10

Consider an ideal gas of molecules with chemical potential μ in thermal and chemical equilibrium with a surface with N sites. Each surface site could adsorb one molecule, and the energy of the adsorbed molecule is $-\epsilon$. The adsorbed atoms are not free to move on the surface.

(i) Show that if there are N_s adsorbed molecules, the partition function for the surface can be written as

$$Z_s = \frac{N!}{N_s!(N-N_s)!} \exp[\beta N_s \epsilon]$$

and the Grand-Canonical partition function is given by

$$\Xi_s = (1 + \exp[\beta(\epsilon + \mu)])^N$$

where μ is the common chemical potential.

(ii) Hence, find an expression for the average number of adsorbed particles $\overline{N}_s$.

(iii) Show that the covering ratio $\Theta = N_s/N$ is given by

$$\Theta = \frac{P}{P + P_0(T)}$$

where

$$P_0(T) = \left(\frac{k_B T}{\lambda^3}\right) \exp[-\beta\epsilon]$$

Problem 5.11

Consider the molecules of an ideal gas of chemical potential μ that have been adsorbed onto a surface with N sites. Each site could be occupied by either zero, one or two molecules. If two molecules occupy the same surface site, the molecules may bind and undergo harmonic oscillations with frequency ω_0. The Hamiltonian is given by

$$H = \sum_{i=1}^{N}\left[-\epsilon n_i + \frac{n_i(n_i - 1)}{2!}\sum_{m_i=0}^{\infty}\hbar\omega_0\left(m_i + \frac{1}{2}\right)\right]$$

where n_i is the number of molecules that occupy site i and m_i are the number of quanta in the oscillator when site i is occupied by two molecules.

(i) Show that the average number of molecules per site, $\overline{n}$, is given by

$$\overline{n} = \frac{2\sinh\frac{\beta\hbar\omega_0}{2} + 2\exp[\beta(\epsilon + \mu)]}{2\sinh\frac{\beta\hbar\omega_0}{2}(\exp[-\beta(\epsilon + \mu)] + 1) + \exp[\beta(\epsilon + \mu)]}$$

(ii) Show that the probability of two molecules occupying the same site is given by the expression

$$\frac{\exp[\beta(\epsilon + \mu)]}{2\sinh\frac{\beta\hbar\omega_0}{2}(\exp[-\beta(\epsilon + \mu)] + 1) + \exp[\beta(\epsilon + \mu)]}$$

Can the average number of pairs of molecules adsorbed onto any site exceed the average number of single adsorbed molecules?

Problem 5.12

Consider an ideal gas of N_v particles with mass m held at temperature T. The molecules move freely in a d-dimensional volume $V_d = L^d$. Consider a surface of the volume of surface area $A_s = L^{d-1}$, on which N_s molecules are adsorbed. The adsorbed molecules move freely on the surface. The total number of molecules is $N = N_v + N_s$. The energy of the adsorbed molecules is given by

$$\epsilon(\underline{p}) = \sum_{i=1}^{d-1} \frac{p_i^2}{2m} - \epsilon$$

The molecules are all indistinguishable.

(i) Show that the partition functions Z_v for the d-dimensional gas and Z_s for the adsorbed gas are given by

$$Z_v = \frac{1}{N_v!} \left(\frac{V_d}{\lambda^d} \right)^{N_v}$$

$$Z_s = \frac{1}{N_s!} \left(\frac{A_s \, \exp[\beta\epsilon]}{\lambda^{d-1}} \right)^{N_s}$$

(ii) Find the expressions for the chemical potentials μ_v and μ_s in terms of N_v and N_s.

Assume that the gas and the adsorbed gas are in chemical equilibrium.

(iii) Show that the number of adsorbed molecules per unit area N_s/A_s is given by

$$\frac{N_s}{A_s} = \frac{N \, \lambda}{A_s \, \lambda + V_d \, \exp[-\beta\epsilon]}$$

(iv) If the d-dimensional volume V_d is increased but the surface A_s, N and T are kept fixed, how does the density of adsorbed molecules N_s/A_s change?

Problem 5.13

Consider a gas of Hydrogen atoms at an extremely high temperature, such that ionization may occur with non-negligible probability. The ionized electrons form an independent gas cloud which coexists with the gas of ionized atoms and neutral atoms. The Debye screening length is assumed to be large such that screening of the Coulomb interaction is negligible. The reaction is

$$H \longleftrightarrow p^+ + e^-$$

The problem is to derive the relative ionization density $\chi = \frac{N_p}{N_H + N_p}$ where N_H is the number of neutral hydrogen atoms and N_p is the number of ionized Hydrogen atoms.

(i) Consider the neutral Hydrogen atoms to be in their ground states. Consider the product of the partition functions $Z_H Z_p Z_e$ for $N_H = N - N_p$ neutral Hydrogen atoms and N_p protons and $N_e = N_p$ electrons. The systems are in thermal equilibrium. The total Free-Energy is a function of the independent quantities N and the total number of protons or electrons N_p.

$$F(T, V, N, N_p) = F_H(T, V, N - N_p) + F_p(T, V, N_p) + F_e(T, V, N_p)$$

The number of ionized Hydrogen atoms N_p in the equilibrium state is given by the value that minimizes $F(T, V, N, N_p)$. Hence, show that

$$\mu_H = \mu_p + \mu_e$$

(ii) By treating all three species of particles as classical ideal gasses and using the Grand-Canonical Ensemble, find expressions for N_H, N_p and N_e. In calculating N_p, be sure to include the ionization energy.

(iii) By forming the ratio of $N_e N_p$ to N_H and using the relation between the chemical potentials, find the Saha equation

$$\frac{N_e N_p}{N_H} = V \left(\frac{m_e k_B T}{2\pi \hbar^2} \right)^{\frac{3}{2}} \exp[-\beta R]$$

where R is the Rydberg constant.

(iv) Setting $N_e = N_p$ and $N_H = N - N_e$, find the degree of ionization

$$\chi = \frac{N_p}{N_H + N_p}$$

Problem 5.14

In this problem the Saha equation is to be re-derived using the principle of detailed balance and the Einstein transition coefficients. We consider recombination of an electron with an ionized Hydrogen atoms in which a photon is emitted resulting in the formation of a neutral Hydrogen atom.

The rate at which the ionized Hydrogen atom and the electron combine along with the spontaneous and stimulated emission of a photon is given by

$$\frac{1}{\tau} = \rho_p \rho_e(p) \left(A_{e,p} + B_{e,p} \, u(\omega) \right)$$

where ρ_p and $\rho_e(p)$ are, respectively, the densities of the ionized Hydrogen and electrons, $u(\omega)$ is the photon energy density. The energy density is given by

$$u(\omega) = \frac{\omega^2}{\pi^2 c^3} \frac{\hbar\omega}{\exp[\beta\hbar\omega] - 1}$$

The quantities A and B are Einstein coefficients which are independent of temperature and frequency. The rate of photo-ionization is given by

$$\frac{1}{\tau} = \rho_H \, B_H \, u(\omega)$$

In equilibrium the two transition rates are equal. The ratios of the Einstein coefficients are given by

$$\frac{B_H}{B_{e,p}} = \frac{p^2}{2\pi^2 \hbar^3}$$

$$\frac{A_{e,p}}{B_{e,p}} = \frac{\hbar\omega^3}{\pi^2 c^3}$$

(i) Equate the two transition rates to show that

$$\frac{\rho_p \rho_e(p)}{\rho_H} = \frac{B_H \, u(\omega)}{A_{e,p} + B_{e,p} \, u(\omega)}$$

(ii) Hence, using energy conservation $\hbar\omega = R + e(p)$ and the Maxwell distribution for $\rho_e(p)$, show that

$$\frac{\rho_p \rho_e}{\rho_H} = \left(\frac{m_e k_B T}{2\pi\hbar^2}\right)^{\frac{3}{2}} \exp[-\beta R]$$

Problem 5.15

Consider an ideal gas composed of a mixture of A atoms, B atoms and AB molecules. The particles undergo a reversible chemical reaction

$$A + B \leftrightarrow AB$$

The density of atoms A, B and molecules AB are, respectively, ρ_A, ρ_B and ρ_{AB}.

(i) Show that the densities obey the law of mass action

$$\frac{\rho_{AB}}{\rho_A \rho_B} = VK(T)$$

where V is the volume occupied by the gas.

(ii) Express $K(T)$ in terms of single atom or molecule partition functions. Consider the following reaction

$$2A + B \leftrightarrow A_2 B$$

(iii) Derive the law of mass action for this reaction.

Problem 5.16

For an electromagnetic cavity in equilibrium with matter, at what temperature T does stimulated emission of electromagnetic radiation of frequency ω dominate over spontaneous emission?

Chapter 6

Quantum Statistical Mechanics

"*Quantum Statistical Mechanics*" describes the thermodynamic properties of macroscopically large many-particle quantum systems.

6.1   Quantum Microstates and Measurements

In Quantum Mechanics a microscopic state of a many-particle system is represented by a vector in Hilbert space

$$|\Psi\rangle$$

Any two states $|\Psi\rangle$ and $|\Phi\rangle$ in Hilbert space have an inner product, which is given by a complex number

$$\langle\Phi|\Psi\rangle = \langle\Psi|\Phi\rangle^* \tag{6.1}$$

The states are normalized to unity

$$\langle\Phi|\Phi\rangle = 1 \tag{6.2}$$

A set of states $|n\rangle$ form an orthonormal set if their inner product satisfies

$$\langle n|m\rangle = \delta_{n\cdot m} \tag{6.3}$$

where $\delta_{n,m}$ is the Kronecker delta function. An orthonormal set is complete if any arbitrary state can be expanded as

$$|\Psi\rangle = \sum_n C_n|n\rangle \tag{6.4}$$

where the expansion coefficients C_n are complex numbers, which are found as

$$C_n = \langle n|\Psi\rangle \tag{6.5}$$

Thus, an arbitrary state can be expanded as

$$|\Psi\rangle = \sum_n |n\rangle\langle n|\Psi\rangle \tag{6.6}$$

Hence, we have the completeness condition

$$\hat{I} = \sum_n |n\rangle\langle n| \tag{6.7}$$

Using the completeness condition, the normalization condition can be written as

$$\langle\Psi|\Psi\rangle = \sum_n \langle\Psi|n\rangle\langle n|\Psi\rangle = 1 \tag{6.8}$$

On choosing the complete set as the set of coordinate eigenstates, this condition reduces to

$$1 = \prod_{i=1}^{3N}\left\{\int dq_i\right\}|\Psi(q_1, q_2, q_3, \ldots, q_{3N})|^2 \tag{6.9}$$

Physical observables $A_j(\{p_i, q_i\})$ are represented by Hermitean operators $A_j(\{\hat{p}_i, \hat{q}_i\})$. If the Poisson Bracket of two classical observables A_j and A_k is represented by

$$\{A_j, A_k\}_{PB}$$

then the Poisson Bracket of two quantum operators is represented by the commutator of the operators divided by $i\hbar$

$$\{A_j, A_k\}_{PB} = \frac{1}{i\hbar}[\hat{A}_j, \hat{A}_k] \tag{6.10}$$

In particular, since the Poisson Bracket for canonically conjugate coordinates and momenta are given by

$$\{p_i, q_j\}_{PB} = -\delta_{i,j} \tag{6.11}$$

then, one has the commutation relations

$$[\hat{p}_i, \hat{q}_j] = -i\hbar\delta_{i,j} \tag{6.12}$$

The possible values of a measurement of A on a systems are the eigenvalues a_n found from the eigenvalue equation

$$\hat{A}|a_n\rangle = a_n|a_n\rangle \tag{6.13}$$

If a measurement of $\hat{A}$ on a system results in the value a_n, then immediately after the measurement the system is definitely known to be in a state which is an eigenstate of A with eigenvalue a_n. The number of eigenstates corresponding to the same eigenvalue a_n is known as the degeneracy $\mathcal{D}_n$. The degenerate eigenstates $|a_{n,\alpha}\rangle$ can be orthonormalized. Since the observable quantities are represented by Hermitian operators, the eigenstates form a complete set and the eigenvalues a_n are real.

It is possible to know with certainty the simultaneous values of measurements of A_j and A_k on a state if the operators $\hat{A}_j$ and $\hat{A}_k$ commute

$$[\hat{A}_j, \hat{A}_k] = 0 \tag{6.14}$$

In which case it is possible to find simultaneous eigenstates of $\hat{A}_j$ and $\hat{A}_k$. A state is completely determined if it is an eigenstate of a maximal set of mutually commuting operators.

If a system is definitely in a state $|\Psi\rangle$, it is in a "*Pure State*". The probability that a measurement of A on the state $|\Psi\rangle$ will yield the result a_n is given by

$$P(a_n) = \sum_{\gamma=1}^{\mathcal{D}_{a_n}} |\langle a_{n,\gamma}|\Psi\rangle|^2 \tag{6.15}$$

where the sum is over the number of $\mathcal{D}_n$-fold degenerate eigenstates[1] $|a_{n,\gamma}\rangle$ that correspond to the eigenvalue a_n. Thus, the average value $\overline{A}$ of the measurement of A on a pure state $|\Psi\rangle$ is given by

$$\begin{aligned}
\overline{A} &= \sum_n P_{a_n} a_n \\
&= \sum_{n,\gamma} \langle\Psi|a_{n,\gamma}\rangle a_n \langle a_{n,\gamma}|\Psi\rangle \\
&= \sum_{n,\gamma} \langle\Psi|\hat{A}|a_{n,\gamma}\rangle \langle a_{n,\gamma}|\Psi\rangle \\
&= \langle\Psi|\hat{A}|\Psi\rangle
\end{aligned} \tag{6.16}$$

In the coordinate representation, the average can be expressed as

$$\overline{A} = \prod_{i=1}^{3N} \left\{ \int dq_i \right\} \Psi^*(q_1, q_2, \ldots, q_{3N})\, A(\{\hat{p}_i, \hat{q}_i\})\, \Psi(q_1, q_2, \ldots, q_{3N}) \tag{6.17}$$

---

[1]The states are assumed to have been orthogonalized, so that the eigenstates of A form a complete orthonormal set.

In the time interval in which no measurements are performed, a state $|\Psi\rangle$ evolves according to the equation

$$i\hbar \frac{\partial}{\partial t} |\Psi\rangle = \hat{H}|\Psi\rangle \tag{6.18}$$

Thus, the time evolution of the state $|\Psi\rangle$ in our closed system is given by

$$|\Psi(t)\rangle = \exp\left[-\frac{i}{\hbar}\hat{H}t\right]|\Psi\rangle \tag{6.19}$$

6.2 The Density Operator and Thermal Averages

In an ensemble, a macroscopic state may correspond to numerous microscopic states $|\Psi_\alpha\rangle$ each of which are assigned probabilities p_α. Such macroscopic states are known as "*Mixed States*". The ensemble average of A is given by the weighted average

$$\begin{aligned}
\overline{A} &= \sum_\alpha p_\alpha \langle \Psi_\alpha|\hat{A}|\Psi_\alpha\rangle \\
&= \sum_\alpha \sum_{n,m} p_\alpha \langle \Psi_\alpha|n\rangle\langle n|\hat{A}|m\rangle\langle m|\Psi_\alpha\rangle \\
&= \sum_{n,m} \langle n|\hat{A}|m\rangle \sum_\alpha p_\alpha \langle m|\Psi_\alpha\rangle\langle \Psi_\alpha|n\rangle
\end{aligned} \tag{6.20}$$

On defining the (probability) "*Density Operator*" or "*Statistical Operator*" $\hat{\rho}$ via

$$\hat{\rho} = \sum_\alpha p_\alpha |\Psi_\alpha\rangle\langle \Psi_\alpha| \tag{6.21}$$

the average can be represented as

$$\begin{aligned}
\overline{A} &= \sum_{n,m} \langle n|\hat{A}|m\rangle\langle m|\hat{\rho}|n\rangle \\
&= \sum_n \langle n|\hat{A}\hat{\rho}|n\rangle \\
&= \text{Trace } \hat{A}\hat{\rho}
\end{aligned} \tag{6.22}$$

where the last line defines the Trace over a complete set of states [33].[2]

---

[2]Gleason's Theorem assures us that the only possible measure of probability on a Hilbert space has the form of a density operator.

Since the probabilities p_α are normalized, the density operator satisfies

$$\text{Trace } \hat{\rho} = \sum_n \langle n|\hat{\rho}|n\rangle$$

$$= \sum_n \sum_\alpha p_\alpha \langle n|\Psi_\alpha\rangle\langle\Psi_\alpha|n\rangle$$

$$= \sum_\alpha p_\alpha \sum_n \langle\Psi_\alpha|n\rangle\langle n|\Psi_\alpha\rangle$$

$$= \sum_\alpha p_\alpha \langle\Psi_\alpha|\Psi_\alpha\rangle$$

$$= \sum_\alpha p_\alpha$$

$$= 1 \qquad (6.23)$$

Thus, the Trace of the density operator is unity.

The density operator satisfies the positivity condition

$$\langle\Phi|\hat{\rho}|\Phi\rangle = \sum_\alpha p_\alpha |\langle\Phi|\Psi_\alpha\rangle|^2$$

$$\geq 0 \qquad (6.24)$$

for any $|\Phi\rangle$, since the probabilities p_α must be positive definite.[3] An equivalent condition is that the eigenvalues of $\hat{\rho}$ are positive definite.

In any basis, other than the basis composed of energy eigenstates of $\hat{H}$, there is no compelling reason for the density operator to be diagonal. For example, if the eigenstates $|a_n\rangle$ of an operator $\hat{A}$ are chosen as a basis chosen and $[\hat{A}, \hat{H}] \neq 0$, then the density operator will have off-diagonal matrix elements $\langle a_n|\hat{\rho}|a_m\rangle$. However, the density operator is a Hermitean operator and it is normalized to unity, independent of any choice made of the basis. It is also independent of any choice of phases assigned to the basis states, since under a gauge transformation

$$|\Psi'_\alpha\rangle = \exp[+i\chi_\alpha]|\Psi_\alpha\rangle$$

$$\langle\Psi'_\alpha| = \exp[-i\chi_\alpha]\langle\Psi_\alpha| \qquad (6.25)$$

so $\hat{\rho}$ is gauge independent. For a system in equilibrium, we know that the density operator must have the symmetry

$$\langle a_m|\hat{\rho}|a_n\rangle = \langle a_n|\hat{\rho}|a_m\rangle^* \qquad (6.26)$$

---

[3]For a Hermitean matrix, Sylvester's condition is a necessary and sufficient condition that the matrix is positive definite. Sylvester's condition has been generalized to positive semidefinite matrices. Alternatively, one could use Gershgorin's Circle Theorem to determine whether the matrix has a negative eigenvalue.

The symmetry is due to the Hermitean nature of the operator and also because equilibrium is a dynamical equilibrium. That is, the system may make transitions between the states $|a_m\rangle$ and $|a_n\rangle$ but the value of a_n is not building up with time since the transitions from $|a_m\rangle$ to $|a_n\rangle$ are, on average, balanced by transitions back from $|a_n\rangle$ to $|a_m\rangle$. This is known as the "*Principle of Detailed Balance*".

6.2.1 *Pure and Mixed States*

In a basis where the density operator is not diagonal, one can identify the density operators that represent pure states as those for which

$$\hat{\rho}^2 = \hat{\rho} \tag{6.27}$$

Thus the density operator for a pure system is idempotent. All other density operators describe mixed states. For a mixed state, one has

$$\text{Trace } \hat{\rho}^2 < 1 \tag{6.28}$$

This follows by considering a density operator in the form

$$\hat{\rho} = \sum_n p_n |n\rangle\langle n| \tag{6.29}$$

where $1 > p_n > 0$ and $\sum_n p_n = 1$, since

$$\text{Trace } \hat{\rho}^2 = \sum_n p_n^2$$

$$< \sum_n p_n$$

$$< 1 \tag{6.30}$$

The quantity Trace $\hat{\rho}^2$ is known as the "*Purity*" of the ensemble.

Example: Density Operators: Mixed and Pure

An operator which acts on the spin states of a system composed of two spin one-half particles is written in the spin-rotationally symmetric form

$$\hat{\rho} = \frac{1}{4} [\hat{I} + \underline{\hat{\sigma}}_1 \cdot \underline{\hat{\sigma}}_2] \tag{6.31}$$

If the operator was a density operator it would describe a mixed state, since

$$\hat{\rho}^2 = \frac{1}{4} \hat{I} \tag{6.32}$$

as can be shown by using

$$\{\hat{\sigma}_\alpha^i, \hat{\sigma}_\alpha^j\}_+ = 2\delta_{i,j}$$
$$[\hat{\sigma}_\alpha^i, \hat{\sigma}_\alpha^j]_- = 2i \sum_k \epsilon^{i,j,k} \hat{\sigma}_\alpha^k \tag{6.33}$$

In the representation where arbitrary states are expressed in terms of the orthonormal basis states $|\phi_i\rangle$ given by $|\uparrow,\uparrow\rangle$, $|\uparrow,\downarrow\rangle$, $|\downarrow,\uparrow\rangle$, $|\downarrow,\downarrow\rangle$ as

$$|\psi\rangle = \sum_i C_i |\phi_i\rangle \tag{6.34}$$

the operator takes the form

$$\hat{\rho} = \frac{1}{2} \begin{pmatrix} 1 & 0 & 0 & 0 \\ 0 & 0 & 1 & 0 \\ 0 & 1 & 0 & 0 \\ 0 & 0 & 0 & 1 \end{pmatrix} \tag{6.35}$$

The matrix has a triply-degenerate eigenvalue of $\lambda = \frac{1}{2}$ and a non-degenerate eigenvalue of $\lambda = -\frac{1}{2}$. Since $\hat{\rho}$ contains a negative eigenvalue, it is not a density operator as density operators must be positive semi-definite.

The spin-rotation symmetric operator $\hat{\rho}$ that acts on the states of two spin one-half particles

$$\hat{\rho} = \frac{1}{4}\left[\hat{I} + \frac{1}{3}\,\underline{\hat{\sigma}}_1 \cdot \underline{\hat{\sigma}}_2\right] \tag{6.36}$$

represents a density operator for a mixed state. Its eigenvalues, λ consists of the non-degenerate value of $\lambda = 0$ and the triply-degenerate values $\lambda = \frac{1}{3}$. Since the non-zero eigenvalues have the maximal degeneracy, $\hat{\rho}$ represents a "*Maximally-Mixed*" state.

A pure state, $|\psi\rangle$, has the density operator given by

$$\hat{\rho} = |\psi\rangle\langle\psi| \tag{6.37}$$

Since states are assumed to be properly normalized, the pure state density operator are idempotent since

$$\hat{\rho}^2 = \hat{\rho} \tag{6.38}$$

The eigenvalue equation of the density operator has the form

$$\hat{\rho}|\chi\rangle = \lambda|\chi\rangle \tag{6.39}$$

Using the eigenvalue equation and idempotent nature of $\hat{\rho}$, one finds that

$$\lambda^2|\chi\rangle = \lambda|\chi\rangle \tag{6.40}$$

so the eigenvalues satisfy

$$\lambda(\lambda - 1) = 0 \tag{6.41}$$

The eigenvalue equation for the eigenstate $|\chi\rangle$ with eigenvalue $\lambda = 1$ reduces to

$$|\psi\rangle\langle\psi|\chi\rangle = |\chi\rangle \tag{6.42}$$

one finds that the eigenstate is the pure state $|\chi\rangle = |\psi\rangle$.

The density operator

$$\hat{\rho} = \begin{pmatrix} \cos^2\theta & -i\sin\theta\cos\theta \\ i\sin\theta\cos\theta & \sin^2\theta \end{pmatrix} \tag{6.43}$$

represented in the basis $|\phi_1\rangle$ and $|\phi_2\rangle$ describes a pure state. The pure state is the eigenstate of $\hat{\rho}$ that has an eigenvalue unity, and is identified as

$$|\psi\rangle = \cos\theta\,|\phi_1\rangle + i\sin\theta|\phi_2\rangle \tag{6.44}$$

6.2.2 Time-Dependence of the Density Operator

The time-dependence of the density operator can be inferred from the time-dependence of the basis states

$$\hat{\rho}(t) = \sum_\alpha p_\alpha |\Psi_\alpha(t)\rangle\langle\Psi_\alpha(t)|$$

$$= \sum_\alpha p_\alpha \exp\left[-\frac{i}{\hbar}\hat{H}t\right] |\Psi_\alpha\rangle\langle\Psi_\alpha| \exp\left[+\frac{i}{\hbar}\hat{H}t\right] \tag{6.45}$$

This shows that the time evolution of the density operator has the form of a unitary transformation. From the above expression, one finds the equation of motion for the density operator is given by

$$i\hbar\frac{\partial\hat{\rho}}{\partial t} = [\hat{H}, \hat{\rho}] \tag{6.46}$$

or, equivalently

$$i\hbar\frac{d\hat{\rho}}{dt} = i\hbar\frac{\partial\hat{\rho}}{\partial t} + [\hat{\rho}, \hat{H}] = 0 \tag{6.47}$$

This last expression could have been derived directly from the Poisson Bracket equation of motion for the classical probability density by Canonical Quantization.

6.2.3 *Density Operator in Various Ensembles*

If a system described by a density operator is in equilibrium, then the density operator should have no explicit time-dependence

$$\frac{\partial}{\partial t} \hat{\rho} = 0 \tag{6.48}$$

This requires that the Hamiltonian and density operator commute

$$[\hat{\rho}, \hat{H}] = 0 \tag{6.49}$$

Thus, the equilibrium density operator can be expressed as a function of the Hamiltonian since it corresponds to a conserved quantity.

In the Micro-Canonical Ensemble, all the states $|\Psi_{n,\gamma}\rangle$ in the ensemble must be energy eigenstates belonging to the same energy eigenvalue $E = E_n$

$$\hat{H}|\Psi_{n,\gamma}\rangle = E_n|\Psi_{n,\gamma}\rangle \tag{6.50}$$

The number of these degenerate eigenstates is denoted by N_Γ. Therefore, in an equilibrium state, the probabilities are given by

$$p_\gamma = \frac{1}{N_\Gamma} \tag{6.51}$$

which is equivalent to the hypothesis of equal a priori probabilities. Thus, the density operator in the Micro-Canonical can be written as

$$\hat{\rho}_{mc} = \frac{1}{N_\Gamma} \sum_{\gamma=1}^{N_\Gamma} |\Psi_{n,\gamma}\rangle\langle\Psi_{n,\gamma}| \tag{6.52}$$

On defining the "*von Neumann Entropy*", S, in terms of the density operator by

$$S = -k_B \, \text{Trace} \, \hat{\rho} \, \ln \hat{\rho} \tag{6.53}$$

the entropy is evaluated as

$$S = -k_B \sum_\gamma p_\gamma \ln p_\gamma$$
$$= k_B \, \ln N_\Gamma \tag{6.54}$$

in agreement with our previous notation. If the energy E of the Micro-Canonical Ensemble corresponds to a non-degenerate state, it is a pure state and has $N_\Gamma = 1$, therefore, the entropy vanishes. This observation is in accordance with the universal constant value of entropy, demanded by Nernst's law in the $T \to 0$, as being defined as zero.

Since the set of all energy eigenstates is complete, in the Canonical Ensemble the density operator is given by

$$\hat{\rho}_c = \frac{1}{Z_N} \exp[-\beta \hat{H}_N] \tag{6.55}$$

where the partition function is given by the normalization condition on $\hat{\rho}_c$

$$Z_N = \text{Trace } \exp[-\beta \hat{H}_N] \tag{6.56}$$

If the Trace is evaluated using a complete basis set of energy eigenstates $|\Psi_\alpha\rangle$, the result for the partition function reduces to

$$Z_N = \sum_\alpha \exp[-\beta \hat{E}_\alpha] \tag{6.57}$$

where the sum runs over all the degenerate states for each energy.

In the Grand-Canonical Ensemble, one is working in a Hilbert space with a variable number of particles (*"Fock Space"*). In this case, one has

$$\hat{\rho}_{gc} = \frac{1}{\Xi} \exp[-\beta(\hat{H} - \mu \hat{N})] \tag{6.58}$$

in which the Grand-Canonical Partition Function is given by

$$\Xi = \text{Trace } \exp[-\beta(\hat{H} - \mu \hat{N})] \tag{6.59}$$

where $\hat{N}$ is the number operator. If the particle number is conserved so that both $\hat{N}$ and $\hat{H}$ can be diagonalized simultaneously, then the partition function can be reduced to the form

$$\Xi = \sum_{N=0}^{\infty} \exp[\beta \mu N] \, Z_N \tag{6.60}$$

where the N are the eigenvalues of the number operator.

6.2.4 *Reduced Density Operators and Partial Traces*

In all Ensembles, except the Micro-Canonical Ensemble, the system is considered to be in contact with a larger system with which it can exchange energy or energy and particles. The density operators used to describe these Ensembles only pertain to the system and not the system and the environment. The reduction of the description to the system is useful, since for a quantum universe composed of a system, a macroscopic quantum measuring device and its environment, it is not only impractical but also impossible to know the precise state of the entire universe. The environmental degrees of freedom are thought as having have been Traced out. This leads to the concept of a *"Partial Trace"*.

Let the set of degrees of freedom of the system S be denoted by n_S and the degrees of freedom of the environment E be denoted by n_E, and the density operator of the combined system and environment be denoted by $\hat{\rho}_{S+E}$. A *"Reduced Density Operator"* $\hat{\rho}_S$ which represents the system can be found by performing a partial Trace, that is, a Trace over the states of the environmental degrees of freedom

$$\hat{\rho}_S = \text{Trace}_{n_E} \ \hat{\rho}_{S+E} \tag{6.61}$$

The density operator of the system ρ_S is still normalized since

$$\text{Trace}_{n_S} \ \hat{\rho}_S = \text{Trace}_{n_S, n_E} \ \hat{\rho}_{S+E}$$
$$= 1 \tag{6.62}$$

The reduced density operator yields the correct expectation values for operators $\hat{A}_S$ which act on the system since

$$\langle \hat{A}_S \rangle = \text{Trace}_{n_S, n_E} \ \hat{\rho}_{S+E} \ \hat{A}_S$$
$$= \text{Trace}_{n_S} \text{Trace}_{n_E} \ \hat{\rho}_{S+E} \ \hat{A}_S$$
$$= \text{Trace}_{n_S} \ \hat{\rho}_S \ \hat{A}_S \tag{6.63}$$

The interactions with the environment, by assumption, lead to the above forms of the density operators for the various Ensembles. However, in practice the density operators for the Ensembles are determined by minimizing the Free-Energy of the system subject to phenomenological constraints and do not involve consideration of the details of the interactions between the system its environment. The terms in the statistical operator off-diagonal in the environment variables are eliminated by taking the partial Trace. Nevertheless, the reduced density operator incorporates the environment's diagonal elements and, thus, give rise to the same set of "probabilities" as the combined density operator. The reduction of the density operator is a non-unitary process and it introduces an element of randomness.

Example: Reduced Density Operators

Consider an atom which has two energy levels E_0 and E_1, where $E_1 > E_0$, that can emit and absorb photons. The state of the electromagnetic cavity is denoted by $|n\rangle$ where n is the number of photons. The composite system is found to be in the pure state

$$|\psi\rangle = \frac{1}{\sqrt{2}} \left(|E_0\rangle + \cos\theta |E_1\rangle \right) |n = 0\rangle - \frac{i}{\sqrt{2}} \sin\theta |E_0\rangle |n = 1\rangle \tag{6.64}$$

This state has the property that no photons have been emitted when the atom is found to be in the excited state $|E_1\rangle$. The density operator of the composite system is given by

$$\hat{\rho}(t) = |\psi\rangle\langle\psi|$$

$$= \frac{1}{2} \begin{pmatrix} 1 & \cos\theta & i\sin\theta & 0 \\ \cos\theta & \cos^2\theta & i\cos\theta\sin\theta & 0 \\ -i\sin\theta & -i\cos\theta\sin\theta & \sin^2\theta & 0 \\ 0 & 0 & 0 & 0 \end{pmatrix} \qquad (6.65)$$

using the notation that if the pure state is written in terms of an orthonormal basis as

$$|\psi\rangle = \sum_{i=1}^{4} a_i|\phi_i\rangle \qquad (6.66)$$

the density matrix is given by

$$\begin{pmatrix} a_1\,a_1^* & a_1\,a_2^* & a_1\,a_3^* & a_1\,a_4^* \\ a_2\,a_1^* & a_2\,a_2^* & a_2\,a_3^* & a_2\,a_4^* \\ a_3\,a_1^* & a_3\,a_2^* & a_3\,a_3^* & a_3\,a_4^* \\ a_4\,a_1^* & a_4\,a_2^* & a_4\,a_3^* & a_4\,a_4^* \end{pmatrix}$$

The reduced density operator for the atom is given by

$$\begin{pmatrix} |a_1|^2 + |a_3|^2 & a_1\,a_2^* + a_3\,a_4^* \\ a_2\,a_1^* + a_4\,a_3^* & |a_2|^2 + |a_4|^2 \end{pmatrix}$$

which is evaluated as

$$\hat{\rho}_{atom}(t) = \frac{1}{2} \begin{pmatrix} 1 + \sin^2\theta & \cos\theta \\ \cos\theta & \cos^2\theta \end{pmatrix} \qquad (6.67)$$

The reduced density operator for the electromagnetic cavity is given by

$$\begin{pmatrix} |a_1|^2 + |a_2|^2 & a_1\,a_3^* + a_2\,a_4^* \\ a_3\,a_1^* + a_4\,a_2^* & |a_3|^2 + |a_4|^2 \end{pmatrix}$$

which becomes

$$\hat{\rho}_{photon}(t) = \frac{1}{2} \begin{pmatrix} 1 + \cos^2\theta & i\sin\theta \\ -i\sin\theta & \sin^2\theta \end{pmatrix} \qquad (6.68)$$

The above reduced density operators represent mixed states.

Given a density operator, it is a conceptually straightforward task to predict outcomes of future experiments on mixed states. The question of how the mixed states were formed is more complex. A subsystem of a bi-partite system which is initially in a pure state can be found to be in a mixed state after a partial Trace has been performed. Generally, a density matrix neither has a unique decomposition, nor do the basis states have to be orthogonal.

Example: Non-Uniqueness of the Basis States of a Density Operator

Consider a pure state of a spin one-half particle. Any pure state of the spin can be represented as

$$|\eta\rangle = \cos\frac{\theta}{2}\,\exp\left[-i\,\frac{\varphi}{2}\right]\left|+\frac{1}{2}\right\rangle + \sin\frac{\theta}{2}\,\exp\left[+i\,\frac{\varphi}{2}\right]\left|-\frac{1}{2}\right\rangle \quad (6.69)$$

where $(1,\theta,\varphi)$ are polar coordinates of a point in the unit sphere. The density operator for the pure state can be expressed as

$$|\eta\rangle\langle\eta| = \frac{1}{2}\,[\hat{I} + \cos\theta\,\hat{\sigma}_z + \sin\theta\,(\cos\varphi\,\hat{\sigma}_x + \sin\varphi\,\hat{\sigma}_y)]$$

$$= \frac{1}{2}\,[\hat{I} + \underline{\eta}\cdot\hat{\underline{\sigma}}] \quad (6.70)$$

where $\underline{\eta}$ is a unit vector in the direction (θ,φ).

The density operator of a mixed state constructed from orthogonal basis states can be expressed as

$$\hat{\rho} = \left(\frac{1+s}{2}\right)|\eta\rangle\langle\eta| + \left(\frac{1-s}{2}\right)|-\eta\rangle\langle-\eta|$$

$$= \frac{1}{2}\,[\hat{I} + s\underline{\eta}\cdot\hat{\underline{\sigma}}] \quad (6.71)$$

where $\underline{\eta}$ is a unit vector and $1 \geq s$. Since the Pauli matrices and the identity span the space of two by two matrices, this is the most general form of an operator for spin of one-half particle. It can represent the density operator, as it is normalized when $1 \geq s$. For the case where $s = 0$, the density operator has lost all the information about the specific choice of orthogonal basis states.

One may also construct a density operator with the same form from the non-orthogonal basis states $|\gamma\rangle$ and $|\delta\rangle$ as

$$\hat{\rho} = \frac{1}{2}\,|\delta\rangle\langle\delta| + \frac{1}{2}\,|\gamma\rangle\langle\gamma|$$

$$= \frac{1}{2}\left[\hat{I} + \frac{1}{2}\,(\underline{\delta}+\underline{\gamma})\cdot\hat{\underline{\sigma}}\right] \quad (6.72)$$

where $\underline{\gamma}$ and $\underline{\delta}$ are unit vectors. If $\underline{\delta}$ and $\underline{\gamma}$ are parallel the state is a pure state, otherwise it is a mixed state.

Many constructs of mixed state density operators yield the exact same mixed state density operator. For example, on choosing $\underline{\eta} = \hat{e}_z$ for the orthogonal basis, one can obtain the same density operator with the non-orthogonal basis by choosing the directions of the basis vectors as being in the same plane as and having the same projections on the z-axis, $\theta_\gamma = \theta_\delta$ with $\varphi_\delta = \pi - \varphi_\gamma$ and then setting $\cos\theta_\gamma = s$. Hence, the decomposition of a density operator is not unique. Furthermore, the diagonal elements can only be interpreted as probabilities of an experimental outcome if the basis states correspond to the eigenstates of the measurement operator. This is an example of the "*Ambiguity of Mixtures*" which is the notion that, when quantities are mixed, information about their origins is lost.

Example: Correlations Between a Pair of Spin One-half Particles

The density operator for the spin degrees of freedom for two spin one-half particles can be written in a spin-rotation symmetric form

$$\hat{\rho} = \frac{1}{4}\left[\hat{I} + c\underline{\hat{\sigma}}_1 \cdot \underline{\hat{\sigma}}_2\right] \tag{6.73}$$

If the density operator describes a pure state, then the condition

$$\hat{\rho}^2 = \hat{\rho} \tag{6.74}$$

introduces the constraint

$$4(\hat{I} + c\underline{\hat{\sigma}}_1 \cdot \underline{\hat{\sigma}}_2) = \hat{I} + 2c\underline{\hat{\sigma}}_1 \cdot \underline{\hat{\sigma}}_2 + c^2\underline{\hat{\sigma}}_1 \cdot \underline{\hat{\sigma}}_2\underline{\hat{\sigma}}_1 \cdot \underline{\hat{\sigma}}_2 \tag{6.75}$$

The term quadratic in the scalar product can be reduced

$$\left(\sum_{i=1}^{3}\hat{\sigma}_1^i\hat{\sigma}_2^i\right)\left(\sum_{j=1}^{3}\hat{\sigma}_1^j\hat{\sigma}_2^j\right) = \sum_{i=1}^{3}(\hat{\sigma}_1^i\hat{\sigma}_2^i)^2 + \sum_{i\neq j}\hat{\sigma}_1^i\hat{\sigma}_1^j\hat{\sigma}_2^i\hat{\sigma}_2^j$$

$$= 3\hat{I} - 2\sum_{k}\sigma_1^k\sigma_2^k \tag{6.76}$$

where the last line has been obtained by using

$$\hat{\sigma}_\alpha^i\hat{\sigma}_\alpha^j = \frac{1}{2}\{\hat{\sigma}_\alpha^i, \hat{\sigma}_\alpha^j\}_+ + \frac{1}{2}[\hat{\sigma}_\alpha^i, \hat{\sigma}_\alpha^j]$$

$$\{\hat{\sigma}_\alpha^i, \hat{\sigma}_\alpha^j\}_+ = 2\delta_{i,j}$$

$$[\hat{\sigma}_\alpha^i, \hat{\sigma}_\alpha^j]_- = 2i\sum_{k}\epsilon^{i,j,k}\hat{\sigma}_\alpha^k \tag{6.77}$$

Hence, the constraint that the state is pure leads to a pair of equations

$$4 = 1 + 3c^2$$
$$4c = 2c - 2c^2 \tag{6.78}$$

which are only satisfied for $c = -1$. Therefore, the pure state is described by

$$\hat{\rho} = \frac{1}{4} [\hat{I} - \hat{\underline{\sigma}}_1 \cdot \hat{\underline{\sigma}}_2]$$

$$= \frac{1}{2} \begin{pmatrix} 0 & 0 & 0 & 0 \\ 0 & 1 & -1 & 0 \\ 0 & -1 & 1 & 0 \\ 0 & 0 & 0 & 0 \end{pmatrix} \tag{6.79}$$

which is independent of the choice of the axis of quantization. This operator represents a spin-singlet state since

$$\text{Trace } \hat{\rho} \, (\hat{\underline{\sigma}}_1 + \hat{\underline{\sigma}}_2) = 0 \tag{6.80}$$

The eigenfunction of $\hat{\rho}$ which corresponds to the eigenvalue of 1 is found as

$$\frac{1}{\sqrt{2}} \begin{pmatrix} 0 \\ 1 \\ -1 \\ 0 \end{pmatrix} \tag{6.81}$$

which is recognized as the spin-singlet state. The other three eigenstates have eigenvalues 0 as they represent the spin-triplet states.

If the component of spin $\underline{\sigma}_1$ is measured along the direction $\underline{\eta}_1$ and the component of spin $\underline{\sigma}_2$ is measured along the direction $\underline{\eta}_2$, then the correlation $C(\hat{\eta}_1, \hat{\eta}_2)$ is defined as

$$C(\hat{\eta}_1, \hat{\eta}_2) = \text{Trace } \hat{\rho} \, (\hat{\underline{\sigma}}_1 \cdot \hat{\eta}_1) (\hat{\underline{\sigma}}_2 \cdot \hat{\eta}_2)$$

$$= \frac{1}{4} \text{Trace } (\hat{I} - \hat{\underline{\sigma}}_1 \cdot \hat{\underline{\sigma}}_2)(\hat{\underline{\sigma}}_1 \cdot \hat{\eta}_1)(\hat{\underline{\sigma}}_2 \cdot \hat{\eta}_2) \tag{6.82}$$

which on using the identity

$$\text{Trace } \hat{\underline{\sigma}} \, (\hat{\underline{\sigma}} \cdot \hat{\eta}) = 2\hat{\eta} \tag{6.83}$$

yields the correlation as

$$C(\hat{\eta}_1, \hat{\eta}_2) = -\hat{\eta}_1 \cdot \hat{\eta}_2 \tag{6.84}$$

Hence, in a spin-singlet state, the directions of the spins are anti-correlated. The probabilities, $P(\pm\hat{\eta}_1, \pm\hat{\eta}_2)$, that measurement will result in the observation of the spin directions as $(\pm\hat{\eta}_1, \pm\hat{\eta}_2)$ are given by the diagonal matrix elements of the density operator evaluated between the spin eigenstates $|\pm\hat{\eta}_1\rangle|\pm\hat{\eta}_2\rangle$

$$P(\pm\hat{\eta}_1,\pm\hat{\eta}_2) = \langle\pm\hat{\eta}_1|\langle\pm\hat{\eta}_2|\hat{\rho}|\pm\hat{\eta}_2\rangle|\pm\hat{\eta}_1\rangle$$
$$= \frac{1}{4}\left(1-(\pm)(\pm)\,\hat{\eta}_1\cdot\hat{\eta}_2\right) \tag{6.85}$$

The diagonal probabilities vanish in the limit $\hat{\eta}_1 \to \hat{\eta}_2$ indicating the perfect predictability of the spins' anti-correlations. The same result could have been obtained by noting that the single-spin operators $\hat{P}_i(\pm\hat{\eta})$ given by

$$\hat{P}_i(\pm\hat{\eta}_i) = \frac{1}{2}\left(\hat{I}\pm\underline{\sigma}_i\cdot\hat{\eta}_i\right) \tag{6.86}$$

projects onto the eigenstate of the $\pm\underline{\sigma}_i\cdot\hat{\eta}_i$. Therefore, the joint probability P is found by projecting the density operator onto the subspace of spin eigenstates $|\pm\hat{\eta}_1\rangle|\pm\hat{\eta}_2\rangle$ and then taking the trace

$$P(\pm\hat{\eta}_1,\pm\hat{\eta}_2) = \text{Trace}\,\hat{\rho}\,\frac{1}{2}\left(\hat{I}\pm\underline{\sigma}_1\cdot\hat{\eta}_1\right)\frac{1}{2}\left(\hat{I}\pm\underline{\sigma}_2\cdot\hat{\eta}_2\right)$$
$$= \frac{1}{4}\left(1-(\pm)(\pm)\,\hat{\eta}_1\cdot\hat{\eta}_2\right) \tag{6.87}$$

Consider a bi-partite system with sub-systems A and B and a pure state which can be written as

$$|\psi\rangle = \sum_{n_A,m_B} C^{A,B}_{n_A,m_B}\,|n_A\rangle|m_B\rangle \tag{6.88}$$

where n_A and m_B are quantum numbers for system A and B separately. This pure state is a *"Factorizable State"* if

$$C^{A,B}_{n_A,m_B} = C^A_{n_A}\,C^B_{m_B} \tag{6.89}$$

Hence, a factorizable state can be expressed as

$$|\psi\rangle = \sum_{n_A} C^A_{n_A}\,|n_A\rangle\sum_{m_B}C^B_{m_B}|m_B\rangle \tag{6.90}$$

The density operator $\rho_{A,B}$ for a factorizable state can then be written as the product

$$\hat{\rho}_{A,B} = \sum_{n_A,n_A'} C^A_{n_A'}{}^*C^A_{n_A}|n_A\rangle\langle n_A'|\sum_{m_B,m_A'}C^B_{m_B'}{}^*C^B_{m_B}|m_B\rangle\langle m_B'|$$
$$= \hat{\rho}_A\hat{\rho}_B \tag{6.91}$$

In the case of factorizable states, the partial Trace over B is evaluated as

$$\text{Trace}_B \, \hat{\rho}_{A,B} = \hat{\rho}_A \sum_{r_B} \sum_{m_B, n'_B} C^B_{m'_B} * C^B_{m_B} \langle r_B | m_B \rangle \langle m'_B | r_B \rangle$$

$$= \hat{\rho}_A \qquad (6.92)$$

Hence, the partial Trace yields the factor $\hat{\rho}_A$. The state described by the reduced density operator is a pure state $|\psi^A\rangle$,

$$|\psi^A\rangle = \sum_{n_A} C^A_{n_A} \, |n_A\rangle \qquad (6.93)$$

When expressed in terms of $|\psi^A\rangle$ the density operator for the factorizable pure state is given by

$$\hat{\rho}_A = |\psi^A\rangle\langle\psi^A| \qquad (6.94)$$

There are also other pure states that are not factorizable but which contain correlations between the two subsystems. Consider each subsystem to have two degrees of freedom, 0 and 1, then the state

$$|\psi\rangle = \left(\frac{|1_A\rangle|0_B\rangle + |0_A\rangle|1_B\rangle}{\sqrt{2}} \right) \qquad (6.95)$$

is a pure state which is not factorizable. The state describes correlations between the subsystems A and B. For example, if a measurement shows that the A sub-system has the quantum number 1, then it is known that the B subsystem has quantum number 0, and vice versa. On performing the partial Trace over the sub-system B, one obtains

$$\hat{\rho}_A = \sum_{m_B} \langle m_B | \hat{\rho}_{A,B} | m_B \rangle$$

$$= \left(\frac{|1_A\rangle\langle 1_A| + |0_A\rangle\langle 0_A|}{2} \right) \qquad (6.96)$$

which clearly represents a mixed state. Pure states that become mixed after a partial Trace has been performed are called *"Entangled States"*. On taking the partial Trace some information has been lost, since states are assigned finite probabilities of occurrence.

The subsystems A and B of a system are defined to be *"Statistically Independent"* if the density operator is factorizable

$$\hat{\rho} = \hat{\rho}_A \hat{\rho}_B \qquad (6.97)$$

The density operators $\hat{\rho}_A$ and $\hat{\rho}_B$ can be obtained by performing partial traces

$$\hat{\rho}_A = \text{Trace}_B \, \hat{\rho}$$
$$\hat{\rho}_B = \text{Trace}_A \, \hat{\rho} \tag{6.98}$$

where each Trace is evaluated as a sum over a complete set of states for the subsystem.

The von Neumann entropies of the subsystems are defined as

$$S_A = -k_B \, \text{Trace}_A \, [\hat{\rho}_A \, \ln \hat{\rho}_A] \tag{6.99}$$

and

$$S_B = -k_B \, \text{Trace}_B \, [\hat{\rho}_B \, \ln \hat{\rho}_B] \tag{6.100}$$

For statistically independent systems the total entropy is the sum of the von-Neumann entropies of the subsystems

$$S = S_A + S_B \tag{6.101}$$

For two subsystems which are not statistically independent, one can also define the density operators for the subsystems by the partial traces

$$\hat{\rho}_A = \text{Trace}_B \, \hat{\rho}$$
$$\hat{\rho}_B = \text{Trace}_A \, \hat{\rho} \tag{6.102}$$

where each Trace is evaluated as a sum over a complete set of states for the subsystem. In this case, one can prove that the entropy satisfies the triangle inequality [34]

$$S_A + S_B \geq S \geq |S_A - S_B| \tag{6.103}$$

The von Neumann entropy is different from the Shannon entropy. The Shannon entropy of a composite system can never be less than the entropy of any one component, but this is not true for von Neumann entropy. For example, the von Neumann entropy of a pure state is zero, whereas, if the pure state is entangled, the entropy of the components are non-zero and must be equal. For a composite system in which the components have different von Neumann entropies, the von Neumann entropy of the joint system must be greater than the magnitude of difference between the von Neumann entropies of the subsystems.

Possible Route to the Increase of Entropy

It has been suggested that entanglement may be the cause of the increase of entropy with time. Consider a bipartite system comprised of a system and its environment, where the initial states of the bipartite system are expressed in terms of separable states

$$|\alpha_S, \alpha_E\rangle = |\alpha_S\rangle|\alpha_E\rangle \tag{6.104}$$

For separable states, the initial density operator is expressible as

$$\hat{\rho}(0) = \sum_{\alpha_S, \alpha_E} p_{\alpha_S, \alpha_E} |\alpha_S\rangle|\alpha_E\rangle\langle\alpha_S|\langle\alpha_E| \tag{6.105}$$

In general, as time evolves, the bipartite system will not remain separable if the composite system has a Hamiltonian of the form

$$\hat{H} = \hat{H}_S + \hat{H}_E + \hat{H}_{int} \tag{6.106}$$

where the three terms represent the Hamiltonian for the system, environment and their interaction. The time-evolution of the density operator is simply a unitary transformation, but the resulting operator will not longer be separable as a result of the interaction

$$\hat{\rho}(t) = \sum_{\alpha_S, \alpha_E} p_{\alpha_S, \alpha_E} |\alpha_S, \alpha_E(t)\rangle\langle\alpha_S, \alpha_E(t)| \tag{6.107}$$

That is, the system will become entangled with the environment, so inserting two completeness relations leads to the following expression for the density operator

$$\hat{\rho}(t) = \sum_{\mu_S, \mu_E} \sum_{\nu_S, \nu_E} \sum_{\alpha_S, \alpha_E} p_{\alpha_S, \alpha_E} \langle\mu_S, \mu_E|\alpha_S, \alpha_E(t)\rangle\langle\alpha_S, \alpha_E(t)|\nu_S, \nu_E\rangle$$
$$\times |\mu_S, \mu_E\rangle\langle\nu_S, \nu_E| \tag{6.108}$$

which has-off diagonal terms. Therefore, due to the interaction, the density matrix is not expected to remain diagonal.

As time evolves, it is expected that correlations between the system and its environment will increase. A measurement on the system at time t will have the effect that the off-diagonal matrix elements of the density operator are lost. The information about the correlations of the system entangled with its environment is lost. Furthermore, after the measurement, all the information one has pertain to the system and one remains indifferent to the state of the environment. After the measurement, one may construct a new density operator that maximizes the entropy. The entropy after the measurement is greater than the initial entropy. This can be thought of that the ignorance of the environment has been transferred to the system as a result of the entanglement.

The above viewpoint is the subject of skepticism. Some believe that, since the entire universe and the observer may form a closed system, the universe should have unitary evolution. The entire universe could even be in a pure state. In the formulation described above, the observer and the collapse of the wave function play central roles. The observer, apparently, is removed from the time-evolution of the universe. An alternate viewpoint [35] suggests that the observer may just simply perceive that entropy is increasing. The suggestion is that apparent time-asymmetry in quantum mechanics results from asking time-asymmetric questions, and that questions and experiments can be framed in a manner such that the results are time-symmetric.

Entropy is Simply Perceived to Increase whenever Correlations are Neglected

The process of performing a partial Trace discards quantum correlations and may produce the perception that entropy increases. This can be illustrated by the following Gedanken Experiment.

Consider an electron in an excited state of an atom which can emit a photon. The initial electron is assumed to be in a pure state which is a linear superposition of the spin up and spin down states

$$|\psi_i\rangle = \alpha|\uparrow\rangle + \beta|\downarrow\rangle \qquad (6.109)$$

where

$$|\alpha|^2 + |\beta|^2 = 1 \qquad (6.110)$$

The electronic state is assumed to decay by the emission of a photon. If the electron is spin up, the emitted photon is assumed to have positive helicity. Whereas, if the initial electron is spin-down, the emitted photon is assumed to have a negative helicity. Thus, the final state is given by

$$|\psi_f\rangle = \alpha|\uparrow, +\rangle + \beta|\downarrow, -\rangle \qquad (6.111)$$

The photon leaves the subsystem and becomes part of a remote environment. If the quantity $\hat{A}$ were to be measured on the initial electronic state, the expectation value would be

$$\langle\psi_i|\hat{A}|\psi_i\rangle = |\alpha|^2\langle\uparrow|\hat{A}|\uparrow\rangle + |\beta|^2\langle\downarrow|\hat{A}|\downarrow\rangle$$
$$+ \alpha^*\beta\langle\uparrow|\hat{A}|\downarrow\rangle + \beta^*\alpha\langle\downarrow|\hat{A}|\uparrow\rangle \qquad (6.112)$$

On the other-hand, a measurement of $\hat{A}$ on the final state would yield the expectation value of

$$
\begin{aligned}
\langle \psi_f | \hat{A} | \psi_f \rangle &= |\alpha|^2 \langle \uparrow, + | \hat{A} | \uparrow, + \rangle + |\beta|^2 \langle \downarrow, - | \hat{A} | \downarrow, - \rangle \\
&\quad + \alpha^* \beta \langle \uparrow, + | \hat{A} | \downarrow, - \rangle + \beta^* \alpha \langle \downarrow, - | \hat{A} | \uparrow, + \rangle \\
&= |\alpha|^2 \langle \uparrow | \hat{A} | \uparrow \rangle + |\beta|^2 \langle \downarrow | \hat{A} | \downarrow \rangle
\end{aligned}
\tag{6.113}
$$

where the interference terms vanish due to the orthogonality of the photon's helicity states. The emission of the photon has resulted in the phenomenon of "*Decoherence*".

If the photon in the final state can never be observed again, then one may perform a partial Trace. The partial Trace yields a mixed state where the classical probabilities for the electron being in the respective spin states are given by

$$
\begin{aligned}
p_\uparrow &= |\alpha|^2 \\
p_\downarrow &= |\beta|^2
\end{aligned}
\tag{6.114}
$$

Hence, the von Neumann entropy of the mixed state is given by

$$
\begin{aligned}
S &= -k_B \left(p_\uparrow \ln p_\uparrow + p_\downarrow \ln p_\downarrow \right) \\
&= -k_B \left(|\alpha|^2 \ln |\alpha|^2 + |\beta|^2 \ln |\beta|^2 \right)
\end{aligned}
\tag{6.115}
$$

Thus, one may argue that the entropy of the final state is perceived to have increased, due to the performance of the partial Trace. The partial Trace was justified on the basis that the emitted photon is assumed to have been removed to the remote environment and will never be observable again. This assumption is clearly time-asymmetric.

6.2.5 *Thermalization of Closed Quantum Systems*

In 1929 von Neumann [36] pointed out that thermalization does not necessarily require an increase of entropy. This can be inferred by considering a pure state that is made from a linear superposition of states with distinct energy eigenvalues E_α

$$
| \psi \rangle = \sum_\alpha C_\alpha | E_\alpha \rangle
\tag{6.116}
$$

where normalization requires

$$
\sum_\alpha |C_\alpha|^2 = 1
\tag{6.117}
$$

The pure state has an intrinsic energy distribution given by

$$P(E) = \langle \psi | \delta(E - H) | \psi \rangle$$
$$= \sum_{\alpha} |C_{\alpha}|^2 \delta(E - E_{\alpha}) \tag{6.118}$$

and has an average energy

$$\langle \psi | \hat{H} | \psi \rangle = \sum_{\alpha} |C_{\alpha}|^2 E_{\alpha} \tag{6.119}$$

The energy distribution can be characterized by its higher-order cumulants. For a system that can be described by the thermodynamic limit, the energy eigenvalue spectrum can be thought of as being quasi-continuous. The quantal pure state evolves with time according to

$$|\psi(t)\rangle = \sum_{\alpha} C_{\alpha} \, \exp[-iE_{\alpha}t] \, |E_{\alpha}\rangle \tag{6.120}$$

The thermodynamic average of an operator $\hat{A}$ in this state may be defined as the long time average

$$\langle \psi | \hat{A} | \psi \rangle = \lim_{T \to \infty} \frac{1}{T} \int_{-T}^{0} dt' \, \langle \psi(t') | \hat{A} | \psi(t') \rangle$$
$$= \lim_{T \to \infty} \frac{1}{T} \int_{-T}^{0} dt' \sum_{\alpha, \beta} C_{\beta}^* \, C_{\alpha} \, \exp\left[i \, \frac{t'}{\hbar} \, (E_{\beta} - E_{\alpha}) \right]$$
$$\times \langle E_{\beta} | \hat{A} | E_{\alpha} \rangle$$
$$\to \sum_{\alpha} |C_{\alpha}|^2 \langle E_{\alpha} | \hat{A} | E_{\alpha} \rangle \tag{6.121}$$

Thus, in the energy representation, the time-average has randomized the phases of the off-diagonal components of the density operator. That is, all correlations between the components with different energies have been lost due to decoherence [37, 38]. In this quantal ergodic picture, thermal equilibration has occurred without a change of entropy since the system has remained in its pure state.[4] However, if a thermodynamic measurement is interpreted in terms of a static ensemble with a diagonal equilibrium density operator, the information about the state could be thought of as being encoded in the values of the expansion coefficients. That is, the energy probability distribution is governed by

$$P(E) = \sum_{\alpha} |C_{\alpha}|^2 \delta(E - E_{\alpha}) \tag{6.122}$$

---

[4]Quantum Statistical Mechanics can be formulated on the basis of the "*Principle of Equal a priori Probabilities*" and the "*Principle of Random a priori Phases*".

which, if interpreted in terms of the distribution function for a Canonical Ensemble, would imply

$$\sum_\alpha |C_\alpha|^2 \, \delta(E - E_\alpha) \sim \rho(E) \, \frac{\exp[-\beta E]}{Z} \tag{6.123}$$

where $\rho(E)$ describes the density of states of the N-particle system. Of course, since the energy distribution is very sharply peaked, it is indistinguishable from the energy-distribution of the Micro-Canonical Ensemble. If the probabilistic aspects of the interpretation is pursued further, it will be perceived that the equilibrium state has a non-zero value for the entropy. This version of the "*Quantal Ergodic Hypothesis*" should be contrasted with "*Eigenstate Thermalization Hypothesis*" [39, 40] which maintains that, for local, translationally invariant operators of few-body systems, the information about the thermal equilibrium state is encoded in the matrix elements of the operators between the energy eigenstates and not the expansion coefficients.[5] Basically, the hypothesis asserts that consecutive energy states are separated by very small energy differences, so the expectation values of an operator between eigenstates are smoothly varying functions of energy, but the correlations between the expectation values diminish exponentially for large energy separations or for large system sizes. That is, in this interpretation, one expects that

$$\langle E_\alpha | \hat{A} | E_\alpha \rangle \tag{6.124}$$

is sharply-peaked for E_α in an infinitesimal range around a fixed E and so would give agreement with the value calculated with the Micro-Canonical Ensemble. Although there is evidence supporting the eigenstate thermalization hypothesis, a rigorous proof is lacking.

6.3 Indistinguishable Particles

Any labelling of indistinguishable particles is unphysical, by definition. Since any measurement of an observable A will produce results that are independent of the choice of labelling, the operators must be symmetric under any permutation of the labels. Hence, every physical operator $A(\{\underline{p}_i, \underline{r}_i\})$, including the Hamiltonian, must be a symmetric function of

---

[5]This alternate hypothesis is not without merit, since a single eigenvector of an $N \times N$ Hermitean matrix contains $2(N-1)$ pieces of information, whereas the eigenvalue only contains one piece of information. Hence, potentially there could be many correlations embedded in the expectation values of certain classes of operators.

the particle position and momenta vectors $(\hat{p}_i, r_i)$ of the N particles. Any permutation of the set of N particles can be represented in terms of the successive interchanges of pairs of particles. The pair of particles labelled as $(\hat{p}_i, r_i)$ and $(\hat{p}_j, r_j)$ are interchanged by the transposition operator $\hat{P}_{i,j}$. The transposition operator $\hat{P}_{i,j}$ is Hermitean and unitary. In the coordinate representation, the transposition operator has the effect

$$\hat{P}_{i,j}\ \Psi(r_1, r_2, \ldots, r_i, \ldots, r_j, \ldots, r_N) = \Psi(r_1, r_2, \ldots, r_j, \ldots, r_i, \ldots, r_N)$$

$$(6.125)$$

Since any physical operator must be invariant under the permutation of any two particles, one has

$$\hat{P}_{i,j}\ A(\{\underline{p}_i, r_i\})\ \hat{P}_{i,j}^{-1} = A(\{\underline{p}_i, r_i\}) \qquad (6.126)$$

Hence, every $\hat{P}_{i,j}$ commutes[6] with every physical operator including the Hamiltonian

$$[\hat{P}_{i,j}, \hat{A}] = 0 \qquad (6.127)$$

Therefore, each transposition operator can be diagonalized simultaneously together with any complete set of compatible physical operators.

The eigenvalues of the transposition operator are defined as p

$$\hat{P}_{i,j}|\Phi_p\rangle = p|\Phi_p\rangle \qquad (6.128)$$

However, as two successive interchanges of the labels i and j leaves the state unchanged, one has

$$\hat{P}_{i,j}^2 = \hat{\mathcal{I}} \qquad (6.129)$$

Thus, the eigenvalues of the transposition operators must satisfy

$$\hat{P}_{i,j}^2|\Phi_p\rangle = p^2|\Phi_p\rangle$$
$$= |\Phi_p\rangle \qquad (6.130)$$

which leads to the solutions for the eigenvalues

$$p = \pm 1 \qquad (6.131)$$

which are constants of motion. Every permutation $\mathcal{P}$ can be expressed as a product of transposition operators

$$\hat{\mathcal{P}} = \prod_{i,j} \hat{P}_{i,j} \qquad (6.132)$$

---

[6]Note that distinct transposition operators $P_{i,j}$ and $P_{j,k}$ do not commute. Therefore, one cannot find a complete set of basis states that are simultaneous eigenstates of all permutation operators. However, it is possible to find specific states that are eigenstates of all permutation operators.

The order of the permutation $n_{\mathcal{P}}$ is the number of transpositions in the decomposition of the permutation $\mathcal{P}$. The decomposition of a permutation is not unique, so the order of the permutation is only defined modulo 2. The eigenvalue of the permutation operator is equal to the product of the eigenvalues of the transposition operators in its decomposition.

Example: Simultaneous Eigenstates of the Permutation Operators

Consider the cyclic permutation $\hat{\mathcal{P}}_{1,2,3}$ that exchanges the indices $(1, 2, 3)$ to $(3, 2, 1)$.

$$\hat{\mathcal{P}}_{1,2,3} \, (1, 2, 3) = (3, 1, 2) \tag{6.133}$$

This permutation operator can be decomposed into a product of transposition operators

$$\hat{\mathcal{P}}_{1,2,3} = \hat{P}_{1,2} \, \hat{P}_{1,3} \tag{6.134}$$

as can be seen from its action

$$
\begin{aligned}
\hat{\mathcal{P}}_{1,2,3} \, (1, 2, 3) &= \hat{P}_{1,2} \, \hat{P}_{1,3} \, (1, 2, 3) \\
&= \hat{P}_{1,2} \, (3, 2, 1) \\
&= (3, 1, 2)
\end{aligned}
\tag{6.135}
$$

However, one also has the equivalent decompositions

$$
\begin{aligned}
\hat{\mathcal{P}}_{1,2,3} &= \hat{P}_{2,3} \, \hat{P}_{1,2} \\
\hat{\mathcal{P}}_{1,2,3} &= \hat{P}_{1,3} \, \hat{P}_{2,3}
\end{aligned}
\tag{6.136}
$$

The eigenvalues of the transposition operators $\hat{P}_{i,j}$ are denoted by $p_{i,j}$ and the eigenvalues of the cyclic permutation operator are denoted by $p_{1,2,3}$. Simultaneous eigenstates of all the permutation operators must have eigenvalues that satisfy

$$p_{1,2,3} = p_{1,2} \, p_{1,3} = p_{1,2} \, p_{2,3} = p_{1,3} \, p_{2,3} \tag{6.137}$$

which implies $p_{1,3} = p_{2,3} = p_{1,2}$. Hence, all the transposition operators must have the same eigenvalues, so the states are either completely symmetric or anti-symmetric under a transposition. Furthermore, since the $p_{i,j} = \pm 1$, one always has $p_{1,2,3} = 1$.

Since every physical operator is symmetric under any permutation of its labels, it must commute with every permutation operator so one may construct simultaneous eigenstates [41]. Moreover, if it is postulated that one may construct simultaneous eigenstates of every permutation operator then, since any permutation can be expressed as a product of transpositions, all transposition operators must have the same eigenvalue p. Hence, in three-dimensions, the physical states must be either completely symmetric or completely antisymmetric. The *"Symmetrization Postulate"* can be stated that every physical realizable state is either totally symmetric or totally antisymmetric under transpositions.

Since the real space probability density given by

$$|\Psi(\underline{r}_1, \underline{r}_2, \ldots, \underline{r}_i, \ldots, \underline{r}_j, \ldots, \underline{r}_N)|^2 \tag{6.138}$$

is observable, and not the wave function, measurements on two states differing only by permutations of the particle labels will yield results that have identical distributions.

6.3.1 *Fermions and Bosons*

Particles are known as *"Fermions"* if their wave functions are antisymmetric under the interchange of any pair of particles, whereas the particles are called *"Bosons"* if the wave function is symmetric under a single interchange.

In certain two-dimensional systems, states with mixed symmetry can occur [42]. The exotic particles with mixed symmetries are known as *"Anyons"*, and they obey fractional statistics.

Fermions and Fermi-Dirac Statistics

Particles are fermions, if their wave function is antisymmetric under the interchange of any pair of particles

$$\Psi(\underline{r}_1, \underline{r}_2, \ldots, \underline{r}_i, \ldots, \underline{r}_j, \ldots, \underline{r}_N)$$
$$= -\Psi(\underline{r}_1, \underline{r}_2, \ldots, \underline{r}_j, \ldots, \underline{r}_i, \ldots, \underline{r}_N) \tag{6.139}$$

Examples of fermions are given by electrons, neutrinos, quarks, protons and neutrons, and Helium^3 atoms.

Bosons and Bose-Einstein Statistics

Particles are bosons if their wave function is symmetric under the interchange of any pair of particles

$$\Psi(\underline{r}_1, \underline{r}_2, \ldots, \underline{r}_i, \ldots, \underline{r}_j, \ldots, \underline{r}_N) = \Psi(\underline{r}_1, \underline{r}_2, \ldots, \underline{r}_j, \ldots, \underline{r}_i, \ldots, \underline{r}_N)$$

$$(6.140)$$

Examples of bosons are given by photons, gluons, phonons, and Helium^4 atoms.

One can represent an arbitrary N-particle state with the required symmetry as a linear superposition of a complete set of orthonormal N-particle basis states Φ. These many-particle basis states are composed as a properly symmetrized product of single-particle wave functions $\phi_\alpha(\underline{r})$, which form a complete orthonormal set

$$\int d^3\underline{r} \; \phi_\beta^*(\underline{r}) \; \phi_\alpha(\underline{r}) = \delta_{\beta,\alpha}$$

$$\sum_\alpha \phi_\alpha^*(\underline{r}') \; \phi_\alpha(\underline{r}) = \delta^3(\underline{r} - \underline{r}') \qquad (6.141)$$

The many-particle basis states $\Phi_{\alpha_1, \alpha_2, \ldots, \alpha_N}(\underline{r}_1, \underline{r}_2, \ldots, \underline{r}_N)$ are composed from a symmetrized linear superposition of N single-particle states

$$\Phi_{\alpha_1, \alpha_2, \ldots, \alpha_N}(\underline{r}_1, \underline{r}_2, \ldots, \underline{r}_N)$$

$$= \aleph^{-1} \sum_{\mathcal{P}} (\pm 1)^{n_\mathcal{P}} \; \hat{\mathcal{P}}_\mathcal{P} \; \phi_{\alpha_1}(\underline{r}_1) \; \phi_{\alpha_2}(\underline{r}_2) \; \cdots \; \phi_{\alpha_N}(\underline{r}_N) \qquad (6.142)$$

where the sum runs over all $N!$ permutations of the particle indices, $\hat{\mathcal{P}}_\mathcal{P}$ is the permutation operator which switches the indices, and $n_\mathcal{P}$ is the order of the permutation. For boson wave functions, the positive sign holds, and the minus sign holds for fermions.

The Pauli Exclusion Principle

The "*Pauli Exclusion Principle*" states that a single-particle state cannot contain more than one fermion. That is, any physically acceptable many-particle state describing identical fermions $\Phi_{\alpha_1, \ldots, \alpha_i, \ldots, \alpha_j \ldots}$ must have $\alpha_i \neq \alpha_j$. Otherwise, if $\alpha_i = \alpha_j$, then one has

$$\hat{P}_{i,j} \; \Phi_{\alpha_1, \ldots, \alpha_i, \ldots, \alpha_i, \ldots}(\underline{r}_1, \ldots, \underline{r}_i, \ldots, \underline{r}_j, \ldots) = \Phi_{\alpha_1, \ldots, \alpha_i, \ldots, \alpha_i, \ldots}(\underline{r}_1, \ldots, \underline{r}_i, \ldots, \underline{r}_j, \ldots)$$

$$(6.143)$$

since interchanging the factors $\phi_{\alpha_i}(\underline{r}_i)$ and $\phi_{\alpha_i}(\underline{r}_j)$ does not change the sign of the wave function. Furthermore, since the many-particle state is an eigenstate of $\hat{P}_{i,j}$ with eigenvalue -1, one also has

$$\hat{P}_{i,j}\Phi_{\alpha_1,..,\alpha_i,...,\alpha_i,..}(\underline{r}_1,..,\underline{r}_i,...,\underline{r}_j,..) = -\Phi_{\alpha_1,..,\alpha_i,...,\alpha_i,..}(\underline{r}_1,..,\underline{r}_i,...,\underline{r}_j,..)$$
(6.144)

Therefore, on equating the right-hand sides one finds that

$$\Phi_{\alpha_1,...,\alpha_i,...,\alpha_i,..}(\underline{r}_1,..,\underline{r}_i,...,\underline{r}_j,..) = 0 \qquad (6.145)$$

which indicates that the state where two fermions occupy the same single-particle eigenstate does not exist.

Example: The Wave Function Symmetries of Two Identical Particles

Fermions

The basis wave function for a system containing two identical fermions at positions $\underline{r}_1$ and $\underline{r}_2$, in which one fermion is occupying a state labeled by the set of single-particle quantum numbers α and the other fermion occupies a state labeled by the quantum numbers β is denoted by $\Phi_{\alpha,\beta}(\underline{r}_1,\underline{r}_2)$. Due to the antisymmetry of fermionic wave functions under a permutation of the labels 1 and 2, the basis wave functions are written as

$$\Phi_{\alpha,\beta}(\underline{r}_1,\underline{r}_2) = \left(\frac{\phi_\alpha(\underline{r}_1)\,\phi_\beta(\underline{r}_2) - \phi_\alpha(\underline{r}_2)\,\phi_\beta(\underline{r}_1)}{\sqrt{2!}}\right) \qquad (6.146)$$

or equivalently as the Slater determinant

$$\Phi_{\alpha,\beta}(\underline{r}_1,\underline{r}_2) = \frac{1}{\sqrt{2!}} \begin{vmatrix} \phi_\alpha(\underline{r}_1) & \phi_\alpha(\underline{r}_2) \\ \phi_\beta(\underline{r}_1) & \phi_\beta(\underline{r}_2) \end{vmatrix} \qquad (6.147)$$

The two fermion basis wave functions are antisymmetric in α and β and, therefore, vanish when $\alpha = \beta$ in accord with Pauli's Exclusion Principle. A general wave function of the two-fermion system can be expressed as a linear superposition of the two-fermion basis wave functions.

Bosons

The basis wave function for a system containing two identical bosons at positions $\underline{r}_1$ and $\underline{r}_2$, in which a boson is occupying a state labeled by the set of single-particle quantum numbers α and the other particle occupies a state labeled by the quantum numbers β is denoted by $\Phi_{\alpha,\beta}(\underline{r}_1,\underline{r}_2)$. Due to the symmetry of bosonic wave functions under permutations of the labels 1 and 2, the basis wave function is written as

$$\Phi_{\alpha,\beta}(\underline{r}_1,\underline{r}_2) = \left(\frac{\phi_\alpha(\underline{r}_1)\ \phi_\beta(\underline{r}_2) + \phi_\alpha(\underline{r}_2)\ \phi_\beta(\underline{r}_1)}{\sqrt{2!}} \right) \tag{6.148}$$

if $\alpha \neq \beta$. If $\alpha = \beta$, the basis wave function can be expressed as

$$\Phi_{\alpha,\alpha}(\underline{r}_1,\underline{r}_2) = \phi_\alpha(\underline{r}_1)\ \phi_\alpha(\underline{r}_2) \tag{6.149}$$

It should be noted that the basis wave function with $\alpha = \beta$ can be found from the case with $\alpha \neq \beta$ by setting $\alpha = \beta$ and by dividing by an extra factor of $\sqrt{n_\alpha!}$ where $n_\alpha = 2$ is the number of bosons occupying the state α.

6.3.2 *The Occupation Number Representation*

Instead of labeling the many-particle basis states by the eigenvalues $\alpha_1, \alpha_2, \ldots, \alpha_N$ we can specify the number of times each single-particle eigenstate is occupied. The number of times that a specific one-particle state ϕ_α occurs is denoted by n_α which is called the "*Occupation Number*". Specifying the occupation numbers $n_{\alpha_1}, n_{\alpha_2}, \ldots$, uniquely specifies the many-particle states $\Phi_{n_{\alpha_1},n_{\alpha_2},\ldots}$. For a system with N particles, the sum of the occupation numbers is just equal to the total number of particles

$$\sum_\alpha n_\alpha = N \tag{6.150}$$

Due to the different symmetries under particle interchange, quantum mechanical particles obey "*Quantum Statistics*". For fermions, the Pauli exclusion principle limits n_α to have values of either 0 or 1. For bosons, n_α can have any positive integer value, including zero.

The orthonormality relation

$$\prod_{i=1}^{N} \left\{ \int d^3\underline{r}_i \right\} \Phi^*_{\{n_\beta\}}(\{\underline{r}_i\})\ \Phi_{\{n_\alpha\}}(\{\underline{r}_i\}) = \delta_{n_{\beta_1},n_{\alpha_1}}\ \delta_{n_{\beta_2},n_{\alpha_2}} \cdots \tag{6.151}$$

leads to the identification of the normalization constant $\aleph$. For fermions, one has

$$\aleph = \sqrt{N!} \tag{6.152}$$

where $N!$ is just the number of terms in the wave function. For bosons, the normalization is given by

$$\aleph = \sqrt{N! \prod_\alpha n_\alpha!} \tag{6.153}$$

where it should be noted that 0! is defined as unity.

An arbitrary many-particle state Ψ can be expressed in terms of the set of basis states as

$$\Psi(\underline{r}_1, \underline{r}_2, \ldots, \underline{r}_N) = \sum_{\{n_\alpha\}} C(n_{\alpha_1}, n_{\alpha_2}, \ldots) \, \Phi_{n_{\alpha_1}, n_{\alpha_2}, \ldots}(\underline{r}_1, \underline{r}_2, \ldots, \underline{r}_N) \tag{6.154}$$

where the expansion coefficients $C(n_{\alpha_1}, n_{\alpha_2}, \ldots)$ play the role of the wave function in the occupation number representation.

6.4 "The Spin-Statistics Theorem" and Composite Particles

The "*Spin-Statistics Theorem*" has its origins in Quantum Field Theory, and states that fermions have half odd integer spins and that bosons have integer spins. The theorem was first proposed by Markus Fierz [43] and later proved by Wolfgang Pauli [44]. Rather than prove the theorem, we shall be content to show that if the Spin-Statistics holds for elementary particles then it will also hold for composite particles. We shall also outline an observation which might be turned into a proof of the Theorem.

Two indistinguishable composite particles are permuted if all the elementary particles composing one composite particles are interchanged with the corresponding constituent particles of the other. When two identical composite particle each of which is composed of n_F elementary fermions and n_B elementary bosons are interchanged, the wave function will change

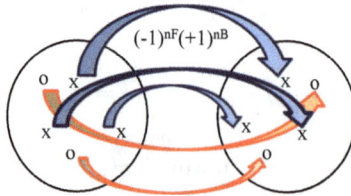

Fig. 6.1 Composite particles are interchanged when their constituent particles are interchanged. For a system with n_F fermions and n_B bosons, the interchange changes the wave function representing the pair of composite particles by a factor of $(-1)^{n_F}(+1)^{n_B}$.

by a factor of

$$(-1)^{n_F} \, (+1)^{n_B}$$

Hence, if a composite particle contains an odd number of fermions, the composite particle will be a fermion since the wave function of two such identical composite particles is antisymmetric under the interchange of the particles. On the other hand, if a composite particle contains an even number of fermions, the composite particle will be a boson since the wave function of two such identical composite particles will be symmetric under the interchange of the particles.

The above result is consistent with the application of the Spin-Statistics Theorem. A composite particle containing n_F fermions and n_B bosons will have a spin composed of n_F half-odd integers and n_B integers. If n_F is odd, the resulting spin will be a half-odd integer, whereas if n_F is even the resulting spin will be integer.

Thus, a composite particle containing an odd number of fermions n_F will have a half-odd integer spin and the wave function of a pair of identical composite particles with odd n_F will be antisymmetric under their interchange. Likewise, a composite particle containing an even number of fermions will have an integer spin, and the wave function of a pair of identical composite particles with even n_F will be symmetric under the interchange of the composite particles. Hence, the Spin-Statistics Theorem will be true for identical composite particles if it is true for their elementary constituents.

Example: The Isotopes of He

He^3 has two protons, a neutron and two electrons. Therefore, He^3 is a fermion.

He^4 has an extra neutron. Thus, it contains two protons, two neutrons and two electrons. Therefore, He^4 is a boson.

The difference in the quantum statistics of the two isotopes results in their having very different properties at low temperatures, although they are chemically similar. For example, their phase diagrams are very different and He^4 exhibits the phenomenon of superfluidity.

The idea behind the Spin-Statistics Theorem is that fields must transform under Lorentz transformations according to the spin of the particles that they describe. The generators of the transformations include boost generators and the generators of rotations. It is noted that rotations about an angle 2π are equivalent to the identity for fields with integer spins, such as scalar or vector fields. However, this is not true for half odd-integer spin fields represented by spinors where a rotation by 2π changes the sign of the spinor. This can be seen by examining the operator $\hat{\mathcal{R}}(\varphi)$ that rotates a spin of one-half about the z-axis by an angle φ. The operator only acts on the two-component spinor and is given by

$$
\begin{aligned}
\hat{\mathcal{R}}(\varphi) &= \exp\left[-\frac{i}{\hbar}\,\varphi \hat{S}^z\right] \\
&= \exp\left[-i\,\frac{\varphi}{2}\,\sigma^z\right] \\
&= \cos\frac{\varphi}{2}\,\hat{I} - i\sin\frac{\varphi}{2}\,\sigma^z
\end{aligned}
\tag{6.155}
$$

and where σ^z is the z-th Pauli spin matrix. Substitution of $\varphi = 2\pi$ leads to

$$
\hat{\mathcal{R}}(2\pi) = -\hat{I}
\tag{6.156}
$$

which shows that a rotation through 2π switches the sign of the wave function of a spin one-half particle. For general values of S, the angle-dependence of spin rotation operator is spanned by the functions $\cos(s_z\varphi)$ and $\sin(s_z\varphi)$, where s_z are the eigenvalues of $\hat{S}^z$. This together with the observation that $\hat{\mathcal{R}}(0) \equiv \hat{I}$ constitutes a proof that the spinor wave functions of half odd-integer spin particles change sign under rotation by 2π, but there is no sign change for integer spin particles.

Consider the expression involving the product of two field operators $\hat{\psi}$, with arbitrary spin, given by

$$
(\hat{\mathcal{R}}(\pi)\,\hat{\psi}(\underline{r}))\,\hat{\psi}(-\underline{r})
\tag{6.157}
$$

where $\underline{r}$ is assumed to lie in the $z = 0$ plane and the matrix $\hat{\mathcal{R}}(\pi)$ acts on the spinor $\hat{\psi}(\underline{r})$ but not $\hat{\psi}(-\underline{r})$. The operator $\hat{\mathcal{R}}(\pi)$ rotates the spin of $\hat{\psi}(\underline{r})$ by π about the z-axis. The product of field operators describes the probability amplitude for finding two identical particles, one at $\underline{r}$ and the other at $-\underline{r}$ which have spins that are rotated by π relative to each other. Now consider a rotation of this configuration about the z-axis through an angle π. Under this rotation, the particles at the two points $\underline{r}$ and $-\underline{r}$

switch places, and the spins of the two field operators are rotated through an angle of π. Thus, the product of the two field operators transforms as

$$(\hat{\mathcal{R}}(2\pi)\ \hat{\psi}(-\underline{r}))\ (\hat{\mathcal{R}}(\pi)\ \hat{\psi}(\underline{r})) \tag{6.158}$$

For integer spins, the above product is equivalent with

$$\hat{\psi}(-\underline{r})\ (\hat{\mathcal{R}}(\pi)\ \hat{\psi}(\underline{r})) \tag{6.159}$$

and for half odd-integer spins, the product is equivalent to

$$-\hat{\psi}(-\underline{r})\ (\hat{\mathcal{R}}(\pi)\ \hat{\psi}(\underline{r})) \tag{6.160}$$

Therefore, the identical particles with half odd-integer spins that are associated with two field operators have been interchanged by a rotation, but the interchange does involve a change of sign. There is no change in sign associated with the interchange of integer spin particles. This implies that integer spin particles are bosons and half-odd integer spin particles are fermions.

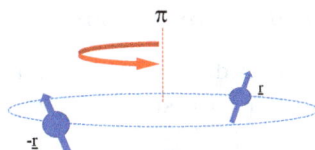

Fig. 6.2   A particle with half-integer spin is added at position $-\underline{r}$ which is followed by the addition of an identical particle at $\underline{r}$ with a spin direction which is obtained from the spin direction of the first particle by a rotation through π about the z-axis. When the entire configuration is rotated about the center of mass through π, the interchange of particles leads back to the initial state but is associated with a change of sign.

The above consideration is only a plausibility argument and is not a proof of the Spin-Statistics Theorem. A proof of the Spin-Statistics Theorem would require that the following assumptions hold true:

(i) The theory has a Lorentz invariant Lagrangian.
(ii) The vacuum is Lorentz invariant.
(iii) The particle is a localized excitation which is not connected to any other object.
(iv) The particle propagates with finite mass.
(v) The particle is a real excitation, so that states involving the particle have positive definite norms.

The above plausibility argument introduced the idea of associating a particle at a point is space $\underline{r}$ with an operator $\hat{\psi}(\underline{r})$, such that

$$\hat{\psi}(\underline{r})\,\hat{\psi}(\underline{r}') = \pm\hat{\psi}(\underline{r}')\,\hat{\psi}(\underline{r}) \qquad (6.161)$$

where the choice of sign depends on the spin of the particle. The properties of the operators are developed over the next two sections.

6.5 Second Quantization

"*Second Quantization*" does not involve further quantization, it merely is a book keeping device which allows one to construct wave functions for an $(N+1)$ particle system out of the wave functions of an N particle system. The second quantum formalism is of great utility for systems in which the number of particles is not conserved. Furthermore, for physical processes in which only a finite number of particles change their states and the majority are unchanged, second quantization allows one to focus attention on the particles that do change states.

6.5.1 *Second Quantization for Bosons*

For Bose particles, one can introduce a set of operators $\hat{a}_\alpha^\dagger$ which are defined by their action on the number basis states

$$\hat{a}_\alpha^\dagger|\Phi_{n_{\alpha_1},\ldots,n_\alpha,\ldots}\rangle = \sqrt{n_\alpha + 1}|\Phi_{n_{\alpha_1},\ldots,n_\alpha+1,\ldots}\rangle \qquad (6.162)$$

where α represents the quantum numbers which describes a single-particle state. The Hermitean conjugate operator is found to satisfy

$$\hat{a}_\alpha|\Phi_{n_{\alpha_1},\ldots,n_\alpha,\ldots}\rangle = \sqrt{n_\alpha}|\Phi_{n_{\alpha_1},\ldots,n_\alpha-1,\ldots}\rangle \qquad (6.163)$$

The "*Creation Operator*" $\hat{a}_\alpha^\dagger$ adds an additional particle to the quantum level α, and the "*Annihilation Operator*" $\hat{a}_\alpha$ annihilates a particle in the state α. If in the initial state Φ, $n_\alpha = 0$ then application of the annihilation operator yields zero.

One can define the "*Number Operator*" $\hat{n}_\alpha$ as the combination

$$\hat{n}_\alpha = \hat{a}_\alpha^\dagger\hat{a}_\alpha \qquad (6.164)$$

which, according to the definitions of the creation and annihilation operator, satisfies the equation

$$\hat{n}_\alpha|\Phi_{n_{\alpha_1},\ldots,n_\alpha,\ldots}\rangle = n_\alpha|\Phi_{n_{\alpha_1},\ldots,n_\alpha,\ldots}\rangle \qquad (6.165)$$

Hence, the occupation number basis states are eigenstate of $\hat{n}_\alpha$ where the eigenvalue is the occupation number of the single-particle state, n_α. Hence, the number operator represents a measurement of the number of particles in a quantum level α. The total number of particles in the system, N, corresponds to the eigenvalues of the operator

$$\hat{N} = \sum_\alpha \hat{n}_\alpha$$
$$= \sum_\alpha \hat{a}_\alpha^\dagger \hat{a}_\alpha \qquad (6.166)$$

The boson creation and annihilation operators satisfy the commutation relations

$$[\hat{a}_\alpha, \hat{a}_\beta^\dagger] = \delta_{\alpha,\beta} \qquad (6.167)$$

and

$$[\hat{a}_\alpha^\dagger, \hat{a}_\beta^\dagger] = 0$$
$$[\hat{a}_\alpha, \hat{a}_\beta] = 0 \qquad (6.168)$$

as can be easily shown. For example, the diagonal elements of the first commutator yields

$$(\hat{a}_\alpha \hat{a}_\alpha^\dagger - \hat{a}_\alpha^\dagger \hat{a}_\alpha)|\Phi_{n_{\alpha_1}...n_\alpha...}\rangle = ((n_\alpha + 1) - n_\alpha)|\Phi_{n_{\alpha_1}...n_\alpha...}\rangle$$
$$= 1|\Phi_{n_{\alpha_1}...n_\alpha...}\rangle \qquad (6.169)$$

and the off-diagonal elements yield zero.

One can create any arbitrary basis state by the repeated action of the creation operators on the vacuum state. If the vacuum state is denoted as $|0\rangle$, the basis states can be expressed in the second quantized form

$$|\Phi_{n_{\alpha_1},n_{\alpha_2}...}\rangle = \prod_{\alpha_i} \left\{ \frac{(a_{\alpha_i}^\dagger)^{n_{\alpha_i}}}{\sqrt{n_{\alpha_i}!}} \right\} |0\rangle \qquad (6.170)$$

where the product runs over all the single-particle quantum numbers α. The wave function of a state is defined as its position representation,

$$\Phi_{n_{\alpha_1},n_{\alpha_2}...}(\underline{r}_1, \underline{r}_2, \cdots \underline{r}_N) = \langle \underline{r}_1, \underline{r}_2, \ldots, \underline{r}_N | \Phi_{n_{\alpha_1},n_{\alpha_2}...}\rangle \qquad (6.171)$$

therefore, the permutation symmetry of the wave function under permutation is ensured if the creation operators commute. Since the annihilation operators are the Hermitean conjugates of the creation operators, they must also commute.

Any operator can also be expressed in terms of the creation and annihilation operators. For example, the Hamiltonian of a system of interacting particles can be written as

$$\hat{H} = \sum_{\alpha;\alpha'} \langle \alpha'|\hat{H}_0|\alpha\rangle \hat{a}_{\alpha'}^\dagger \hat{a}_\alpha + \frac{1}{2!} \sum_{\alpha,\beta;\alpha',\beta'} \langle \alpha'\beta'|\hat{V}_{int}|\alpha\beta\rangle \hat{a}_{\alpha'}^\dagger \hat{a}_{\beta'}^\dagger \hat{a}_\beta \hat{a}_\alpha \qquad (6.172)$$

where

$$\langle \alpha' | \hat{H}_0 | \alpha \rangle = \int d^3\underline{r} \; \phi_{\alpha'}^*(\underline{r}) \left[\frac{\hat{p}^2}{2m} + V_0(\underline{r}) \right] \phi_\alpha(\underline{r}) \tag{6.173}$$

denotes the matrix elements of the single-particle energy and

$$\langle \alpha' \beta' | \hat{V}_{int} | \alpha \beta \rangle = \int d^3\underline{r} \int d^3\underline{r}' \; \phi_{\alpha'}^*(\underline{r}) \; \phi_{\beta'}^*(\underline{r}') \; V_{int}(\underline{r}, \underline{r}') \; \phi_\beta(\underline{r}') \; \phi_\alpha(\underline{r}) \tag{6.174}$$

represent the matrix elements of the two-body interaction terms.

6.5.2 *Second Quantization for Fermions*

For fermi particles, one can formally introduce a set of operators $\hat{c}_\alpha^\dagger$ in which α represents a set of quantum numbers that describe a single-particle state. It is important to note that a convention must be adopted which governs the order in which the labeled occupation numbers appear in the number basis states $|\Phi_{n_{\alpha_1},\dots,n_\alpha,\dots}\rangle$ and that the same convention be used consistently. The creation operators are defined by

$$\hat{c}_\alpha^\dagger |\Phi_{n_{\alpha_1},\dots,n_\alpha,\dots}\rangle = (-1)^{M_\alpha} \sqrt{n_\alpha + 1} \; |\Phi_{n_{\alpha_1},\dots,n_\alpha+1,\dots}\rangle \tag{6.175}$$

where M_α is a positive integer which, for antisymmetric states Φ, is defined by

$$M_\alpha = \sum_{\alpha_i < \alpha} n_{\alpha_i} \tag{6.176}$$

which counts the number of particles occupying the levels α_i that appear before n_α in the ordered list of occupation numbers. The Hermitean conjugate operator $\hat{c}_\alpha$ is then found to satisfy

$$\hat{c}_\alpha |\Phi_{n_{\alpha_1},\dots,n_\alpha,\dots}\rangle = (-1)^{M_\alpha} \sqrt{n_\alpha} \; |\Phi_{n_{\alpha_1},\dots,n_\alpha-1,\dots}\rangle \tag{6.177}$$

The phase factor is either +1 or -1, depending on whether the total number of particles in the single-particle states α_i that appear in the ordered list before α is even or odd.

Example: Two-Fermion States

If $|0\rangle$ is the vacuum state in which all fermion occupation numbers are zero, then one can express the state in which a single fermion is occupying a single-particle level labeled by the set of quantum numbers α as

$$|\alpha\rangle = \hat{c}_\alpha^\dagger |0\rangle \tag{6.178}$$

which has the wave function

$$\langle \underline{r}_1 | \alpha \rangle = \phi_\alpha(\underline{r}_1) \tag{6.179}$$

A two-particle state can be created by adding another fermion, but this time to the single-particle state β

$$|\beta, \alpha\rangle = \hat{c}_\beta^\dagger \, \hat{c}_\alpha^\dagger |0\rangle \tag{6.180}$$

The two-fermion state has the wave function $\Phi_{\alpha,\beta}(\underline{r}_1, \underline{r}_2)$

$$\langle \underline{r}_1, \underline{r}_2 | \beta, \alpha \rangle = \Phi_{\alpha,\beta}(\underline{r}_1, \underline{r}_2) \tag{6.181}$$

If the two-fermion state is created by adding the particles in the reverse order, one obtains

$$|\alpha, \beta\rangle = \hat{c}_\alpha^\dagger \, \hat{c}_\beta^\dagger |0\rangle \tag{6.182}$$

which has the wave function

$$\langle \underline{r}_1, \underline{r}_2 | \alpha, \beta \rangle = \Phi_{\beta,\alpha}(\underline{r}_1, \underline{r}_2) \tag{6.183}$$

Since fermionic wavefunctions are antisymmetric under the permutation of two particles, one has

$$\Phi_{\beta,\alpha}(\underline{r}_1, \underline{r}_2) = -\Phi_{\alpha,\beta}(\underline{r}_1, \underline{r}_2) \tag{6.184}$$

Therefore, one infers that the fermion creation operators must anticommute

$$\hat{c}_\alpha^\dagger \, \hat{c}_\beta^\dagger = -\hat{c}_\beta^\dagger \, \hat{c}_\alpha^\dagger \tag{6.185}$$

By taking the Hermitean conjugate of the above relation, one finds that the annihilation operators must also anticommute.

Due to the antisymmetric nature of the many-body wave functions $\Phi_{\{n_{\alpha_i}\}}$, the fermion creation and annihilation operators must satisfy the anti-commutation relations

$$\{\hat{c}_\alpha, \hat{c}_\beta^\dagger\} = \delta_{\alpha,\beta} \tag{6.186}$$

and

$$\{\hat{c}_\alpha^\dagger, \hat{c}_\beta^\dagger\} = 0$$
$$\{\hat{c}_\alpha, \hat{c}_\beta\} = 0 \tag{6.187}$$

where the anti-commutator of two operators $\hat{A}$ and $\hat{B}$ are defined by

$$\{\hat{A}, \hat{B}\} = \hat{A}\hat{B} + \hat{B}\hat{A} \tag{6.188}$$

The anti-commutation relations restrict the occupation numbers n_α for a single-particle state α to zero or unity, as required by the Pauli exclusion principle. This can be seen since the anti-commutation relation yields

$$\hat{c}_\alpha \hat{c}_\alpha \left| \Phi_{n_{\alpha_1}, \ldots, n_\alpha, \ldots} \right\rangle = -\hat{c}_\alpha \hat{c}_\alpha \left| \Phi_{n_{\alpha_1}, \ldots, n_\alpha, \ldots} \right\rangle$$
$$= 0 \qquad (6.189)$$

which shows that two identical creation operators acting on a number state annihilates the state. Hence, on using the definition of the annihilation operator, one finds

$$\sqrt{(n_\alpha - 1)\, n_\alpha} \left| \Phi_{n_{\alpha_1}, \ldots, n_\alpha - 2, \ldots} \right\rangle = 0 \qquad (6.190)$$

so either $n_\alpha = 1$ or $n_\alpha = 0$. Incorporating this restriction into the definition of the creation operator yields

$$\hat{c}_\alpha^\dagger \left| \Phi_{n_{\alpha_1}, \ldots, n_\alpha, \ldots} \right\rangle = (-1)^{M_\alpha} \left| \Phi_{n_{\alpha_1}, \ldots, n_\alpha + 1, \ldots} \right\rangle \qquad \text{for } n_\alpha = 0$$
$$= 0 \quad \text{for } n_\alpha = 1 \qquad (6.191)$$

and similarly for the annihilation operator

$$\hat{c}_\alpha \left| \Phi_{n_{\alpha_1}, \ldots, n_\alpha, \ldots} \right\rangle = (-1)^{M_\alpha} \left| \Phi_{n_{\alpha_1}, \ldots, n_\alpha - 1, \ldots} \right\rangle \qquad \text{for } n_\alpha = 1$$
$$= 0 \quad \text{for } n_\alpha = 0 \qquad (6.192)$$

The occupation number operator is defined as

$$\hat{n}_\alpha = \hat{c}_\alpha^\dagger \hat{c}_\alpha \qquad (6.193)$$

which is seen to satisfy the eigenvalue equation

$$\hat{n}_\alpha \left| \Phi_{n_{\alpha_1}, \ldots, n_\alpha, \ldots} \right\rangle = n_\alpha \left| \Phi_{n_{\alpha_1}, \ldots, n_\alpha, \ldots} \right\rangle \qquad (6.194)$$

similar to the case for bosons. The total number operator, $\hat{N}$, is then defined as

$$\hat{N} = \sum_\alpha \hat{n}_\alpha$$
$$= \sum_\alpha \hat{c}_\alpha^\dagger \hat{c}_\alpha \qquad (6.195)$$

If the vacuum state is denoted by $|0\rangle$, one can express any arbitrary basis state in terms of the repeated action of creation operators on the vacuum state

$$\left| \Phi_{n_{\alpha_1}, n_{\alpha_2} \ldots} \right\rangle = \prod_{\alpha_i} \left\{ (\hat{c}_{\alpha_i}^\dagger)^{n_{\alpha_i}} \right\} |0\rangle \qquad (6.196)$$

where the ordered product runs over all the single-particle quantum numbers α_i. The phase factor

$$(-1)^{M_\alpha}$$

which appears in the definition of the actions of either the creation or annihilation operator on the number basis state, is recognized as just the number of anticommutations that are required to move the operator from the front of the expression to a position in which it has the correct order.

Any operator can also be expressed in terms of the creation and annihilation operators. For example, the Hamiltonian for a system of interacting fermions can be written as

$$
\hat{H} = \sum_{\alpha;\alpha'} \langle \alpha' | \hat{H}_0 | \alpha \rangle \hat{c}_{\alpha'}^\dagger \hat{c}_\alpha
$$
$$
+ \frac{1}{2!} \sum_{\alpha,\beta;\alpha',\beta'} \langle \alpha'\beta' | \hat{V}_{int} | \alpha\beta \rangle \hat{c}_{\alpha'}^\dagger \hat{c}_{\beta'}^\dagger \hat{c}_\beta \hat{c}_\alpha \tag{6.197}
$$

where

$$
\langle \alpha' | \hat{H}_0 | \alpha \rangle = \int d^3\underline{r} \; \phi_{\alpha'}^*(\underline{r}) \left[\frac{\hat{p}^2}{2m} + V_0(\underline{r}) \right] \phi_\alpha(\underline{r}) \tag{6.198}
$$

is the matrix elements of the single-particle energy and $\langle \alpha'\beta' | \hat{V}_{int} | \alpha\beta \rangle$ represents the matrix elements of the two-body interaction between products of two single-fermion wave functions.

$$
\langle \alpha'\beta' | \hat{V}_{int} | \alpha\beta \rangle = \int d^3\underline{r} \int d^3\underline{r}' \; \phi_{\alpha'}^*(\underline{r}) \, \phi_{\beta'}^*(\underline{r}') \, V_{int}(\underline{r},\underline{r}') \, \phi_\beta(\underline{r}') \, \phi_\alpha(\underline{r}) \tag{6.199}
$$

As required by Hermiticity, the creation operators appear in the reverse order of the annihilation operators.

Coherent States

Since the occupation number is unrestricted, an unusual type of state is allowed for bosons. We shall focus our attention on one single-particle quantum level, and shall drop the index α which labels the level. A "*Coherent State*" $|a_\varphi\rangle$ is defined as an eigenstate of the annihilation operator [31]

$$
\hat{a} | a_\varphi \rangle = a_\varphi | a_\varphi \rangle \tag{6.200}
$$

where a_φ is a complex number. For example, the vacuum state or ground state is an eigenstate of the annihilation operator, in which case $a_\varphi = 0$.

The coherent state [32] has the form of an infinite linear superposition of eigenstates of the number operator with eigenvalues n

$$|a_\varphi\rangle = \sum_{n=0}^{\infty} C_n |n\rangle \tag{6.201}$$

On substituting this form into the definition of the coherent state

$$\hat{a}|a_\varphi\rangle = \sum_n C_n \hat{a}|n\rangle$$

$$= a_\varphi \sum_n C_n |n\rangle \tag{6.202}$$

and using the property of the annihilation operator, one has

$$\sum_n C_n \sqrt{n}|n-1\rangle = a_\varphi \sum_n C_n |n\rangle \tag{6.203}$$

On taking the matrix elements of this equation with the state $\langle m|$, and using the orthonormality of the eigenstates of the number operator, one finds

$$C_{m+1}\sqrt{m+1} = a_\varphi C_m \tag{6.204}$$

Hence, on iterating downwards, the C_m are determined up to a normalization constant as

$$C_m = \left(\frac{a_\varphi^m}{\sqrt{m!}}\right) C_0 \tag{6.205}$$

and the coherent state can be expressed as

$$|a_\varphi\rangle = C_0 \sum_{n=0}^{\infty} \left(\frac{a_\varphi^n}{\sqrt{n!}}\right) |n\rangle \tag{6.206}$$

The normalization constant C_0 can be found from

$$1 = C_0^* C_0 \sum_{n=0}^{\infty} \left(\frac{a_\varphi^{n*} a_\varphi^n}{n!}\right) \tag{6.207}$$

by noting that the sum exponentiates to yield

$$1 = C_0^* C_0 \ \exp[a_\varphi^* a_\varphi] \tag{6.208}$$

so, on choosing the phase of C_0, one has

$$C_0 = \exp\left[-\frac{1}{2} a_\varphi^* a_\varphi\right] \tag{6.209}$$

From this, it can be shown that if the number of bosons in a coherent state is measured, the result n will occur with a probability given by

$$P(n) = \frac{(a_\varphi^* a_\varphi)^n}{n!} \exp[-a_\varphi^* a_\varphi] \tag{6.210}$$

Thus, the boson statistics are governed by a Poisson distribution. Furthermore, the quantity $a_\varphi^* a_\varphi$ is the average number of bosons $\bar{n}$ present in the coherent state.

The coherent states can be written in a more compact form. Since the state with occupation number n can be written as

$$|n\rangle = \frac{(\hat{a}^\dagger)^n}{\sqrt{n!}} |0\rangle \tag{6.211}$$

the coherent state can also be expressed as

$$|a_\varphi\rangle = \exp\left[-\frac{1}{2} a_\varphi^* a_\varphi\right] \sum_{n=0}^{\infty} \frac{(a_\varphi \hat{a}^\dagger)^n}{n!} |0\rangle \tag{6.212}$$

or on summing the series as an exponential

$$|a_\varphi\rangle = \exp\left[-\frac{1}{2} a_\varphi^* a_\varphi\right] \exp[a_\varphi \hat{a}^\dagger] |0\rangle \tag{6.213}$$

Thus the coherent state is an infinite linear superposition of states with different occupation numbers, each coefficient in the linear superposition has a specific phase relation with every other coefficient.

The above equation represents a transformation between number operator states and the coherent states. The inverse transformation can be found by expressing a_φ as a magnitude a and a phase φ

$$a_\varphi = a \exp[i\varphi] \tag{6.214}$$

The number states can be expressed in terms of the coherent states via the inverse transformation

$$|n\rangle = \frac{\sqrt{n!}}{a^n} \exp\left[+\frac{1}{2} a^2\right] \int_0^{2\pi} \frac{d\varphi}{2\pi} \exp[-in\varphi] |a_\varphi\rangle \tag{6.215}$$

by integrating over the phase φ of the coherent state. Since the set of occupation number states is complete, the set of coherent states must also span Hilbert space. In fact, the set of coherent states is over-complete.

A number of systems do have states which have properties that closely resemble coherent states, such as the photon states in a laser or such as the superfluid condensate of He^4 at low temperatures. Although they do not have the precise mathematical form of the coherent states investigated by Glauber, the approximate states are characterized by having sufficiently large fluctuations in their occupation numbers so that the expectation value of the annihilation operator is well-defined.

6.6   Field Quantization

Second quantization can be achieved from single-particle quantum me-
chanics simply by replacing the wave functions $\psi(\underline{r})$ and $\psi^*(\underline{r})$ by "*Field
Operators*" $\hat{\psi}(\underline{r})$ and $\hat{\psi}^\dagger(\underline{r})$ that are defined by

$$\hat{\psi}(\underline{r}) = \sum_\alpha \phi_\alpha(\underline{r})\, \hat{a}_\alpha$$

$$\hat{\psi}^\dagger(\underline{r}) = \sum_\alpha \phi_\alpha^*(\underline{r})\, \hat{a}_\alpha^\dagger \tag{6.216}$$

in which the $\phi_\alpha(\underline{r})$ are a complete orthonormal set of single-particle wave
functions. The field operator and its Hermitean conjugate, respectively,
have the effect of annihilating and creating a particle at position $\underline{r}$. The
inverse transform is found by using the orthonormality of the single-particle
basis functions $\phi_\alpha(\underline{r})$

$$\hat{a}_\alpha = \int d^3\underline{r}\; \phi_\alpha^*(\underline{r})\, \hat{\psi}(\underline{r})$$

$$\hat{a}_\alpha^\dagger = \int d^3\underline{r}\; \phi_\alpha(\underline{r})\, \hat{\psi}^\dagger(\underline{r}) \tag{6.217}$$

For fermions, we can define $\phi_\alpha(\underline{r})$ as a two-component spinor so that the
field operators can also be represented as spinor operators $\hat{\psi}_\sigma(\underline{r})$ with single-
particle basis states $\phi_{m,\sigma}(\underline{r})$.

The field operators satisfy equal-time commutation or anti-commutation
relations which follow from the type of commutation relations satisfied by
the creation and annihilation relations. For example, the boson/fermion
field creation and annihilation operators satisfy the commutation/anti-
commutation relations

$$\hat{\psi}(\underline{r})\,\hat{\psi}^\dagger(\underline{r}') \pm \hat{\psi}^\dagger(\underline{r}')\,\hat{\psi}(\underline{r}) = \sum_{\alpha,\beta} \phi_\alpha(\underline{r})\phi_\beta^*(\underline{r}')\,(\hat{a}_\alpha\hat{a}_\beta^\dagger \pm \hat{a}_\beta^\dagger\hat{a}_\alpha)$$

$$= \sum_{\alpha,\beta} \phi_\alpha(\underline{r})\phi_\beta^*(\underline{r}')\,\delta_{\alpha,\beta}$$

$$= \sum_\alpha \phi_\alpha(\underline{r})\phi_\alpha^*(\underline{r}')$$

$$= \delta^3(\underline{r} - \underline{r}') \tag{6.218}$$

The completeness relation for the single-particle wave functions has been
used to obtain the last line. Similarly, the other two commutation/anti-

commutation relations are found to be

$$\hat{\psi}^\dagger(\underline{r})\,\hat{\psi}^\dagger(\underline{r}') \pm \hat{\psi}^\dagger(\underline{r}')\,\hat{\psi}^\dagger(\underline{r}) = 0$$
$$\hat{\psi}(\underline{r})\,\hat{\psi}(\underline{r}') \pm \hat{\psi}(\underline{r}')\,\hat{\psi}(\underline{r}) = 0 \qquad (6.219)$$

The commutation/anti-commutation relations can be extended to unequal times and, for Lorentz invariant theories, is seen to be an expression of causality. For equal times, there is no possibility that a measurement made at position $\underline{r}$ can affect the physics at position $\underline{r}' \neq \underline{r}$ since the disturbance can only travel as fast as the speed of light.

Since the field operators create or annihilate particles at position $\underline{r}$, they provide a natural link between the wave function and second quantized representations. For example, the second quantized N-particle basis state (with the set of occupation numbers $\{n_\alpha\}$) that is expressed by

$$|\Phi_{\{n_\alpha\}}\rangle = \prod_j \left\{ \frac{(\hat{a}_{\alpha_j}^\dagger)^{n_{\alpha_j}}}{\sqrt{n_{\alpha_j}!}} \right\} |0\rangle \qquad (6.220)$$

where the occupation numbers are subject to the constraint

$$N = \sum_j n_{\alpha_j} \qquad (6.221)$$

has a real space basis wave function which is generated by

$$\Phi_{\{n_\alpha\}}(\underline{r}_1, \underline{r}_2, \dots \underline{r}_N) = \langle \underline{r}_1, \underline{r}_2, \dots \underline{r}_N | \Phi_{\{n_\alpha\}} \rangle$$
$$= \frac{(\pm 1)^{N-1}}{\sqrt{N!}} \langle 0| \prod_{i=1}^N \{\hat{\psi}(\underline{r}_i)\} \prod_j \left\{ \frac{(\hat{a}_{\alpha_j}^\dagger)^{n_{\alpha_j}}}{\sqrt{n_{\alpha_j}!}} \right\} |0\rangle$$

$$(6.222)$$

as can be shown combinatorially. The above relation leads to the identification of the field operators as the creators of the N-particle position basis states

$$|\underline{r}_1, \underline{r}_2, \dots, \underline{r}_N\rangle = \frac{1}{\sqrt{N!}}\,\hat{\psi}^\dagger(\underline{r}_N)\,\hat{\psi}^\dagger(\underline{r}_{N-1}) \dots \hat{\psi}^\dagger(\underline{r}_1)\,|0\rangle \qquad (6.223)$$

The link between the real space basis wave functions and the field operators can be used to express the real space representation of matrix elements of operators in terms of the second quantized basis states. This process produces the field quantized form of operators. As an example, consider the matrix elements of a single-particle operator $\hat{A}_1$

$$\hat{A}_1 = \sum_{i=1}^N \hat{A}_1(\underline{r}_i) \qquad (6.224)$$

On expressing the wave functions in terms of field operators, the expectation value can be written as

$$\langle \Phi_{\{n_\beta\}} | \hat{A}_1 | \Phi_{\{n_\alpha\}} \rangle = \frac{1}{N!} \prod_{k=1}^{N} \left\{ \int d^3 \underline{r}_k \right\} \langle \Phi_{\{n_\beta\}} | \left(\prod_{j=1}^{N} \hat{\psi}(\underline{r}_j) \right)^{\dagger}$$

$$\times \sum_{i=1}^{N} \hat{A}_1(\underline{r}_i) \left(\prod_{j=1}^{N} \hat{\psi}(\underline{r}_j) \right) | \Phi_{\{n_\alpha\}} \rangle \qquad (6.225)$$

It should be noted that since the operator $\hat{A}_1$ has been positioned symmetrically and is surrounded by equal numbers of creation (annihilation) and field operators acting on the vacuum state, the projection onto the vacuum is redundant and has been omitted. The operator $\hat{A}_1(\underline{r}_i)$ only acts on the field operator $\hat{\psi}(\underline{r}_i)$ and should be kept next to it. For each term in the sum over i, the pair of field operators are commuted/anticommuted until $\hat{\psi}^{\dagger}(\underline{r}_i)$ and $\hat{\psi}(\underline{r}_i)$ are positioned to the extreme left and right of the strings of field operators, respectively. Since the re-ordering involves an even number of commutations or anticommutations, the signs of the matrix elements do not change. The integration over $\underline{r}_1$ may be performed leading to the total number operator

$$\int d^3 \underline{r}_1 \, \hat{\psi}^{\dagger}(\underline{r}_1) \, \hat{\psi}(\underline{r}_1) = \sum_{\alpha,\beta} \int d^3 \underline{r}_1 \, \phi_\beta^*(\underline{r}_1) \, \phi_\alpha(\underline{r}_1) \, \hat{a}_\beta^{\dagger} \hat{a}_\alpha$$

$$= \sum_{\alpha} \hat{a}_\alpha^{\dagger} \hat{a}_\alpha$$

$$= \hat{N} \qquad (6.226)$$

In arriving at the above expression, we have used the orthonormality of the single-particle basis functions. The expectation value of the total number operator is unity. The integration over $\underline{r}_2$ also generates a factor of the total number operator $\hat{N}$ which, due to the reduction in the number of field operators on the right, produces an eigenvalue of 2. The integration is performed $(N-1)$ times, where at each step the eigenvalue of the number operator is increased by one, until one arrives at the expression

$$\langle \Phi_{\{n_\beta\}} | \hat{A}_1 | \Phi_{\{n_\alpha\}} \rangle = \frac{(N-1)!}{N!} \sum_{i=1}^{N} \int d^3 \underline{r}_i$$

$$\times \langle \Phi_{\{n_\beta\}} | \hat{\psi}^{\dagger}(\underline{r}_i) \hat{A}_1(\underline{r}_i) \hat{\psi}(\underline{r}_i) | \Phi_{\{n_\alpha\}} \rangle$$

$$= \langle \Phi_{\{n_\beta\}} | \int d^3 \underline{r} \hat{\psi}^{\dagger}(\underline{r}) \, \hat{A}_1(\underline{r}) \, \hat{\psi}(\underline{r}) | \Phi_{\{n_\alpha\}} \rangle \qquad (6.227)$$

Hence, the expectation value has been expressed in terms of the field quantized operators as

$$\hat{A}_1 = \int d^3\underline{r}\ \hat{\psi}^\dagger(\underline{r})\ \hat{A}_1(\underline{r})\ \hat{\psi}(\underline{r}) \qquad (6.228)$$

which is to be sandwiched between states that are expressed as products of creation (annihilation) operators acting on the vacuum state. This process can also be applied to yield the field quantized expressions for two-particle operators, and results in

$$\hat{A}_2 = \frac{1}{2!}\int d^3\underline{r}\int d^3\underline{r}'\ \hat{\psi}^\dagger(\underline{r})\ \hat{\psi}^\dagger(\underline{r}')\ \hat{A}_2(\underline{r},\underline{r}')\ \hat{\psi}(\underline{r}')\ \hat{\psi}(\underline{r}) \qquad (6.229)$$

Due to their locality, the field operators can be used to represent the density operator in position space $\hat{\rho}(\underline{r})$

$$\hat{\rho}(\underline{r}) = \hat{\psi}^\dagger(\underline{r})\ \hat{\psi}(\underline{r}) \qquad (6.230)$$

The number operator is given by the integral of the density operator over all space

$$\hat{N} = \int d^3\underline{r}\ \hat{\psi}^\dagger(\underline{r})\ \hat{\psi}(\underline{r})$$

$$= \sum_\alpha \hat{a}_\alpha^\dagger \hat{a}_\alpha \qquad (6.231)$$

The result is recognized as the sum of the number operators for the single particle states.

Noether's Theorem provides the corresponding expression for the second-quantized probability current density operator. For non-relativistic particles, the current density is expressed in terms of the field operators by

$$\hat{\underline{j}}(\underline{r}) = \frac{\hbar}{2im}\ [\hat{\psi}^\dagger(\underline{r})\ \underline{\nabla}\hat{\psi}(\underline{r}) - (\underline{\nabla}\hat{\psi}^\dagger(\underline{r}))\ \hat{\psi}(\underline{r})] \qquad (6.232)$$

which reduces to the second quantized form

$$\hat{\underline{j}}(\underline{r}) = \frac{\hbar}{2im}\ \sum_{\alpha,\beta}(\phi_\beta^*(\underline{r})\underline{\nabla}\phi_\alpha(\underline{r}) - (\underline{\nabla}\phi_\beta^*(\underline{r}))\ \phi_\alpha(\underline{r}))\ \hat{a}_\beta^\dagger \hat{a}_\alpha \qquad (6.233)$$

When multiplied by the charge, this expression forms the paramagnetic component of the electrical current density, but the diamagnetic component of the current density has to be obtained by using the minimal coupling procedure that incorporates the vector potential.

The field quantized expression for the Hamiltonian is given by

$$\hat{H} = \hat{H}_0 + \hat{H}_{int} \tag{6.234}$$

where the non-interacting Hamiltonian $\hat{H}_0$ is expressed as

$$\hat{H}_0 = \int d^3\underline{r}\ \hat{\psi}^\dagger(\underline{r}) \left[-\frac{\hbar^2}{2m}\nabla^2 + V_0(\underline{r}) \right] \hat{\psi}(\underline{r}) \tag{6.235}$$

and the pairwise Coulomb interaction as

$$\hat{H}_{int} = \frac{1}{2!} \int d^3\underline{r} \int d^3\underline{r}'\ \hat{\psi}^\dagger(\underline{r})\ \hat{\psi}^\dagger(\underline{r}')\ V_{int}(\underline{r} - \underline{r}')\ \hat{\psi}(\underline{r}')\ \hat{\psi}(\underline{r}) \tag{6.236}$$

On expressing the field operators as sums of single-particle basis state operators, the above expressions reduce to the previously discussed forms of second quantized Hamiltonians.

6.7   Fermion and Boson Distributions

The Fermi-Dirac and Bose-Einstein distribution functions can be derived by using the statistical operator $\hat{\rho}$ and second quantization.

In the Heisenberg representation, the time dependence of an annihilation operator is given by

$$\hat{a}_\alpha(t) = \exp\left[+\frac{i}{\hbar}\hat{H}t \right] \hat{a}_\alpha \exp\left[-\frac{i}{\hbar}\hat{H}t \right] \tag{6.237}$$

The Heisenberg equation of motion for the annihilation operator has the form

$$i\hbar\,\frac{\partial}{\partial t}\hat{a}_\alpha(t) = [\hat{a}_\alpha(t), \hat{H}(t)] \tag{6.238}$$

For non-interacting particles which are described by the Hamiltonian

$$\hat{H} = \sum_\beta \epsilon_\beta \hat{a}_\beta^\dagger \hat{a}_\beta \tag{6.239}$$

the equation of motion simplifies to

$$i\hbar\,\frac{\partial}{\partial t}\hat{a}_\alpha(t) = \sum_\beta \epsilon_\beta\ [\hat{a}_\alpha(t), \hat{a}_\beta^\dagger(t)\ \hat{a}_\beta(t)]$$

$$= \sum_\beta \epsilon_\beta \hat{a}_\beta(t)\ \delta_{\alpha,\beta}$$

$$= \epsilon_\alpha \hat{a}_\alpha(t) \tag{6.240}$$

where the second line results by using either fermionic or bosonic commutation relations. After solving the differential equation, one finds the time-dependence of the annihilation operator is simply given by

$$\hat{a}_\alpha(t) = \exp\left[-\frac{i}{\hbar}\,\epsilon_\alpha t\right]\hat{a}_\alpha \qquad (6.241)$$

for both fermions and bosons. Likewise, the time-dependence of the creation operator is found to be given by the Hermitean conjugate expression

$$\hat{a}_\alpha^\dagger(t) = \exp\left[+\frac{i}{\hbar}\,\epsilon_\alpha t\right]\hat{a}_\alpha^\dagger \qquad (6.242)$$

The above relations can be extended to imaginary times $t = i\beta$.

In the Grand-Canonical Ensemble, where the statistical operator is given by

$$\hat{\rho} = \frac{1}{\Xi}\,\exp[-\beta(\hat{H} - \mu\hat{N})] \qquad (6.243)$$

the second-quantized thermal averaged occupation number is given by

$$
\begin{aligned}
\langle\hat{n}_\alpha\rangle &= \frac{1}{\Xi}\,\mathrm{Trace}(\exp[-\beta(\hat{H} - \mu\hat{N})]\,\hat{a}_\alpha^\dagger\hat{a}_\alpha) \\
&= \frac{1}{\Xi}\,\mathrm{Trace}(\exp[-\beta(\hat{H} - \mu\hat{N})]\,\hat{a}_\alpha^\dagger\,\exp[+\beta(\hat{H} - \mu\hat{N})] \\
&\quad \times \exp[-\beta(\hat{H} - \mu\hat{N})]\,\hat{a}_\alpha) \\
&= \exp[-\beta(\epsilon_\alpha - \mu)]\,\frac{1}{\Xi}\,\mathrm{Trace}\,(\hat{a}_\alpha^\dagger\,\exp[-\beta(\hat{H} - \mu\hat{N})]\,\hat{a}_\alpha) \\
&= \exp[-\beta(\epsilon_\alpha - \mu)]\,\frac{1}{\Xi}\,\mathrm{Trace}\,(\exp[-\beta(\hat{H} - \mu\hat{N})]\,\hat{a}_\alpha\hat{a}_\alpha^\dagger) \qquad (6.244)
\end{aligned}
$$

where we have used the expression for the imaginary time dependence of the creation operator in the second line and the cyclic invariance of the Trace in the third line. On using the fermion and boson commutation relations, the thermally averaged occupation number reduces to

$$\langle\hat{n}_\alpha\rangle = \exp[-\beta(\epsilon_\alpha - \mu)]\,\frac{1}{\Xi}\,\mathrm{Trace}\,(\exp[-\beta(\hat{H} - \mu\hat{N})]\,[1 \pm \hat{a}_\alpha^\dagger\hat{a}_\alpha]) \qquad (6.245)$$

where the plus sign pertains to boson and the negative sign to fermions. Therefore, the thermally averaged occupation numbers are determined self-consistently from the equation

$$\langle\hat{n}_\alpha\rangle = \exp[-\beta(\epsilon_\alpha - \mu)]\,(1 \pm \langle\hat{n}_\alpha\rangle) \qquad (6.246)$$

The solutions for the quantum distribution functions are

$$\langle \hat{n}_\alpha \rangle = \frac{1}{\exp[\beta\,(\epsilon_\alpha - \mu)] \mp 1} \tag{6.247}$$

which corresponds to the Bose-Einstein and Fermi-Dirac distribution functions.

6.8   Correlation Functions

Non-interacting classical particles do not show any correlations between themselves. However, since quantum mechanical particles obey "*Quantum Statistics*" they do exhibit correlations.

A "*Single-Particle Correlation Function*" $C_1(r, r')$ can be defined in terms of the field operators by

$$C_1(r, r') = \langle \hat{\psi}^\dagger(r)\,\hat{\psi}(r') \rangle \tag{6.248}$$

The correlation function describes the probability amplitude that the annihilation of a particle at position r' followed by the creation of a particle at r leads to the recovery of the initial state. The correlation function is evaluated as

$$\begin{aligned}
C_1(r, r') &= \sum_{\alpha,\beta} \phi_\beta^*(r)\,\phi_\alpha(r')\langle \hat{a}_\beta^\dagger \hat{a}_\alpha \rangle \\
&= \sum_{\alpha,\beta} \phi_\beta^*(r)\,\phi_\alpha(r')\,\delta_{\alpha,\beta}\langle \hat{n}_\alpha \rangle \\
&= \sum_\alpha \phi_\alpha^*(r)\,\phi_\alpha(r')\langle \hat{n}_\alpha \rangle
\end{aligned} \tag{6.249}$$

The form of the single-particle correlations has a similar form for fermion or bosons, and only differ due to the different explicit expressions for the thermally averaged occupation number.

A two-particle or "*Pair Correlation Function*" $C_2(r, r')$ can be defined in terms of the field operator by

$$\begin{aligned}
C_2(r, r') &= \langle \hat{\psi}^\dagger(r)\hat{\psi}^\dagger(r')\hat{\psi}(r')\hat{\psi}(r) \rangle \\
&= \sum_{\alpha,\beta,\alpha',\beta'} \phi_{\alpha'}^*(r)\phi_{\beta'}^*(r')\phi_\beta(r')\phi_\alpha(r)\langle \hat{a}_{\alpha'}^\dagger\,\hat{a}_{\beta'}^\dagger\,\hat{a}_\beta\hat{a}_\alpha \rangle
\end{aligned} \tag{6.250}$$

The pair correlation function describes the probability density that if a particle is found at position r another particle will be found at position r'. For convenience of notation, we shall define the zero of energy as μ. The expectation value of the products of creation and annihilation operators can be evaluated as

$$\langle \hat{a}^\dagger_{\alpha'} \, \hat{a}^\dagger_{\beta'} \, \hat{a}_\beta \hat{a}_\alpha \rangle = \frac{1}{\Xi} \text{ Trace } \exp[-\beta \hat{H}] \, \hat{a}^\dagger_{\alpha'} \, \hat{a}^\dagger_{\beta'} \, \hat{a}_\beta \hat{a}_\alpha$$

$$= \exp[-\beta \epsilon_{\alpha'}] \frac{1}{\Xi} \text{ Trace } \exp[-\beta \hat{H}] \, \hat{a}^\dagger_{\beta'} \, \hat{a}_\beta \hat{a}_\alpha \hat{a}^\dagger_{\alpha'}$$

$$= \exp[-\beta \epsilon_{\alpha'}] \frac{1}{\Xi} \text{ Trace } \exp[-\beta \hat{H}] \, \hat{a}^\dagger_{\beta'} \, \hat{a}_\beta \, [\delta_{\alpha,\alpha'} \pm \hat{a}^\dagger_{\alpha'} \, \hat{a}_\alpha]$$

$$\text{(6.251)}$$

where the last line is obtained for boson or fermion operators by using, respectively, the commutation or anti-commutation relations. Hence,

$$\langle \hat{a}^\dagger_{\alpha'} \, \hat{a}^\dagger_{\beta'} \, \hat{a}_\beta \hat{a}_\alpha \rangle = \exp[-\beta \epsilon_{\alpha'}] \left[\langle \hat{n}_\beta \rangle \, \delta_{\beta,\beta'} \, \delta_{\alpha,\alpha'} \right.$$

$$\left. \pm \frac{1}{\Xi} \text{ Trace } \exp[-\beta \hat{H}] \, \hat{a}^\dagger_{\beta'} \, \hat{a}_\beta \hat{a}^\dagger_{\alpha'} \, \hat{a}_\alpha \right]$$

$$= \exp[-\beta \epsilon_{\alpha'}] \left[\langle \hat{n}_\beta \rangle \delta_{\beta,\beta'} \, \delta_{\alpha,\alpha'} \right.$$

$$\left. \pm \frac{1}{\Xi} \text{ Trace } \exp[-\beta \hat{H}] \, \hat{a}^\dagger_{\beta'} \, [\delta_{\alpha',\beta} \pm \hat{a}^\dagger_{\alpha'} \, \hat{a}_\beta] \, \hat{a}_\alpha \right]$$

$$= \exp[-\beta \epsilon_{\alpha'}] \, [\langle \hat{n}_\beta \rangle \, \delta_{\beta,\beta'} \, \delta_{\alpha,\alpha'} \pm \langle \hat{n}_\alpha \rangle \, \delta_{\alpha,\beta'} \, \delta_{\beta,\alpha'}]$$

$$+ \frac{\exp[-\beta \epsilon_{\alpha'}]}{\Xi} \text{ Trace } \exp[-\beta \hat{H}] \, \hat{a}^\dagger_{\beta'} \, \hat{a}^\dagger_{\alpha'} \, \hat{a}_\beta \hat{a}_\alpha$$

$$= \exp[-\beta \epsilon_{\alpha'}] \, [\langle \hat{n}_\beta \rangle \, \delta_{\beta,\beta'} \, \delta_{\alpha,\alpha'} \pm \langle \hat{n}_\alpha \rangle \, \delta_{\alpha,\beta'} \, \delta_{\beta,\alpha'}]$$

$$\pm \exp[-\beta \epsilon_{\alpha'}] \langle \hat{a}^\dagger_{\alpha'} \, \hat{a}^\dagger_{\beta'} \, \hat{a}_\beta \hat{a}_\alpha \rangle \qquad \text{(6.252)}$$

Thus, the expectation value of the operators is given by the expression

$$\langle \hat{a}^\dagger_{\alpha'} \, \hat{a}^\dagger_{\beta'} \, \hat{a}_\beta \hat{a}_\alpha \rangle = \langle \hat{n}_\alpha \rangle \langle \hat{n}_\beta \rangle [\delta_{\beta,\beta'} \, \delta_{\alpha,\alpha'} \pm \delta_{\alpha,\beta'} \, \delta_{\beta,\alpha'}] \qquad \text{(6.253)}$$

Using the above expression for the expectation value, one finds that the two-particle correlation function reduces to

$$C_2(\underline{r},\underline{r}') = \langle \hat{\psi}^\dagger(\underline{r}) \, \hat{\psi}^\dagger(\underline{r}') \, \hat{\psi}(\underline{r}') \, \hat{\psi}(\underline{r}) \rangle$$

$$= \sum_{\alpha,\beta,\alpha'\beta'} \langle \hat{n}_\alpha \rangle \langle \hat{n}_\beta \rangle [\delta_{\beta,\beta'} \, \delta_{\alpha,\alpha'} \pm \delta_{\alpha,\beta'} \, \delta_{\beta,\alpha'}]$$

$$\times \phi^*_{\alpha'}(\underline{r}) \, \phi^*_{\beta'}(\underline{r}') \, \phi_\beta(\underline{r}') \, \phi_\alpha(\underline{r})$$

$$= \sum_{\alpha,\beta} \langle \hat{n}_\alpha \rangle \langle \hat{n}_\beta \rangle$$

$$\times [|\phi_\alpha(\underline{r})|^2 |\phi_\beta(\underline{r}')|^2 \pm \phi^*_\alpha(\underline{r}) \phi_\beta(\underline{r}) \, \phi^*_\beta(\underline{r}') \phi_\alpha(\underline{r}')] \qquad \text{(6.254)}$$

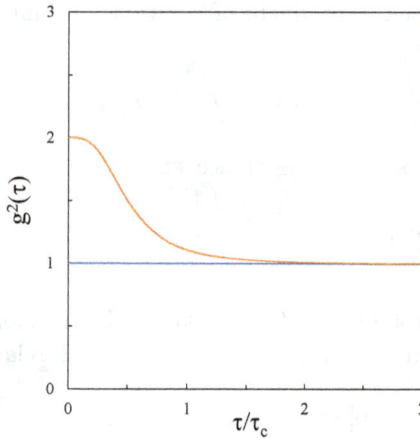

Fig. 6.3   The temporal correlations of the intensity fluctuations of radiation emitted by a thermal source. The probability of observing a second photon is increased by a factor of two just after one photon has been observed. The temporal correlations are related to the spatial correlation function of Eq. (6.254) with the positive sign for Bose-Einstein Statistics. The red line corresponds to the uniform intensity expected from coherent radiation.

Apart from the explicit forms of the average occupation numbers, the two-particle correlation function for fermions and bosons differ due to the sign of the last term in the brackets. The boson two-particle correlation function can show a two-fold increase as $\underline{r} \rightarrow \underline{r}'$, as was found by Hanbury Brown and Twiss [45]. The Hanbury Brown and Twiss experiment showed that photons emitted from a thermal source exhibited bunching due to Bose-Einstein Statistics. The two-fermion correlation function for electrons with parallel spins is expected to vanish in the limit $|\underline{r} - \underline{r}'| \rightarrow 0$, due to the existence of an "*Exchange Correlation Hole*" caused by the Pauli exclusion principle. However, there is no exchange hole for electrons with anti-parallel spins.

6.9 Problems

Problem 6.1

Consider the construction of a density operator $\hat{\rho}$ in the Micro-Canonical Ensemble. In the energy representation, the density operator has two important properties

$$\text{Trace } \hat{\rho} = 1$$

$$\langle E_m|\hat{\rho}|E_n\rangle = \delta_{n,m}\langle E_n|\hat{\rho}|E_n\rangle$$

where the degenerate energy eigenstates $|E_n\rangle$ have been orthogonalized.

Construct the density operator formed in the basis set of eigenstates $|a_n\rangle$ of a Hermitean operator $\hat{A}$ that commutes with $\hat{H}$. In particular, one may form an Ensemble of N normalized pure states $|\psi_n^\alpha\rangle$ expressed as linear superpositions

$$|\psi^\alpha\rangle = \sum_n C_n^\alpha |a_n\rangle$$

where C_n^α are complex coefficients. Using the assumption of equal a priori probabilities, the matrix elements of the density operator can be written as

$$\langle a_m|\hat{\rho}|a_n\rangle = \frac{1}{N}\sum_{\alpha=1}^{N} C_m^\alpha C_n^{\alpha *}$$

(i) Prove that the density operator has a Trace of unity.
(ii) Show that for the density operator to be diagonal, one must assume that the phases of C_n^α must be randomized. That is, the pure states $|\psi^\alpha\rangle$ have the phases of their components randomized.
This suggests that the commonly used Quantum Statistical Ensembles not only assume the "Principle of Equal a priori Probabilities" but also the *"Principle of a priori Random Phases"*.

Problem 6.2

Consider a quantum particle with energy E which is constrained to move in one dimension. A general state can be expressed as

$$|\psi\rangle = C_+|+k\rangle + C_-|-k\rangle$$

in the momentum representation. The density operator of an ensemble of N such systems can be expressed as

$$\hat{\rho} = \frac{1}{N} \sum_{i=1}^{N} |\psi_i\rangle\langle\psi_i|$$

(i) Find an expression for the probability $P(x)$ of finding a particle in the interval dx around x

$$P(x) = \langle x|\hat{\rho}|x\rangle$$

(ii) What condition has to be placed on the coefficients $C_\pm^i$ of the ensemble if the particle is equally likely to be found at any position x?

(iii) Show that this condition is satisfied if the phases of the states in the ensemble are to be averaged over.

Problem 6.3

Density operators are frequently assumed to be diagonal, due to the assumption that the states in the probabilistic interpretation of the density operator are linear superpositions in which the coefficients have random phases. That is, the phases of C_n^m in the expansion of the m-th state in the ensemble

$$|\psi^m\rangle = \sum_{n=1}^{n=N_d} C_n^m |\phi_n\rangle$$

are randomly distributed. Then, on averaging over the random phases, one finds that the density operator

$$\overline{C_n^* \, C_{n'}} = \rho_{n,n} \, \delta_{n,n'}$$

only has diagonal matrix elements. However, if one assumes that at an initial time the density operator only has diagonal elements between the basis states

$$\langle\phi_{n'}|\hat{\rho}|\phi_n\rangle = \langle\phi_n|\hat{\rho}|\phi_n\rangle\delta_{n,n'}$$

then, generally, the density operator will not remain diagonal.

(i) By evaluating the time-derivative of the off-diagonal matrix elements of the density operator's equation of motion

$$i\hbar \, \frac{\partial}{\partial t} \, \hat{\rho} + [\hat{\rho}, \hat{H}] = 0$$

show that if the initial density operator only has diagonal elements, then if it is to remain diagonal at all times one must have

$$((\langle \phi_n | \hat{\rho} | \phi_n \rangle - \langle \phi_{n'} | \hat{\rho} | \phi_{n'} \rangle)) \langle \phi_{n'} | \hat{H} | \phi_n \rangle = 0$$

Thus, either the probabilities of the two states are, a priori, equal

$$\langle \phi_n | \hat{\rho} | \phi_n \rangle = \langle \phi_{n'} | \hat{\rho} | \phi_{n'} \rangle$$

or the two states are not connected by the unitary time-evolution operator to first-order in the Hamiltonian

$$\langle \phi_{n'} | \hat{H} | \phi_n \rangle = 0$$

The density operator for the Canonical Ensemble in the basis of energy eigenstates does remain diagonal, since either sets of degenerate eigenstates are assigned equal a priori probabilities or sets of energy eigenstates corresponding to different eigenvalues are orthogonal.

Problem 6.4

Consider an N-dimensional Hilbert space.
(i) Show that it takes a minimum of $N^2 - 1$ measurements to completely describe a general density operator.
(ii) Show that it takes $2(N - 1)$ measurements to completely specify the density operator for a system which is known to be pure.
(iii) Show that it takes an additional $2(N - 2)$ measurements to completely specify a second distinct pure state. Hence, rationalize the result of (i).

Problem 6.5

Show that the statement Trace $\hat{\rho}^2 = 1$, is equivalent to the statement that the von Neumann entropy vanishes, $S = 0$.

Problem 6.6

Consider the expectation value of a time-independent operator $\hat{A}_S$ in the Schrödinger representation with the density operator $\hat{\rho}$ using the Canonical Ensemble.
(i) Express the expectation value of $\hat{A}_H(t)$ at time t using the Heisenberg representation.
(ii) Show that the expectation value is time-independent.

Problem 6.7

Consider the density operator for a spin in an external magnetic field $\underline{B}$ governed by the Hamiltonian

$$\hat{H} = -\frac{g}{2}\,\mu_B\,\underline{\sigma}\cdot\underline{B}$$

(i) Determine the time-dependence of the spin-polarization $\underline{P}$.

(ii) Use the normalization of the density operator and its Hermitean nature to prove that $\hat{\rho}$ must have the form

$$\hat{\rho} = \frac{1}{2}\,(1+\underline{a}\cdot\underline{\sigma})$$

and identify $\underline{a}$ with the induced polarization.

(iii) Hence evaluate the thermal average of the equation of motion

$$\frac{\partial \underline{P}}{\partial t} = g\mu_B\,(\underline{P}\wedge\underline{B})$$

Problem 6.8

Consider the density operator $\hat{\rho}$ for a system with angular momentum $j = \frac{1}{2}$ which when expressed in terms of the eigenstates of j^z is given by

$$\hat{\rho} = \frac{1}{4}\,|+\rangle\langle+| + \frac{3}{4}\,|-\rangle\langle-|$$

(i) Show that this represents a mixed state.

(ii) Determine the expectation values of j^x, j^y and j^z.

Problem 6.9

(i) Find a general expression for the density operator $\hat{\rho}$ for the Canonical Ensemble for a single spin one-half particle with an energy given by

$$\hat{H} = -\mu_B\underline{B}\cdot\underline{\sigma}$$

by using the operator identity

$$\exp[\beta\mu_B\underline{B}\cdot\underline{\sigma}] = \cosh\beta\mu_B B + \frac{\underline{B}\cdot\underline{\sigma}}{B}\,\sinh\beta\mu_B B$$

(ii) Determine the average value of σ_z if the magnetic field is oriented along the positive z-axis.

(iii) Determine the average value of σ_x if the magnetic field B is oriented along the positive z-direction.

(iv) Show that the unitary transformation

$$\hat{U} = \frac{1}{\sqrt{2}} \begin{pmatrix} 1 & 1 \\ -1 & 1 \end{pmatrix}$$

transforms the density operator between the basis where σ_z and the basis where σ_x is diagonal.

Problem 6.10

Glassy materials are often modeled as two level systems [46]. In these models, an atom can either be in state A denoted by

$$\begin{pmatrix} 1 \\ 0 \end{pmatrix}$$

or in state B represented by

$$\begin{pmatrix} 0 \\ 1 \end{pmatrix}$$

The atom can tunnel between these two states. It is proposed that the Hamiltonian of the system can be represented by

$$\hat{H} = \begin{pmatrix} 0 & -\Delta \\ \Delta & 0 \end{pmatrix}$$

(i) What conditions have to be imposed on Δ?

(ii) Without diagonalizing $\hat{H}$, show that the density operator can be expressed as

$$\hat{\rho} = \frac{1}{Z} \begin{pmatrix} \cosh \beta|\Delta| & +i \sinh \beta|\Delta| \\ -i \sinh \beta|\Delta| & \cosh \beta|\Delta| \end{pmatrix}$$

(iii) Hence, show that

$$Z = 2 \cosh \beta|\Delta|$$

(iv) Determine the average value of $\hat{H}$.

(v) Determine the heat capacity of the model.

Problem 6.11

Consider the density operator $\hat{\rho}$ for a two-state system. Let the density operator be represented by

$$\hat{\rho} = \frac{1}{2} \begin{pmatrix} 1 & 1 - 2p \\ 1 - 2p & 1 \end{pmatrix}$$

where p is a continuous parameter with the range $1 > p > 0$.
(i) Calculate the von Neumann entropy $S(p)$ corresponding to the above density operator.
(ii) Plot the entropy as a function of p and give a physical interpretation to the plot.
Consider a density operator ρ which is expressed as a linear superposition of two density matrices

$$\hat{\rho} = \theta \hat{\rho}_1 + (1 - \theta) \, \hat{\rho}_2$$

where $1 > \theta > 0$, where

$$\rho_1 = \frac{1}{2} \begin{pmatrix} 1 & -1 \\ -1 & 1 \end{pmatrix}$$

and

$$\rho_2 = \frac{1}{2} \begin{pmatrix} 1 & 1 \\ 1 & 1 \end{pmatrix}$$

(iii) Show that the equality in von Neumann's Mixing Theorem[7] holds

$$S(\rho) \geq \theta S(\rho_1) + (1 - \theta) \, S(\rho_2)$$

if and only if, either $\theta = 1$ or $\theta = 0$.
(iv) Show that the inequality holds otherwise.
Hint: The inequality can be shown to be satisfied if $(1 - \frac{\theta}{2})^{(1 - \frac{\theta}{2})} (2\theta)^{\frac{\theta}{2}} < 1$. Hence, find the value of the argument at its maximum value.

Problem 6.12

A matrix in the j_z basis is given by

$$\frac{1}{4} \begin{pmatrix} 2 & 1 & 1 \\ 1 & 1 & 0 \\ 1 & 0 & 1 \end{pmatrix}$$

---

[7]The theorem is based on the concavity of the entropy. The entropy has the form of $-x \ln x$ where x is in the range $1 > x > 0$. The second derivative is $-1/x$, and so it is a concave function.

(i) Could this matrix represent a density operator?
(ii) If so, find the average value of $\bar{j}_z$.
(iii) Find the average value of

$$(\hat{j}_z - \bar{j}_z)^2$$

Problem 6.13

(i) Show that the most general form of an operator in the Hilbert space with angular momentum $j = 1$, is

$$\hat{O} = \sum_{i,j} b_{i,j}\; \hat{J}_i \hat{J}_j$$

with nine independent coefficients. Hence, show that the most general form of a density operator for a spin-one system can be expressed as

$$\hat{\rho} = \frac{1}{3}\,\hat{I} + \underline{a} \cdot \underline{\hat{J}} + \sum_{i,j} b_{i,j}\left[\frac{1}{2}\,(\hat{J}_i\hat{J}_j + \hat{J}_j\hat{J}_i) - \frac{2}{3}\,\delta_{i,j}\,\hat{I}\right]$$

where $\sum_i b_{i,i} = \frac{1}{2}$.
(ii) Show that the density operator

$$\hat{\rho} = \frac{1}{2}\begin{pmatrix} 1 & 0 & i \\ 0 & 0 & 0 \\ -i & 0 & 1 \end{pmatrix}$$

is a pure state which corresponds to an eigenstate of the quadrupole moment operator $Q_{x,y}$

$$Q_{x,y} = \left(\frac{\hat{J}_x\hat{J}_y + \hat{J}_y\hat{J}_x}{2}\right)$$

$$= \frac{1}{2}\begin{pmatrix} 0 & 0 & -i \\ 0 & 0 & 0 \\ i & 0 & 0 \end{pmatrix}$$

and that this state has no dipole moment

$$\text{Trace }\underline{\hat{J}}\hat{\rho} = 0$$

Problem 6.14

Consider a system which has statistically independent sub-systems A and B. Statistically independent systems are defined as having a factorizable density operator

$$\hat{\rho} = \hat{\rho}_A \, \hat{\rho}_B$$

The von Neumann entropies of the subsystems are defined as

$$S_A = -k_B \, \text{Trace}_A \, [\hat{\rho}_A \, \ln \hat{\rho}_A]$$

and

$$S_B = -k_B \, \text{Trace}_B \, [\hat{\rho}_B \, \ln \hat{\rho}_B]$$

where the Trace over A is evaluated as a sum over a complete set of states for the subsystem A, and a similar definition holds for subsystem B.

(i) Show that, for statistically independent sub-systems, the von Neumann entropies are additive

$$S = S_A + S_B$$

(ii) Find an example of a quantum system in a pure state for which $S = 0$ but, nevertheless, it has subsystems that are not statistical independent where S_A and S_B are greater than zero.

(iii) The density operator for a pure state of a bi-partite system can be reduced into the density operators $\hat{\rho}_A$ and $\hat{\rho}_B$ for its two subsystems. Show that the von Neumann entropies of the subsystems must be equal.

$$S_A = S_B$$

Hint: Express the pure state as a Schmidt decomposition.

Problem 6.15

Consider a pure state composed of two spin one-half particles

$$|\psi\rangle = \frac{1}{\sqrt{2}} \, (|+, -\rangle + |-, +\rangle)$$

where the z-axis is chosen as the axis of quantization.

(i) Construct the density operator.

(ii) Form the reduced density operator and find the von Neumann entropy.

(iii) Express the pure state in terms of the basis in which the axis of quantization is chosen to be the (θ, ϕ) direction and hence find the density operator in the new basis.

(iv) Show that the reduced density operator is independent of the basis.

Problem 6.16

Show that the von Neumann entropy is constant, due to the unitary nature of the time evolution operator.

Problem 6.17

Determine whether the following particles are Bosons or Fermions.
(i) α-particle
(ii) ^{3}He
(iii) H_2 molecule
(iv) ^{6}Li$^+$ ion
(v) ^{7}Li$^+$ ion
(vi) Rb^{87}

Problem 6.18

Consider a quantum mechanical system composed of a few fermions with spin S enclosed in a volume V. The particles have a spin-independent dispersion relation

$$\epsilon(p) = \frac{p^2}{2m}$$

The system is held at constant temperature T.
(i) Show that, in the position representation, the spin Trace of the density operator for a single fermion has matrix elements given by

$$\langle \underline{r}'|\hat{\rho}|\underline{r}\rangle = \frac{(2S+1)}{Z_1} \int \frac{d^3\underline{p}}{(2\pi\hbar)^3} \exp\left[-\beta\frac{p^2}{2m}\right] \exp\left[\frac{i}{\hbar}\underline{p}\cdot(\underline{r}'-\underline{r})\right]$$

$$= \frac{(2S+1)}{Z_1} \frac{\exp\left[-\pi\left(\frac{r-r'}{\lambda}\right)^2\right]}{\lambda^3}$$

where λ is the thermal de Broglie wavelength.
(ii) Show by combining spin angular momenta, that two spin S fermions can form $(S+1)(2S+1)$ states that are symmetric under the exchange of the spin components of the wave functions and $S(2S+1)$ states that are antisymmetric under the exchange of spin. Show that the spatial components of these wave functions must be, respectively, odd and even under the exchange of the spatial coordinates.

(iii) Show that, in the position representation with spin-symmetric states, the two-particle density operator has non-vanishing spin matrix elements given by

$$\langle \underline{r}_1', \underline{r}_2' | \hat{\rho} | \underline{r}_1, \underline{r}_2 \rangle = \frac{1}{2! Z_2} \left(\frac{\exp\left[-\pi \left(\frac{r_1 - r_1'}{\lambda} \right)^2 \right]}{\lambda^3} \frac{\exp\left[-\pi \left(\frac{r_2 - r_2'}{\lambda} \right)^2 \right]}{\lambda^3} \right.$$

$$\left. - \frac{\exp\left[-\pi \left(\frac{r_1 - r_2'}{\lambda} \right)^2 \right]}{\lambda^3} \frac{\exp\left[-\pi \left(\frac{r_2 - r_1'}{\lambda} \right)^2 \right]}{\lambda^3} \right)$$

(iv) Show that, in the position representation with spin-antisymmetric states, the two-particle density operator has non-zero matrix elements given by

$$\langle \underline{r}_1', \underline{r}_2' | \hat{\rho} | \underline{r}_1, \underline{r}_2 \rangle = \frac{1}{2! Z_2} \left(\frac{\exp\left[-\pi \left(\frac{r_1 - r_1'}{\lambda} \right)^2 \right]}{\lambda^3} \frac{\exp\left[-\pi \left(\frac{r_2 - r_2'}{\lambda} \right)^2 \right]}{\lambda^3} \right.$$

$$\left. + \frac{\exp\left[-\pi \left(\frac{r_1 - r_2'}{\lambda} \right)^2 \right]}{\lambda^3} \frac{\exp\left[-\pi \left(\frac{r_2 - r_1'}{\lambda} \right)^2 \right]}{\lambda^3} \right)$$

(v) Show that by performing the Trace over the spins, the diagonal matrix element of the two-particle density operator is

$$\langle \underline{r}_1, \underline{r}_2 | \hat{\rho} | \underline{r}_1, \underline{r}_2 \rangle = \frac{(2S+1)}{2! Z_2 \, \lambda^6} \left((2S+1) - \exp\left[-2\pi \left(\frac{r_1 - r_2}{\lambda} \right)^2 \right] \right)$$

(vi) Hence, show that

$$Z_2 = \frac{Z_1^2}{2!} \left[1 - \frac{1}{2^{\frac{3}{2}} Z_1} \right]$$

(vii) Generalize your result to show that the leading-order correction to the classical results for N fermions is given by

$$Z_N \approx \frac{Z_1^N}{N!} \left[1 - \frac{N(N-1)}{2! 2^{\frac{3}{2}} Z_1} + \dots \right]$$

Explain your reasoning.

(viii) Calculate the correction to the internal energy and the specific heat. Estimate the temperature at which the classical approximation breaks down.

Problem 6.19

Consider a quantum mechanical system composed of a few bosons with spin S enclosed in a volume V. The particles have a dispersion relation

$$\epsilon(p) = \frac{p^2}{2m}$$

that is independent of the spin. The system is held at constant temperature T. It is important to note that for a two-particle system and with $\alpha \neq \beta$, the two-particle states (α, β) and (β, α) are merely different representations of a unique state.

(i) Evaluate the single-particle density operator in the position representation. Hence, determine the partition function Z_1 for a system containing a single particle.

(ii) Show that the partition Z_2 for a two-particle ideal boson gas is given by

$$Z_2 = \frac{Z_1^2}{2!} \left[1 + \frac{1}{2^{\frac{3}{2}} Z_1} \right]$$

(iii) Estimate the temperature below which the classical approximation breaks down for an ideal gas of N bosons.

Problem 6.20

The density operator for a quantum mechanical Simple Harmonic Oscillator, in the Canonical Ensemble can be expressed as

$$\hat{\rho} = \sum_{n=0}^{\infty} |n\rangle\langle n| \, \exp\left[-\beta\hbar\omega\left(n + \frac{1}{2} \right) \right]$$

where $|n\rangle$ is the n-th energy eigenstate. In the position representation, the n-th eigenstate of a quantum mechanical harmonic oscillator is given by

$$\langle q|n\rangle = \left(\frac{m\omega}{\pi\hbar} \right)^{\frac{1}{4}} \frac{H_n(\sqrt{\frac{m\omega}{\hbar}}q)}{2^n \, n!} \, \exp\left[-\frac{m\omega \, q^2}{2\hbar} \right]$$

where $H_n(x)$ are the Hermite polynomials. The Hermite polynomials are given by

$$H_n(x) = (-1)^n \, \exp[+x^2] \left(\frac{d^n}{dx^n} \right) \exp[-x^2]$$

$$= \frac{\exp[+x^2]}{\sqrt{\pi}} \int_{-\infty}^{\infty} du(-i2u)^n \, \exp[-u^2 + 2ixu]$$

where the last line has been obtained by representing the exponential term as a Gaussian integral, and then taking the derivative wrt x. Show that, in the position representation the density matrix has the position matrix elements

$$\langle q'|\hat{\rho}|q\rangle = \frac{1}{Z}\sqrt{\frac{m\omega}{2\pi\hbar \, \sinh\beta\hbar\omega}}$$

$$\times \exp\left[-\frac{m\omega}{\hbar}\left(\left(\frac{q+q'}{2}\right)^2\tanh\frac{\beta\hbar\omega}{2}+\left(\frac{q-q'}{2}\right)^2\coth\frac{\beta\hbar\omega}{2}\right)\right]$$

Problem 6.21

The density operator for a one-dimensional harmonic oscillator has position matrix elements which are given by

$$\langle q'|\hat{\rho}|q\rangle = \frac{1}{Z}\sqrt{\frac{m\omega_0}{2\pi\hbar \, \sinh\beta\hbar\omega_0}}$$

$$\times \exp\left[-\frac{m\omega_0}{\hbar}\left(\left(\frac{q+q'}{2}\right)^2\tanh\frac{\beta\hbar\omega_0}{2}\right.\right.$$

$$\left.\left.+\left(\frac{q-q'}{2}\right)^2\coth\frac{\beta\hbar\omega_0}{2}\right)\right]$$

(i) Determine the partition function Z.
(ii) Determine the form of the matrix elements of $\hat{\rho}$ in the momentum representation.
(iii) Examine the high temperature limit of the matrix elements. Do the matrix elements of the density operator reduce to the corresponding factors in the classical probability density?

Problem 6.22

A Hydrogen molecule consists of two protons which are fermions with spin one-half. The nuclear wave functions must be antisymmetric under exchange. That is, the wave functions have either an antisymmetric spin wave function and symmetric spatial wave function or a symmetric spin wave function and antisymmetric spatial wave function.
(i) Show that for a pair of nucleons with spin S, one can construct $S(2S+1)$ antisymmetric spin wave functions and one can construct $(S+1)(2S+1)$ symmetric spin wave functions.

(ii) Provide an argument which shows that symmetric spatial nuclear wave functions correspond to even angular momentum l and the antisymmetric spatial nuclear wave functions correspond to odd values of the angular momentum l.

(iii) Show that, in chemical equilibrium, the partition function for the pair of protons should be written as

$$Z = S(2S+1)\ Z_{S,even} + (S+1)(2S+1)\ Z_{S,odd}$$

where, if the system can be considered as a rigid rotator, the spatial factor of the partition function is given by

$$Z_S = \sum_l (2l+1)\ \exp\left[-\beta\ \frac{l(l+1)\ \hbar^2}{2I}\right]$$

in which l runs over either even or odd values.

(iv) Show that these expressions can be combined to yield the equilibrium partition function

$$Z = \frac{(2S+1)}{2} \sum_l [(2S+1) - (-1)^l](2l+1)\ \exp\left[-\beta\ \frac{l(l+1)\ \hbar^2}{2\ I}\right]$$

For protons with $S = \frac{1}{2}$, one finds that there are two types of states for the Hydrogen molecules, which are known as either para-Hydrogen or ortho-Hydrogen.

At room temperature, ortho and para Hydrogen are expected to be in equilibrium and be in the ratio of $(S+1)$ to S, that is three to one. At lower temperatures, the relative abundance persists. The reason is that the system is not in chemical equilibrium. Ortho-Hydrogen is unstable at low temperatures and converts to para-Hydrogen, but the rate of decay is extremely slow since there is no natural radiation mode that can de-excite the molecule. This leads to a large deviation between the equilibrium value of the energy or heat capacity from its equilibrium value.

(v) Find the equilibrium value of the low temperature rotational specific heat for a H_2 gas and compare this with the non-equilibrium value.

Ortho and para Oxygen molecules O_2 also exist. The molecules exists in triplet and singlet states, which arise from the two unpaired electrons. The triplet state has a paramagnetic moment of $S = 1$ and is the ground state as expected from Hund's rules. The singlet state is a diamagnetic excited state.

Problem 6.23

The rotational Raman spectra of a diatomic molecule involving identical atoms of spin S exhibits lines indicative of rotation with energy

$$\epsilon = \frac{\hbar^2 l(l+1)}{2I}$$

where l is the angular momentum and I is the moment of inertia. The rotational Raman spectrum consists of two series of alternating lines.

(i) Show that the ratios of the intensities of the two series is given by $S/(S+1)$.

(ii) Show that the series reaches maxima near the angular momentum l_m which is given by

$$(2l_m + 1)^2 = 4k_B T I$$

Problem 6.24

The number operator $\hat{n}_\alpha$ for a single-particle quantum state α is defined as

$$\hat{n}_\alpha = a_\alpha^\dagger a_\alpha$$

where the creation operators, $a_\alpha^\dagger$, and the annihilation operators, a_α, either satisfy commutation or anticommutation relations.

(i) Show that the number operators satisfy the following commutation relations with the creation and annihilation operators

$$[a_\alpha^\dagger, \hat{n}_\beta] = -a_\alpha^\dagger \delta_{\alpha,\beta}$$
$$[a_\alpha, \hat{n}_\beta] = a_\alpha \delta_{\alpha,\beta}$$

(ii) Show that the eigenstates of the number operator $\hat{n}_\alpha$ generally can be written as

$$|n_\alpha\rangle = \frac{(a_\alpha^\dagger)^{n_\alpha}}{\sqrt{n_\alpha!}} |0\rangle$$

where n_α is the eigenvalue and $|0\rangle$ is the vacuum state defined by

$$a_\alpha |0\rangle = 0$$

(iii) Show, generally, that the following identity holds

$$a_\alpha^\dagger a_\beta^\dagger a_\beta a_\alpha = \hat{n}_\alpha(\hat{n}_\beta - \delta_{\alpha,\beta})$$

which can be used to count the number of pairs of particles.

(iv) Show that the following matrix elements can be evaluated as

$$\langle n_\alpha | a_\alpha^\dagger a_\alpha^\dagger a_\alpha a_\alpha | n_\alpha \rangle = \pm n_\alpha(n_\alpha - 1)$$

where the plus sign refers to bosons and the minus sign corresponds to fermions. Using this result show that for fermions, which obey anticommutation relations, the only allowed values of the eigenvalues of the number operator are $n_\alpha = 0$ and $n_\alpha = 1$.

Problem 6.25

(i) For a system of identical bosons, evaluate the matrix elements

$$\langle n_\alpha | (a_\alpha^\dagger + a_\alpha)^m | n_\alpha \rangle$$

for $m = 0$, 1 and 2.
(ii) Evaluate the commutation relation

$$\left[a_\alpha, \frac{U}{2!} \sum_{\gamma \neq \delta} a_\gamma^\dagger a_\gamma a_\delta^\dagger a_\delta \right]$$

for the cases of fermion and boson operators.

Problem 6.26

A quantum mechanical harmonic oscillator is in a coherent state $|a_\phi\rangle$.
(i) Determine the expectation value of the position operator $\hat{x}(t)$ given by

$$\hat{x}(t) = \sqrt{\frac{\hbar}{2m\omega_0}} \left(a \ \exp[-i\omega_0 t] + a^\dagger \ \exp[+i\omega_0 t] \right)$$

in a coherent state.
(ii) Determine the expectation value of the momentum operator $\hat{p}(t)$ in the coherent state $|a_\phi\rangle$, where

$$\hat{p}(t) = -i\hbar \sqrt{\frac{m \ \omega_0}{2\hbar}} \left(a \ \exp[-i\omega_0 t] - a^\dagger \ \exp[+i\omega_0 t] \right)$$

(iii) Compare these results with the time-dependence of the position and momentum that of a classical harmonic oscillator.

Problem 6.27

The coherent states yield expectation values of an operator equal to the classical value, if the operator is normal ordered. Operators are normal ordered if, in each term, creation operators appear on the left side of the expression and annihilation operators appear on the right. The expectation values of normal ordered operators vanish for the vacuum state since an annihilation operator acting on the vacuum ket yields zero and the creation operator acting on the vacuum bra also produces zero.

(i) Evaluate the expectation value of the normal ordered operator

$$a^\dagger a^\dagger a^\dagger aa$$

for a coherent state $|a_\phi\rangle$ where $a^\dagger$ and a are boson creation and annihilation operators.

(ii) Find the expectation value of the operators

$$a^\dagger a^\dagger aa^\dagger a \equiv a^\dagger a^\dagger a^\dagger aa + a^\dagger a^\dagger a$$

$$a^\dagger aaa^\dagger \equiv a^\dagger a^\dagger aa + 2a^\dagger a$$

$$aa^\dagger \equiv a^\dagger a + 1$$

in the coherent state $|a_\phi\rangle$ and the creation and annihilation operators satisfy bosonic commutation relations. From your results, it is seen that the expectation value for a quantum coherent state differs from the classical result since, in the classical limit, operators commute and can be replaced by numbers.

Problem 6.28

(i) Show that the projection onto a coherent state can be expressed in terms of the normal ordered operator

$$|\alpha\rangle\langle\alpha| = : \exp[-(a^\dagger - \alpha^*)(a - \alpha)] :$$

where $: \hat{A} :$ denotes normal ordering in which creation operators appear on the left of annihilation operators.

Problem 6.29

(i) Show that the distribution of the number of quanta n in a coherent state characterized by the amplitude a_ϕ, $P(n)$ is given by

$$P(n) = \frac{|a_\phi|^{2n}}{n!} \exp[-|a_\phi|^2]$$

(ii) Show that the average number of quanta is given by

$$\bar{n} = |a_\phi|^2$$

(iii) Hence, show that the number of quanta have a Poisson distribution function

$$P(n) = \frac{\bar{n}^n}{n!} \exp[-\bar{n}]$$

Problem 6.30

Consider the Hamiltonian with fixed l

$$\hat{H} = -\frac{\hbar^2}{2m} \frac{\partial^2}{\partial r^2} + \frac{m\omega^2 \, r^2}{2} + \frac{\hbar^2 l(l+1)}{2mr^2}$$

The operators $b_l^\dagger$ and b_l are defined as

$$b_l = \sqrt{\frac{\hbar}{2m\omega}} \left(\frac{\partial}{\partial r} - \frac{l}{r} \right) + \sqrt{\frac{m\omega}{2\hbar}} \, r$$

$$b_l^\dagger = \sqrt{\frac{\hbar}{2m\omega}} \left(-\frac{\partial}{\partial r} - \frac{l}{r} \right) + \sqrt{\frac{m\omega}{2\hbar}} \, r$$

(i) Show that the operators satisfy bosonic commutation relations

$$[b_l, b_l^\dagger] = 1 + \frac{l}{r}$$

$$[b_l, b_{-l}] = -\frac{l}{r^2}$$

$$[b_l, b_{-l}^\dagger] = 1$$

(ii) Show that the fixed-l Hamiltonian can be expressed as

$$\hat{H}_l = \hbar\omega \left(b_l^\dagger b_l + l - \frac{1}{2} \right)$$

Problem 6.31

The operators $a_x^\dagger$, a_x, $a_y^\dagger$, a_y obey Bose-Einstein commutation relations and represent plane polarized waves. The phase difference between the polarizations is φ. Consider the operators

$$\hat{S}_0 = a_x^\dagger a_x + a_y^\dagger a_y$$

$$\hat{S}_1 = a_x^\dagger a_x - a_y^\dagger a_y$$

$$\hat{S}_2 = a_x^\dagger a_y \, \exp[+i\varphi] + a_y^\dagger a_x \, \exp[-i\varphi]$$

$$\hat{S}_3 = -i(a_x^\dagger a_y \, \exp[+i\varphi] - a_y^\dagger a_x \, \exp[-i\varphi])$$

which are potential spin-one Casimir and angular momentum operators.
(i) Show that the operators satisfy the commutation relations

$$[\hat{S}_i, \hat{S}_j] = 2i \sum_k \xi^{i,j,k} \, \hat{S}_k$$

$$[\hat{S}_0, \hat{S}_i] = 0$$

for $i = 1, 2, 3$.

Chapter 7

Fermi-Dirac Statistics

7.1   Non-Interacting Fermions

An ideal gas of non-interacting fermions is described by the Hamiltonian $\hat{H}$ which is given by the sum

$$\hat{H} = \sum_{\alpha} \epsilon_{\alpha}\, \hat{n}_{\alpha} \tag{7.1}$$

where $\hat{n}_{\alpha}$ represents the occupation number of the α-th single-particle energy level. This is just the sum of the contributions from each particle, grouped according to the energy levels that they occupy. Likewise, the operator $\hat{N}$ representing the total number of particles is given by

$$\hat{N} = \sum_{\alpha} \hat{n}_{\alpha} \tag{7.2}$$

where the sum is over the single-particle energy levels. Hence, for non-interacting fermions, the density operator is diagonal in the occupation number representation.

The Grand-Canonical Partition Function Ξ is given by

$$\Xi = \text{Trace } \exp[-\beta(\hat{H} - \mu\hat{N})] \tag{7.3}$$

where the Trace is over a complete set of microscopic states with variable N for the entire system. A convenient basis is given by the N-particle states $|\Phi_{n_{\alpha_1}, n_{\alpha_2},...}\rangle$, since both $\hat{H}$ and $\hat{N}$ are diagonal in this basis. The Trace reduces to the sum of all configurations $\{n_{\alpha}\}$, and since the total number of particles N is also being summed over, the Trace is unrestricted. Therefore, the Trace can be evaluated by summing over all possible values of the eigenvalues n_{α} for each consecutive value of α. That is

$$\text{Trace } \{...\} \equiv \sum_{n_{\alpha_1}=0}^{1} \sum_{n_{\alpha_2}=0}^{1} \sum_{n_{\alpha_3}=0}^{1} \cdots \{...\} \tag{7.4}$$

Since the exponential term in Ξ reduces to the form of the exponential of a sum of independent terms, it can be written as the product of exponential factors

$$\exp[-\beta(\hat{H} - \mu\hat{N})] = \exp\left[-\beta \sum_\alpha (\epsilon_\alpha - \mu)\, \hat{n}_\alpha\right]$$

$$= \prod_\alpha \exp[-\beta\, (\epsilon_\alpha - \mu)\, \hat{n}_\alpha] \qquad (7.5)$$

Therefore, the Trace in the expression for Ξ can be reduced to

$$\Xi = \prod_\alpha \left\{ \sum_{n_\alpha=0}^{1} \exp[-\beta(\epsilon_\alpha - \mu)\, n_\alpha] \right\} \qquad (7.6)$$

where the sum is over all the occupation numbers n_α of either zero or unity as is allowed by Fermi-Dirac statistics. On performing the sum of the geometric series, one obtains

$$\sum_{n_\alpha=0}^{1} \exp[-\beta(\epsilon_\alpha - \mu)n_\alpha] = 1 + \exp[-\beta(\epsilon_\alpha - \mu)] \qquad (7.7)$$

Therefore, the Grand-Canonical Partition Function is given by

$$\Xi = \prod_\alpha \{1 + \exp[-\beta(\epsilon_\alpha - \mu)]\} \qquad (7.8)$$

The Grand-Canonical Potential Ω is found from

$$\Xi = \exp[-\beta\Omega] \qquad (7.9)$$

which, on taking the logarithm, yields

$$-\beta\Omega = \sum_\alpha \ln\,[1 + \exp[-\beta(\epsilon_\alpha - \mu)]] \qquad (7.10)$$

or

$$\Omega = -k_B T \sum_\alpha \ln\,[1 + \exp[-\beta(\epsilon_\alpha - \mu)]] \qquad (7.11)$$

On introducing the density of single-particle states $\rho(\epsilon)$ defined as

$$\rho(\epsilon) = \sum_\alpha \delta(\epsilon - \epsilon_\alpha) \qquad (7.12)$$

the summation can be transformed into an integral over ϵ

$$\Omega = -k_B T \int_{-\infty}^{\infty} d\epsilon\, \rho(\epsilon)\, \ln\,[1 + \exp[-\beta(\epsilon - \mu)]] \qquad (7.13)$$

where the density of states is bounded from below by ϵ_0. After evaluating the integral, one may find all the thermodynamic properties of the system from Ω.

7.2 The Fermi-Dirac Distribution Function

The average number of particles in the system $\overline{N}$ can be determined from Ω via the relation

$$\overline{N} = -\left(\frac{\partial\Omega}{\partial\mu}\right)_{T,V} \tag{7.14}$$

which is evaluated as

$$\overline{N} = k_B T \int_{-\infty}^{\infty} d\epsilon \, \rho(\epsilon) \, \frac{\beta \, \exp[-\beta(\epsilon - \mu)]}{1 + \exp[-\beta(\epsilon - \mu)]}$$

$$= \int_{-\infty}^{\infty} d\epsilon \, \rho(\epsilon) \, \frac{1}{\exp[\beta(\epsilon - \mu)] + 1} \tag{7.15}$$

or equivalently

$$\overline{N} = \sum_{\alpha} \frac{1}{\exp[\beta(\epsilon_\alpha - \mu)] + 1} \tag{7.16}$$

but, by definition, one has

$$\overline{N} = \sum_{\alpha} \overline{n}_\alpha \tag{7.17}$$

Therefore, the function $f(\epsilon_\alpha)$ defined by

$$f(\epsilon_\alpha) = \frac{1}{\exp[\beta(\epsilon_\alpha - \mu)] + 1} \tag{7.18}$$

represents the average number of particles $\overline{n}_\alpha$ in a quantum level with a single-particle energy ϵ_α. The function $f(\epsilon)$ is the Fermi-Dirac distribution function. The Fermi-Dirac distribution varies as

$$f(\epsilon) \approx \exp[-\beta(\epsilon - \mu)] \tag{7.19}$$

for $\epsilon - \mu > k_B T$. However, for $\mu - \epsilon > k_B T$, the Fermi-Dirac distribution function tends to unity

$$f(\epsilon) \approx 1 - \exp[\beta(\epsilon - \mu)] \tag{7.20}$$

and falls off rapidly from 1 to 0 at $\epsilon = \mu$ where it takes on the value

$$f(\mu) = \frac{1}{2} \tag{7.21}$$

The range of ϵ over which the function differs from either 1 or 0 is governed by $k_B T$.

At zero temperature, all the states with single-particle energies less than μ are occupied and all states with energies greater than μ are empty. The value of $\mu(T = 0)$ is called the Fermi-energy and is denoted by ϵ_F. Note

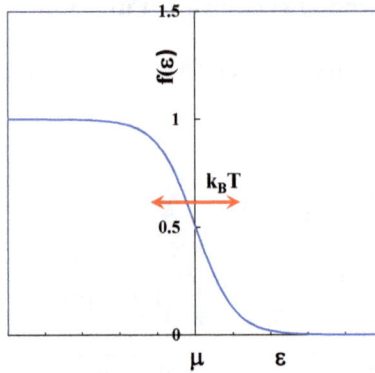

Fig. 7.1   The Fermi-Dirac Distribution function.

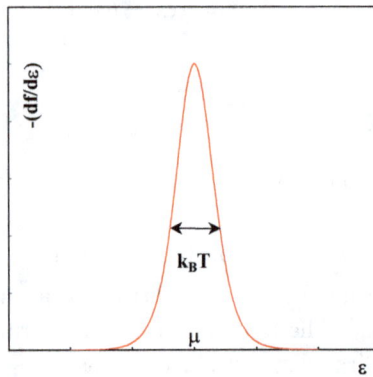

Fig. 7.2   The negative energy derivative of Fermi-Dirac distribution function.

that since

$$\int_{-\infty}^{\infty} d\epsilon \, \frac{\partial f(\epsilon)}{\partial \epsilon} = -1 \tag{7.22}$$

and as $T \to 0$ one has

$$\frac{\partial f(\epsilon)}{\partial \epsilon} \approx 0 \qquad \text{if } |\epsilon - \mu| > k_B T \tag{7.23}$$

then

$$-\frac{\partial f(\epsilon)}{\partial \epsilon} \sim \delta(\epsilon - \mu) \tag{7.24}$$

as it resembles the Dirac delta function.

7.3 Thermodynamic Properties

The thermodynamic energy U coincides with the average energy $\overline{E}$. The thermodynamic energy can be obtained from the Grand-Canonical Potential Ω via its definition

$$\Omega = U - TS - \mu \overline{N} \tag{7.25}$$

which, together with

$$S = -\left(\frac{\partial \Omega}{\partial T}\right)_{\mu, V}$$

$$\overline{N} = -\left(\frac{\partial \Omega}{\partial \mu}\right)_{T, V} \tag{7.26}$$

can be inverted to yield

$$U = \Omega + TS + \mu \overline{N}$$

$$= \Omega - T \left(\frac{\partial \Omega}{\partial T}\right)_{\mu, V} - \mu \left(\frac{\partial \Omega}{\partial \mu}\right)_{T, V} \tag{7.27}$$

On substituting

$$\Omega = -k_B T \int_{-\infty}^{\infty} d\epsilon \, \rho(\epsilon) \, \ln\left[1 + \exp[-\beta(\epsilon - \mu)]\right] \tag{7.28}$$

one obtains

$$
\begin{aligned}
U &= k_B T^2 \int_{-\infty}^{\infty} d\epsilon \, \rho(\epsilon) \, \frac{(\epsilon - \mu)\left(-\frac{\partial \beta}{\partial T}\right) \exp[-\beta(\epsilon - \mu)]}{1 + \exp[-\beta(\epsilon - \mu)]} \\
&\quad + \mu k_B T \int_{-\infty}^{\infty} d\epsilon \, \rho(\epsilon) \, \frac{\beta \, \exp[-\beta(\epsilon - \mu)]}{1 + \exp[-\beta(\epsilon - \mu)]} \\
&= \int_{-\infty}^{\infty} d\epsilon \, \rho(\epsilon) \, \frac{\epsilon}{\exp[\beta(\epsilon - \mu)] + 1} \\
&= \int_{-\infty}^{\infty} d\epsilon \, \rho(\epsilon) \, \epsilon \, f(\epsilon) \\
&= \sum_{\alpha} \epsilon_\alpha \, f(\epsilon_\alpha) \\
&= \sum_{\alpha} \epsilon_\alpha \, \overline{n}_\alpha \tag{7.29}
\end{aligned}
$$

which shows that the thermodynamic energy for a system of particles is just the average energy, since the average energy of a system of non-interacting particles is just the sum over the average energy for each energy level. This reinforces the interpretation of $f(\epsilon)$ as the average number of fermions in a single-particle state with energy ϵ.

The entropy S is determined from the equation

$$S = -\left(\frac{\partial \Omega}{\partial T}\right)_{\mu, V} \tag{7.30}$$

which yields

$$
\begin{aligned}
S = k_B \sum_\alpha & \ln\left[1 + \exp[\beta(\mu - \epsilon_\alpha)]\right] \\
& - \sum_\alpha \frac{(\mu - \epsilon_\alpha)}{T} \frac{\exp[\beta(\mu - \epsilon_\alpha)]}{1 + \exp[\beta(\mu - \epsilon_\alpha)]} \\
= -k_B \sum_\alpha & \ln\left[1 - f(\epsilon_\alpha)\right] \\
& + k_B \sum_\alpha \left(\frac{\mu - \epsilon_\alpha}{T}\right) f(\epsilon_\alpha)
\end{aligned} \tag{7.31}
$$

However, one may rewrite the factor

$$\left(\frac{\mu - \epsilon_\alpha}{k_B T}\right) \tag{7.32}$$

as

$$
\begin{aligned}
\left(\frac{\mu - \epsilon_\alpha}{k_B T}\right) &= \ln\left[\exp[\beta(\mu - \epsilon_\alpha)]\right] \\
&= -\ln\left[1 - f(\epsilon_\alpha)\right] + \ln\left[f(\epsilon_\alpha)\right]
\end{aligned} \tag{7.33}
$$

Therefore, on combining the above expressions, one finds that the entropy of the non-interacting fermion gas can be expressed as

$$S = -k_B \sum_\alpha \left((1 - f(\epsilon_\alpha)) \ln\left[1 - f(\epsilon_\alpha)\right] + f(\epsilon_\alpha) \ln\left[f(\epsilon_\alpha)\right]\right) \tag{7.34}$$

The entropy has the form of

$$S = -k_B \sum_\gamma p_\gamma \ln p_\gamma \tag{7.35}$$

where $p_\gamma = f(\epsilon_\alpha)$ is the probability that the α-th level is occupied and $p_\gamma = 1 - f(\epsilon_\alpha)$ is the probability that the level is empty. This form of the entropy follows naturally from the assumption that the non-interacting particles are statistically independent together with the Shannon definition of entropy.

7.4 The Equation of State

The equation of state for a gas of non-interacting fermions can be found from Ω by noting that

$$\Omega = -PV \tag{7.36}$$

The equation of state can be obtained directly when the single-particle density of states $\rho(\epsilon)$ has the form of a simple power law

$$\rho(\epsilon) = C \, \epsilon^\alpha \qquad \text{for } \epsilon \geq 0$$
$$= 0 \qquad \text{otherwise} \tag{7.37}$$

where α is a constant. The Grand-Canonical Potential Ω is found as

$$\Omega = -PV$$
$$= -k_B T \int_{-\infty}^{\infty} d\epsilon \, \rho(\epsilon) \, \ln\left[1 + \exp[-\beta(\epsilon - \mu)]\right] \tag{7.38}$$

Hence, one has

$$\frac{PV}{k_B T} = C \int_0^{\infty} d\epsilon \, \epsilon^\alpha \, \ln\left[1 + \exp[-\beta(\epsilon - \mu)]\right]$$
$$= \frac{C}{\alpha + 1} \int_0^{\infty} d\epsilon \, \frac{d\epsilon^{(\alpha+1)}}{d\epsilon} \, \ln\left[1 + \exp[-\beta(\epsilon - \mu)]\right] \tag{7.39}$$

On integrating by parts, one obtains

$$\frac{PV}{k_B T} = \frac{C}{\alpha + 1} \left[\epsilon^{\alpha+1} \, \ln\left(1 + \exp[-\beta(\epsilon - \mu)]\right)\right]\big|_0^\infty$$
$$+ \frac{C}{\alpha + 1} \int_0^{\infty} d\epsilon \, \epsilon^{(\alpha+1)} \, \frac{\beta \, \exp[-\beta(\epsilon - \mu)]}{1 + \exp[-\beta(\epsilon - \mu)]} \tag{7.40}$$

The boundary terms vanish, since the density of states vanishes at the lower limit of integration $\epsilon = 0$ and the logarithmic factor vanishes exponentially when $\epsilon \to \infty$. Thus, on canceling a factor of β, one finds

$$PV = \frac{1}{\alpha + 1} \int_0^{\infty} d\epsilon \, C \, \epsilon^{(\alpha+1)} \, \frac{1}{\exp[\beta(\epsilon - \mu)] + 1}$$
$$= \frac{1}{\alpha + 1} \int_{-\infty}^{\infty} d\epsilon \, \rho(\epsilon) \, \epsilon \, \frac{1}{\exp[\beta(\epsilon - \mu)] + 1}$$
$$= \frac{U}{\alpha + 1} \tag{7.41}$$

That is, the equation of state for the system of non-interacting fermions is found as

$$PV = \frac{U}{\alpha + 1} \tag{7.42}$$

For $\alpha = \frac{1}{2}$, the relation is identical to that found for the classical ideal gas.

In fact, the high-temperature limit of the equation of state for the system of non-interacting fermions can be evaluated as

$$
\begin{aligned}
\Omega &= -PV \\
&= -k_B T \int_0^\infty d\epsilon\, \rho(\epsilon)\, \ln\left[1 + \exp[-\beta(\epsilon - \mu)]\right] \\
&\approx -k_B T \int_0^\infty d\epsilon\, \rho(\epsilon)\, \exp[-\beta(\epsilon - \mu)] \\
&\approx -k_B T \int_0^\infty d\epsilon\, \rho(\epsilon)\, \frac{1}{\exp[\beta(\epsilon - \mu)] + 1} \\
&\approx -\overline{N}\, k_B T
\end{aligned}
\tag{7.43}
$$

since at high-temperatures $\mu < 0$ and since we have assumed that the single-particle density of states is zero below $\epsilon = 0$. Therefore, we have re-derived the ideal gas law from the high temperature limit of a set of non-interacting particles which obey Fermi-Dirac statistics.

7.5 The Chemical Potential

We have seen that, at high temperatures, the equation of state of a gas of particles obeying Fermi-Dirac statistics reduces to the equation of state for particles obeying Classical statistics if

$$
\exp[\beta\mu] \ll 1 \tag{7.44}
$$

(where we have restricted $\epsilon \geq 0$) since under these conditions the Fermi-Dirac distribution function reduces to the Boltzmann distribution function. The above restriction is also consistent with our previous discussion of the ideal gas, where

$$
\exp[\beta\mu(T)] = \left(\frac{V}{\lambda^3}\right) \tag{7.45}
$$

We shall examine the temperature dependence of the chemical potential and show that Fermi-Dirac statistics is similar to classical Maxwell-Boltzmann statistics at sufficiently high temperatures.

For a system with a fixed number of fermions, $\overline{N}$, governed by a condition like electrical neutrality, then the chemical potential is temperature dependent and $\mu(T)$ is found as the solution of the equation

$$
\overline{N} = \int_0^\infty d\epsilon\, \rho(\epsilon)\, f(\epsilon) \tag{7.46}
$$

For large and negative values of $\mu(T)$, one can expand the Fermi-Dirac distribution function

$$\overline{N} = \int_0^\infty d\epsilon \, \rho(\epsilon) \, \frac{\exp[-\beta(\epsilon - \mu)]}{1 + \exp[-\beta(\epsilon - \mu)]}$$

$$= \int_0^\infty d\epsilon \, \rho(\epsilon) \, \exp[-\beta(\epsilon - \mu)] \sum_{n=0}^\infty (-1)^n \, \exp[-n\beta(\epsilon - \mu)]$$

$$= \int_0^\infty d\epsilon \, \rho(\epsilon) \sum_{n=1}^\infty (-1)^{n+1} \, \exp[n\beta(\mu - \epsilon)]$$

$$= \sum_{n=1}^\infty (-1)^{n+1} \, \exp[n\beta\mu] \int_0^\infty d\epsilon \, \rho(\epsilon) \, \exp[-n\beta\epsilon] \qquad (7.47)$$

On substituting for the chemical potential in terms of the fugacity z

$$z = \exp[\beta\mu] \qquad (7.48)$$

and on introducing an expression for the density of states

$$\rho(\epsilon) = C \, \epsilon^\alpha \qquad (7.49)$$

one has

$$\overline{N} = C \sum_{n=1}^\infty (-1)^{n+1} \, z^n \int_0^\infty d\epsilon \, \epsilon^\alpha \, \exp[-n\beta\epsilon] \qquad (7.50)$$

On changing the variable of integration from ϵ to the dimensionless variable x where x is defined as

$$x = n\beta\epsilon \qquad (7.51)$$

the expression for the average number of particles has the form

$$\overline{N} = C(k_B T)^{\alpha+1} \int_0^\infty dx \, x^\alpha \, \exp[-x] \sum_{n=1}^\infty (-1)^{n+1} \, \frac{z^n}{n^{\alpha+1}}$$

$$= C(k_B T)^{\alpha+1} \, \Gamma(\alpha+1) \sum_{n=1}^\infty (-1)^{n+1} \, \frac{z^n}{n^{\alpha+1}} \qquad (7.52)$$

where $\Gamma(x)$ is the Gamma function. Since one can re-write the above equation as

$$\left(\frac{\overline{N}}{\Gamma(\alpha+1) \, C(k_B T)^{\alpha+1}}\right) = \sum_{n=1}^\infty (-1)^{n+1} \, \frac{z^n}{n^{\alpha+1}} \qquad (7.53)$$

one can see that, for fixed $\overline{N}$ and high T, z must be small. In the case where $z \ll 1$, which occurs for sufficiently high temperatures, one may only retain the first term in the power series for z. This leads to the solution for z and, hence, the chemical potential

$$z \approx \left(\frac{\overline{N}}{\Gamma(\alpha+1) \ C(k_B T)^{\alpha+1}} \right) \tag{7.54}$$

or alternatively

$$\mu(T) \approx -k_B T \ \ln \left[\frac{\Gamma(\alpha+1) \ C(k_B T)^{\alpha+1}}{\overline{N}} \right] \tag{7.55}$$

which illustrates that $\mu(T)$ must be temperature dependent. Furthermore, we see that at sufficiently high temperatures, only the first term in the expansion of the Fermi-Dirac distribution contributes. In this limit, Fermi-Dirac statistics reduces to classical Maxwell-Boltzmann statistics. We shall see later that something similar happens in the high temperature limit with Bose-Einstein statistics.

One also sees that on decreasing the temperature downwards, starting from the high temperature limit, then z increases. Therefore, one must retain an increasing number of terms in the expansion

$$\left(\frac{\overline{N}}{\Gamma(\alpha+1) \ C(k_B T)^{\alpha+1}} \right) = \sum_{n=1}^{\infty} (-1)^{n+1} \ \frac{z^n}{n^{\alpha+1}} \tag{7.56}$$

if one wants to determine the chemical potential accurately for lower temperatures. The reversion of the series is only practical if $z \leq 1$, above which the Fermi-Dirac gas is said to be non-degenerate. For temperatures below the degeneracy temperature, at which $\mu = 0$ and therefore $z = 1$, the chemical potential must be found by other methods.

7.6 The Sommerfeld Expansion

Many thermodynamic properties of a system of electrons can be written in the form

$$A(T) = \int_{-\infty}^{\infty} d\epsilon \ \Phi(\epsilon) \ f(\epsilon) \tag{7.57}$$

where $\Phi(\epsilon)$ is some function of the single-particle energy ϵ multiplied by the single-particle density of states $\rho(\epsilon)$ and $f(\epsilon)$ is the Fermi-function.

Integrals of this form can be evaluated very accurately, if the temperature T is of the order of room temperature $T \sim 300$ K, which corresponds to an energy scale of

$$k_B T \sim \frac{1}{40} \ \text{eV} \tag{7.58}$$

and if the typical energy scale for $\Phi(\epsilon)$ is between 1 and 10 eV. Typical electronic scales are given by the binding energy of an electron in a Hydrogen atom $\sim$13.6 eV or the total band width in a transition metal which is about 10 eV. In Na, the chemical potential μ measured from the bottom of the valence band density of states is about 3 eV and is about 12 eV for Al. Clearly, under ambient conditions in a metal one has

$$\mu \gg k_B T \qquad (7.59)$$

so a metal can usually be thought as being below its degeneracy temperature.

The Sommerfeld expansion [47] expresses integrals of the form

$$A(T) = \int_{-\infty}^{\infty} d\epsilon \ \Phi(\epsilon) \ f(\epsilon) \qquad (7.60)$$

in terms of a sum of the $T = 0$ limit of the integral and a power series of $k_B T/\mu$.

As a first approximation, one can estimate $A(T)$ as

$$A(T) \sim \int_{-\infty}^{\mu(T)} d\epsilon \ \Phi(\epsilon) \ + \ 0 \qquad (7.61)$$

since at $T = 0$, one can write the Fermi-function as

$$f(\epsilon) = 1 \qquad \text{for} \ \ \epsilon < \mu(T)$$
$$f(\epsilon) = 0 \qquad \text{for} \ \ \epsilon > \mu(T) \qquad (7.62)$$

We would like to obtain a better approximation, which reflects the temperature dependence of the Fermi-function $f(\epsilon)$. A better approximation for $A(T)$ can be obtained by re-writing the exact expression for $A(T)$ as

$$A(T) = \int_{-\infty}^{\mu(T)} d\epsilon \ \Phi(\epsilon) \ + \int_{\mu(T)}^{\infty} d\epsilon \ 0$$

$$+ \int_{-\infty}^{\mu(T)} d\epsilon \ \Phi(\epsilon) \ (f(\epsilon) - 1) + \int_{\mu(T)}^{\infty} d\epsilon \ \Phi(\epsilon)(f(\epsilon) - 0)$$

$$(7.63)$$

In this we have included the exact corrections to the $T = 0$ approximation to each region of the integral. This is evaluated as

$$
A(T) = \int_{-\infty}^{\mu(T)} d\epsilon \; \Phi(\epsilon) + \int_{-\infty}^{\mu(T)} d\epsilon \; \Phi(\epsilon) \left(\frac{1}{\exp[\beta(\epsilon - \mu)] + 1} - 1 \right)
$$
$$
+ \int_{\mu(T)}^{\infty} d\epsilon \; \Phi(\epsilon) \frac{1}{\exp[\beta(\epsilon - \mu)] + 1}
$$
$$
= \int_{-\infty}^{\mu(T)} d\epsilon \; \Phi(\epsilon) - \int_{-\infty}^{\mu(T)} d\epsilon \; \Phi(\epsilon) \frac{1}{\exp[-\beta(\epsilon - \mu)] + 1}
$$
$$
+ \int_{\mu(T)}^{\infty} d\epsilon \; \Phi(\epsilon) \frac{1}{\exp[\beta(\epsilon - \mu)] + 1} \tag{7.64}
$$

where we have substituted the identity

$$
\frac{1}{\exp[\beta(\epsilon - \mu)] + 1} - 1 = - \frac{\exp[\beta(\epsilon - \mu)]}{\exp[\beta(\epsilon - \mu)] + 1}
$$
$$
= - \frac{1}{1 + \exp[-\beta(\epsilon - \mu)]} \tag{7.65}
$$

in the second term of the first line. The two temperature-dependent correction terms in $A(T)$ involve a function of the form

$$
\frac{1}{\exp[x] + 1} \tag{7.66}
$$

We shall set $x = \beta(\epsilon - \mu(T))$ or

$$
\epsilon = \mu(T) + k_B T x \tag{7.67}
$$

in the second correction term which yields

$$
A(T) = \int_{-\infty}^{\mu(T)} d\epsilon \; \Phi(\epsilon) - \int_{-\infty}^{\mu(T)} d\epsilon \; \Phi(\epsilon) \frac{1}{\exp[-\beta(\epsilon - \mu)] + 1}
$$
$$
+ k_B T \int_0^{\infty} dx \; \Phi(\mu + k_B T x) \frac{1}{\exp[x] + 1} \tag{7.68}
$$

The first correction term can be expressed in terms of the variable $y = -\beta(\epsilon - \mu(T))$ or

$$
\epsilon = \mu(T) - k_B T y \tag{7.69}
$$

and the boundaries of the integration over y run from 0 to ∞.

$$
A(T) = \int_{-\infty}^{\mu(T)} d\epsilon \; \Phi(\epsilon) + k_B T \int_{\infty}^{0} dy \; \Phi(\mu - k_B T y) \frac{1}{\exp[y] + 1}
$$
$$
+ k_B T \int_0^{\infty} dx \; \Phi(\mu + k_B T x) \frac{1}{\exp[x] + 1} \tag{7.70}
$$

Except for the terms $+k_BTx$ and $-k_BTy$ in the arguments of the function Φ, the correction terms would cancel and vanish. On changing the integration variable from y to x in the second term, the integrals can be combined as

$$
A(T) = \int_{-\infty}^{\mu(T)} d\epsilon \ \Phi(\epsilon)
$$
$$
+ k_BT \int_0^{\infty} dx [\Phi(\mu + k_BTx) - \Phi(\mu - k_BTx)] \ \frac{1}{\exp[x] + 1}
$$
$$(7.71)$$

On Taylor expanding the terms in the large square parenthesis, one finds that only the odd terms in k_BTx survive.

$$
A(T) = \int_{-\infty}^{\mu(T)} d\epsilon \ \Phi(\epsilon)
$$
$$
+ 2k_BT \left[\sum_{n=0}^{\infty} \frac{\partial^{2n+1}\Phi(\epsilon)}{\partial \epsilon^{2n+1}} \bigg|_{\mu(T)} \right.
$$
$$
\left. \times \int_0^{\infty} dx \ \frac{(k_BTx)^{2n+1}}{(2n+1)!} \right] \frac{1}{\exp[x] + 1}
$$
$$(7.72)$$

Interchanging the order of the summation and the integration, results in

$$
A(T) = \int_{-\infty}^{\mu(T)} d\epsilon \ \Phi(\epsilon)
$$
$$
+ 2 \sum_{n=0}^{\infty} \frac{(k_BT)^{2n+2}}{(2n+1)!} \frac{\partial^{2n+1}\Phi(\epsilon)}{\partial \epsilon^{2n+1}} \bigg|_{\mu(T)} \int_0^{\infty} dx \ \frac{x^{2n+1}}{\exp[x] + 1}
$$
$$(7.73)$$

where all the derivatives of Φ are to be evaluated at $\mu(T)$. The integrals over x are convergent. One should note that the power series only contains terms of even powers in k_BT. Since the derivatives of Φ such as

$$
\frac{\partial^{2n+1}\Phi(\epsilon)}{\partial \epsilon^{2n+1}}
$$
$$(7.74)$$

have the dimensions of Φ/μ^{2n+1}, one might think of this expansion as being in powers of the dimensionless quantity

$$
\left(\frac{k_BT}{\mu} \right)^2
$$
$$(7.75)$$

which is assumed to be much smaller than unity. Therefore, the series could be expected to be rapidly convergent.

The integral

$$\int_0^\infty dx \, \frac{x^{2n+1}}{\exp[x] + 1} \tag{7.76}$$

is convergent since it is just the area under the curve that varies as x^{2n+1} for small x and vanishes exponentially as $\exp[-x] \, x^{2n+1}$ for large x. The

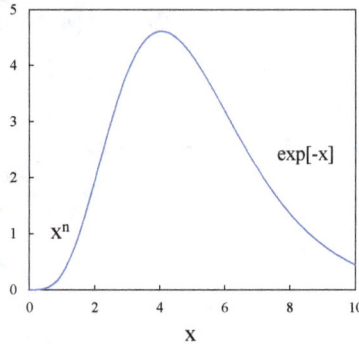

Fig. 7.3   The integrand $\frac{x^n}{\exp[x]+1}$.

integral can be evaluated by considering I_m given by

$$I_m = \int_0^\infty dx \, \frac{x^m}{\exp[x] + 1} \tag{7.77}$$

and noting that since $x > 0$ then $\exp[-x] < 1$. Therefore, by rewriting the integral as

$$I_m = \int_0^\infty dx \, \frac{x^m}{\exp[x] \, (1 + \exp[-x])}$$

$$= \int_0^\infty dx \, \frac{x^m \, \exp[-x]}{(1 + \exp[-x])} \tag{7.78}$$

one can expand the integral in powers of $\exp[-x]$

$$I_m = \int_0^\infty dx \, x^m \, \exp[-x] \sum_{l=0}^\infty (-1)^l \, \exp[-lx]$$

$$= \sum_{l=0}^\infty (-1)^l \int_0^\infty dx \, x^m \, \exp[-(l+1)x] \tag{7.79}$$

On changing the variable of integration from x to y where

$$y = (l+1) \, x \tag{7.80}$$

one has

$$I_m = \sum_{l=0}^{\infty} \frac{(-1)^l}{(l+1)^{m+1}} \int_0^{\infty} dy \; y^m \; \exp[-y] \tag{7.81}$$

The integral $\int_0^{\infty} dy \; y^m \; \exp[-y]$ can be evaluated by successive integration by parts. That is,

$$\int_0^{\infty} dy \; y^m \; \exp[-y] = -\int_0^{\infty} dy \; y^m \; \frac{\partial}{\partial y} \exp[-y]$$

$$= -y^m \; \exp[-y]\big|_0^{\infty} + \int_0^{\infty} dy \; m \; y^{m-1} \; \exp[-y] \tag{7.82}$$

The boundary term vanishes like y^m near $y = 0$ and vanishes like $\exp[-y]$ when $y \to \infty$. Therefore,

$$\int_0^{\infty} dy \; y^m = m \int_0^{\infty} dy \; y^{m-1} \; \exp[-y]$$

$$= m! \int_0^{\infty} dy \; \exp[-y]$$

$$= m! \tag{7.83}$$

Thus, we have

$$I_m = m! \sum_{l=0}^{\infty} \frac{(-1)^l}{(l+1)^{m+1}} \tag{7.84}$$

However, since the Riemann zeta function ζ is defined by

$$\zeta(m+1) = \sum_{l=0}^{\infty} \frac{1}{(l+1)^{m+1}} \tag{7.85}$$

one has

$$I_m = m! \left(1 - \frac{2}{2^{m+1}}\right) \zeta(m+1) \tag{7.86}$$

Therefore, the Sommerfeld expansion takes the form

$$A(T) = \int_{-\infty}^{\mu(T)} d\epsilon \; \Phi(\epsilon)$$

$$+ 2 \sum_{n=0}^{\infty} \left(1 - \frac{1}{2^{(2n+1)}}\right) \zeta(2(n+1))(k_B T)^{2n+2} \frac{\partial^{2n+1}\Phi(\epsilon)}{\partial \epsilon^{2n+1}}\bigg|_{\mu(T)} \tag{7.87}$$

which can be written as

$$A(T) = \int_{-\infty}^{\mu(T)} d\epsilon \; \Phi(\epsilon)$$

$$+ 2 \sum_{n=1}^{\infty} \left(1 - \frac{1}{2^{(2n-1)}} \right) \zeta(2n)(k_B T)^{2n} \left. \frac{\partial^{2n-1}\Phi(\epsilon)}{\partial\epsilon^{2n-1}} \right|_{\mu(T)}$$

$$(7.88)$$

where $\mu(T)$ is the value of the chemical potential at temperature T and the Riemann ζ function has the values

$$\zeta(2) = \frac{\pi^2}{6}$$

$$\zeta(4) = \frac{\pi^4}{90}$$

$$\zeta(6) = \frac{\pi^6}{945} \tag{7.89}$$

etc. Thus, at sufficiently low temperatures, one expects that one might be able to approximate $A(T)$ by the Sommerfeld expansion

$$A(T) \approx \int_{-\infty}^{\mu(T)} d\epsilon \; \Phi(\epsilon)$$

$$+ \frac{\pi^2}{6} (k_B T)^2 \left. \frac{\partial\Phi(\epsilon)}{\partial\epsilon} \right|_{\mu(T)} + \dots \tag{7.90}$$

7.7   The Low-Temperature Specific Heat

The condition of electrical neutrality determines the number of electrons in a metal and keeps the number constant. The number of electrons $\overline{N}$ is given by

$$\overline{N} = \int_{-\infty}^{\infty} d\epsilon \; \rho(\epsilon) \; f(\epsilon) \tag{7.91}$$

which can be approximated by the first few terms in the Sommerfeld expansion

$$\overline{N} = \int_{-\infty}^{\mu(T)} d\epsilon \; \rho(\epsilon) + \frac{\pi^2}{6} (k_B T)^2 \left. \frac{\partial\rho}{\partial\epsilon} \right|_{\mu} + \dots \tag{7.92}$$

This yields the temperature dependence of $\mu(T)$. Since $\overline{N}$ is independent of temperature

$$\frac{\partial N}{\partial T} = 0$$

$$= \frac{\partial}{\partial T} \int_{-\infty}^{\mu(T)} d\epsilon \, \rho(\epsilon) + \frac{2\pi^2}{6} k_B(k_B T) \left.\frac{\partial \rho}{\partial \epsilon}\right|_\mu + \dots$$

$$0 = \frac{\partial \mu}{\partial T} \rho(\mu) + \frac{2\pi^2}{6} k_B(k_B T) \left.\frac{\partial \rho}{\partial \epsilon}\right|_\mu + \dots \tag{7.93}$$

Thus, we have found that

$$\frac{\partial \mu}{\partial T} = -\frac{2\pi^2}{6} k_B(k_B T) \left.\frac{\frac{\partial \rho}{\partial \epsilon}}{\rho}\right|_\mu \tag{7.94}$$

which implies that the derivative of μ w.r.t. T has the opposite sign to $\frac{\partial \rho}{\partial \epsilon}$. Thus, if $\mu(T = 0)$ is just below a peak in $\rho(\epsilon)$ then the integral expression for N runs over a range of ϵ which avoids the peak. If μ did not decrease with increasing T, then at finite temperatures, the peak when multiplied by the tail of the Fermi-function could give an extra contribution to $\overline{N}$. This

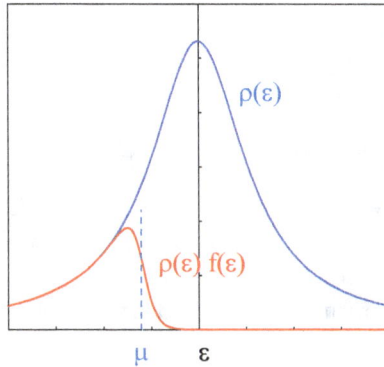

Fig. 7.4   The density of states $\rho(\epsilon)$ and the density of states weighted by the Fermi-Dirac distribution function $\rho(\epsilon)f(\epsilon)$.

increase must be offset by moving $\mu(T)$ down from ϵ_F, so the contribution from the tail at the peak is smaller and is offset by the smaller area under the curve to $\mu(T)$. Similar reasoning applies for the increase in $\mu(T)$ if ϵ_F is located just above a peak in $\rho(\epsilon)$. However, if ϵ_F is located at the top of a symmetric peak in the density of states, then the chemical potential should not depend on temperature.

The internal energy can also be expressed as

$$\overline{E} = \int_{-\infty}^{\infty} d\epsilon \, \rho(\epsilon) \, \epsilon \, f(\epsilon) \tag{7.95}$$

which can be approximated by the first few terms in the Sommerfeld expansion

$$\overline{E} = \int_{-\infty}^{\mu(T)} d\epsilon \, \rho(\epsilon) \, \epsilon + \frac{\pi^2}{6} \, (k_B T)^2 \, \left. \frac{\partial(\epsilon\rho)}{\partial\epsilon} \right|_{\mu} + \cdots \tag{7.96}$$

The specific heat is given by the temperature derivative of the internal energy at fixed V

$$
\begin{aligned}
C_V &= \left(\frac{\partial \overline{E}}{\partial T} \right)_V \\
&= \frac{\partial}{\partial T} \int_{-\infty}^{\mu(T)} d\epsilon \, \epsilon \, \rho(\epsilon) + \frac{2\pi^2}{6} \, k_B(k_B T) \, \left. \frac{\partial(\epsilon\rho)}{\partial\epsilon} \right|_{\mu} + \cdots \\
&= \frac{\partial\mu}{\partial T} \, \mu \, \rho(\mu) + \frac{2\pi^2}{6} \, k_B(k_B T) \, \left. \frac{\partial(\epsilon\rho)}{\partial\epsilon} \right|_{\mu} + \cdots
\end{aligned} \tag{7.97}
$$

On substituting for $\frac{\partial\mu}{\partial T}$, one finds

$$
\begin{aligned}
C_V &= -\frac{\pi^2}{3} \, k_B(k_B T) \, \mu \, \left. \frac{\partial\rho}{\partial\epsilon} \right|_{\mu} + \frac{2\pi^2}{6} \, k_B(k_B T) \, \left. \frac{\partial(\epsilon\rho)}{\partial\epsilon} \right|_{\mu} + \cdots \\
&= \frac{\pi^2}{3} \, k_B(k_B T) \, \rho(\mu) + O(T^3)
\end{aligned} \tag{7.98}
$$

since on expanding $\frac{\partial\epsilon\rho}{\partial\epsilon}$, one finds the term $\epsilon\frac{\partial\rho}{\partial\epsilon}$ cancels with the term $-\mu\frac{\partial\rho}{\partial\epsilon}$ coming from the temperature dependence of the chemical potential. Hence, the low-temperature electronic specific heat at constant volume is linearly proportional to temperature and the coefficient involves the density of states at the Fermi-energy.

$$C_V = \frac{\pi^2}{3} \, k_B(k_B T) \, \rho(\mu) \tag{7.99}$$

The above result is in contrast with the specific heat of a classical gas, which is given by

$$C_v = \frac{3}{2} \, Nk_B \tag{7.100}$$

The result found using quantum statistical mechanics

$$C_V = \frac{\pi^2}{3} \, k_B(k_B T) \, \rho(\mu) \tag{7.101}$$

is consistent with Nernst's law as

$$C_V = T \left(\frac{\partial S}{\partial T}\right)_T \qquad (7.102)$$

vanishes as $T \to 0$ since S vanishes as $T \to 0$. This occurs since the quantum ground state is unique so that S/N vanishes as $T \to 0$. The uniqueness occurs since the lowest energy single-particle states are all occupied by one electron, in accordance with the Pauli exclusion principle. Since there is no degeneracy, $S = 0$.

The specific heat of the electron gas is proportional to T. This can be understood by considering the effect of the Pauli exclusion principle. In a classical gas, where there is no exclusion principle, on supplying the thermal energy to the system, on average each particle acquires a kinetic energy of $\frac{3}{2} k_B T$. Hence, the excitation energy of the system is proportional to $k_B T$ and

$$\overline{E} = \frac{3}{2} N k_B T \qquad (7.103)$$

so

$$C_v = \left(\frac{\partial E}{\partial T}\right)_V$$
$$= \frac{3}{2} N k_B \qquad (7.104)$$

For fermions, if one supplies the thermal energy to the system, only the electrons within $k_B T$ of the Fermi-energy can be excited. An electron in an energy level far below ϵ_F cannot be excited by $k_B T$ since the final state with higher energy is already occupied. Thus, the Pauli exclusion principle forbids it to be excited. However, electrons within $k_B T$ of the Fermi-energy can be excited. The initial state is occupied, but the final state is above the Fermi-energy and can accept the excited electron.

Only the electrons within $k_B T$ of the Fermi-energy can be excited. The number of these electrons is approximately given by

$$\rho(\epsilon_F) \, k_B T \qquad (7.105)$$

Each of these electrons can be excited by the thermal energy $k_B T$, so the increase in the systems energy is given by

$$\Delta E = \rho(\epsilon_F)(k_B T)^2 \qquad (7.106)$$

Hence, the specific heat is estimated as

$$C_V = \frac{\Delta E}{T}$$
$$= \rho(\epsilon_F) \, k_B^2 T \qquad (7.107)$$

which shows that the linear T dependence is due to the Pauli exclusion principle.

Similar arguments apply to other thermodynamic properties or transport coefficients of the electron gas. The electrons in the states far below ϵ_F are inert since they cannot be excited by the small energies involved in the process, as their electrons cannot move up in energy because the desired final states are already occupied. The Pauli exclusion principle blocks these states from participating in processes which involve low excitation energies. Thus, they don't participate in electrical conduction, etc. These processes are all dominated by the states near ϵ_F, hence they depend on the density of states evaluated at the Fermi-energy $\rho(\mu)$ or its derivatives $\frac{\partial \rho}{\partial \mu}$, etc. Therefore, it may be useful to find other experimental properties which can be used to measure $\rho(\mu)$.

7.8 The Pauli Paramagnetic Susceptibility

The Pauli paramagnetic susceptibility provides an alternate measure of the single-particle density of states at the Fermi-energy. The effect of the Pauli exclusion principle limits the number of electrons that are allowed to flip their spins to an energy interval of the order of $\mu_B H$ around the Fermi-energy.

Consider a gas of non-interacting electrons, each of which carries a spin $S = \frac{1}{2}$. These spins couple to an applied magnetic field H^z aligned along the z-axis, via the anomalous Zeeman interaction

$$\hat{H}_{int} = -g\mu_B\, S^z\, H^z \tag{7.108}$$

where $g = 2$ is the gyromagnetic ratio. Hence, in the presence of the field, the single-electron energy levels become

$$\epsilon_\alpha - \mu_B H^z \qquad \text{for} \quad S^z = +\frac{1}{2}$$

$$\epsilon_\alpha + \mu_B H^z \qquad \text{for} \quad S^z = -\frac{1}{2} \tag{7.109}$$

Therefore, the Grand-Canonical potential is given by

$$-\beta\Omega = \sum_{\alpha,\pm} \ln\left[1 + \exp[-\beta(\epsilon_\alpha \mp \mu_B H^z - \mu)]\right] \tag{7.110}$$

which can be written as an integral over the (zero-field) density of states

$$-\beta\Omega = \frac{1}{2}\sum_{\pm} \int_{-\infty}^{\infty} d\epsilon\, \rho(\epsilon)\, \ln\left[1 + \exp[-\beta(\epsilon \mp \mu_B H^z - \mu)]\right] \tag{7.111}$$

where the factor $\frac{\rho}{2}$ is the density of states per spin direction (for zero applied fields). The magnetization is defined as

$$M^z = -\left(\frac{\partial \Omega}{\partial H^z}\right) \tag{7.112}$$

which yields

$$M^z = \frac{1}{2} \sum_{\pm} (\pm) \, \mu_B \int_{-\infty}^{\infty} d\epsilon \, \rho(\epsilon) \, \frac{\exp[-\beta(\epsilon \mp \mu_B H^z - \mu)]}{1 + \exp[-\beta(\epsilon \mp \mu_B H^z - \mu)]}$$

$$= \frac{1}{2} \sum_{\pm} (\pm) \, \mu_B \int_{-\infty}^{\infty} d\epsilon \, \rho(\epsilon) \, f(\epsilon \mp \mu_B H^z) \tag{7.113}$$

This can simply be interpreted as μ_B times the excess of spin-up electrons over down-spin electrons

$$M^z = \sum_{\alpha,\pm} (\pm) \, \mu_B \, \overline{n}_{\alpha,\pm} \tag{7.114}$$

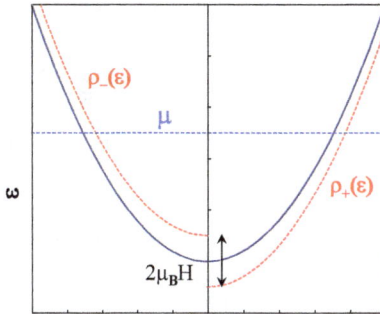

Fig. 7.5 The spin-split single-electron density of states (red), in the presence of a finite magnetic field. Due to the Pauli principle, the field can only realign the spins of electrons which have energies within $\mu_B H^z$ of the Fermi-energy.

which vanishes when $H^z \to 0$. The differential susceptibility is defined as

$$\chi^{z,z} = \left(\frac{\partial M^z}{\partial H^z}\right)_{H^z=0} \tag{7.115}$$

which measure the linear field dependence of M. The susceptibility is evaluated as

$$\chi^{z,z} = -\frac{1}{2} \sum_{\pm} \mu_B^2 \int_{-\infty}^{\infty} d\epsilon \, \rho(\epsilon) \, \frac{\partial f}{\partial \epsilon} (\epsilon \mp \mu_B H^z) \tag{7.116}$$

which remains finite in the limit $H^z \to 0$. In the limit of zero field, the susceptibility simplifies to

$$\chi^{z,z} = -\frac{1}{2} \sum_{\pm} \mu_B^2 \int_{-\infty}^{\infty} d\epsilon \, \rho(\epsilon) \, \frac{\partial f}{\partial \epsilon}$$

$$= -\mu_B^2 \int_{-\infty}^{\infty} d\epsilon \, \rho(\epsilon) \, \frac{\partial f}{\partial \epsilon} \tag{7.117}$$

In the limit of zero temperature, one has

$$-\frac{\partial f}{\partial \epsilon} = \delta(\epsilon - \mu) \tag{7.118}$$

therefore, one finds

$$\chi^{z,z} = \mu_B^2 \, \rho(\mu) \tag{7.119}$$

Hence, the ratio of the specific heat and the susceptibility given by

$$\frac{\mu_B^2 \, C_V}{k_B^2 T \, \chi^{z,z}} = \frac{\pi^2}{3} \tag{7.120}$$

This relation is independent of $\rho(\mu)$ and provided a check on the theory. It is satisfied for all simple metals and for most of the early transition metals. Thus the low temperature limit of $\chi^{z,z}$ is a measure of the density of states at the Fermi-energy.

The leading temperature dependent corrections to χ can be obtained from the Sommerfeld expansion. The susceptibility is given by

$$\chi^{z,z} = -\mu_B^2 \int_{-\infty}^{\infty} d\epsilon \, \rho(\epsilon) \, \frac{\partial f}{\partial \epsilon} \tag{7.121}$$

and on integrating by parts, one obtains

$$\chi^{z,z} = -\mu_B^2 \, \rho(\epsilon) \, f(\epsilon)|_{-\infty}^{\infty} + \mu_B^2 \int_{-\infty}^{\infty} d\epsilon \, \frac{\partial \rho}{\partial \epsilon} \, f(\epsilon)$$

$$= \mu_B^2 \int_{-\infty}^{\infty} d\epsilon \, \frac{\partial \rho}{\partial \epsilon} \, f(\epsilon) \tag{7.122}$$

since the boundary terms vanish. On using the Sommerfeld expansion, one obtains the result

$$\chi^{z,z} = \mu_B^2 \, \rho(\mu) + \mu_B^2 \, \frac{\pi^2}{6} \, (k_B T)^2 \left[\frac{\partial^2 \rho}{\partial \epsilon^2} - \frac{1}{\rho} \left(\frac{\partial \rho}{\partial \epsilon} \right)^2 \right] \Bigg|_{\epsilon_F} + \dots \tag{7.123}$$

The corrections aren't important unless $k_B T \sim \rho/\frac{\partial \rho}{\partial \epsilon}$, i.e. the temperature is of the order of the energy over which ρ varies. Thus, $\chi^{z,z}$ is approximately temperature independent.

7.9 The High-Temperature Susceptibility

The high-temperature limit of the susceptibility

$$\chi^{z,z} = -\mu_B^2 \int_{-\infty}^{\infty} d\epsilon \, \rho(\epsilon) \, \frac{\partial f}{\partial \epsilon} \tag{7.124}$$

can be found by using the high-temperature approximation

$$f(\epsilon) \approx \exp[-\beta(\epsilon - \mu)] \tag{7.125}$$

which yields the approximation

$$\chi^{z,z} \sim \beta \, \mu_B^2 \int_{-\infty}^{\infty} d\epsilon \, \rho(\epsilon) \, \exp[-\beta(\epsilon - \mu)] \tag{7.126}$$

which is evaluated as

$$\chi^{z,z} \sim \beta \mu_B^2 \, \overline{N} \tag{7.127}$$

on utilizing the expression for the number of electrons

$$\overline{N} = \int_{-\infty}^{\infty} d\epsilon \, \rho(\epsilon) \, \exp[-\beta(\epsilon - \mu)] \tag{7.128}$$

Hence, the Pauli paramagnetic susceptibility turns over into a Curie law at sufficiently high temperatures. Since the high temperature variation

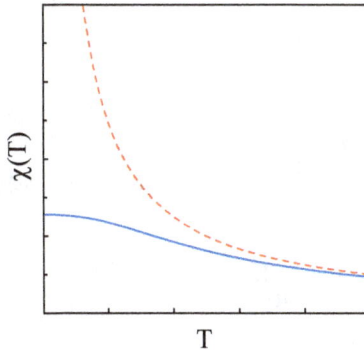

Fig. 7.6 The temperature dependence of the Pauli paramagnetic susceptibility $\chi(T)$ (blue curve). The high temperature Curie law is shown by the dashed red curve.

first happens when $k_B T \approx \mu$, the high temperature limit first applies at temperatures of the order of $T \approx 12{,}000$ K.

7.10   The Pressure of a Fermion Gas

The Sommerfeld expansion can also be used to calculate the temperature-dependence of the pressure for a gas of non-interacting fermions at low temperatures. Starting from the expression for the Grand-Canonical potential Ω and integrating by parts, one finds that

$$\Omega = -PV$$
$$= -\frac{2}{3} \int_0^\infty d\epsilon \, \rho(\epsilon) \, \epsilon \, \frac{1}{\exp[\beta \, (\epsilon - \mu)] + 1} \tag{7.129}$$

where the density of states (including a factor of 2 for both spin directions) is given by

$$\rho(\epsilon) = \frac{V}{2\pi^2} \left(\frac{2m}{\hbar^2}\right)^{\frac{3}{2}} \sqrt{\epsilon} \tag{7.130}$$

Hence, on substituting for the single-particle density of states and on canceling a factor of V, the pressure is expressed as

$$P = \frac{1}{3\pi^2} \left(\frac{2m}{\hbar^2}\right)^{\frac{3}{2}} \int_0^\infty d\epsilon \, \epsilon^{\frac{3}{2}} \, f(\epsilon) \tag{7.131}$$

where $f(\epsilon)$ is the Fermi-Dirac distribution function. On using the Sommerfeld expansion, one obtains the approximate expression

$$P \approx \frac{1}{3\pi^2} \left(\frac{2m}{\hbar^2}\right)^{\frac{3}{2}} \left[\int_0^{\mu(T)} d\epsilon \, \epsilon^{\frac{3}{2}} + \frac{\pi^2}{6} (k_B T)^2 \, \frac{3}{2} \, \mu(T)^{\frac{1}{2}} + \ldots\right] \tag{7.132}$$

which is evaluated as

$$P \approx \frac{1}{3\pi^2} \left(\frac{2m}{\hbar^2}\right)^{\frac{3}{2}} \left[\frac{2}{5} \, \mu(T)^{\frac{5}{2}} + \frac{\pi^2}{4} (k_B T)^2 \, \mu(T)^{\frac{1}{2}} + \ldots\right] \tag{7.133}$$

The temperature dependence of the chemical potential is approximately given by

$$\mu(T) \approx \epsilon_F \left[1 - \frac{\pi^2}{12} \left(\frac{k_B T}{\epsilon_F}\right)^2 + \ldots\right] \tag{7.134}$$

Hence, we obtain the final result

$$P \approx \frac{1}{3\pi^2} \left(\frac{2m}{\hbar^2}\right)^{\frac{3}{2}} \left[\frac{2}{5} \, \epsilon_F^{\frac{5}{2}} + \frac{\pi^2}{6} (k_B T)^2 \, \epsilon_F^{\frac{1}{2}} + \ldots\right]$$
$$\approx \frac{2}{15\pi^2} \left(\frac{2m}{\hbar^2}\right)^{\frac{3}{2}} \epsilon_F^{\frac{5}{2}} \left[1 + \frac{5\pi^2}{12} \left(\frac{k_B T}{\epsilon_F}\right)^2 + \ldots\right] \tag{7.135}$$

Thus, a gas of non-interacting particles which obey Fermi-Dirac statistics exerts a finite pressure in the limit $T \to 0$. This can be understood, since at $T = 0$, the particles occupy states with finite momenta up to the Fermi-energy and, therefore, they collide with the container's walls giving rise to pressure. This is in direct contrast with the behavior found for a classical ideal gas and, as we shall see later, is also in contrast with the pressure of a non-interacting gas of bosons at low temperatures.

7.11 Fluctuations in the Occupation Numbers

If one considers a group of energy levels ϵ_α with large enough degeneracy, then it is possible to consider the statistics of the occupation number for α.

The average occupation number of the energy levels ϵ_α, is given by

$$
\begin{aligned}
\bar{n}_\alpha &= \frac{1}{\Xi} \text{ Trace } \hat{n}_\alpha \, \exp[-\beta(\hat{H} - \mu\hat{N})] \\
&= \frac{1}{\Xi} \text{ Trace } n_\alpha \, \exp\left[-\beta \sum_{\alpha'} (\epsilon_{\alpha'} - \mu) \, n_{\alpha'}\right] \\
&= -k_B T \frac{\partial}{\partial \epsilon_\alpha} \ln \Xi
\end{aligned}
\tag{7.136}
$$

where the derivative is w.r.t. the energy level ϵ_α. As expected, this is given by

$$
\begin{aligned}
\bar{n}_\alpha &= \frac{1}{\exp[\beta(\epsilon_\alpha - \mu)] + 1} \\
&= f(\epsilon_\alpha)
\end{aligned}
\tag{7.137}
$$

which is just the Fermi-Dirac distribution function.

The mean squared fluctuations around this average is given by

$$
\begin{aligned}
\bar{n}_\alpha^2 - \bar{n}_\alpha^2 &= \left(-\frac{1}{\beta} \frac{\partial}{\partial \epsilon_\alpha}\right)^2 \ln \Xi \\
&= \left(-\frac{1}{\beta} \frac{\partial}{\partial \epsilon_\alpha}\right) \bar{n}_\alpha \\
&= f(\epsilon_\alpha) \, (1 - f(\epsilon_\alpha))
\end{aligned}
\tag{7.138}
$$

The r.m.s. number fluctuation is reduced from the classical value of $\bar{n}_\alpha$. In fact, due to the Pauli exclusion principle, the fluctuations are only non-zero in an energy width of $k_B T$ around the Fermi-energy. The reduction in the fluctuation of fermion occupation numbers is in strong contrast to the fluctuations that are found for particles that obey Bose-Einstein Statistics.

The specific heat provides a measure of the fluctuations in the energy divided by $k_B T^2$ and, for a non-interacting gas of fermions, the fluctuations in the occupation numbers of each single-particle energy level can be considered as being independent. Hence, the linear-T dependence of the low-temperature specific heat is seen to be a consequence of the fluctuations of the gas of fermions. Likewise, the Pauli paramagnetic susceptibility is related to β times the fluctuations of the magnetization which, in turn, can also be related to the fluctuations in the occupation numbers. Since the fluctuations are restricted to within $k_B T$ of the Fermi-energy, the Pauli susceptibility is approximately constant. As both the susceptibility and the specific heat both involve the fluctuations in the occupations numbers, it is expected that the "*Sommerfeld Ratio*", defined by the low-temperature limit of

$$R_S = \frac{\pi^2}{3}\left(\frac{k_B^2 T}{\mu_B^2}\right)\frac{\chi}{C} \qquad (7.139)$$

is approximately unity. For many metals in which the electron-electron interactions are weak, the experimentally determined ratio is found to be close to unity. Even for the strongly correlated materials known as heavy-fermion systems, such as $CeAl_3$ or UBe_{13} in which the linear-T term in specific heat is up to 1000 times larger than is expected from non-interacting electrons, the experimentally determined Sommerfeld ratio is still close to unity.

7.12 The Pauli Spin Susceptibility: Revisited

The Pauli spin susceptibility can expressed in terms of the fluctuations of the spin magnetization. That is, the susceptibility is expressed in terms of integrals over a spatially-dependent spin-spin correlation function

$$\chi^{z,z} = \beta \int d^3\underline{r} \int d^3\underline{r}' [\langle \hat{M}^z(\underline{r})\hat{M}^z(\underline{r}')\rangle - \langle \hat{M}^z(\underline{r})\rangle\langle \hat{M}^z(\underline{r}')\rangle] \qquad (7.140)$$

where the second-quantized magnetization density operator is expressed in terms of the field operators as

$$\hat{\underline{M}}(\underline{r}) = \mu_B \, \hat{\psi}^\dagger(\underline{r}) \, \underline{\sigma} \, \hat{\psi}(\underline{r}) \qquad (7.141)$$

and in which we have used the conduction electron value of the gyromagnetic ratio $g = 2$. For a metal, the field operators can be expanded in terms of Bloch functions labelled by the Bloch wave vectors $\underline{k}$. For a paramagnet, the average value of the magnetization vanishes in the absence of a field. The vanishing of the magnetization follows from

$$
\int d^3\underline{r} \langle \hat{M}^z(r) \rangle = \mu_B \sum_{\underline{k},\sigma;\underline{k}',\sigma'} \int d^2\underline{r} \phi_{\underline{k},\sigma}^*(r) \sigma^z \phi_{\underline{k}',\sigma'}(\underline{r}) \langle \hat{c}_{\underline{k},\sigma}^\dagger \hat{c}_{\underline{k}',\sigma'} \rangle
$$

$$
= \mu_B \sum_{\underline{k},\sigma;\underline{k}',\sigma'} \sigma \langle \hat{c}_{\underline{k},\sigma}^\dagger \hat{c}_{\underline{k},\sigma} \rangle \delta_{\underline{k},\underline{k}'} \delta_{\sigma,\sigma'}
$$

$$
= \mu_B \sum_{\underline{k},\sigma} \sigma \langle \hat{n}_{\underline{k},\sigma} \rangle
$$

$$
= 0 \tag{7.142}
$$

where the last line is zero due to the sum over spins and since the thermally averaged occupation numbers are spin-independent. Therefore, the paramagnetic susceptibility is proportional to the term

$$
\int d^3\underline{r} \int d^3\underline{r}' \langle \hat{M}^z(\underline{r}) \hat{M}^z(\underline{r}') \rangle = \mu_B^2 \sum_{\underline{k},\sigma;\underline{k}',\sigma'} \sigma\sigma' \langle \hat{c}_{\underline{k},\sigma}^\dagger \hat{c}_{\underline{k},\sigma} \hat{c}_{\underline{k}',\sigma'}^\dagger \hat{c}_{\underline{k}',\sigma'} \rangle
$$

$$
= \mu_B^2 \sum_{\sigma,\underline{k},\underline{k}'} [\langle \hat{c}_{\underline{k},\sigma}^\dagger \hat{c}_{\underline{k},\sigma} \hat{c}_{\underline{k}',\sigma}^\dagger \hat{c}_{\underline{k}',\sigma} \rangle
$$

$$
- \langle \hat{n}_{\underline{k},\sigma} \rangle \langle \hat{n}_{\underline{k}',\bar{\sigma}} \rangle] \tag{7.143}
$$

The expectation value of the product of the four equal-spin creation and annihilation operators is evaluated as

$$
\langle \hat{c}_{\underline{k},\sigma}^\dagger \hat{c}_{\underline{k},\sigma} \hat{c}_{\underline{k}',\sigma}^\dagger \hat{c}_{\underline{k}',\sigma} \rangle = \text{Trace} \left[\frac{\exp[-\beta(\hat{H} - \mu\hat{N})]}{\Xi} \right.
$$

$$
\left. \times \hat{c}_{\underline{k},\sigma}^\dagger \hat{c}_{\underline{k},\sigma} \hat{c}_{\underline{k}',\sigma}^\dagger \hat{c}_{\underline{k}',\sigma} \right]
$$

$$
= \exp[-\beta(\epsilon_{\underline{k}} - \mu)] \text{ Trace} \left[\hat{c}_{\underline{k},\sigma}^\dagger \right.
$$

$$
\left. \times \frac{\exp[-\beta(\hat{H} - \mu\hat{N})]}{\Xi} \hat{c}_{\underline{k},\sigma} \hat{c}_{\underline{k}',\sigma}^\dagger \hat{c}_{\underline{k}',\sigma} \right]
$$

$$
= \exp[-\beta(\epsilon_{\underline{k}} - \mu)] \text{ Trace} \left[\frac{\exp[-\beta(\hat{H} - \mu\hat{N})]}{\Xi} \right.
$$

$$
\left. \times \hat{c}_{\underline{k},\sigma} \hat{c}_{\underline{k}',\sigma}^\dagger \hat{c}_{\underline{k}',\sigma} \hat{c}_{\underline{k},\sigma}^\dagger \right] \tag{7.144}
$$

where we have used the commutation relation

$$
\exp[-\beta(\hat{H} - \mu\hat{N})] \hat{c}_{\underline{k},\sigma}^\dagger = \hat{c}_{\underline{k},\sigma}^\dagger \exp[-\beta(\epsilon_{\underline{k},\sigma} - \mu)] \exp[-\beta(\hat{H} - \mu\hat{N})] \tag{7.145}
$$

and the cyclic invariance of the Trace. On transposing $\hat{c}^\dagger_{\underline{k},\sigma}$ with $\hat{c}_{\underline{k}',\sigma}$ one obtains

$$
\begin{aligned}
\langle \hat{c}^\dagger_{\underline{k},\sigma} \hat{c}_{\underline{k},\sigma} \hat{c}^\dagger_{\underline{k}',\sigma} \hat{c}_{\underline{k}',\sigma} \rangle &= \exp[-\beta(\epsilon_{\underline{k}} - \mu)] \, \text{Trace} \left[\frac{\exp[-\beta(\hat{H} - \mu\hat{N})]}{\Xi} \right. \\
&\quad \left. \times \hat{c}_{\underline{k},\sigma} \hat{c}^\dagger_{\underline{k}',\sigma} (\delta_{\underline{k},\underline{k}'} - \hat{c}^\dagger_{\underline{k},\sigma} \hat{c}_{\underline{k}',\sigma}) \right] \\
&= \exp[-\beta(\epsilon_{\underline{k}} - \mu)] \langle \hat{c}_{\underline{k},\sigma} \hat{c}^\dagger_{\underline{k}',\sigma} \rangle \delta_{\underline{k},\underline{k}'} \\
&\quad - \exp[-\beta(\epsilon_{\underline{k}} - \mu)] \langle \hat{c}_{\underline{k},\sigma} \hat{c}^\dagger_{\underline{k}',\sigma} \hat{c}^\dagger_{\underline{k},\sigma} \hat{c}_{\underline{k}',\sigma} \rangle \\
&= \exp[-\beta(\epsilon_{\underline{k}} - \mu)] \langle \hat{c}_{\underline{k},\sigma} \hat{c}^\dagger_{\underline{k}',\sigma} \rangle \delta_{\underline{k},\underline{k}'} \\
&\quad + \exp[-\beta(\epsilon_{\underline{k}} - \mu)] \langle \hat{c}_{\underline{k},\sigma} \hat{c}^\dagger_{\underline{k},\sigma} \hat{c}^\dagger_{\underline{k}',\sigma} \hat{c}_{\underline{k}',\sigma} \rangle
\end{aligned}
\tag{7.146}
$$

In obtaining the last line, we have anticommuted the two creation operators. Using the anticommutation relation

$$
\{ \hat{c}^\dagger_{\underline{k},\sigma}, \hat{c}^\dagger_{\underline{k},\sigma} \}_+ = 0
\tag{7.147}
$$

one finds that

$$
\begin{aligned}
\langle \hat{c}^\dagger_{\underline{k},\sigma} \hat{c}_{\underline{k},\sigma} \hat{c}^\dagger_{\underline{k}',\sigma} \hat{c}_{\underline{k}',\sigma} \rangle &= \exp[-\beta(\epsilon_{\underline{k}} - \mu)][\langle \hat{c}_{\underline{k},\sigma} \hat{c}^\dagger_{\underline{k}',\sigma} \rangle \delta_{\underline{k},\underline{k}'} + \langle \hat{n}_{\underline{k}',\sigma} \rangle] \\
&\quad - \exp[-\beta(\epsilon_{\underline{k}} - \mu)] \langle \hat{c}^\dagger_{\underline{k},\sigma} \hat{c}_{\underline{k},\sigma} \hat{c}^\dagger_{\underline{k}',\sigma} \hat{c}_{\underline{k}',\sigma} \rangle
\end{aligned}
\tag{7.148}
$$

where the expectation values of the four fermion operators are identical. Furthermore, since

$$
\langle \hat{c}_{\underline{k},\sigma} \hat{c}^\dagger_{\underline{k}',\sigma} \rangle \delta_{\underline{k},\underline{k}'} = (1 - \langle \hat{n}_{\underline{k},\sigma} \rangle) \, \delta_{\underline{k},\underline{k}'}
\tag{7.149}
$$

and since

$$
\langle \hat{n}_{\underline{k},\sigma} \rangle = \frac{\exp[-\beta(\epsilon_{\underline{k}} - \mu)]}{1 + \exp[-\beta(\epsilon_{\underline{k}} - \mu)]}
\tag{7.150}
$$

one concludes that the equal spin expectation value is given by

$$
\langle \hat{c}^\dagger_{\underline{k},\sigma} \hat{c}_{\underline{k},\sigma} \hat{c}^\dagger_{\underline{k}',\sigma} \hat{c}_{\underline{k}',\sigma} \rangle = \langle \hat{n}_{\underline{k},\sigma} \rangle [\langle \hat{n}_{\underline{k}',\sigma} \rangle + (1 - \langle \hat{n}_{\underline{k},\sigma} \rangle) \, \delta_{\underline{k},\underline{k}'}]
\tag{7.151}
$$

Since the occupation numbers are spin-independent at zero field, the low-temperature susceptibility is given by

$$
\begin{aligned}
\chi^{z,z} &= 2\beta\mu_B^2 \sum_{\underline{k}} \langle \hat{n}_{\underline{k}} \rangle (1 - \langle \hat{n}_{\underline{k}} \rangle) \\
&= -2\mu_B^2 \int d\epsilon \, N\rho(\epsilon) \left(\frac{df}{d\epsilon} \right) \\
&\approx 2\mu_B^2 \, N\rho(\mu)
\end{aligned}
\tag{7.152}
$$

where $f(\epsilon)$ is the Fermi-Dirac distribution and $\rho(\epsilon)$ is the single-electron density of states per spin. In the last line we have used the low-temperature approximation

$$-\left(\frac{df}{d\epsilon}\right) \approx \delta(\epsilon - \mu) \qquad (7.153)$$

It should also be noted that the factor of β, which for isolated spins leads to a Curie susceptibility, has been compensated by a factor of $k_B T$ due to the Pauli exclusion principle that restricts the magnetic fluctuations to within $k_B T$ of the metal's Fermi-energy. We conclude that the Pauli susceptibility obtained from consideration of the spatial correlation functions agrees with the susceptibility calculated by using the Grand-Canonical potential.

7.13 The Hartree-Fock Approximation

The "*Hartree-Fock Approximation*" is used to find the best energy eigenvalues and eigenstates of an interacting N-electron system, if the many-body wave function is restricted to have the form of a single Slater determinant [48]. The approximation is based on the Rayleigh-Ritz variational principle for the energy eigenstates of a Hamiltonian.

The Hamiltonian of the interacting system of electrons is given by the operator

$$\hat{H} = \int d^3\underline{r}\hat{\psi}^\dagger(\underline{r}) \left[-\frac{\hbar^2}{2m}\nabla^2 + V_0(\underline{r}) \right] \hat{\psi}(\underline{r})$$

$$+ \frac{1}{2!} \int d^3\underline{r} \int d^3\underline{r}'\hat{\psi}^\dagger(\underline{r})\hat{\psi}^\dagger(\underline{r}')V_{int}(\underline{r},\underline{r}')\hat{\psi}(\underline{r}')\hat{\psi}(\underline{r}) \qquad (7.154)$$

The above expressions reduce to the second quantized form

$$\hat{H} = \sum_{\alpha,\alpha'} \hat{c}^\dagger_{\alpha'}\hat{c}_\alpha \int d^3\underline{r}\phi^*_{\alpha'}(\underline{r}) \left[-\frac{\hbar^2}{2m}\nabla^2 + V_0(\underline{r}) \right] \phi_\alpha(\underline{r})$$

$$+ \frac{1}{2!} \sum_{\alpha,\beta,\alpha',\beta'} \hat{c}^\dagger_{\alpha'}\hat{c}^\dagger_{\beta'}\hat{c}_\beta\hat{c}_\alpha$$

$$\times \int d^3\underline{r} \int d^3\underline{r}'\phi^*_{\alpha'}(\underline{r})\phi^*_{\beta'}(\underline{r}')V_{int}(\underline{r},\underline{r}')\phi_\beta(\underline{r}')\phi_\alpha(\underline{r}) \qquad (7.155)$$

The expectation values of the operator products for a state which consists of a single Slater determinant are given by the expressions

$$\langle \hat{c}^{\dagger}_{\alpha'} \hat{c}_{\alpha} \rangle = n_{\alpha} \delta_{\alpha,\alpha'}$$

$$\langle \hat{c}^{\dagger}_{\alpha'} \hat{c}^{\dagger}_{\beta'} \hat{c}_{\beta} \hat{c}_{\alpha} \rangle = n_{\alpha} n_{\beta} [\delta_{\beta,\beta'} \delta_{\alpha,\alpha'} - \delta_{\alpha,\beta'} \delta_{\beta,\alpha'}] \tag{7.156}$$

where the occupation numbers n_{α}'s are just unity or zero, depending on whether or not the single particle basis function $\phi_{\alpha}(\underline{r})$ is included in the chosen Slater determinant. Hence, the Hartree-Fock energy is given by the functional

$$H_{HF}[\{\phi_{\alpha}(\underline{r})\}] = \sum_{\alpha} n_{\alpha} \int d^3 \underline{r} \phi^*_{\alpha}(\underline{r}) \left[-\frac{\hbar^2}{2m} \nabla^2 + V_0(\underline{r}) \right] \phi_{\alpha}(\underline{r})$$

$$+ \frac{1}{2!} \sum_{\alpha,\beta} n_{\alpha} n_{\beta}$$

$$\times \int d^3 \underline{r} \int d^3 \underline{r}' \phi^*_{\alpha}(\underline{r}) \phi^*_{\beta}(\underline{r}') V_{int}(\underline{r}, \underline{r}') \phi_{\beta}(\underline{r}') \phi_{\alpha}(\underline{r})$$

$$- \frac{1}{2!} \sum_{\alpha,\beta} n_{\alpha} n_{\beta}$$

$$\times \int d^3 \underline{r} \int d^3 \underline{r}' \phi^*_{\beta}(\underline{r}) \phi^*_{\alpha}(\underline{r}') V_{int}(\underline{r}, \underline{r}') \phi_{\beta}(\underline{r}') \phi_{\alpha}(\underline{r}) \tag{7.157}$$

On representing the eigenvalues as α by (m, σ) and the corresponding spinors as

$$\phi_{\alpha}(\underline{r}) = \phi_m(\underline{r}) \chi_{\sigma}$$

$$\phi^*_{\alpha}(\underline{r}) = \phi^*_m(\underline{r}) \chi^{\dagger}_{\sigma} \tag{7.158}$$

and β by (m', σ') so that

$$\phi_{\beta}(\underline{r}) = \phi_{m'}(\underline{r}) \chi_{\sigma'}$$

$$\phi^*_{\beta}(\underline{r}) = \phi^*_{m'}(\underline{r}) \chi^{\dagger}_{\sigma'} \tag{7.159}$$

one sees that the last term in the functional H_{HF} vanishes for spin-independent potential unless $\sigma = \sigma'$. The first term of the energy functional is just the energy of non-interacting electrons, whereas the second term represents the Coulomb interaction energy due to the charge densities of pairs of electrons. The last term represents the "*Exchange Interaction*" which is caused by the Fermi-Dirac statistics and it only acts on pairs of electrons which have spins that point in the same direction. Thus, the energy functional can be expressed as

$$H_{HF}[\{\phi_m(\underline{r})\}] = \sum_{m,\sigma} n_{m,\sigma} \int d^3\underline{r}\, \phi_m^*(\underline{r}) \left[-\frac{\hbar^2}{2m}\nabla^2 + V_0(\underline{r}) \right] \phi_m(\underline{r})$$

$$+ \frac{1}{2!} \sum_{m,\sigma,m',\sigma'} n_{m,\sigma} n_{m',\sigma'}$$

$$\times \int d^3\underline{r} \int d^3\underline{r}' |\phi_{m'}(\underline{r}')|^2 V_{int}(\underline{r},\underline{r}') |\phi_m(\underline{r})|^2$$

$$- \frac{1}{2!} \sum_{m,m',\sigma} n_{m,\sigma} n_{m',\sigma}$$

$$\times \int d^3\underline{r} \int d^3\underline{r}' \phi_m^*(\underline{r}')\phi_{m'}(\underline{r}') V_{int}(\underline{r},\underline{r}') \phi_{m'}^*(\underline{r})\phi_m(\underline{r})$$

$$(7.160)$$

The functional is to be minimized with respect to variations of $\phi_m(\underline{r})$ subject to the constraint that the single-particle basis functions are normalized

$$\int d^3\underline{r}\, |\phi_m(\underline{r})|^2 = 1 \qquad (7.161)$$

The minimization subject to the constraints is achieved by using Lagrange's method of undetermined multipliers. Lagrange's method consists of minimizing the auxiliary functional

$$I[\{\phi_m(\underline{r})\}] = H_{HF}[\{\phi_m(\underline{r})\}] + \sum_{m,\sigma} \epsilon_m n_{m,\sigma} \left(1 - \int d^3\underline{r}\, |\phi_m(\underline{r})|^2 \right)$$

$$(7.162)$$

where ϵ_m are the undetermined multipliers, and it can be seen that the last term vanishes if all the single-particle basis functions are normalized.

The variational procedure assumes that the extremal state exists and is unique. Then one may consider functions that are distorted from the extremal function by arbitrarily small amounts and with various shapes

$$\phi_m(\underline{r}) \to \phi_m(\underline{r}) + \lambda\delta\phi_m(\underline{r}) \qquad (7.163)$$

The extremal condition requires that the change in the functional due to any arbitrary deviation must be zero. Hence, on substituting the distorted basis state in the functional and by expanding in powers of λ, one requires that the first-order term in λ must vanish

$$\lambda\delta^1 I[\{\phi_m(\underline{r})\}] = 0 \qquad (7.164)$$

Since the basis functions are complex functions and the real and imaginary parts are independent, one could vary the real and the imaginary parts separately. However, since the functional is real, variation wrt the ϕ_m and their complex conjugates ϕ_m^* are equivalent. Varying the functional wrt ϕ_m^* yields

$$\delta^1 I[\{\phi_m(\underline{r})\}] = \int d^3\underline{r}\,\delta\phi_m^*(\underline{r}) \left[\left(-\frac{\hbar^2}{2m}\nabla^2 + V_0(\underline{r}) - \epsilon_m \right) \phi_m(\underline{r}) \right.$$

$$+ \sum_{m',\sigma'} n_{m',\sigma'} \int d^3\underline{r}'|\phi_{m'}(\underline{r}')|^2 V_{int}(\underline{r},\underline{r}')\phi_m(\underline{r})$$

$$\left. - \sum_{m'} n_{m',\sigma} \int d^3\underline{r}'\,\phi_{m'}^*(\underline{r}')\phi_{m'}(\underline{r})V_{int}(\underline{r},\underline{r}')\phi_m(\underline{r}') \right]$$

$$(7.165)$$

Since $\delta^1 I$ vanishes for any function $\delta\phi_m^*(\underline{r})$, the deviation may be chosen as a dirac delta function located at any point in space. The integration over the delta function leads to the conclusion that the function multiplying $\delta\phi_m^*$ must vanish at every point in space. This conclusion yields the "*Hartree-Fock Equations*" for the optimized basis functions

$$\epsilon_m \phi_m(\underline{r}) = \left[-\frac{\hbar^2}{2m}\nabla^2 + V_0(\underline{r}) \right.$$

$$\left. + \sum_{m',\sigma'} n_{m',\sigma'} \int d^3\underline{r}'|\phi_{m'}(\underline{r}')|^2 V_{int}(\underline{r},\underline{r}') \right] \phi_m(\underline{r})$$

$$- \int d^3\underline{r}' \left[\sum_{m'} n_{m',\sigma}\phi_{m'}^*(\underline{r}')\phi_{m'}(\underline{r})V_{int}(\underline{r},\underline{r}') \right] \phi_m(\underline{r}')$$

$$(7.166)$$

The first two lines have the form of a one-electron energy eigenvalue equation which includes the external potential V_0 potential which is supplemented by the potential due to the Coulomb interaction due to the other electrons. The energy eigenvalue is identified with the Lagrange undetermined multiplier ϵ_m. The last line describes an integral operator which has no classical analogue since it is due to antisymmetric nature of the many-electron wave function.

The Hartree-Fock equations are a set of N coupled equations that must be solved self-consistently. The equations can be solved by iteration, where the iterations are stopped at a stage where the basis functions calculated in the ultimate step coincide with the basis functions found from the previous iteration. It is important to note that the resulting states need not have the same symmetry as the Hamiltonian. In such a case, the Hartree-Fock approximation provides an indication that the system may have undergone a phase transition at which the symmetry of the system has been broken spontaneously and the total energies of the *"Broken Symmetry"* states are lower than that of the symmetric state. The Hartree-Fock approximation has proved useful in the description of states of matter with broken symmetry, such as ferromagnetic states.

The Hartree-Fock approximation for the energy of the many-body state is given by

$$
E_{HF} = \sum_{m,\sigma} n_{m,\sigma} \int d^3\underline{r}\phi_m^*(\underline{r}) \left[-\frac{\hbar^2}{2m}\nabla^2 + V_0(\underline{r}) \right] \phi_m(\underline{r})
$$

$$
+ \frac{1}{2!} \sum_{m,\sigma,m',\sigma'} n_{m,\sigma}n_{m',\sigma'}
$$

$$
\times \int d^3\underline{r} \int d^3\underline{r}' |\phi_{m'}(\underline{r}')|^2 V_{int}(\underline{r},\underline{r}') |\phi_m(\underline{r})|^2
$$

$$
- \frac{1}{2!} \sum_{m,m',\sigma} n_{m,\sigma}n_{m',\sigma}
$$

$$
\times \int d^3\underline{r} \int d^3\underline{r}' \phi_m^*(\underline{r}')\phi_{m'}(\underline{r}') V_{int}(\underline{r},\underline{r}') \phi_{m'}^*(\underline{r})\phi_m(\underline{r})
$$

$$(7.167)$$

or equivalently by

$$
E_{HF} = \sum_{m,\sigma} n_{m,\sigma}\epsilon_m
$$

$$
- \frac{1}{2!} \sum_{m,\sigma,m',\sigma'} n_{m,\sigma}n_{m',\sigma'}
$$

$$
\times \int d^3\underline{r} \int d^3\underline{r}' |\phi_{m'}(\underline{r}')|^2 V_{int}(\underline{r},\underline{r}') |\phi_m(\underline{r})|^2
$$

$$
+ \frac{1}{2!} \sum_{m,m',\sigma} n_{m,\sigma}n_{m',\sigma}
$$

$$
\times \int d^3\underline{r} \int d^3\underline{r}' \phi_m^*(\underline{r}')\phi_{m'}(\underline{r}') V_{int}(\underline{r},\underline{r}') \phi_{m'}^*(\underline{r})\phi_m(\underline{r})
$$

$$(7.168)$$

which indicates that ϵ_m might be considered as a quasi-particle energy that includes the energy due to the interactions with the other electrons. The other terms in the Hartree-Fock energy simply reflects that the interaction is between pairs of electrons and that interaction has been counted twice in the sum of the quasi-particle energies ϵ_m. If the basis states $\phi_m(\underline{r})$ do not change significantly when the integers $n_{m,\sigma}$ are changed, then one expects that the change in the direct Coulomb and exchange energies will be negligible. In the absence of such relaxation, "*Koopman's Theorem*" [49] shows that the energy for removing an electron from state (m, σ) is $-\epsilon_m$.

An alternative to the Hartree-Fock approximation which is sometimes used is density functional theory [50] which, if the density functional was known, would lead to the exact ground-state energy and density. Unfortunately, only approximations to the exact density functional exist. Furthermore, the Kohn-Sham method [51] of mapping density functional theory to a single-electron eigenvalue equation leads to single-electron energy eigenvalues and eigenfunctions that, if considered physical, would imply a Slater determinant wave function that is inferior to the Slater determinant uniquely determined by the Hartree-Fock approximation.

A general many-electron state cannot be expressed as a single Slater determinant, but must be expressed as a linear superposition of Slater determinants with complex coefficients. The correlations of such many-electron states are embedded in the expansion coefficients. The BCS approximation for the wave function for a superconducting state is built of a linear superposition of Slater determinants, but it is a linear superposition of states that contain different total numbers of electrons.

7.14 Problems

Problem 7.1

For an ideal gas of fermions, the probability distribution function $p_\alpha(n)$ for finding n particles in the α-th single-particle quantum state is given by

$$p_\alpha(n) = \left(\frac{1}{1 + \exp[-\beta(\epsilon_\alpha - \mu)]}\right) \exp[-n\beta(\epsilon_\alpha - \mu)]$$

where $n = 0$ or 1.

(i) Show that the average number of particles in the α-th level is given by

$$\bar{n}_\alpha = \left(\frac{\exp[-\beta(\epsilon_\alpha - \mu)]}{1 + \exp[-\beta(\epsilon_\alpha - \mu)]}\right)$$

(ii) Hence, show that $p_\alpha(n)$ can be written as

$$p_\alpha(n) = (1 - \bar{n}_\alpha)\left(\frac{\bar{n}_\alpha}{1 - \bar{n}_\alpha}\right)^n$$

(iii) Find the moment and cumulant generating functions. Hence, find the first few cumulants.

(iv) Calculate the Shannon entropy.

Problem 7.2

Consider a gas of identical fermions which undergo collisions, $\alpha, \beta \to \alpha', \beta'$. The interaction can be written as

$$\hat{H}_{int} = \frac{1}{2!}\sum_{\alpha,\beta,\alpha',\beta'}\langle\alpha',\beta'|\hat{V}|\beta,\alpha\rangle c_{\alpha'}^\dagger c_{\beta'}^\dagger c_\beta c_\alpha$$

where the creation and annihilation operators satisfy anticommutation relations.

(i) Using the Fermi-Golden rule, derive an expression for the rate at which particles in a single particle quantum state α with energy ϵ_α will be scattered out of that state. Express your result in terms of the probabilities p_α that a single-particle quantum state will be occupied by a particle.

(ii) Using the Fermi-Golden rule, determine an expression for the rate at which particles will be scattered into state α.

(iii) Show that the "Principle of Detailed Balance" holds if the p_α are replaced by the Fermi-Dirac distribution functions $f(\epsilon_\alpha)$.

Problem 7.3

Consider a two-level system with energies 0 and ϵ, which can be occupied by spinless fermions. The system is in equilibrium with a heat bath at temperature T and chemical potential μ.

(i) Show that the Grand-Canonical partition function Ξ can be expressed as

$$\Xi = 1 + \exp[\beta\mu] + \exp[\beta(\mu - \epsilon)] + \exp[\beta(2\mu - \epsilon)]$$

(ii) Evaluate the average energy and the average number of particles in the single-particle state with energy ϵ.

(iii) Show that the results are compatible with the Fermi-Dirac distribution function.

Problem 7.4

Consider an ideal gas of N electrons in a d-dimensional volume $V_d = L^d$, held at zero temperature and subject to an external magnetic field B. Neglecting Landau quantization of the orbital motion, the single-electron energies are given by

$$\epsilon(\underline{p}) = \sum_{i=1}^{d} \frac{p_i^2}{2m} \pm \mu_B \, B$$

(i) Find an expression for the chemical potential μ when $B = 0$.

(ii) When B is non-zero, find an expression for the Fermi-momenta $p_{F,\uparrow}$ for the up-spin electrons and $p_{F,\downarrow}$ for the down-spin electrons.

(iii) Find the energies $U_\uparrow$ for the gas of up-spin and $U_\downarrow$ for the down-spin electrons.

(iv) Determine the leading order terms of the expansion of the total energy of the electronic system $U = U_\uparrow + U_\downarrow$ in powers of B. Hence, determine the $T = 0$ expression for the Pauli paramagnetic susceptibility χ.

Problem 7.5

Assume that the energy of a non-interacting fermion in a gravitational field g is adequately described by the W.K.B. expression

$$E(n, p_\perp) = \frac{p_\perp^2}{2m} + mg \left(\frac{9\pi^2 \hbar^2}{8m^2 g} \right)^{\frac{1}{3}} \left(n + \frac{3}{4} \right)^{\frac{3}{2}}$$

where $n = 0, 1, 2 \dots$.

(i) Calculate the single-particle density of states $\rho(E)$.

(ii) Calculate the Grand-Canonical potential for an ideal gas of fermions which is confined in a cylinder of cross-sectional area A and is subject to a gravitational field. Does this reduce to the classical expression at high temperatures?

(iii) Determine expressions for the internal energy U and the number of particles N in the system.

(iv) Determine the low-temperature specific heat.

Problem 7.6

Graphene is a two-dimensional array of carbon atoms arranged on a honeycomb lattice. The electronic structure shows the existence of Dirac cones located at two inequivalent points on the Brillouin zone boundary. In the vicinity of these points, the electronic dispersion relation has the form

$$\epsilon^2 - c^2 \hbar^2 k^2 = 0$$

(i) Show that the one-electron density of states is given by

$$\rho(\epsilon) \approx 2 \; \frac{A}{N} \; \frac{|\epsilon|}{\pi c^2 \hbar^2}$$

where A/N is the area per unit cell. For undoped graphene, the chemical potential coincides with the energy where the density of states is zero.

(ii) Show that the difference in energy of the system between $T = 0$ and temperature T, $\Delta U(T)$, can be written as

$$\Delta U(T) = 2 \int_0^\infty d\epsilon \; \epsilon \; \rho(\epsilon) \; f(\epsilon)$$

(iii) Hence, find an exact expression for the specific heat.

Problem 7.7

Consider an ultra-relativistic gas of fermions with dispersion relation

$$\epsilon(\underline{p}) = c \sqrt{\sum_{i=1}^{d} p_i^2}$$

The system is held at $T = 0$ and the fermion number density per spin is given by ρ.

(i) Determine the Fermi-Energy μ for the system.

(ii) Determine the total energy, per spin, of the gas of fermions.

Problem 7.8

An intrinsic semi-conductor, such as Ge or Si, have band gaps $2\,\Delta$ of the order of an eV. The density of states shows a gap between the occupied valence band states and the empty conduction band states. The conduction band states have a dispersion relation

$$\epsilon_c(\underline{p}) = \Delta + \frac{p^2}{2m_c}$$

and the valence band states are described by

$$\epsilon_v(\underline{p}) = -\Delta - \frac{p^2}{2m_v}$$

Consider the process of creation and annihilation of electron hole pairs. The annihilation process may occur by emission of radiation

$$e + h \rightarrow \gamma$$

in addition to other processes. The stoichiometric coefficients for this reaction are $\nu_c = 1$ and $\nu_h = 1$ and $\nu_\gamma = -1$. For the above reaction to be in chemical equilibrium, the Gibbs Free-Energy must satisfy $dG \equiv 0$, or

$$dG = \mu_c\,dN_c + \mu_h\,dN_h = 0$$

since one can formally define $\mu_\gamma = 0$. Furthermore, since $dN_c = dN_h$, one finds that

$$\mu_c + \mu_h = 0$$

The relation $\mu_c = -\mu_h$ could also be obtained from the fact that the probability of a hole excitation with excitation energy ϵ, $p_h(\epsilon)$ is equal to probability that the electronic state with energy $-\epsilon$ is unoccupied

$$p_h(\epsilon) = 1 - f(-\epsilon)$$

$$= \frac{1}{\exp[\beta(\epsilon + \mu)] + 1}$$

Since holes obey Fermi-Dirac statistics, one identifies μ_h as $\mu_h = -\mu$.

(i) Determine the Grand-Canonical Partition function Ξ_c for the conduction band. Hence, show that the average number of electrons in the conduction band, $\overline{N}_c$, is given by

$$\overline{N}_c = 2V \int \frac{d^3p}{(2\pi\hbar)^3} \frac{1}{\exp\left[\beta\left(\Delta + \frac{p^2}{2m_c} - \mu\right)\right] + 1}$$

(ii) Similarly, show that number of holes in the valence band is given by

$$\overline{N}_h = 2V \int \frac{d^3p}{(2\pi\hbar)^3} \frac{1}{\exp\left[\beta\left(\Delta + \frac{p^2}{2m_v} + \mu\right)\right] + 1}$$

Evaluate these expressions in limit $(\Delta \mp \mu) \gg k_B T$.

(iii) Hence, in this limit, show that

$$\overline{N}_c = \overline{N}_h$$
$$\approx 2 \left(\frac{V}{\sqrt{\lambda_c^3 \lambda_v^3}}\right) \exp[-\beta\Delta]$$

(iv) Hence, in the same limit, determine the chemical potential μ

$$\mu \approx \frac{3}{4} k_B T \ln \left(\frac{m_v}{m_c}\right)$$

The above expression shows that the chemical potential is pinned to the middle of the gap.

Problem 7.9

Consider an n-type semiconductor, in which the presence of a number N_D of impurities produces N_D bound states at an energy Δ below the bottom of the conduction band. The large interaction between pairs of electrons in the number of localized levels prevents them from being occupied by more than one electron. At zero temperature, the N_D bound states are assumed to be occupied by exactly one electron. At non-zero temperatures, electrons are thermally activated from the impurity levels to the conduction band.

(i) Using the Grand-Canonical Ensemble, assuming the energy is given by $E = -n_D \Delta$ where n_D is the number of electrons in the donor levels, show that the Grand-Canonical Partition function for the impurity levels Ξ_D, is given by

$$\Xi_D = [1 + 2 \ \exp[\beta(\Delta + \mu)]]^{N_D}$$

(ii) Hence, show that

$$n_D = \frac{N_D}{1 + \frac{1}{2} \ \exp[-\beta(\Delta + \mu)]}$$

or, equivalently

$$2 \frac{(N_D - n_D)}{n_D} = \exp[-\beta(\Delta + \mu)]$$

(iii) Calculate the number of conduction electrons, n_c, and show that for $\mu \gg k_B T$, it can be reduced to

$$n_c \approx 2 \frac{V}{\lambda^3} \exp[\beta\mu]$$

(iv) Eliminating μ, show that the number of conduction electrons is given by the solution of the quadratic equation

$$n_c^2 \approx (N_D - n_c) \frac{V}{\lambda^3} \exp[-\beta\Delta]$$

Problem 7.10

Consider an n-type semiconductor, in which the presence of a number N_D of impurities produces N_D bound states at an energy Δ below the bottom of the conduction band. At zero temperature, there are N_D electrons in the bound states. At non-zero temperatures, electrons are thermally activated from the impurity levels to the conduction band.

(i) Using the Grand-Canonical Ensemble, assuming the energy is given by $E = -n_D \, \Delta$ where n_D is the number of electrons in the donor levels, show that the Grand-Canonical Partition function for the impurity levels Ξ_D, is given by

$$\Xi_D = [1 + \exp[\beta(\Delta + \mu)]]^{2N_D}$$

(ii) Hence, show that

$$n_D = \frac{2N_D}{1 + \exp[-\beta(\Delta + \mu)]}$$

or, equivalently

$$\frac{(2N_D - n_D)}{n_D} = \exp[-\beta(\Delta + \mu)]$$

(iii) Calculate the number of conduction electrons, n_c, and show that for $\mu \gg k_B T$, it can be reduced to

$$n_c \approx 2 \frac{V}{\lambda^3} \exp[\beta\mu]$$

(iv) Eliminating μ, show that the number of conduction electrons is given by

$$(N_D + n_c) \, n_c \approx 2(N_D - n_c) \frac{V}{\lambda^3} \exp[-\beta\Delta]$$

Problem 7.11

Consider a gas of N fermions confined in an anisotropic harmonic potential in d-dimensions

$$V = \sum_{i=1}^{d} \frac{m\omega_i^2}{2} q_i^2$$

(i) By considering all the states bounded by a hyperplane with intercepts $(E/\hbar\omega_i)$ on the axes, show that the density of states can be approximated by

$$\rho(E) \approx \frac{2S+1}{(d-1)!} \frac{E^{d-1}}{\prod_i \hbar\omega_i}$$

Under what circumstances is this approximation reasonable?

(ii) Hence, show that the number of fermions N is given by

$$N = \frac{2S+1}{(d-1)!} \int_0^\infty dx \, \frac{x^{d-1}}{z^{-1} \exp[x] + 1} \prod_{i=1}^{d} \left(\frac{k_B T}{\hbar\omega_i} \right)$$

where z is the fugacity. Evaluate the integral in the low-temperature limit.

Problem 7.12

A star with one solar mass will burn out when the nuclear reactions stop for lack of fuel and, due to the gravitational force of the protons, collapse into a white dwarf star. The system then consists of a degenerate neutral gas of He nuclei and electrons. Due to the high density, the electron gas should be considered as relativistic and, since the Fermi-energy is quite large, can be treated as being in thermal equilibrium at $T = 0$.

(i) Assume the star has radius R, mass M and has a uniform density. Show that the gravitational energy is given by

$$U_G(M, R) = -\frac{3GM^2}{5R}$$

where G is the gravitational constant.

(ii) Assume that there are equal numbers of electrons and nuclei. Derive an expression for the total kinetic energy $U_K(M, R)$ of the electron gas, assuming that it is non-relativistic. Can the kinetic energy of the nuclei be neglected?

(iii) From (ii) extremalize the total energy $U(M,R) = U_G(M,R) + U_K(M,R)$ w.r.t. R, to determine the mass-radius relation of the white dwarf star.

(iv) Assume that there are equal numbers of electrons and nuclei. Derive an expression for the total kinetic energy $U_K(M,R)$ of the electron gas, assuming that it is ultra-relativistic.

(v) From (iv), calculate $U(M,R)$ and provide an argument that indicates that the white dwarf star will become unstable due to collapse for a mass above a critical value [52]. The critical value of the mass is known as the Chandrasekhar limit [53].

Problem 7.13

A white dwarf star above the Chandrasekhar limit will collapse since the pressure from the electron gas cannot balance the gravitational force. The white dwarf collapses to such high densities that the protons and electrons react to form neutrons and neutrinos. The resulting star is comprised of only neutrons, since the neutrinos escape from the star. Assume that the gravitational energy is given by

$$U_G(M,R) = -\frac{3GM^2}{5R}$$

where G is the gravitational constant.

(i) Find the mass-radius relation for a non-relativistic neutron star, if the neutrons can be treated as being non-relativistic.

(ii) Treat the neutrons as being ultra-relativistic and determine the critical mass M_c, above which, the neutron star collapses and forms a black hole.

Problem 7.14

(i) In order to describe relativistic Fermi-gasses, show that the number of particles, energy and pressure can be expressed as

$$N = V\left(\frac{m^3 c^3}{\pi^2 \hbar^3}\right) \int_0^\infty d\theta \frac{\sinh^2 \theta \cosh \theta}{\exp[-\beta\mu + \beta mc^2 \cosh \theta] + 1}$$

$$U = V\left(\frac{m^4 c^5}{\pi^2 \hbar^3}\right) \int_0^\infty d\theta \frac{\sinh^2 \theta \cosh^2 \theta}{\exp[-\beta\mu + \beta mc^2 \cosh \theta] + 1}$$

$$P = \left(\frac{m^4 c^5}{3\pi^2 \hbar^3}\right) \int_0^\infty d\theta \frac{\sinh^4 \theta}{\exp[-\beta\mu + \beta mc^2 \cosh \theta] + 1}$$

where θ is defined through the momentum by

$$p = mc\sinh\theta$$

(ii) Setting

$$\mu = mc^2\cosh\theta_0$$

show that, in the limit $T \to 0$, the integrals are evaluated as

$$N = V\left(\frac{m^3c^3}{3\pi^2\hbar^3}\right)\sinh^3\theta_0$$

$$U = V\left(\frac{m^4c^5}{32\pi^2\hbar^3}\right)[\sinh 4\theta_0 - 4\theta_0]$$

$$P = \left(\frac{m^4c^5}{24\pi^2\hbar^3}\right)\left[\frac{1}{4}\sinh 4\theta_0 - 2\sinh 2\theta_0 + 3\theta_0\right]$$

Problem 7.15

Consider an electron/positron/photon plasma. The electrons and positrons share the same relativistic dispersion relation

$$\epsilon(p) = \sqrt{p^2c^2 + m^2c^4}$$

The electrons and positrons can be created and annihilated via the emission of photons

$$e^- + e^+ \to 2\gamma$$

(i) Determine an expression if the number of electron is N_e at $T = 0$ and find an expression for the chemical potential in terms of the mass m.
(ii) If the gas is ultra-relativistic, show that excess of electrons over positrons is given by

$$(N_{e^-} - N_{e^+}) = 2\frac{4\pi V}{(2\pi\hbar)^3}\int_0^\infty dp\, p^2 \left[\frac{1}{\exp[\beta(cp - \mu)] + 1}\right.$$

$$\left. - \frac{1}{\exp[\beta(cp + \mu)] + 1}\right]$$

Evaluate this expression using contour integration and show that it is exactly equal to

$$2\frac{4\pi V}{3}\left(\frac{k_B T}{2\pi\hbar c}\right)^3\left[\pi^2\left(\frac{\mu}{k_B T}\right) + \left(\frac{\mu}{k_B T}\right)^3\right]$$

Hint: Set $z = \exp[\beta cp]$ and integrate

$$\int_C dz \ \frac{\ln^3 z}{z^2 + 2z \cosh \beta\mu + 1}$$

over a contour which consists of small radius circle around the origin, parallel lines excluding the branch cut on the real positive axis and the contour is completed by a circular contour at infinity.

(iii) In the ultra-relativistic limit, determine expressions for the average number of electrons, positrons and photons for the neutral plasma. Also determine an expression for the average energy.

(iv) What is the ratio of the number of photons to positrons? What is the ratio of the energy of the photon gas to the energy of the gas of positrons?

Problem 7.16

Consider an ideal quark-gluon plasma held in a d-dimensional volume. The particles can be treated as ultra-relativistic.

(i) Determine the degeneracy factors $\mathcal{D}_Q$ for the quarks, antiquarks and $\mathcal{D}_G$ for the gluons.

(ii) Determine an expression for density of quarks in terms of the chemical potential. Evaluate the integral under the condition that the gas is sufficiently degenerate so that one may use the Sommerfeld expansion. Estimate the magnitude of the possible error.

(iii) Evaluate the number of antiquarks approximately assuming that the magnitude of μ/T is sufficiently large.

(iv) Determine the gluon density.

(v) If there are equal numbers of quarks and anti-quarks, what is the ratio of the numbers of gluons to quarks? What is the ratio of the quark to gluon partial pressures?

Problem 7.17

A container is separated into two compartments by a sliding piston. Two non-relativistic fermi-gasses with spin S_L and S_R which have the same mass are placed in the left and right compartments, respectively. Show that the ratio of the equilibrium densities at $T = 0$ is given by

$$\frac{\rho_L}{\rho_R} = \sqrt{\frac{(2S_L + 1)}{(2S_R + 1)}}$$

Problem 7.18

Edmund Stoner created a collective model for itinerant magnetism. The model consists of a degenerate band of itinerant electrons described by a dispersion relation $\epsilon_{\underline{k}}$. The electrons within the i-th unit cell interact via the pair-wise interaction

$$\hat{H}_{int} = \frac{U}{2} \sum_{i,\alpha,\beta} c^\dagger_{i,\alpha} c^\dagger_{i,\beta} c_{i,\beta} c_{i,\alpha}$$

where α and β denote the degeneracy indices of the $\mathcal{D}$-fold degenerate states. (Usually $\mathcal{D} = 2$ corresponding to a spin degeneracy.) The quantity U represents a screened Coulomb interaction strength. The pair-wise interaction between electrons in states $\alpha \neq \beta$ is written in terms of the average number of electrons per site $\overline{n}_{i,\alpha}$ and its fluctuation $\Delta n_{i,\alpha}$, via

$$\hat{n}_{i,\alpha} = \overline{n}_{i,\alpha} + \Delta \hat{n}_{i,\alpha}$$

The product of the operators take the form

$$\hat{n}_\alpha \hat{n}_\beta = \overline{n}_{i,\alpha} \overline{n}_{i,\beta} + \overline{n}_{i,\alpha} \Delta \hat{n}_{i,\beta} + \Delta \hat{n}_{i,\alpha} \overline{n}_{i,\beta} + \Delta \hat{n}_{i\alpha} \Delta \hat{n}_{i,\beta}$$

On neglecting the correlations between the fluctuations, the interaction term can be written as

$$\hat{H}_{int} = \frac{U}{2} \sum_{i,\alpha \neq \beta} (\overline{n}_{i,\alpha} \hat{n}_{i,\beta} + \hat{n}_{i,\alpha} \overline{n}_{i,\beta} - \overline{n}_{i\alpha} \overline{n}_{i,\beta})$$

where the constant term prevents over-counting the contribution of the interaction to the total energy. The average occupation numbers are to be treated as variational parameters. The occupation numbers will be assumed to be independent of the position i. It can be shown that the energy of a single particle with degeneracy label α is shifted by an amount $U \sum_{\beta \neq \alpha} \overline{n}_\beta$. The variational Grand-Canonical Potential Ω is written as

$$\Omega = \sum_{\underline{k},\alpha} \left(\epsilon_{\underline{k}} + U \sum_{\beta \neq \alpha} \overline{n}_\beta - \mu \right) f_{\underline{k},\alpha} - \frac{U}{2} \sum_{i,\alpha \neq \beta} \overline{n}_\alpha \overline{n}_\beta$$
$$+ k_B T \sum_{\underline{k},\alpha} (f_{\underline{k},\alpha} \ln f_{\underline{k},\alpha} + (1 - f_{\underline{k},\alpha}) \ln (1 - f_{\underline{k},\alpha}))$$

where $f_{\underline{k},\alpha}$ is the occupation number.

(i) Minimizing the expression for Ω with respect to $f_{\underline{k},\alpha}$. Hence. find an explicit form for $f_{\underline{k},\alpha}$.

(ii) Minimize the expression for Ω w.r.t the set of $\bar{n}_\alpha$, to identify $\bar{n}_\alpha$ in terms of the $f_{\underline{k},\alpha}$.

$$\bar{n}_\alpha = \frac{1}{N} \sum_{\underline{k}} f_{\underline{k},\alpha}$$

These equations form a set of self-consistency equations that need to be solved.

In the state with unbroken symmetry, all the states are degenerate, so $\bar{n}_\alpha = \bar{n}_\beta$ for all α and β. Hence

$$\bar{n}_\alpha = \frac{n_e}{\mathcal{D}}$$

where n_e is the total number of electrons per unit cell.

Now assume that the symmetry is broken and the degeneracy $\mathcal{D}$ is lifted. Assume that the states are separated into two sets with equal degeneracies $\frac{\mathcal{D}}{2}$. The occupation numbers of the two sets can be written as

$$\bar{n}_+ = \frac{n_e + \delta n}{\mathcal{D}}$$

$$\bar{n}_- = \frac{n_e - \delta n}{\mathcal{D}}$$

(iii) Expand the set of self-consistency equations to first-order in the deviations. Show the equations only have trivial solutions $\delta n = 0$ when

$$1 > -U \int d\epsilon \, \rho(\epsilon) \frac{\partial f}{\partial \epsilon}$$

and a non-trivial solution is possible whenever

$$1 < -U \int d\epsilon \, \rho(\epsilon) \frac{\partial f}{\partial \epsilon}$$

where $\rho(\epsilon)$ is the one-electron density of states and $f(\epsilon)$ is the Fermi-Dirac distribution function.

(iv) Expand the Grand-Canonical potential Ω to quadratic order in δn and show that the non-trivial solutions are more stable than the trivial solutions.

Problem 7.19

The BCS pairing interaction is a simplified interaction in which only pairs of electrons near the Fermi-surface with antiparallel spins and zero center of mass momentum are considered. The Hamiltonian has the form

$$\hat{H} - \mu \hat{N} = \sum_{\underline{k},\sigma} (\epsilon(\underline{k}) - \mu) c^\dagger_{\underline{k},\sigma} c_{\underline{k},\sigma} - \sum_{\underline{k},\underline{k}'} V c^\dagger_{\underline{k}',\uparrow} c^\dagger_{-\underline{k}',\downarrow} c_{-\underline{k},\downarrow} c_{\underline{k},\uparrow}$$

where V is an attractive potential. The BCS s-wave order parameter is defined by

$$\Delta = \sum_{\underline{k}} V \langle c_{-\underline{k},\downarrow} c_{\underline{k},\uparrow} \rangle$$

$$\Delta^* = \sum_{\underline{k}} V \langle c^{\dagger}_{\underline{k},\uparrow} c^{\dagger}_{-\underline{k},\downarrow} \rangle$$

(i) Show that the mean-field Hamiltonian is given by

$$\hat{H} = \sum_{\underline{k}} ((\epsilon(\underline{k}) - \mu)(c^{\dagger}_{\underline{k},\uparrow} c_{\underline{k},\uparrow} + c^{\dagger}_{-\underline{k},\downarrow} c_{-\underline{k},\downarrow})$$

$$- \Delta c^{\dagger}_{\underline{k},\uparrow} c^{\dagger}_{-\underline{k},\downarrow} - \Delta^* c_{-\underline{k},\downarrow} c_{\underline{k},\uparrow}) + \frac{|\Delta|^2}{V}$$

The last term prevents overcounting the interaction in the energy.

(ii) Assuming a variational ground state of the form

$$|\psi\rangle = \prod_{\underline{k}} (u_{\underline{k}} + v_{\underline{k}} c^{\dagger}_{-\underline{k},\downarrow} c^{\dagger}_{\underline{k},\uparrow}) |0\rangle$$

where $|0\rangle$ is the vacuum state and $u_{\underline{k}}$ and $v_{\underline{k}}$ satisfy the normalization condition

$$|u_{\underline{k}}|^2 + |v_{\underline{k}}|^2 = 1$$

determine an expression for the variational energy

$$E = \langle \psi | \hat{H} | \psi \rangle$$

(iii) By varying $u_{\underline{k}}^*$ and $v_{\underline{k}}^*$ and Δ^* show that, with a suitable choice of gauge for Δ,

$$|v_{\underline{k}}|^2 = \frac{1}{2} \left(1 - \frac{(\epsilon(\underline{k}) - \mu)}{\sqrt{(\epsilon(\underline{k}) - \mu)^2 + \Delta^2}} \right)$$

$$|u_{\underline{k}}|^2 = \frac{1}{2} \left(1 + \frac{(\epsilon(\underline{k}) - \mu)}{\sqrt{(\epsilon(\underline{k}) - \mu)^2 + \Delta^2}} \right)$$

$$\Delta = \sum_{\underline{k}} V u_{\underline{k}}^* v_{\underline{k}}$$

extremalize the mean-field energy and satisfy the normalization condition. The equation for Δ can be written in the form

$$\Delta = \frac{V}{2} \sum_{\underline{k}} \frac{\Delta}{\sqrt{(\epsilon(\underline{k}) - \mu)^2 + \Delta^2}}$$

which is the BCS gap equation that has to be solved self-consistently.

Chapter 8

Bose-Einstein Statistics

8.1 Non-Interacting Bosons

An ideal gas of non-interacting bosons [54, 55] is described by the Hamiltonian $\hat{H}$ which is given by the sum

$$\hat{H} = \sum_{\alpha} \epsilon_\alpha \, \hat{n}_\alpha \tag{8.1}$$

where $\hat{n}_\alpha$ represents the occupation number of the α-th single-particle energy level. This is just the sum of the contributions from each particle, grouped according to the energy levels that they occupy. Likewise, the operator $\hat{N}$ representing the total number of particles is given by

$$\hat{N} = \sum_{\alpha} \hat{n}_\alpha \tag{8.2}$$

where the sum is over the single-particle energy levels. Hence, for non-interacting bosons, the density operator is diagonal in the occupation number representation.

The Grand-Canonical Partition Function Ξ is given by

$$\Xi = \text{Trace } \exp[-\beta(\hat{H} - \mu\hat{N})] \tag{8.3}$$

where the Trace is over a complete set of microscopic states with variable N for the entire system. A convenient basis is given by the N-particle states $|\Phi_{n_{\alpha_1}, n_{\alpha_2}, \dots}\rangle$, since both $\hat{H}$ and $\hat{N}$ are diagonal in this basis. The Trace reduces to the sum of all configurations $\{n_\alpha\}$, and since the total number of particles N is also being summed over, the Trace is unrestricted. Therefore, the Trace can be evaluated by summing over all possible values of the eigenvalues n_α for each consecutive value of α. That is

$$\text{Trace } \{\dots\} \equiv \sum_{n_{\alpha_1}=0}^{\infty} \sum_{n_{\alpha_2}=0}^{\infty} \sum_{n_{\alpha_3}=0}^{\infty} \dots \{\dots\} \tag{8.4}$$

387

Since the exponential term in Ξ reduces to the form of the exponential of a sum of independent terms, it can be written as the product of exponential factors

$$\exp[-\beta(\hat{H} - \mu\hat{N})] = \exp\left[-\beta \sum_\alpha (\epsilon_\alpha - \mu)\, \hat{n}_\alpha\right]$$

$$= \prod_\alpha \exp[-\beta(\epsilon_\alpha - \mu)\, \hat{n}_\alpha] \qquad (8.5)$$

Therefore, the Trace in the expression for Ξ can be reduced to

$$\Xi = \prod_\alpha \left\{ \sum_{n_\alpha=0}^{\infty} \exp[-\beta(\epsilon_\alpha - \mu)\, n_\alpha] \right\} \qquad (8.6)$$

where the sum is over all the occupation numbers n_α allowed by Bose-Einstein statistics. The summation is of the form of a geometric series, as can be seen by introducing the variable x_α defined by

$$x_\alpha = \exp[-\beta(\epsilon_\alpha - \mu)] \qquad (8.7)$$

so

$$\sum_{n_\alpha=0}^{\infty} \exp[-\beta(\epsilon_\alpha - \mu)\, n_\alpha] = \sum_{n_\alpha=0}^{\infty} x_\alpha^{n_\alpha} \qquad (8.8)$$

This geometric series converges if $x_\alpha \leq 1$, i.e.

$$\exp[-\beta(\epsilon_\alpha - \mu)] \leq 1 \qquad (8.9)$$

which requires that $\epsilon_\alpha > \mu$. This condition has to hold for all ϵ_α, so μ must be smaller than the lowest single-particle energy level ϵ_0. Therefore, we require that

$$\epsilon_0 > \mu \qquad (8.10)$$

On performing the sum of the geometric series, one obtains

$$\sum_{n_\alpha=0}^{\infty} \exp[-\beta(\epsilon_\alpha - \mu)\, n_\alpha] = \frac{1}{1 - \exp[-\beta(\epsilon_\alpha - \mu)]} \quad \text{where } \epsilon_0 > \mu \quad (8.11)$$

Therefore, the Grand-Canonical Partition Function is given by

$$\Xi = \prod_\alpha \left\{ \frac{1}{1 - \exp[-\beta(\epsilon_\alpha - \mu)]} \right\} \qquad (8.12)$$

The Grand-Canonical Potential Ω is found from

$$\Xi = \exp[-\beta\Omega] \qquad (8.13)$$

which, on taking the logarithm, yields

$$-\beta\Omega = \sum_\alpha \ln \left[\frac{1}{1 - \exp[-\beta(\epsilon_\alpha - \mu)]} \right] \tag{8.14}$$

or

$$\Omega = k_B T \sum_\alpha \ln \left[1 - \exp[-\beta(\epsilon_\alpha - \mu)] \right] \tag{8.15}$$

On introducing the density of single-particle states $\rho(\epsilon)$ defined as

$$\rho(\epsilon) = \sum_\alpha \delta(\epsilon - \epsilon_\alpha) \tag{8.16}$$

the summation can be transformed into an integral over ϵ

$$\Omega = k_B T \int_{-\infty}^{\infty} d\epsilon \, \rho(\epsilon) \, \ln \left[1 - \exp[-\beta(\epsilon - \mu)] \right] \tag{8.17}$$

where the density of states goes to zero below ϵ_0. After evaluating the integral, one may find all the thermodynamic properties of the system from Ω.

8.2 The Bose-Einstein Distribution Function

The average number of particles in the system $\overline{N}$ can be determined from Ω via the relation

$$\overline{N} = -\left(\frac{\partial \Omega}{\partial \mu} \right)_{T,V} \tag{8.18}$$

which is evaluated as

$$\begin{aligned}
\overline{N} &= k_B T \int_{-\infty}^{\infty} d\epsilon \, \rho(\epsilon) \, \frac{\beta \, \exp[-\beta(\epsilon - \mu)]}{1 - \exp[-\beta(\epsilon - \mu)]} \\
&= \int_{-\infty}^{\infty} d\epsilon \, \rho(\epsilon) \, \frac{1}{\exp[\beta(\epsilon - \mu)] - 1}
\end{aligned} \tag{8.19}$$

or equivalently

$$\overline{N} = \sum_\alpha \frac{1}{\exp[\beta(\epsilon_\alpha - \mu)] - 1} \tag{8.20}$$

By definition, one has

$$\overline{N} = \sum_\alpha \overline{n}_\alpha \tag{8.21}$$

therefore, the function $N(\epsilon)$ defined by

$$N(\epsilon) = \frac{1}{\exp[\beta(\epsilon - \mu)] - 1} \tag{8.22}$$

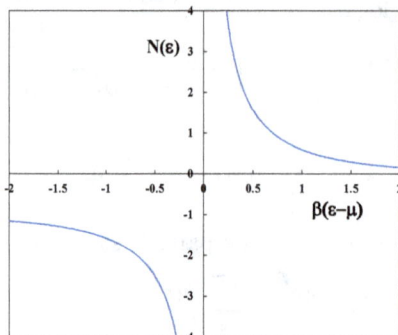

Fig. 8.1 The Bose-Einstein Distribution function.

represents the average number of particles $\overline{n}$ in a quantum level with a single-particle energy ϵ. The function $N(\epsilon)$ is the Bose-Einstein distribution function. The Bose-Einstein distribution varies as

$$N(\epsilon) \approx \exp[-\beta(\epsilon - \mu)] \tag{8.23}$$

for $\epsilon - \mu > k_B T$. For low energies $\mu < \epsilon < \mu + k_B T$, the Bose-Einstein distribution varies as

$$N(\epsilon) \approx \frac{k_B T}{\epsilon - \mu} - \frac{1}{2} \tag{8.24}$$

which can become arbitrarily large.

The Bose-Einstein distribution function enters the expressions for other thermodynamic quantities, such as the average energy. The energy can be found from the expression for the entropy S. The entropy can be found from the infinitesimal relation

$$d\Omega = -S \, dT - P \, dV - N \, d\mu \tag{8.25}$$

which yields

$$
\begin{aligned}
S &= -\left(\frac{\partial \Omega}{\partial T}\right)_{V,\mu} \\
&= -\frac{\Omega}{T} + k_B \beta \int_{-\infty}^{\infty} d\epsilon \, \rho(\epsilon) \, \frac{(\epsilon - \mu)}{\exp[\beta(\epsilon - \mu)] - 1}
\end{aligned} \tag{8.26}
$$

On recalling the expression for $\overline{N}$, one has

$$S = -\frac{(\Omega + \mu \overline{N})}{T} + \frac{1}{T} \int_{-\infty}^{\infty} d\epsilon \, \rho(\epsilon) \, \epsilon \, \frac{1}{\exp[\beta(\epsilon - \mu)] - 1} \tag{8.27}$$

On using the definition

$$\Omega = U - TS - \mu \overline{N} \tag{8.28}$$

one finds that the thermodynamic energy is given by

$$U = \int_{-\infty}^{\infty} d\epsilon \, \rho(\epsilon) \, \epsilon \, \frac{1}{\exp[\beta(\epsilon - \mu)] - 1} \tag{8.29}$$

or alternately as

$$U = \sum_{\alpha} \epsilon_{\alpha} \, \frac{1}{\exp[\beta(\epsilon_{\alpha} - \mu)] - 1} \tag{8.30}$$

The thermodynamic energy U should be compared with the expression for the average energy of the non-interacting particles

$$\overline{E} = \sum_{\alpha} \epsilon_{\alpha} \, \overline{n}_{\alpha} \tag{8.31}$$

This comparison reconfirms our identification of the average number of particles in the α-th energy level as

$$\overline{n}_{\alpha} = \frac{1}{\exp[\beta(\epsilon_{\alpha} - \mu)] - 1} \tag{8.32}$$

In general, the Bose-Einstein distribution function is defined as

$$N(\epsilon_{\alpha}) = \frac{1}{\exp[\beta(\epsilon_{\alpha} - \mu)] - 1} \tag{8.33}$$

where $\epsilon \geq \mu$ and corresponds to the thermal average of the occupation number of a quantum level α with single-particle energy ϵ. Since the occupation numbers can have the values $n_{\alpha} = 0, 1, 2, \ldots \infty$, one has

$$\infty \geq n_{\alpha} \geq 0 \tag{8.34}$$

so the average value must also satisfy the same inequalities

$$\infty \geq N(\epsilon_{\alpha}) \geq 0 \tag{8.35}$$

The positivity of $N(\epsilon_{\alpha})$ requires that $\epsilon_{\alpha} \geq \mu$. However, there does exist a possibility that, in thermal equilibrium, a level can be occupied by an average number of particles that tends to infinity. This possibility requires that the energy of this level is sufficiently close to μ, i.e. $\epsilon_0 \approx \mu$.

For a system with a fixed average number of particles $\overline{N}$, the equation

$$\overline{N} = \int_{-\infty}^{\infty} d\epsilon \, \rho(\epsilon) \, N(\epsilon) \tag{8.36}$$

has to be regarded as an implicit equation for $\mu(T)$. Once $\mu(T)$ has been found, one may then calculate other thermodynamic averages. For example, the average energy of our system of non-interacting particles is given by

$$U = \int_{-\infty}^{\infty} d\epsilon \, \rho(\epsilon) \, \epsilon \, N(\epsilon) \tag{8.37}$$

etc.

8.3   The Equation of State for a Boson Gas

The equation of state for a gas of non-interacting bosons can be found from Ω by noting that

$$\Omega = -PV \tag{8.38}$$

The equation of state can be obtained directly when the single-particle density of states $\rho(\epsilon)$ has the form of a simple power law

$$\rho(\epsilon) = C\epsilon^{\alpha} \qquad \text{for } \epsilon \geq 0$$
$$= 0 \qquad \text{otherwise} \tag{8.39}$$

where α is a constant. The Grand-Canonical Potential Ω is found as

$$\Omega = -PV$$
$$= k_B T \int_{-\infty}^{\infty} d\epsilon \, \rho(\epsilon) \, \ln \left[1 - \exp[-\beta(\epsilon - \mu)] \right] \tag{8.40}$$

Hence, one has

$$-\frac{PV}{k_B T} = C \int_0^{\infty} d\epsilon \, \epsilon^{\alpha} \, \ln \left[1 - \exp[-\beta(\epsilon - \mu)] \right]$$
$$= \frac{C}{\alpha+1} \int_0^{\infty} d\epsilon \, \frac{d\epsilon^{(\alpha+1)}}{d\epsilon} \, \ln \left[1 - \exp[-\beta(\epsilon - \mu)] \right] \tag{8.41}$$

On integrating by parts, one obtains

$$-\frac{PV}{k_B T} = \frac{C}{\alpha+1} \left[\epsilon^{\alpha+1} \, \ln \left(1 - \exp[-\beta(\epsilon - \mu)] \right) \right] \Big|_0^{\infty}$$
$$- \frac{C}{\alpha+1} \int_0^{\infty} d\epsilon \, \epsilon^{(\alpha+1)} \, \frac{\beta \, \exp[-\beta(\epsilon - \mu)]}{1 - \exp[-\beta(\epsilon - \mu)]} \tag{8.42}$$

The boundary terms vanish, since the density of states vanishes at the lower limit of integration $\epsilon = 0$ and the logarithmic factor vanishes exponentially when $\epsilon \to \infty$. Thus, on canceling a factor of β, one finds

$$
\begin{aligned}
PV &= \frac{1}{\alpha+1} \int_0^\infty d\epsilon \, C \, \epsilon^{(\alpha+1)} \, \frac{1}{\exp[\beta(\epsilon-\mu)]-1} \\
&= \frac{1}{\alpha+1} \int_{-\infty}^\infty d\epsilon \, \rho(\epsilon) \, \epsilon \, \frac{1}{\exp[\beta(\epsilon-\mu)]-1} \\
&= \frac{U}{\alpha+1} \tag{8.43}
\end{aligned}
$$

That is, the equation of state for an ideal gas of bosons is found as

$$
PV = \frac{U}{\alpha+1} \tag{8.44}
$$

The same method was used to find the equation of state for an ideal gas of fermions. The result is the same. Therefore, the equation of state holds true, independent of the quantum statistics used. The equation of state must also apply to the classical ideal gas, as the ideal gas can be considered as the high-temperature limiting form of the ideal quantum gasses.

8.4 The Fugacity at High Temperatures

For a system with a fixed number of particles $\overline{N}$, the equation

$$
\overline{N} = \int_{-\infty}^\infty d\epsilon \, \rho(\epsilon) \, N(\epsilon) \tag{8.45}
$$

implicitly determines $\mu(T)$. For non-relativistic bosons with the dispersion relation

$$
\epsilon_{\underline{k}} = \frac{\hbar^2 \underline{k}^2}{2m} \tag{8.46}
$$

in three-dimensions, the density of states can be calculated by replacing the sum over discrete values of $\underline{k}$ by an integral over a density if points in phase space

$$
\begin{aligned}
\rho(\epsilon) &= \sum_{\underline{k}} \delta(\epsilon - \epsilon_{\underline{k}}) \\
&= \frac{V}{(2\pi)^3} \int d^3\underline{k} \, \delta(\epsilon - \epsilon_{\underline{k}}) \\
&= \frac{V}{2\pi^2} \int_0^\infty dk \, k^2 \, \frac{2m}{\hbar^2} \, \delta\left(\frac{2m\epsilon}{\hbar^2} - k^2\right) \\
&= \frac{V}{2\pi^2} \frac{m}{\hbar^2} \sqrt{\frac{2m\epsilon}{\hbar^2}} \tag{8.47}
\end{aligned}
$$

Thus, the average number of particles is expressed as

$$\overline{N} = \frac{V}{4\pi^2} \left(\frac{2m}{\hbar^2}\right)^{\frac{3}{2}} \int_0^\infty d\epsilon \; \epsilon^{\frac{1}{2}} \frac{1}{z^{-1} \exp[\beta\epsilon] - 1} \tag{8.48}$$

where z is the fugacity is defined as

$$z = \exp[\beta\mu] \tag{8.49}$$

The equation for $\overline{N}$ can be re-written in terms of the dimensionless variable $x = \beta\epsilon$ as

$$\overline{N} = \frac{V}{4\pi^2} \left(\frac{2mk_BT}{\hbar^2}\right)^{\frac{3}{2}} \int_0^\infty dx \; x^{\frac{1}{2}} \frac{1}{z^{-1} \exp[x] - 1} \tag{8.50}$$

which determines the fugacity z as a function of temperature. This equation can be expressed as

$$\frac{2}{\sqrt{\pi}} \int_0^\infty dx \; x^{\frac{1}{2}} \frac{1}{z^{-1} \exp[x] - 1} = \frac{N}{V} \left(\frac{2\pi\hbar^2}{mk_BT}\right)^{\frac{3}{2}} \tag{8.51}$$

or as

$$\frac{1}{\Gamma(\frac{3}{2})} \int_0^\infty dx \; x^{\frac{1}{2}} \frac{1}{z^{-1} \exp[x] - 1} = \frac{N}{V} \left(\frac{2\pi\hbar^2}{mk_BT}\right)^{\frac{3}{2}} \tag{8.52}$$

where

$$\Gamma(\alpha + 1) = \int_0^\infty dx \; x^\alpha \exp[-x] \tag{8.53}$$

This type of integral appears frequently in the evaluation of other quantities [56]. We shall denote the integral $I_{\alpha+1}(z)$ as

$$I_{\alpha+1}(z) = \frac{1}{\Gamma(\alpha+1)} \int_0^\infty dx \; x^\alpha \frac{1}{z^{-1} \exp[x] - 1} \tag{8.54}$$

where $z < 1$ and $x > 1$. The integrand can be expanded as

$$I_{\alpha+1}(z) = \frac{1}{\Gamma(\alpha+1)} \int_0^\infty dx \; x^\alpha \frac{z \exp[-x]}{1 - z \exp[-x]}$$

$$= \frac{1}{\Gamma(\alpha+1)} \int_0^\infty dx \; x^\alpha \sum_{m=1}^\infty z^m \exp[-mx] \tag{8.55}$$

On transforming the variable of integration to $y = mx$ one obtains

$$I_{\alpha+1}(z) = \frac{1}{\Gamma(\alpha+1)} \int_0^\infty dy \; y^\alpha \exp[-y] \sum_{m=1}^\infty \frac{z^m}{m^{\alpha+1}}$$

$$= \sum_{m=1}^\infty \frac{z^m}{m^{\alpha+1}} \tag{8.56}$$

For small z, $z \ll 1$, one only needs to retain the first terms so

$$I_{\alpha+1}(z) \approx z \qquad (8.57)$$

while for z equal to unity, $z = 1$, the integral has the value

$$I_{\alpha+1}(1) = \sum_{m=1}^{\infty} m^{-(\alpha+1)}$$
$$= \xi(\alpha + 1) \qquad (8.58)$$

The equation determining $\overline{N}$ can be expressed in terms of the above set of functions as

$$I_{\frac{3}{2}}(z) = \frac{N}{V} \left(\frac{2\pi\hbar^2}{mk_BT} \right)^{\frac{3}{2}} \qquad (8.59)$$

where $\alpha = \frac{1}{2}$. This equation can be solved graphically if the function $I_{\frac{3}{2}}(z)$ is plotted vs z since the intersection with the line representing the righthand side yields the solution for z. At high temperatures, one finds that the solution for z is much less than unity so that the chemical potential μ is negative. As the temperature decreases, the value of z increases towards its maximum value of one, which corresponds to μ increasing towards zero. The temperature for which $z = 1$ is given by

$$\xi\left(\frac{3}{2}\right) = \frac{N}{V} \left(\frac{2\pi\hbar^2}{mk_BT_c} \right)^{\frac{3}{2}} \qquad (8.60)$$

For temperatures below T_c, the equation for $\overline{N}$ cannot be satisfied and the lowest energy level has to have a macroscopic occupation number. That is for $T < T_c$, the bosons must condense into the lowest energy state, as first predicted by Einstein. For this low temperature range, it is no longer sufficient to use a continuum expression for the density of single-particle state $\rho(\epsilon)$ which fails to give the proper weight for the lowest energy state. That is, the method used for calculating the density of states approximates the sum over states by an integration over the density of points in phase space states. Therefore, it only calculates the average number of points on the constant energy surface. This approximation fails miserably at very low energies, when the number of points on the constant energy surface is low. A better approximation has the form

$$\rho(\epsilon) = \delta(\epsilon) + C\epsilon^{\frac{1}{2}} \qquad (8.61)$$

which explicitly includes a delta function of weight of unity for the lowest energy state and C is an extensive constant.

8.5   Fluctuations in the Occupation Numbers

Since the subject under consideration is that of an ideal gas of bosons, the
bosons are non-interacting. Hence, the probabilities that a single-particle
quantum state α is occupied by n bosons is independent of the occupations
of the other single-particle quantum states. Then, in the Grand-Canonical
Ensemble, the probability $p_\alpha(n)$ that the state with energy ϵ_α is occupied
by n bosons is given by

$$p_\alpha(n) = \frac{\exp[-n\beta(\epsilon_\alpha - \mu)]}{\Xi_\alpha} \qquad (8.62)$$

where Ξ_α is determined from the normalization condition for $p_\alpha(n)$. That
is

$$\sum_{n=0}^{\infty} p_\alpha(n) = \sum_{n=0}^{\infty} \frac{\exp[-n\beta(\epsilon_\alpha - \mu)]}{\Xi_\alpha}$$

$$= \frac{1}{\Xi_\alpha} \left[1 - \exp[-\beta(\epsilon_\alpha - \mu)]\right]^{-1}$$

$$= 1 \qquad (8.63)$$

Therefore, the Grand-Canonical partition function for the α-th state, Ξ_α,
is given by

$$\Xi_\alpha = \left[1 - \exp[-\beta(\epsilon_\alpha - \mu)]\right]^{-1} \qquad (8.64)$$

The resulting probability $p_\alpha(n)$ for the state α to be occupied by n bosons
has the form of a geometric probability

$$p_\alpha(n) = \left[1 - \exp[-\beta(\epsilon_\alpha - \mu)]\right] \exp[-n\beta(\epsilon_\alpha - \mu)] \qquad (8.65)$$

The average occupation number of the α-th single-particle quantum state,
$\overline{n}_\alpha$ is calculated as

$$\overline{n}_\alpha = \sum_{n=0}^{\infty} p_\alpha(n)\, n$$

$$= \frac{\exp[-\beta(\epsilon_\alpha - \mu)]}{1 - \exp[-\beta(\epsilon_\alpha - \mu)]}$$

$$= \frac{1}{\exp[\beta(\epsilon_\alpha - \mu)] - 1} \qquad (8.66)$$

which is the usual expression for the Bose-Einstein distribution. This equation can be inverted to express the exponential

$$\exp[-\beta(\epsilon_\alpha - \mu)] = \frac{\bar{n}_\alpha}{\bar{n}_\alpha + 1} \qquad (8.67)$$

as a rational function of the average occupation number. The probability can then be written entirely in terms of the average occupation number

$$p_\alpha(n) = \frac{1}{1 + \bar{n}_\alpha} \left(\frac{\bar{n}_\alpha}{1 + \bar{n}_\alpha} \right)^n \qquad (8.68)$$

The characteristics of the probability distribution can be found by standard means. The k-th moment, M_k^α, of the probability distribution function is defined as the average of n^k

$$M_k^\alpha = \sum_{n=0}^\infty p_\alpha(n) \, n^k \qquad (8.69)$$

The moment generating function $M_\alpha(t)$ is defined as

$$\begin{aligned} M_\alpha(t) &= \sum_{k=0}^\infty M_k^\alpha \frac{t^k}{k!} \\ &= \sum_{n=0}^\infty p_\alpha(n) \sum_{k=0}^\infty \frac{(nt)^k}{k!} \\ &= \sum_{n=0}^\infty p_\alpha(n) \, \exp[nt] \end{aligned} \qquad (8.70)$$

so the k-th moment M_k^α is given by

$$M_k^\alpha = \left(\frac{\partial^k M_\alpha(t)}{\partial t^k} \right)_{t=0} \qquad (8.71)$$

The moment generating function is evaluated

$$M_\alpha(t) = \frac{1}{1 + \bar{n}_\alpha - \bar{n}_\alpha \, \exp[t]} \qquad (8.72)$$

The cumulant generating function $K_\alpha(t)$ is defined as

$$K_\alpha(t) = \ln M_\alpha(t) \qquad (8.73)$$

so that the cumulants κ_k are the coefficients of a power series

$$K_\alpha(t) = \sum_{k=1}^\infty \kappa_k^\alpha \frac{t^k}{k!} \qquad (8.74)$$

Therefore, the cumulants can be found by direct expansion

$$K_\alpha(t) = -\ln\left(1 + \overline{n}_\alpha - \overline{n}_\alpha \exp[t]\right)$$

$$\approx t\overline{n}_\alpha + \frac{t^2}{2!}\,\overline{n}_\alpha(\overline{n}_\alpha + 1) + \ldots \tag{8.75}$$

which yields the average and the mean-squared fluctuation

$$\overline{n}_\alpha = \overline{n}_\alpha$$

$$\overline{\Delta n_\alpha^2} = \overline{n}_\alpha(\overline{n}_\alpha + 1) \tag{8.76}$$

One can form the relative mean-squared fluctuation of the boson occupation numbers, which is given by

$$\frac{\overline{\Delta n_\alpha^2}}{\overline{n}_\alpha^2} = 1 + \frac{1}{\overline{n}_\alpha} \tag{8.77}$$

Einstein identified the term of order unity as originating from the interference of classical waves and the second term as corresponding to classical particles. The skew and the kurtosis can be found from the higher-order terms in the expansion of the cumulant generating function.

Having found the probability $p_\alpha(n)$ for quantum state α and since the $p_\alpha(n)$ for each state is independent of the other states, one can form the entropy of the entire system as the sum of the entropies for each quantum state. The Shannon entropy can be expressed in terms of the average occupation numbers as

$$S = -k_B \sum_\alpha \left[\sum_{n=0}^\infty p_\alpha(n)\,\ln\,p_\alpha(n)\right]$$

$$= -k_B \sum_\alpha \frac{1}{1 + \overline{n}_\alpha}$$

$$\times \left[\sum_{n=0}^\infty \left(\frac{\overline{n}_\alpha}{1 + \overline{n}_\alpha}\right)^n \left(n\,\ln\left(\frac{\overline{n}_\alpha}{1 + \overline{n}_\alpha}\right) - \ln(1 + \overline{n}_\alpha)\right)\right]$$

$$= -k_B \sum_\alpha \left[\overline{n}_\alpha \ln \overline{n}_\alpha - (1 + \overline{n}_\alpha) \ln(1 + \overline{n}_\alpha)\right] \tag{8.78}$$

which is agreement with the entropy found thermodynamically from the Grand-Canonical potential $\Omega(T, V, \mu)$.

8.6   Bose-Einstein Condensation

The expression for the Bose-Einstein distribution function only makes sense when the chemical potential μ is less than the energy of the single-particle quantum state ϵ. Since this is true for all ϵ and since we have defined the lowest single-particle energy as $\epsilon = 0$, we must require that $\mu < 0$. As was first pointed out by Einstein, if μ approaches zero, then there can be a macroscopic occupation of a single-particle quantum level.

When there is a macroscopic occupation of the lowest energy level, the single particle density of states must explicitly include the lowest energy state. In this case, we need to use the better approximation to the density of states given by

$$\rho(\epsilon) = \delta(\epsilon) + C\epsilon^\alpha \, \Theta(\epsilon) \tag{8.79}$$

where the function $\Theta(\epsilon)$ is the Heaviside step function. The delta function represents the lowest energy state. The second term represents the approximation for the extensive part of the density of states. This expression can also be used at temperatures above T_c since the contribution from the delta function is not extensive and can be ignored.

On using the above expression for the density of states to evaluate the average number of particles, one finds

$$\overline{N} = \int_{-\infty}^{\infty} d\epsilon \, \rho(\epsilon) \, N(\epsilon)$$

$$= \int_{-\infty}^{\infty} d\epsilon \, (\delta(\epsilon) + C\epsilon^\alpha \, \Theta(\epsilon)) \, N(\epsilon)$$

$$= N(0) + \int_0^\infty d\epsilon \, C\epsilon^\alpha \, N(\epsilon) \tag{8.80}$$

where the first term represents the number of particles in the quantum level with zero energy. On changing the variable of integration to $x = \beta\epsilon$, the expression can be re-written as

$$\overline{N} = \frac{1}{z^{-1} - 1} + C(k_B T)^{\alpha+1} \int_0^\infty dx \, \frac{x^\alpha}{z^{-1} \exp[x] - 1}$$

$$= \frac{z}{1 - z} + \Gamma(\alpha + 1) \, C(k_B T)^{\alpha+1} \, I_{\alpha+1}(z) \tag{8.81}$$

This equation determines z when the average number of particles $\overline{N}$ is fixed. The above equation can be interpreted as the sum of the particles in the lowest energy state N_0 and the number of particles in the excited states N_{exc}

$$\overline{N} = N_0 + N_{exc} \tag{8.82}$$

where the number of excited particles is given by

$$N_{exc} = \Gamma(\alpha + 1) \, C(k_B T)^{\alpha+1} \, I_{\alpha+1}(z) \tag{8.83}$$

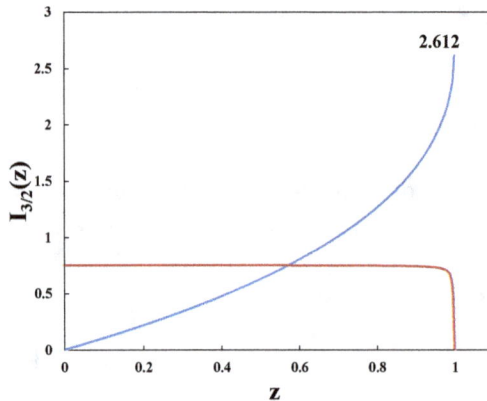

Fig. 8.2 The graphical solution for the fugacity found from plotting both $I_{\frac{3}{2}}(z)$ (blue) and $A[\overline{N} - z/(1-z)]$ (red) versus z. The point of intersection of the curves yields the value of the fugacity. Note that although $I_{\frac{3}{2}}(z)$ has a divergent derivative at $z = 1$, its value there is finite and is given by 2.612.

The number of particles in the condensate N_0 is given by

$$N_0 = \frac{z}{1 - z} \tag{8.84}$$

Above the Condensation Temperature

For $T > T_c$, one has

$$N_{exc} = \overline{N} \tag{8.85}$$

as the explicit equation for $\overline{N}$ determines the value of z to be $z < 1$. Thus, the number of particles in the condensate is not extensive and is negligible, so all the particles can be considered as being in the excited states.

The Condensation Temperature

The Bose-Einstein condensation temperature T_c is evaluated from the condition that $z = 1 - \eta$ where η is an infinitesimal positive quantity

$$\overline{N} = \Gamma(\alpha + 1) \, C(k_B T_c)^{\alpha+1} \, \xi(\alpha + 1) \tag{8.86}$$

which determines the lowest temperature at which the number of the particles in the condensate is still negligibly small ($N_0 \sim \eta^{-1}$). Note that, since $C \propto V$, the condensation temperature T_c depends on the density, or if the number of particles is fixed it depends on the volume $T_c(V)$.

Below the Condensation Temperature

For temperatures below T_c, $T < T_c$, the number of particles in the excited states is temperature dependent and decreases towards zero as T is reduced to zero according to a simple power law.

$$N_{exc} = \Gamma(\alpha + 1)\, C(k_B T)^{\alpha+1}\, \xi(\alpha + 1) \tag{8.87}$$

or equivalently as

$$N_{exc} = \overline{N}\left(\frac{T}{T_c}\right)^{\alpha+1} \tag{8.88}$$

where we have used the equation for the Bose-Einstein condensation temperature to eliminate the constant C. The number of particles in the condensate is defined as

$$N_0 = \overline{N} - N_{exc} \tag{8.89}$$

which is evaluated as

$$N_0 = \overline{N}\left[1 - \left(\frac{T}{T_c}\right)^{\alpha+1}\right] \tag{8.90}$$

that tends to N in the limit of $T \to 0$. In this case, one also has

$$\frac{z}{1-z} = \overline{N}\left[1 - \left(\frac{T}{T_c}\right)^{\alpha+1}\right] \tag{8.91}$$

which determines z as

$$z = \frac{1}{1 - \frac{1}{N}\left[1 - \left(\frac{T}{T_c}\right)^{\alpha+1}\right]^{-1}} \tag{8.92}$$

that confirms that $z \approx 1$.

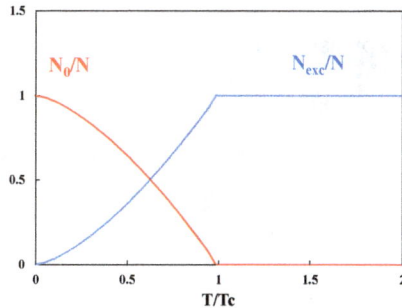

Fig. 8.3 The temperature-dependence of the relative number of condensate particles N_0 and the relative number of excited particles N_{exc}.

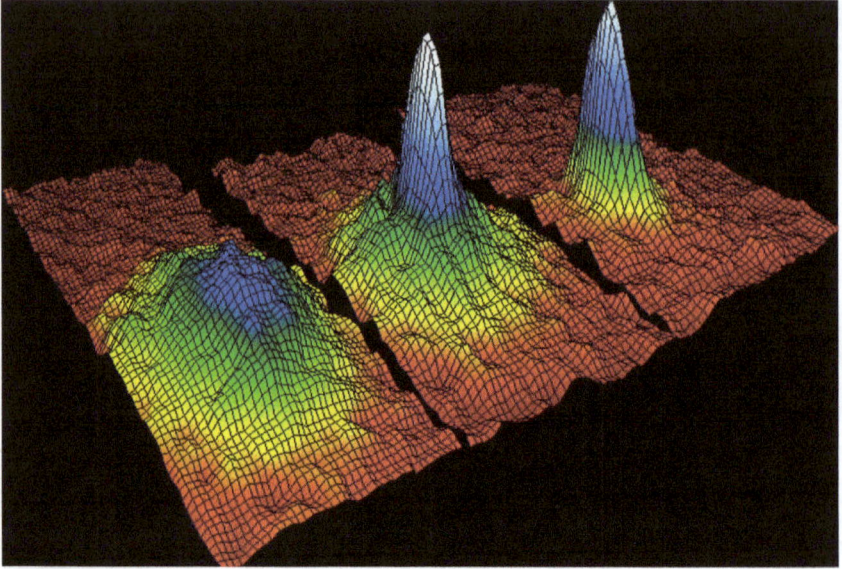

Fig. 8.4 The velocity distribution of a gas of Rb atoms at three different temperatures, showing the evolution of the macroscopic occupation of the lowest energy single-particle state as the temperature is lowered. [M.H. Anderson, J.R. Ensher, M.R. Matthews, C.E. Wieman, and E.A. Cornell, "Observation of Bose-Einstein Condensation in a Dilute Atomic Vapor", Science 269, 198-201 (1995).]courtesy of NIST/JILA/CU-Boulder

8.6.1 *Properties of the Idealized Condensed Phase*

The average energy U of the condensed phase of an ideal gas of bosons is given by

$$U = \int_{-\infty}^{\infty} d\epsilon \; \rho(\epsilon) \; \epsilon \; N(\epsilon) \tag{8.93}$$

which together with expression for the single-particle density of states

$$\rho(\epsilon) = \delta(\epsilon) + C\epsilon^\alpha \; \Theta(\epsilon) \tag{8.94}$$

yields the expression

$$U = \int_{0}^{\infty} d\epsilon \; C\epsilon^\alpha \; \epsilon \; N(\epsilon) \tag{8.95}$$

since the $N(0)$ particles with $\epsilon = 0$ don't contribute to the total energy. Then for $z \approx 1$, the integral reduces to

$$\begin{aligned} U &= C(k_B T)^{\alpha+2} \; \Gamma(\alpha+2) \; I_{\alpha+2}(1) \\ &= C(k_B T)^{\alpha+2} \; \Gamma(\alpha+2) \; \xi(\alpha+2) \end{aligned} \tag{8.96}$$

Then, the specific heat for $T \leq T_c$ can be found from

$$C_{V,N} = \left(\frac{\partial U}{\partial T}\right)_{V,N}$$

$$= k_B \, C(k_B T)^{\alpha+1} \, \Gamma(\alpha+3) \, \xi(\alpha+2) \tag{8.97}$$

which follows a simple power law in T

$$C_{V,N} = \overline{N}k_B \, (\alpha+2)(\alpha+1) \left(\frac{\xi(\alpha+2)}{\xi(\alpha+1)}\right) \left(\frac{T}{T_c}\right)^{\alpha+1} \tag{8.98}$$

The power law variation is simply understood. At $T = 0$ all the particles occupy the state with $\epsilon = 0$. At finite temperatures, the states with energy less than $k_B T$ are occupied. For a density of states proportional to ϵ^α, there are approximately $(k_B T)^{\alpha+1}$ occupied states, each of which carries an energy of approximately $k_B T$. Therefore, the total energy is proportional to $(k_B T)^{\alpha+2}$. Hence, the specific heat is proportional to $(k_B T)^{\alpha+1}$.

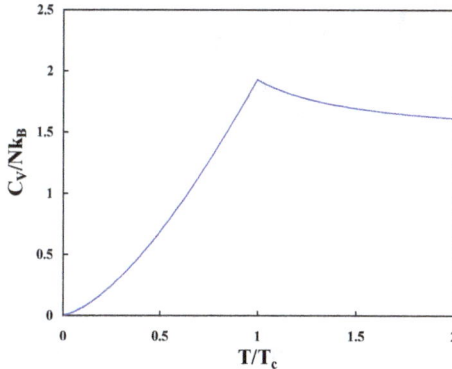

Fig. 8.5 The temperature dependence of the heat capacity for an ideal Bose Gas for $\alpha = \frac{1}{2}$.

The Cusp in the Specific Heat

The specific heat that evaluated in the condensed phase, at a temperature below T_c. At T_c, the expression is evaluated as

$$C_{V,N}(T_c) = \overline{N}k_B(\alpha+2)(\alpha+1) \left(\frac{\xi(\alpha+2)}{\xi(\alpha+1)}\right) \tag{8.99}$$

As shall be shown now, in the normal phase, just above the condensation temperature T_c, the specific heat is given by

$$C_{V,N} = (\alpha + 1)\,\overline{N}k_B \left[(\alpha + 2)\,\frac{\xi(\alpha + 2)}{\xi(\alpha + 1)} - (\alpha + 1)\,\frac{\xi(\alpha + 1)}{\xi(\alpha)}\right]$$

(8.100)

which indicates that it may be discontinuous at T_c. However, for $\alpha = \frac{1}{2}$, $\xi(\alpha)$ diverges, so the last term vanishes. Hence, for $\alpha = \frac{1}{2}$ the specific heat is continuous at T_c and takes the value

$$C_{V,N}(T_c) = \overline{N}k_B\,\frac{15}{4}\left(\frac{\xi(\frac{5}{2})}{\xi(\frac{3}{2})}\right)$$

$$= 1.925\,\overline{N}k_B$$

(8.101)

Thus, the specific heat at T_c exceeds the high temperature classical value of $1.5\,\overline{N}k_B$. In fact there is a cusp in $C_{V,N}$ at T_c.

The existence of a cusp can be inferred from an examination of the general expression for the heat capacity, valid in the normal liquid and the condensed phase.

$$C_{V,N} = \left(\frac{\partial U}{\partial T}\right)_{V,N}$$

$$= k_B C(k_B T)^{\alpha+1}\,\Gamma(\alpha + 2)\left[(\alpha + 2)\,I_{\alpha+2}(z) + T\,\frac{\partial z}{\partial T}\,\frac{\partial}{\partial z}I_{\alpha+2}(z)\right]$$

(8.102)

However, since

$$I_{\alpha+1}(z) = \sum_{m=1}^{\infty} \frac{z^m}{m^{\alpha+1}}$$

(8.103)

then the derivative w.r.t. z is simply given by

$$\frac{\partial}{\partial z}I_{\alpha+1}(z) = \frac{1}{z}\,I_\alpha(z)$$

(8.104)

Thus, the specific heat in the normal phase can be expressed as

$$C_{V,N} = Ck_B(k_B T)^{\alpha+1}\,\Gamma(\alpha + 2)\left[(\alpha + 2)\,I_{\alpha+2}(z) + T\,\frac{\partial z}{\partial T}\,\frac{1}{z}\,I_{\alpha+1}(z)\right]$$

(8.105)

which requires knowledge of $\frac{\partial z}{\partial T}$ in the normal phase. In the Bose Condensed phase, $z = 1$ so the derivative vanishes. The value of $\frac{\partial z}{\partial T}$ in the normal phase can be found from the condition

$$\left(\frac{\partial \overline{N}}{\partial T}\right)_V = 0$$

(8.106)

which yields

$$0 = \left(\frac{\partial \overline{N}}{\partial T}\right)_V$$

$$= Ck_B(k_BT)^\alpha \, \Gamma(\alpha+1) \left[(\alpha+1)\, I_{\alpha+1}(z) + T\, \frac{\partial z}{\partial T}\, \frac{1}{z}\, I_\alpha(z)\right]$$

$$(8.107)$$

This is an equation which determines the temperature-dependence of the fugacity

$$\frac{\partial z}{\partial T} = -(\alpha+1)\, \frac{z}{T}\, \frac{I_{\alpha+1}(z)}{I_\alpha(z)} \tag{8.108}$$

Therefore, the specific heat for the normal phase is given by the expressions

$$C_{V,N} = Ck_B(k_BT)^{\alpha+1}\, \Gamma(\alpha+2) \left[(\alpha+2)\, I_{\alpha+2}(z) - (\alpha+1)\, \frac{I_{\alpha+1}^2(z)}{I_\alpha(z)}\right]$$

$$= (\alpha+1)\, \overline{N}k_B \left[(\alpha+2)\, \frac{I_{\alpha+2}(z)}{I_{\alpha+1}(z)} - (\alpha+1)\, \frac{I_{\alpha+1}(z)}{I_\alpha(z)}\right] \tag{8.109}$$

In the high temperature limit, the specific heat can be expanded as

$$C_{V,N} \approx (\alpha+1)\, \overline{N}k_B \left[1 + \frac{\alpha}{2^{\alpha+2}}\, z + \ldots\right] \tag{8.110}$$

which reaches the classical limit as z approaches zero and increases when z increases. If $\alpha = \frac{1}{2}$, the denominator of the last term in the exact expression diverges at the Bose-Einstein condensation temperature where $z = 1$. This occurs as the sum defining $\xi(\frac{1}{2})$ is divergent. The divergence of the $I_{\frac{1}{2}}(z)$ at $z = 1$ causes the last term to vanish and makes the specific heat continuous at T_c. However, the specific heat does have a discontinuity in its slope which is given by

$$\left(\frac{\partial C_{V,N}}{\partial T}\right)\Bigg|_{T_c-\eta}^{T_c+\eta} = 3.66\, \frac{\overline{N}k_B}{T_c} \tag{8.111}$$

A cusp is also seen in the temperature dependence of the specific heat of liquid He^4, which is a signature of the so-called λ transition.

The Pressure in the Condensed Phase

The pressure can be found directly from the Grand-Canonical Potential Ω, as

$$\Omega = -PV \tag{8.112}$$

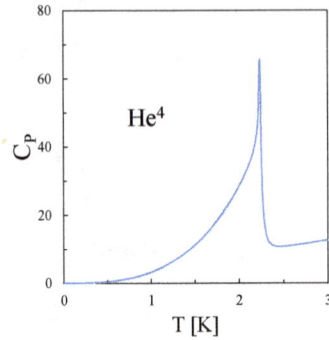

Fig. 8.6   The temperature dependence of the specific heat of He II near the superfluid temperature T_λ. (Schematic)

which is evaluated as

$$\Omega - -PV$$
$$= k_B T \int_{-\infty}^{\infty} d\epsilon \ \rho(\epsilon) \ \ln[1 - \exp[-\beta(\epsilon - \mu)]] \tag{8.113}$$

with

$$\rho(\epsilon) = \delta(\epsilon) + C\epsilon^\alpha \ \Theta(\epsilon) \tag{8.114}$$

This yields

$$- PV = k_B T \ \ln(1 - z) + Ck_B T \int_0^{\infty} d\epsilon \ \epsilon^\alpha \ \ln[1 - z \ \exp[-\beta\epsilon]] \tag{8.115}$$

On changing the integration variable to $x = \beta\epsilon$, one obtains

$$PV = k_B T \ \ln N(0) - k_B T \ \ln z - C(k_B T)^{\alpha+2}$$
$$\times \int_0^{\infty} dx \ x^\alpha \ \ln[1 - z \ \exp[-x]] \tag{8.116}$$

Integrating by parts in the last term, yields

$$PV = k_B T \ \ln N(0) - k_B T \ \ln \ z + \frac{C}{\alpha + 1} \ (k_B T)^{\alpha+2}$$
$$\times \int_0^{\infty} dx \ \frac{x^{\alpha+1}}{z^{-1} \exp[x] - 1} \tag{8.117}$$

where the boundary terms have vanished. Hence, we find the equation of state has the form

$$PV = k_B T \ \ln N(0) - k_B T \ \ln z + \frac{C}{\alpha + 1} \ (k_B T)^{\alpha+2} \ \Gamma(\alpha + 2) \ I_{\alpha+2}(z) \tag{8.118}$$

For $T < T_c$, one has $z = 1$, therefore the expression reduces to

$$PV = k_B T \ln N(0) + \frac{C}{\alpha + 1} (k_B T)^{\alpha+2} \Gamma(\alpha + 2) \xi(\alpha + 2) \qquad (8.119)$$

Since C is proportional to the volume, the first and last terms are extensive, while the logarithmic term is not and can be neglected. This further reduces the equation of state to the form

$$PV = C(k_B T)^{\alpha+2} \Gamma(\alpha + 1) \xi(\alpha + 2) \qquad (8.120)$$

or equivalently

$$P = \frac{C}{V} (k_B T)^{\alpha+2} \Gamma(\alpha + 1) \xi(\alpha + 2) \qquad (8.121)$$

which, as the volume dependence of C cancels with V, one finds that the pressure is independent of volume. The pressure only depends on temperature, for $T < T_c$, Thus, the isotherms become flat on entering the condensed

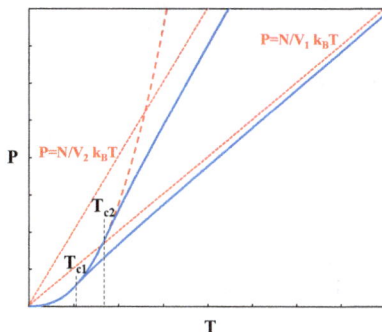

Fig. 8.7 The $P - T$ relations for ideal boson gasses with two different densities (blue) and different T_c's (vertical dashed lines). In the Bose-Einstein condensed phase, the $P - T$ curves collapse onto one curve. The classical asymptotic limiting form for the $P - T$ relations at these two densities are shown by the dashed red lines.

phase. Furthermore, since

$$N_{exc} = \Gamma(\alpha + 1) C(k_B T)^{\alpha+1} \xi(\alpha + 1) \qquad (8.122)$$

one can write

$$PV = \left(\frac{\xi(\alpha + 2)}{\xi(\alpha + 1)} \right) N_{exc} k_B T \qquad (8.123)$$

This makes sense since only the excited particles carry momentum and collide with the walls.

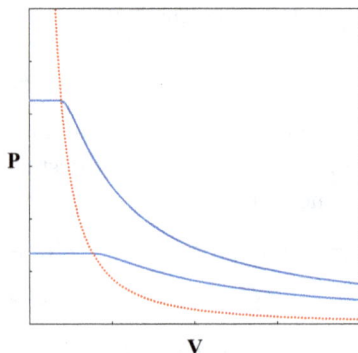

Fig. 8.8 The $P - V$ relations for an ideal boson gas at two different temperatures. Since the number of particles is fixed, the condensation temperature depends on volume $T_c(V) \propto V^{-\frac{3}{2}}$. Thus, the critical pressure P_c varies as $P_c \propto V^{-\frac{5}{3}}$.

The Entropy of the Condensed Phase

The entropy of the condensed phase can be found from

$$S = -\left(\frac{\partial \Omega}{\partial T}\right)_{V,\mu} \tag{8.124}$$

which leads to

$$TS = E - \Omega - \mu N$$
$$= \left(\frac{\alpha + 2}{\alpha + 1}\right) U - \mu \overline{N} \tag{8.125}$$

In the Bose condensed phase, $\mu = 0$. Hence, with

$$U = (\alpha + 1) \left(\frac{\xi(\alpha + 2)}{\xi(\alpha + 1)}\right) N_{exc} \, k_B T \tag{8.126}$$

one finds that the entropy is given by

$$S = (\alpha + 2) \left(\frac{\xi(\alpha + 2)}{\xi(\alpha + 1)}\right) k_B \, N_{exc} \tag{8.127}$$

This implies that only the excited particles are disordered and contribute to the entropy.

8.6.2 *Long-Range Order in Bose-Einstein Condensates*

The single-particle correlation function $C_1(\underline{r}, \underline{r}')$ for non-interacting bosons is given by

$$C_1(\underline{r}, \underline{r}') = \langle \hat{\psi}^\dagger(\underline{r}) \hat{\psi}(\underline{r}') \rangle$$
$$= \sum_\alpha \phi_\alpha^*(\underline{r}) \phi_\alpha(\underline{r}') \langle \hat{n}_\alpha \rangle \qquad (8.128)$$

For a translational invariant and isotropic system, the correlation function only depends on R where $\underline{R} = \underline{r} - \underline{r}'$. For example, with plane wave energy eigenstates, the correlation function is evaluated as

$$C_1(R) = \frac{1}{V} \sum_k \frac{\exp[i\,\underline{k} \cdot \underline{R}]}{\exp[\beta(\epsilon(k) - \mu)] - 1} \qquad (8.129)$$

At $R = 0$, the correlation function reduces to the uniform density, $\rho = \frac{N}{V}$, and can be shown to decease quadratically in R for short distances. We shall consider a three-dimensional system where the boson dispersion relation is isotropic and has the form $\epsilon(k) = \frac{\hbar^2 k^2}{2m}$. On replacing the sum by an integration, one finds the correlation function shows an initial quadratic decrease

$$C_1(R) = 4\pi \int_0^\infty \frac{dk}{(2\pi)^3} k^2 \left(\frac{\sin kR}{kR} \right) \frac{1}{\exp[\beta(\epsilon(k) - \mu)] - 1}$$
$$\sim \frac{N}{V} \left(1 - \frac{1}{3!} k_0^2 R^2 + \dots \right) \qquad (8.130)$$

The long distance behavior is dramatically different in a Bose condensed phase from a non-condensed phase, due to the occupation N_0 of the state with $k = 0$. On retaining the contribution from the $k = 0$ state, the correlation function becomes

$$C_1(R) = \frac{N_0}{V} + 4\pi \int_0^\infty \frac{dk}{(2\pi)^3} k^2 \left(\frac{\sin kR}{kR} \right) \frac{1}{\exp[\beta(\epsilon(k) - \mu)] - 1}$$
$$(8.131)$$

For $T_c > T$, one has $\mu = 0$ and N_0 is a macroscopic quantity, so for long distances, the correlation is dominated by small values of k

$$C_1(R) \sim \frac{N_0}{V} + \left(\frac{mk_B T}{\pi^2 \hbar^2} \right) \int_0^\infty dk \left(\frac{\sin kR}{kR} \right)$$
$$\sim \frac{N_0}{V} + \left(\frac{mk_B T}{2\pi^2 \hbar^2} \right) \frac{1}{R} \int_{-\infty}^\infty dz \frac{\exp[iz]}{iz}$$
$$\sim \frac{N_0}{V} + \left(\frac{mk_B T}{2\pi \hbar^2} \right) \frac{1}{R} \qquad (8.132)$$

which does not vanish as $R \to \infty$ since it saturates at the condensate density. On the other hand, for $T > T_c$, the condensate density vanishes and the chemical potential is negative. Expressing the chemical potential as $\mu = -\frac{\hbar^2 \kappa^2}{2m}$, the asymptotic long distance behavior of the correlation function reduces to

$$
\begin{aligned}
C_1(R) &\sim \left(\frac{mk_BT}{\pi^2\hbar^2}\right) \int_0^\infty dk \; \frac{k^2}{k^2 + \kappa^2} \left(\frac{\sin kR}{kR}\right) \\
&\sim \left(\frac{mk_BT}{2\pi^2\hbar^2}\right) \frac{1}{iR} \int_{-\infty}^\infty dk \; \frac{k}{k^2 + \kappa^2} \; \exp[ikR] \\
&\sim \left(\frac{mk_BT}{2\pi\hbar^2}\right) \frac{\exp[-\kappa R]}{R}
\end{aligned}
\tag{8.133}
$$

which decreases exponentially as $R \to \infty$. The phenomenon in which of $C_1(R)$ tends to a finite value in the limit $R \to \infty$ for a condensate is known as "*Off-Diagonal Long-Range Order*" (ODLRO) [57, 58]. The existence of ODLRO is suggestive that when $R \to \infty$ the correlation function may be separable [59, 60] such that field operators $\hat{\psi}^\dagger(\underline{r})$ and $\hat{\psi}(\underline{r})$ have macroscopic average values that are related to the square root of the condensate density. The idea of ODLRO has also been extended to superconductors by Gorkov in which the bosonic field operator $\hat{\psi}$ is replaced by the electron pair field operators $\hat{\phi}\,\hat{\phi}$ which acts as the superconducting order parameter.

The above discussions relate to non-interacting boson gasses. It is expected that Bose-Einstein condensation may be hindered by interactions,

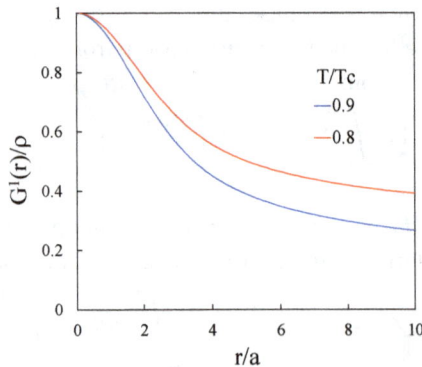

Fig. 8.9   The spatial variation of the single-particle correlation function $C_1(R)$ for temperatures below the Bose-Einstein condensation temperature T_c, as function of distance normalized to the density. The correlation function saturates as $R \to \infty$ to the value of the condensate density.

since the density should resemble the squared modulus of the lowest energy
single particle wave function and exhibit the same non-uniformity in space.
Local interactions are expected to make the fluid's density uniform.

8.7   Superfluidity

Helium has two electrons located in the $1s$ orbitals, so it has a closed atomic
shell and is relatively chemically inert. There does exist a van der Waals
interaction between the He atoms. The interatomic potential consists of
the sum of the short-ranged exponential repulsion and the attractive van
der Waals interaction [61]. The potential has a weak minimum of depth 9
K, at a distance of about 3 Å. Since the atom is relatively light, it doesn't

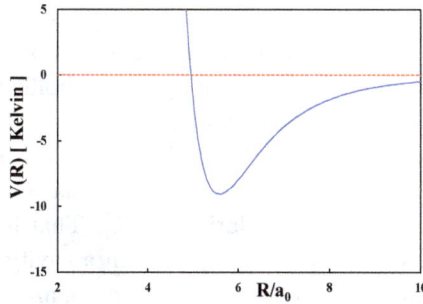

Fig. 8.10   The interaction potential of atomic Helium as a function of radial distance
R/a_0, where a_0 is the Bohr radius.

solidify easily. If the atoms did solidify, with a lattice spacing d, then the
uncertainty in the momentum Δp of any atom would be of the order of
$\hbar/d$. The kinetic energy would be of the order of

$$\frac{\Delta p^2}{2m} \approx \frac{\hbar^2}{2m\,d^2} \tag{8.134}$$

for $d \sim 3$ Å, this energy is greater than the minimum of the potential.
Hence, the lattice would melt. Therefore, He remains a liquid at ambient
pressures. However, He interacts quite strongly, so at pressures of the order
of 25 atmospheres, it can solidify at very low temperatures.

   The isotope He^3 is a fermion and a He^3 gas forms a Fermi-liquid at low
temperatures. On the other hand, He^4 is a boson and obeys Bose-Einstein

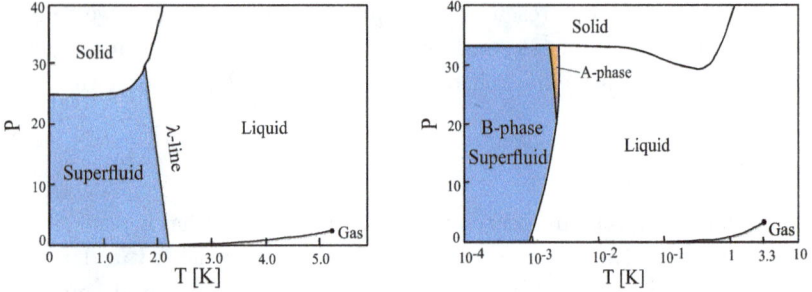

Fig. 8.11   The $P - T$ phase diagrams of the He^4 (left) and the He^3 (right) which obey different quantum statistics. The Bose-system He^4 remains a fluid for pressures below 2.5 MPa. The Liquid-Gas critical point is located at a temperature of 5.2 K. The liquid phase undergoes a further transition from He I (Liquid) to He II (Superfluid), as the temperature is reduced below 2.18 K. Despite the chemical similarities, the Fermi-system He^3 has a completely different phase diagram to He^4.

Statistics. He^4 can be cooled by a process of evaporation. When it is cooled, the Helium becomes turbulent, just like boiling water. However, at a temperature of 2.2 K it suddenly becomes clear and the turbulence disappears. This signals a phase change from the He I phase, to He II.

He II has unusual properties, it flows as if it has a vanishing viscosity, and can flow through narrow capillaries [62,63]. That is, it exhibits super-flow. Furthermore, when the fluid is placed in a cavity containing closely spaced disks that are free to rotate, and the disks are connected to a tor-sional rod and are set into oscillation through some small angles (torsional

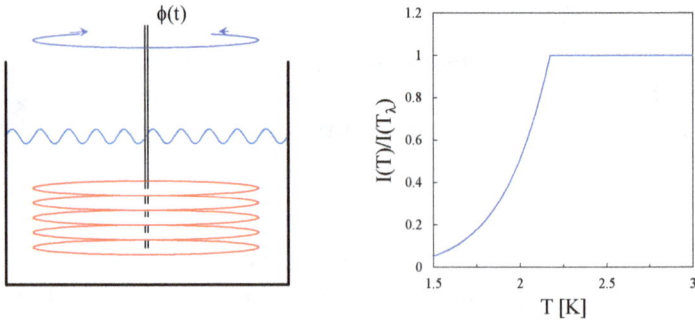

Fig. 8.12   Andronikashvili's 1959 experiment on the moment of inertia of a stack of disks emersed in He^4. Below the lambda transition temperature, T_λ, the moment of inertia due to the fluid $I(T)$ dropped precipitously.

oscillations), then the effective moment of inertia is temperature dependent. The increase in moment of inertia over that of the disks indicates that some fluid is being dragged by the rotating disks. Andronikashvili [64] found that above the critical temperature T_c the fluid was being dragged by the plates, but below T_c the moment of inertia decreased. In fact, the fraction of the fluid which was dragged by the rotating disks was found to vanish proportional to T^4 as T was decreased to zero.[1]

In 1938 Fritz London [65] proposed that the transition to He II is related to Bose-Einstein condensation. The reasons for this proposal was:

(i) The observed critical temperature T_c has the magnitude of 2.18 K, whereas the Bose-Einstein condensation temperature for free bosons is given by

$$k_B T_c = \frac{2\pi\hbar^2}{2m} \left[\xi\left(\frac{3}{2}\right) \frac{V}{N} \right]^{2/3} \tag{8.135}$$

which is calculated as 3.14 K.

(ii) The transition occurs in the boson system He^4 but not in the fermionic system He^3.

(iii) For the Bose condensed phase of an ideal gas, the pressure is independent of the density and only a function of temperature. Hence, pressure variations are directly related to temperature variations, as is true for He II. The specific heat of He II also shows a λ anomaly at T_c, whereas the ideal Bose gas is only expected to exhibit a cusp at T_c. However, lambda-like anomalies in the specific heat are universally found at second-order phase transitions.

Superfluid He^4 differs from the Bose-Einstein condensate in many details, primarily due to the effects of interactions. For example, an ideal Bose condensate would not be a superfluid. However, interactions can change the form of the dispersion relation for the low-energy excitations and that results in He II becoming a superfluid.

---

[1]The T^4 variation is to be contrasted with the $T^{\frac{3}{2}}$ variation of the density of excited particles expected for non-interacting bosons but is closer to the T^3 variation that is expected if the excitations have a "phonon-like" dispersion relation. Landau predicted a T^4 variation using a "phonon-like" dispersion relation in which the extra factor of T is connected with the motion of the plates.

8.7.1   *A Weakly-Interacting Bose Gas*

Consider a gas of weakly-interacting bosons contained in a volume V. The Hamiltonian is described by

$$\hat{H} = \sum_{\alpha;\alpha'} \langle \alpha' | \frac{\hat{p}^2}{2m} | \alpha \rangle \, \hat{a}^\dagger_{\alpha'} \, \hat{a}_\alpha$$

$$+ \frac{1}{2!} \sum_{\alpha,\beta;\alpha',\beta'} \langle \alpha'\beta' | \hat{V}_{int} | \alpha\beta \rangle \, \hat{a}^\dagger_{\alpha'} \, \hat{a}^\dagger_{\beta'} \, \hat{a}_\beta \hat{a}_\alpha \qquad (8.136)$$

where the two-body interaction represents a short-ranged repulsive interaction. The one-body part can be diagonalized by choosing the single-particle wave functions $\phi_\alpha(\underline{r})$ to be momentum eigenstates

$$\phi_{\underline{k}}(\underline{r}) = \frac{1}{\sqrt{V}} \, \exp[i\underline{k} \cdot \underline{r}] \qquad (8.137)$$

The Hamiltonian can be expressed as

$$\hat{H} = \sum_{\underline{k}} \frac{\hbar^2 \underline{k}^2}{2m} \, \hat{a}^\dagger_{\underline{k}} \, \hat{a}_{\underline{k}} + \frac{1}{2! \, V} \sum_{\underline{k},\underline{k}';\underline{q}} V_{int}(\underline{q}) \, \hat{a}^\dagger_{\underline{k}-\underline{q}} \, \hat{a}^\dagger_{\underline{k}'+\underline{q}} \, \hat{a}_{\underline{k}'} \, \hat{a}_{\underline{k}} \qquad (8.138)$$

where the scattering term conserves momentum and $V_{int}(\underline{q})$ is the Fourier Transform of the interaction potential. Therefore, the two-body scattering does not change the total momentum of the system. We shall assume that the potential is sufficiently short-ranged so that the limit, $\lim_{q \to 0} V_{int}(\underline{q}) = V_{int}(0)$, is well-defined.

At sufficiently low temperatures, the bosons are expected to form a condensate. Let the number of particles in the condensate be N_0 and we shall assume that N_0 is much larger than the number of excited particles N_{exc}

$$\overline{N}_{exc} = \sum_{\underline{k}} \overline{\hat{a}^\dagger_{\underline{k}} \hat{a}_{\underline{k}}} \qquad (8.139)$$

so

$$\overline{N}_0 = N - \sum_{\underline{k}} \overline{\hat{a}^\dagger_{\underline{k}} \hat{a}_{\underline{k}}} \qquad (8.140)$$

The condensate contains a large number of bosons with $\underline{k} = 0$ and, therefore can be considered to be a coherent state. The expectation value of the creation and annihilation $\hat{a}^\dagger_0$ and $\hat{a}_0$, respectively, can be replaced by the complex numbers a_0^* and a_0. Thus, the interaction Hamiltonian can be expanded in powers of a_0 or a_0^* as

$$\hat{H} = \sum_{\underline{k}} \frac{\hbar^2 \underline{k}^2}{2m} \hat{a}_{\underline{k}}^\dagger \hat{a}_{\underline{k}} + \frac{1}{2! \, V} |a_0|^4 \, V_{int}(0)$$

$$+ \frac{1}{2! \, V} \sum_{\underline{k}} V_{int}(\underline{k}) \, (a_0^2 \hat{a}_{\underline{k}}^\dagger \hat{a}_{-\underline{k}}^\dagger + a_0^{*2} \hat{a}_{-\underline{k}} \hat{a}_{\underline{k}})$$

$$+ \frac{2}{2! \, V} \sum_{\underline{k}} V_{int}(0) |a_0|^2 \, (\hat{a}_{\underline{k}}^\dagger \hat{a}_{\underline{k}} + \hat{a}_{-\underline{k}}^\dagger \hat{a}_{-\underline{k}}) + \ldots \quad (8.141)$$

Terms cubic in a_0 and a_0^* are forbidden due to the requirement of conservation of momentum. In this expression, we have ignored terms involving more than two excited boson creation or annihilation operators. The above form of the Hamiltonian contains terms involving unbalanced creation and annihilation operators. The term with two creation operators represent processes in which two bosons are scattered out of the condensate and the term with two annihilation operators represents the absorbtion of two bosons into the condensate. On replacing $|a_0|^2$ by $N - \sum_k a_k^\dagger a_k$, one finds

$$\hat{H} = \sum_{\underline{k}} \frac{\hbar^2 \underline{k}^2}{2m} \hat{a}_{\underline{k}}^\dagger \hat{a}_{\underline{k}} + \frac{N^2}{2! \, V} \, V_{int}(0)$$

$$+ \frac{N}{2! \, V} \sum_{\underline{k}} V_{int}(\underline{k}) \, (\exp[+2i\varphi] \, \hat{a}_{\underline{k}}^\dagger \hat{a}_{-\underline{k}}^\dagger + \exp[-2i\varphi] \, \hat{a}_{-\underline{k}} \hat{a}_{\underline{k}})$$

$$+ \frac{N}{2! \, V} \sum_{\underline{k}} V_{int}(0) \, (\hat{a}_{\underline{k}}^\dagger \hat{a}_{\underline{k}} + \hat{a}_{-\underline{k}}^\dagger \hat{a}_{-\underline{k}}) + \ldots \quad (8.142)$$

where φ is a constant phase. When the condensate adopts a phase, the continuous $U(1)$ phase symmetry of the Hamiltonian has been spontaneously broken.

The Hamiltonian can be put in diagonal form [66] by using a suitably chosen unitary transformation $\hat{U}$, so

$$\hat{H}' = \hat{U} \hat{H} \hat{U}^\dagger \quad (8.143)$$

The transformed Hamiltonian $\hat{H}'$ has the same spectrum of eigenvalues as the original Hamiltonian $\hat{H}$. We shall choose the transformation to be of the form

$$\hat{U} = \exp\left[-\sum_{\underline{k}} \Theta_k \, (\hat{a}_{\underline{k}}^\dagger \hat{a}_{-\underline{k}}^\dagger \, \exp[+2i\varphi_k] - \hat{a}_{-\underline{k}} \hat{a}_{\underline{k}} \, \exp[-2i\varphi_k]) \right] \quad (8.144)$$

which is unitary when Θ_k is real. The creation operators transform as

$$\alpha_{\underline{k}}^\dagger = \hat{U} \, \hat{a}_{\underline{k}}^\dagger \, \hat{U}^\dagger$$
$$= \cosh \Theta_k \, \hat{a}_{\underline{k}}^\dagger + \sinh \Theta_k \, \exp[-2i\varphi_k] \, \hat{a}_{-\underline{k}} \qquad (8.145)$$

and the annihilation operators are found to transform as

$$\alpha_{\underline{k}} = \hat{U} \, a_{\underline{k}} \, \hat{U}^\dagger$$
$$= \cosh \Theta_k \, a_{\underline{k}} + \sinh \Theta_k \, \exp[+2i\varphi_k] \, a_{-\underline{k}}^\dagger \qquad (8.146)$$

Since the transformation is unitary, it does not affect the canonical conjugate commutation relations

$$[\hat{\alpha}_{\underline{k}}, \hat{\alpha}_{\underline{k}'}^\dagger] = \delta_{\underline{k},\underline{k}'} \qquad (8.147)$$

and

$$[\hat{\alpha}_{\underline{k}}^\dagger, \hat{\alpha}_{\underline{k}'}^\dagger] = 0$$
$$[\hat{a}_{\underline{k}}, \hat{\alpha}_{\underline{k}'}] = 0 \qquad (8.148)$$

The transformed Hamiltonian is expressed as

$$\hat{H}' = \frac{N^2}{2! \, V} \, V_{int}(0)$$
$$+ \sum_{\underline{k}} \left(\frac{\hbar^2 k^2}{2m} + \frac{N}{V} V_{int}(0) \right)$$
$$\times [\cosh^2 \Theta_k \, \hat{a}_{\underline{k}}^\dagger \hat{a}_{\underline{k}} + \sinh^2 \Theta_k \, \hat{a}_{-\underline{k}} \hat{a}_{-\underline{k}}^\dagger]$$
$$+ \frac{N}{V} \sum_{\underline{k}} V_{int}(\underline{k}) \, \sinh \Theta_k \, \cosh \Theta_k \, \cos[2(\varphi - \varphi_k)]$$
$$\times [\hat{a}_{\underline{k}}^\dagger \hat{a}_{\underline{k}} + \hat{a}_{-\underline{k}} \hat{a}_{-\underline{k}}^\dagger]$$
$$+ \sum_{\underline{k}} \left(\frac{\hbar^2 k^2}{2m} + \frac{N}{V} V_{int}(0) \right) \sinh \Theta_k \, \cosh \Theta_k$$
$$\times [\exp[+2i\varphi_k] \, \hat{a}_{\underline{k}}^\dagger \hat{a}_{-\underline{k}}^\dagger + \exp[-2i\varphi_k] \, \hat{a}_{-\underline{k}} \hat{a}_{\underline{k}}]$$
$$+ \frac{N}{2! \, V} \sum_{\underline{k}} V_{int}(\underline{k})(\cosh^2 \Theta_k + \sinh^2 \Theta_k)$$
$$\times [\exp[+2i\varphi] \, \hat{a}_{\underline{k}}^\dagger \hat{a}_{-\underline{k}}^\dagger + \exp[-2i\varphi] \, \hat{a}_{-\underline{k}} \hat{a}_{\underline{k}}] \qquad (8.149)$$

when written in terms of the original creation and annihilation operators. The terms non-diagonal in the particle creation and annihilation operators can be eliminated by the appropriate choice of Θ_k and φ_k. We shall set φ_k

equal to the phase of the condensate, $\varphi_k = \varphi$. Then, the off-diagonal terms vanish if one chooses Θ_k to satisfy

$$\tanh 2\Theta_k = -\frac{\frac{N}{V} V_{int}(\underline{k})}{\frac{\hbar^2 k^2}{2m} + \frac{N}{V} V_{int}(0)} \tag{8.150}$$

With this choice, the Hamiltonian reduces to

$$\hat{H}' = \frac{N^2}{2! \, V} V_{int}(0) + \frac{1}{2} \sum_{\underline{k}} \left(E(\underline{k}) - \frac{\hbar^2 k^2}{2m} - \frac{N}{V} V_{int}(0) \right)$$

$$+ \sum_{\underline{k}} E(k) \, \hat{a}^\dagger_{\underline{k}} \, \hat{a}_{\underline{k}} \tag{8.151}$$

where the first line represents the ground state energy and the second line represents the energy of the elementary excitations. The energy of the elementary excitation $E(\underline{k})$ can be reduced to

$$E(\underline{k}) = \sqrt{\left(\frac{\hbar^2 k^2}{2m} + \frac{N}{V} V_{int}(0) \right)^2 - \left(\frac{N}{V} V_{int}(\underline{k}) \right)^2} \tag{8.152}$$

This dispersion relation vanishes identically at $k = 0$ and is approximately linear for small k,

$$E(\underline{k}) \approx k \sqrt{\frac{\hbar^2}{m} \frac{N}{V} V_{int}(0)} \tag{8.153}$$

where the excitations have the characteristics of phonons. The excitations are the Goldstone modes (See Section 9.8) associated with the broken gauge symmetry.[2] It should be noted that $V_{int}(0)$ must be positive for this solution to be stable. At higher values of k the dispersion relation reduces to

$$E(\underline{k}) \approx \frac{\hbar^2 k^2}{2m} + \frac{N}{V} V_{int}(0) \tag{8.154}$$

which represents the bare particle dispersion relation together with a constant energy shift due to the interaction with the particles in the condensate.

In summary one observes that, due to the interactions with the particles in the condensate, the dispersion of the elementary excitations has changed from quadratic to being linear. This has the experimental consequence that the specific heat changes from being proportional to $T^{\frac{3}{2}}$ at low temperatures

---

[2]Gauge symmetry is broken since each condensate has a particular value for the phase φ, whereas the theory shows that the energy of the Bose-condensed state is independent of φ, where φ is a continuous variable that lies in the range $2\pi > \varphi > 0$. Hence, the Bose-condensate has broken the continuous phase-symmetry of the Hamiltonian.

to having a T^3 variation. More importantly, the change in the dispersion relation due to the interactions allows the Bose-Einstein Condensate to exhibit superfluidity.

8.7.2 *The Coherent Nature of the Ground State*

The transformation of the creation and annihilation operators with non-zero momentum may have obscured the physics, specially since the unitary transformation does not conserve the number of particles. An appropriate generalization to the case of conserved particles is given by

$$\hat{U} = \exp\left[-\sum_{\underline{k}} \frac{\Theta_{\underline{k}}}{N} \left(\hat{a}^{\dagger}_{\underline{k}} \hat{a}^{\dagger}_{-\underline{k}} \hat{a}_0 \hat{a}_0 - \hat{a}^{\dagger}_0 \hat{a}^{\dagger}_0 \hat{a}_{-\underline{k}} \hat{a}_{\underline{k}} \right) \right] \tag{8.155}$$

where, for convenience, we have set the phase of the condensate to zero. The states of the untransformed system can be obtain from those of the transformed system by the inverse transformation

$$|\Psi\rangle = \hat{U}^{\dagger} |\Psi'\rangle \tag{8.156}$$

In the primed frame, the ground state is an eigenstate of the number operator $\hat{n}_{\underline{k}} = \hat{a}^{\dagger}_{\underline{k}} \hat{a}_{\underline{k}}$ with eigenvalue zero. Hence, in the primed frame, the ground state simply corresponds to N bosons in the condensate

$$|\Psi'\rangle = \frac{1}{\sqrt{N!}} (\hat{a}^{\dagger}_0)^N |0\rangle \tag{8.157}$$

where $|0\rangle$ is the vacuum. Since the vacuum satisfies

$$\hat{U}^{\dagger} |0\rangle = |0\rangle \tag{8.158}$$

one finds that the ground state in the un-transformed system is given by

$$\begin{aligned} |\Psi\rangle &= \frac{1}{\sqrt{N!}} \hat{U}^{\dagger} (\hat{a}^{\dagger}_0)^N \hat{U} \hat{U}^{\dagger} |0\rangle \\ &= \frac{1}{\sqrt{N!}} \hat{U}^{\dagger} (\hat{a}^{\dagger}_0)^N \hat{U} |0\rangle \\ &= \frac{1}{\sqrt{N!}} (\hat{U}^{\dagger} \hat{a}^{\dagger}_0 \hat{a}^{\dagger}_0 \hat{U})^{\frac{N}{2}} |0\rangle \end{aligned} \tag{8.159}$$

Thus, the ground state of the condensate has the form of a product of linear superpositions [67]

$$|\Psi\rangle \sim \frac{1}{\sqrt{N!}} \left(\hat{a}^{\dagger}_0 \hat{a}^{\dagger}_0 - \sum_{\underline{k}} \tanh\Theta_k \, \hat{a}^{\dagger}_{\underline{k}} \hat{a}^{\dagger}_{-\underline{k}} \right)^{\frac{N}{2}} |0\rangle \tag{8.160}$$

in which it is seen that the interaction has scattered pairs of bosons out of the condensate. Conservation of momentum shows that the pairs of particles scattered out of the condensate have zero total momentum. Thus, the number of particles with zero momentum is smaller than the number of particles. The ground state is a form of coherent state, in the sense the number of particles in the condensate is large, as is the number fluctuations. It is also seen that the components of the ground state with different numbers of particles in the condensate have definite phase relationships.

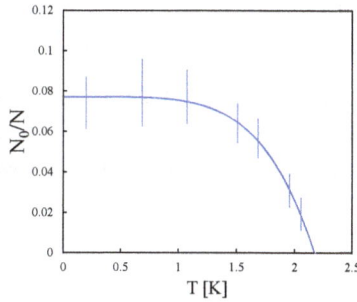

Fig. 8.13   The temperature dependence of the number of particles in the zero momentum states for Helium II (schematic). The condensate fraction at $T = 0$ is reduced from unity. [After E.F. Sears and E.C. Svensson, Phys. Rev. Lett. **43**, 2009, (1979).] [68]

The depletion of the condensate at $T = 0$ can be found more directly by using Nikolai Bogoliubov's approximation. The number of particles in the condensate at $T = 0$ is given by

$$N_0 = N - \sum_{\underline{k} \neq 0} \langle \Psi | \sum_{\underline{k} \neq 0} \hat{a}_{\underline{k}}^\dagger \, \hat{a}_{\underline{k}} \, | \Psi \rangle$$

$$= N - \sum_{\underline{k} \neq 0} \langle \Psi | \, \hat{U}^\dagger \, \hat{U} \sum_{\underline{k} \neq 0} \hat{a}_{\underline{k}}^\dagger \, \hat{a}_{\underline{k}} \, \hat{U}^\dagger \, \hat{U} \, | \Psi \rangle$$

$$= N - \sum_{\underline{k} \neq 0} \sinh^2 \Theta_{\underline{k}} \tag{8.161}$$

The interactions have caused the condensate to be depleted, because the perturbing interactions have virtually mixed the states with zero momentum with pairs of states that have finite $\underline{k}$ values. Thus, even though there may be no excitations present in the $T = 0$ Bose-condensed system, the condensate fraction is less than N. The condensate fraction is evaluated as

$$N_0 = N - \frac{V}{4\pi^2} \left(\frac{2mV_{int}(0) \, N}{\hbar^2 V} \right)^{\frac{3}{2}} I \tag{8.162}$$

where I is a dimensionless quantity given by

$$I = \int_0^\infty dx\; x \left(\frac{(1+x^2)}{\sqrt{2+x^2}} - x \right) \tag{8.163}$$

which is evaluated as $\frac{\sqrt{2}}{3}$. The depletion is found to be proportional to the interaction to the power of $\frac{3}{2}$. The Bogoliubov approximation is only reasonable when the condensate depletion is small. For He^4 with $T \to 0$, the number of particles in the condensate N_0 is approximately $0.1 \times N$.

The states with a single elementary excitation present are proportional to

$$\hat{U}^\dagger\, \hat{a}_{\underline{k}}^\dagger\, \hat{U}\, |\Psi\rangle = (\cosh \Theta_k\, \hat{a}_{\underline{k}}^\dagger - \sinh \Theta_k\, \hat{a}_{-\underline{k}})\, |\Psi\rangle \tag{8.164}$$

Hence, the elementary excitations of the Bose-Einstein condensate are of the form of a linear superposition. The excitations are of the form of a compressional wave. For example, the Fourier Transform of the density operator $\hat{\rho}_q$ is defined as

$$\hat{\rho}_{\underline{q}} = \int d^3\underline{r}\; \exp[i\underline{q} \cdot \underline{r}]\; \hat{\psi}^\dagger(\underline{r})\, \hat{\psi}(\underline{r})$$
$$= \sum_{\underline{k}} \hat{a}_{\underline{k}+\underline{q}}^\dagger\, \hat{a}_{\underline{k}} \tag{8.165}$$

which, at low temperatures, reduces to

$$\hat{\rho}_{\underline{q}} = a_0\, \hat{a}_{\underline{q}}^\dagger + a_0^*\, \hat{a}_{-\underline{q}} \tag{8.166}$$

On performing the Bogoliubov transform, one finds that the excitation is essentially the same as $\hat{a}_{\underline{q}}^\dagger$. Hence, the excitations are of the form of density waves.

Thus, for a Bose-Einstein condensate, the interactions not only produces a change in the dispersion relation of the excitations, but also changes the character of the excitations.

8.7.3 *The Critical Velocity*

The change in the character of the dispersion relation at low energies has the consequence that the condensate of a weakly-interacting Bose-Einstein particles exhibits superfluidity. Superfluidity is the property of flowing through narrow capillaries without exhibiting viscosity. Thus, a superfluid will flow in the absence of any driving forces.

We shall consider the flow of a fluid at zero temperature, flowing with velocity $\underline{v}$ through a capillary. In the primed frame of reference moving with the fluid, the walls of the capillary are moving with velocity $-\underline{v}$. In this primed reference frame, the fluid is considered to initially be in a Bose-Einstein Condensate which carries no momentum. The total energy of the fluid consists of the rest mass energy, Mc^2, and the total momentum is zero. If the viscosity were to be finite, the interaction with the moving capillary walls would cause the fluid to start moving. The change in the state of the initial fluid could only be caused by exciting the internal degrees of freedom of the superfluid. That is, if the viscosity is present, the interaction with the moving capillary walls should produce an excitation in the liquid. In the primed reference frame, the total energy for the condensate in which there is an excitation with momentum $\underline{p}'$ is given by

$$E_T' = Mc^2 + E(\underline{p}') \tag{8.167}$$

In the reference frame where the capillary walls are stationary, this energy is given by the Lorentz transformation

$$E_T = \frac{1}{\sqrt{1 - v/c^2}} \left(E_T' + \underline{v} \cdot \underline{p}' \right)$$

$$\approx Mc^2 + E(\underline{p}') + \underline{p}' \cdot \underline{v} + \frac{M}{2}\, \underline{v}^2 \tag{8.168}$$

and the momentum in the rest frame is given by

$$\underline{p} = \frac{1}{\sqrt{1 - v/c^2}} \left(\underline{p}' + \underline{v}\, \frac{E'}{c^2} \right)$$

$$\approx \underline{p}' + M\underline{v} \tag{8.169}$$

Hence, the energy of the excitation in the stationary reference frame is given by a Doppler shifted value

$$\Delta E = E(\underline{p}') + \underline{p}' \cdot \underline{v} \tag{8.170}$$

and its momentum is $\underline{p}'$. The excitation energy must be negative, if the excitation is to be allowed. The rationale for this is that, since the capillary is at rest and at $T = 0$, it cannot provide the energy necessary to create a positive-energy excitation. On the other hand, since the fluid is moving, it can lose energy by reducing its state of motion and dissipate the excess energy through the creation of positive-energy excitations in the capillary's walls. Hence, we require that the excitation energy is negative

$$E(\underline{p}') + \underline{p}' \cdot \underline{v} < 0 \tag{8.171}$$

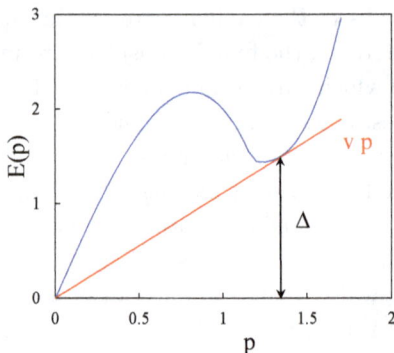

Fig. 8.14   The schematic energy dispersion of HeII, exhibiting phonon like dispersion at low p and the roton minimum with energy Δ. The construct for determining the critical velocity is shown by the red line.

It is possible to satisfy this criterion if $\underline{p}'$ is anti-parallel to $\underline{v}$. After the excitation has occurred, the liquid is expected to have slowed down.[3]

The criterion for dissipation to occur in the fluid is that

$$E(\underline{p}') + \underline{p}' \cdot \underline{v} < 0 \tag{8.173}$$

On assuming v and p' are oppositely directed, the criterion for viscous flow reduces to

$$v > \min E(p')/p' \tag{8.174}$$

for some value of p'. The critical velocity v_c is defined as

$$v_c = \min E(p')/p' \tag{8.175}$$

The "*Landau Criterion*" [69] states that superflow occurs when $v_c > v > 0$ and viscous flow occurs when $v > v_c$. Geometrically, the critical velocity is

---

[3]If one considers the recoil of the fluid, one may derive the criterion entirely in the laboratory frame. In the laboratory frame, the total energy and momentum of the state with the excitation present are given by

$$E_T = \frac{M}{2}(\underline{v} - \Delta \underline{v})^2 + E(\underline{p})$$
$$\underline{P}_T = M(\underline{v} - \Delta \underline{v}) + \underline{p}$$

where the fluid has slowed down by Δv. If one enforces conservation of momentum, $\Delta \underline{P}_T = 0$, and assumes that M is large, then

$$\Delta E_T = E(\underline{p}) - \underline{p} \cdot \underline{v} \tag{8.172}$$

If ΔE_T is negative, the excitation will form spontaneously as the fluid evolves to its equilibrium state.

the minimum value of the slope of a line from the origin to a point p_0 on the curve $E(p)$. This point is given by the solution of

$$\frac{E(p_0)}{p_0} = \frac{dE(p)}{dp}\bigg|_{p_0} \tag{8.176}$$

Any Bose-Einstein condensate with a parabolic dispersion relation $E(p)$ cannot exhibit superflow, since $p_0 = 0$. Therefore, for a Bose-Einstein condensate with a parabolic dispersion relation, $v_c = 0$ so any flow is viscous. For He^4, the theoretically calculated and experimentally measured dispersion relation $E(p)$ is linear at small p but exhibits a minimum at some finite value of p, which is known as the roton minimum. The roton minima is caused by the strong interactions between the particles which are almost strong enough to result in the formation of a crystal. If a crystal were to be formed, the "phonon" dispersion relation should have a periodicity determined by the lattice structure. In fact Feynman[4] has shown that the dispersion relation can be written in terms of the geometric structure factor $S(p)$ via

$$E(p) = \frac{p^2}{2mS(p)} \tag{8.177}$$

For large p, the structure factor has a maximum at a momentum p_0 given by $p_0 \sim \frac{\hbar}{d}$ where d is the inter-particle spacing. The maximum in the structure factor gives rise to the observed roton minimum. However, the observed roton minimum gives a critical velocity of 60 m/sec. This velocity is up to an order of magnitude greater than the critical velocities observed in He II. Feynman and Cohen [70] has shown that the critical velocity is very close to that expected for a single vortex ring.

8.8 The Superfluid Velocity and Vortices

A spatially varying condensate can be characterized by a spatially varying wave function $\psi(r)$ which is given by the creation amplitude for the particles in the condensate. Then the total number of particles in the condensate N_0 is given by

$$N_0 = \int d^3r \, |\psi(r)|^2 \tag{8.178}$$

---

[4]Feynman used the Rayleigh-Ritz principle, using density waves as an ansatz for the excitations. The normalization factor in the denominator is recognized as the structure factor and, after expressing the numerator as a double commutator of $\hat{H}$ with $\hat{\rho}_k^\dagger$ and $\hat{\rho}_k$, the numerator reduces to the kinetic energy.

The spatial variation of the condensate can be described by a phenomeno-logical Landau-Free-Energy

$$F[\psi, \psi^*] = \int d^3\underline{r} \left[\frac{\hbar^2}{2m} |\underline{\nabla} \psi|^2 + V(\underline{r}) |\psi|^2 + V_{int} |\psi|^4 \right] \qquad (8.179)$$

where the last term represents a localized interaction between the parti-cles in the condensate. Minimization of the Free-Energy w.r.t ψ^* yields a Schrödinger-like equation with non-linear terms. Conservation of particles requires that the condensate density ρ and current density $\underline{j}$ must satisfy the continuity equation

$$\left(\frac{\partial \rho}{\partial t} \right) + \underline{\nabla} \cdot \underline{j} = 0 \qquad (8.180)$$

where

$$\rho(\underline{r}) = |\psi|^2 \qquad (8.181)$$

and

$$\underline{j}(\underline{r}) = \frac{\hbar}{2mi} \left(\psi^* \underline{\nabla}\psi - \underline{\nabla}\psi^* \psi \right) \qquad (8.182)$$

If the amplitude of the condensate wave function varies slowly compared to the phase, one may write

$$\psi(\underline{r}) = \sqrt{\frac{N_0}{V}} \exp[i\varphi(\underline{r})] \qquad (8.183)$$

Hence, one finds that the condensate's current density is given by

$$\underline{j}(\underline{r}) = \frac{\hbar}{m} \frac{N_0}{V} \nabla\varphi \qquad (8.184)$$

This allows one to define a superfluid velocity $\underline{v}_s$ via

$$\underline{v}_s = \frac{\hbar}{m} \nabla\varphi \qquad (8.185)$$

which is governed by the spatial variation of the phase of the condensate.

The circulation of the superfluid velocity field is given by the line integral around a closed loop inside the superfluid

$$\oint d\underline{r} \cdot \underline{v}_s = \frac{\hbar}{m} \oint d\underline{r} \cdot \underline{\nabla}\varphi \qquad (8.186)$$

This equation relates the superfluid circulation to the change of the phase of the condensate wave function at the end point of the loop. Since the condensate wave function must be single-valued, the phase at any point must be defined up to a multiple of 2π. Thus, from continuity of the wave function, the circulation must be quantized [71]

$$\oint d\underline{r} \cdot \underline{v}_s = \frac{\hbar}{m} \, n2\pi \qquad (8.187)$$

where n is an integer, $n = 0, \pm 1, \pm 2, \ldots$. The quantization of circulation is a manifestation of the quantum nature of the superfluid. If one now considers a loop in the condensate that is simply connected and shrinks the size of the loop to zero, one finds that the only possible value of the phase quantum number n is zero. Hence, the condensate must be irrotational. On the other hand, if the loop is in a multiply connected region of the condensate, then one may not be able to shrink the loop to zero in which case the quantum numbers n can be non-zero. Vortices with low numbers of circulation quanta are preferred energetically. Thus, the circulation of the superfluid can be non-zero when the loop encloses regions in which the condensate density is zero. This analysis has the experimental consequence that if one starts to rotate a cylindrical vessel containing a superfluid, the superfluid liquid will initially remain at rest. However, if the angular velocity is increased above a critical value, excitations consisting of normal regions called vortices are introduced into the superfluid. A vortex consists of a one-dimensional region of normal fluid (i.e. non-superfluid) that is oriented parallel to the axis of rotation and extends throughout the entire length of the cylinder [See Fig. 8.15]. The vortices are topological excitations. An increase in the angular velocity of the cylinder will result in an increase in the number of vortices that penetrate the superfluid, and the vortex lines will form a two-dimensional array [72]. In a non-rotating superfluid, it is possible to have vortices, if the vortices close up on themselves in loops.

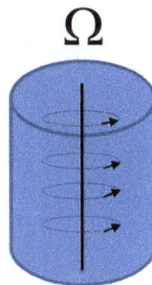

Fig. 8.15 A single vortex excitation in a superfluid contained in a cylinder rotating with angular velocity Ω.

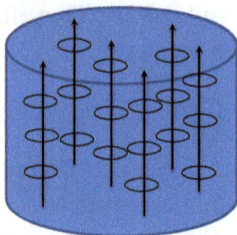

Fig. 8.16   The configuration of vortex excitations in a superfluid contained within a rotating cylinder.

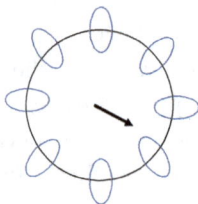

Fig. 8.17   A vortex ring moving through a superfluid.

Such vortex rings have the form of "smoke rings" [See Fig. 8.17]. In a flowing liquid, large vortex rings can be created before the critical velocity for the creation of rotons is reached. Hence, the critical velocity of a superfluid is usually determined by the vortices. If the areal dimension of the system is reduced below the size of the vortex rings, the critical velocity is bounded from above by the critical velocity inferred from the roton excitations.

8.9 Problems

Problem 8.1

Consider a system containing a fixed finite number of bosons N in equilibrium at temperature T, which has two single-particle energy levels, one with energy 0 and the other with energy $\epsilon > 0$.

(i) Find the partition function Z_N in the Canonical Ensemble.

(ii) Show that the average total occupation number of the state with zero energy, $\overline{N}(0)$ is given by

$$N(0) = \frac{N+1}{1 - \exp[-\beta\epsilon(N+1)]} - \frac{1}{1 - \exp[-\beta\epsilon]}$$

(iii) Calculate Z_N and $N(0)$ using Maxwell-Boltzmann Statistics for indistinguishable particles, and plot $N(0)/N$ for Bose-Einstein and Maxwell Statistics as a function of T.

Problem 8.2

Consider an ideal gas of bosons confined in a volume V. The bosons have the dispersion relation

$$\epsilon(\underline{p}) = \frac{p^2}{2m}$$

(i) Show that the Grand-Canonical potential Ω is given by

$$\beta\Omega = V \int \frac{d^3 p}{(2\pi\hbar)^3} \ln(1 - \exp[-\beta \, (\epsilon(\underline{p}) - \mu)])$$

(ii) Expand $\beta\Omega$ in powers of $\exp[\beta\mu]$ and determine the first three non-zero coefficients in the expansion. Hence, find the first two corrections to the ideal gas law.

(iii) Use the definition of the Grand-Canonical partition function Ξ in terms of the N-particle quantity Z_N

$$\Xi = \sum_{N=0}^{\infty} Z_N \, \exp[\beta\mu N]$$

to determine the leading-order corrections to the N-particle partition function from its classical value

$$Z_N \approx \frac{Z_1^N}{N!} \, [1 + \ldots]$$

Problem 8.3

Consider a gas of identical bosons which undergo collisions, $\alpha, \beta \to \alpha', \beta'$. The interaction can be written as

$$\hat{H}_{int} = \frac{1}{2!} \sum_{\alpha,\beta,\alpha',\beta'} \langle \alpha', \beta' | \, \hat{V} | \beta, \alpha \rangle \, a^\dagger_{\alpha'} \, a^\dagger_{\beta'} \, a_\beta \, a_\alpha$$

(i) Using the Fermi-Golden rule, derive an expression for the rate at which particles in a single particle quantum state α with energy ϵ_α will be scattered out of that state. Express your result in terms of the probabilities p_α that a single-particle quantum state will be occupied by a particle.

(ii) Using the Fermi-Golden rule, determine an expression for the rate at which particles will be scattered into state α.

(iii) Show that the "Principle of Detailed Balance" holds if the p_α are replaced by the Bose-Einstein distribution functions $N(\epsilon_\alpha)$.

Problem 8.4

It is postulated that the entropy of a system, S, is given by the expression

$$S = -k_B \sum_\alpha [n_\alpha \ln n_\alpha - (1 + n_\alpha) \ln (1 + n_\alpha)]$$

where n_α is the average occupation number of a single-particle state with energy ϵ_α. Determine the form of n_α subject to the constraints that the average energy of the system is E and that the average number of particles is N.

Problem 8.5

Consider an gas of non-interacting bosons in which the single-particle energies are given by ϵ_α. The gas is held at temperature T and at a chemical potential μ.

(i) Find the joint probability $P(\{n_\alpha\})$ for finding a system in the microscopic state specified by the set of single-particle occupation numbers $\{n_\alpha\}$.

(ii) Find the moment generating function $M_\alpha(T)$ and cumulant generating function $K_\alpha(t)$ for the single-particle state α. Hence, determine the average occupation $\bar{n}_\alpha$ and the mean squared fluctuation of the occupation number.

(iii) Express the joint probability distribution function in terms of $\bar{n}_\alpha$. Hence, evaluate the Shannon entropy S

$$S = -k_B \sum_{\{n_\alpha\}} P(\{n_\alpha\}) \ln P(\{n_\alpha\})$$

Problem 8.6

(i) Using Maxwell's relations, show that

$$\left(\frac{\partial U}{\partial N} \right)_{T,V} = \mu - T \left(\frac{\partial \mu}{\partial T} \right)_{N,V}$$

$$= \frac{\partial}{\partial \beta} (\ln z)$$

where z is the fugacity.

(ii) Show that, in the Bose-Condensed phase, the change in energy for particle addition is given by

$$\left(\frac{\partial U}{\partial N} \right)_{T,V} = 0$$

(iii) Determine an expression for the density of bosons for temperatures above the condensation temperature. Hence show that, above the transition

$$\left(\frac{\partial U}{\partial N} \right)_{T,V} \geq 0$$

Problem 8.7

(i) Show that the specific heat satisfies the thermodynamic relations

$$C_V = VT \left(\frac{\partial^2 P}{\partial T^2} \right)_V - NT \left(\frac{\partial^2 \mu}{\partial T^2} \right)_V$$

$$C_p = -NT \left(\frac{\partial^2 \mu}{\partial T^2} \right)_V$$

(ii) Are these relations are satisfied for an ideal gas of bosons?

Problem 8.8

Consider an ideal gas of bosons confined in a volume $V = L^d$, with dispersion relation

$$\epsilon(p) = \alpha p^s$$

(i) Show that the occupation number fluctuations of the boson gas satisfy

$$\overline{\Delta n(\underline{p})^2} = \overline{n(\underline{p})}\left(\overline{n(\underline{p})} + 1\right)$$

(ii) Show that the occupation number fluctuations in the lowest level diverge as $(k_B T)^2 \, L^{2s}$ when $L \to \infty$.

Problem 8.9

Show that for temperatures just above the critical value T_c for Bose-Einstein condensation, the chemical potential of a three-dimensional Bose-gas can be expressed as

$$\mu \sim -k_B T_c \left[\left(\frac{T_c}{T}\right)^{\frac{3}{2}} - 1\right]^2 \left[\frac{\zeta(3/2)\,\Gamma(3/2)}{\pi}\right]^2$$

Problem 8.10

Consider an ideal gas of bosons with mass m in three dimensions. Derive an approximation for the discontinuity in the temperature derivative of the specific heat at the condensation temperature T_c

$$\Delta\left(\frac{\partial C_V}{\partial T}\right)\bigg|_{T_c}$$

Problem 8.11

Consider an ideal gas of N bosons confined within a d-dimensional volume $V = L^d$. The bosons have a dispersion relation

$$\epsilon(p) = \alpha p^s$$

for $s > 0$. The bosons can also be trapped in a bound state with energy $\epsilon = -\Delta$.

(i) For what value of μ will the bound state be occupied by a macroscopic number of bosons?

(ii) Show that the number of bosons in the continuum states, N_e, can be expressed as

$$\frac{N_e}{V} = \frac{S_d\,\Gamma\left(\frac{d}{s}\right)}{s}\left(\frac{(k_B T/\alpha)^{\frac{1}{s}}}{2\pi\hbar}\right)^d \sum_{n=1}^{\infty} \frac{\exp[n\beta\mu]}{n^{\frac{d}{s}}}$$

(iii) Hence show that, for finite Δ, the gas will condense at a finite temperature for any value of $\frac{d}{s}$. Contrast your result with the limit $\Delta \to 0$.

Problem 8.12

Although a two-dimensional ideal gas of bosons of mass m does not Bose-Einstein condense in the thermodynamic limit, the lowest level may have a macroscopic occupation if the total number of particles N is finite.

(i) Show that there is no limit on the number of bosons that can occupy the excited states at any temperature T since

$$N_e = -\frac{A}{\lambda^2} \ln\left(1 - \exp\left[\frac{\mu}{k_B T}\right]\right)$$

and N_e diverges when $\mu \to 0$.

(ii) Calculate the temperature T_1 at which the lowest-energy single-particle state is occupied by $\frac{N}{10}$ particles.

Problem 8.13

Show that density $\rho_1(\underline{r})$ for non-interacting bosons of mass m in a large three-dimensional volume V is given by the expression

$$\rho_1(\underline{r}) = \frac{1}{V} \sum_{\underline{k}} \exp[i\underline{k} \cdot \underline{r}] \, N(\epsilon(\underline{k}))$$

where $N(x)$ is the Bose-Einstein distribution function.

(i) Find an integral expression for the limit when $T \to T_c$ where T_c is the Bose-Einstein condensation temperature.

(ii) Show that, as $r \to \infty$, and at $T = T_c$

$$\lim_{r \to \infty} \rho_1(r) \to \left(\frac{m k_B T_c}{2\pi \hbar^2}\right) \frac{1}{r}$$

Problem 8.14

Assume that the energy of a bosonic particle in a gravitational field g is adequately described by the W.K.B. expression

$$E(n, p_\perp) = \frac{p_\perp^2}{2m} + mg \left(\frac{9\pi^2 \hbar^2}{8m^2 g}\right)^{\frac{1}{3}} \left(n + \frac{3}{4}\right)^{\frac{3}{2}}$$

where $n = 0, 1, 2 \ldots$ and $p_\perp$ is the momentum in the plane perpendicular to the direction of g.

(i) Calculate the single-particle density of states $\rho(E)$.

(ii) Calculate the Grand-Canonical potential for an ideal gas of bosons which is confined in a cylinder of cross-sectional area A and is subject to a gravitational field.

(iii) Determine expressions for the internal energy U and the number of particles N in the system.

(iv) Determine the temperature, T_c, at which the gas will condense.

Problem 8.15

Consider an ideal gas of N bosons in an isotropic d-dimension harmonic trap

$$V(\underline{r}) = \frac{m\omega_0^2}{2} \sum_{i=1}^{d} q_i^2$$

(i) Show that the degeneracy of a single-particle state with Q quanta in it is given by

$$\mathcal{D}_Q = \binom{d+Q-1}{d}$$

(ii) Determine an expression for the Grand-Canonical potential Ω.

(iii) Determine an expression for the average number of particles N in terms of μ.

(iv) Assume that Q is large w.r.t. d, so one may replace the sums over Q by integrals. Find the temperature, T_c, for which

$$\mu \approx \frac{d}{2} \hbar\omega_0$$

(v) Calculate the number of particles in the state $Q = 0$ for $T_c > T$.

Problem 8.16

Rb^{87} was reported to Bose Condense [73]. The gas was trapped in a three-dimensional harmonic potential with frequency $\omega_0 \approx 750$ Hz.

(i) Evaluate an expression for the density of states $\rho(E)$.

(ii) Find expressions for the average energy E and the average number of particles N at T_c.

(iii) Find an expression for the condensation temperature in terms of the number of particles N and $\hbar\omega$.

For a number density of 2.5×10^{12} per cm^3, the observed condensation temperature was $T_c \approx 170$ nK.

(iv) Using the average value of the energy, estimate the density and, hence, determine T_c. Compare your result for T_c with the experimental value.

Problem 8.17

A relativistic gas of bosonic particles and antiparticles of mass m is held in a d-dimensional volume V and is in thermal equilibrium at temperature T. The excess number of bosons over anti-bosons is in N.

(i) Determine the condition on μ for the bosons to Bose-Einstein condense.
(ii) Determine the condensation temperature T_c. You may express your equation for T_c in terms of a power series.

Problem 8.18

An ideal gas of N bosons of mass m and spin $S = 1$ is contained in a volume V. The gas is subjected to an applied magnetic field B.

(i) Find an expression for the magnetic susceptibility, χ, for $T > T_c$. Determine the high-temperature variation of χ.
(ii) Show that the magnetic susceptibility χ diverges at $T = T_c$.
(iii) By considering the effect of an infinitesimal magnetic field, determine the spontaneous magnetization for $T < T_c$.

Problem 8.19

The problem is to show that the specific heat of a two-dimensional ideal gas of bosons with mass m is identical to the specific heat of an two-dimensional ideal gas of fermions with the same mass. (May's Theorem.) Since the Bose gas with a quadratic dispersion relation does not condense, one has to find a relationship between the integrals for the number of particles in the excited states (and also for the energies) for both types of statistics.

(i) Evaluate the integrals

$$F_\nu(x) = \frac{1}{\Gamma(\nu)} \int_0^\infty dx \, \frac{x^{\nu-1}}{z_F^{-1} \exp[x] + 1}$$

$$G_\nu(x) = \frac{1}{\Gamma(\nu)} \int_0^\infty dx \, \frac{x^{\nu-1}}{z_B^{-1} \exp[x] - 1}$$

for $\nu = 1$.

(ii) Hence, show that the fugacity of the Fermion gas z_F is related to the fugacity of the Boson gas z_B by

$$1 + z_F = \frac{1}{1 - z_B}$$

(iii) By integrating by parts, prove that

$$\frac{\partial G_\nu(z)}{\partial z} = \frac{G_{\nu-1}(z)}{z}$$

(iv) Show that

$$F_2(z_F) = \int_0^{z_F} dz \, \frac{\ln(1 + z)}{z}$$

and derive a similar expression for $G_2(z_B)$.
Thus, show that

$$F_2(z_F) - G_2(z_B) = \frac{1}{2} \ln^2(1 + z_F)$$

(v) By combining the results, show that the fermionic internal energy $U_F(T)$ only differs from the bosonic internal energy $U_B(T)$ by a constant. Therefore, show that the specific heats are equal.

Problem 8.20

Consider a non-relativistic ideal gas of spin one-half fermions of mass m confined in a two-dimensional area A at temperature T. Opposite spin fermions can combine and form a boson of mass $2m$ with binding energy $-\Delta$.

(i) Find expressions for the number of free fermions, N_F and the number of bosons, N_B.

(ii) The fermions and bosons are in chemical equilibrium. Determine the relation between the chemical potentials of the fermions and bosons.

(iii) If the initial number of fermions is N, and since the fermion number is conserved

$$N = N_F + 2N_B$$

Find an expression for the number of bosons at low temperatures.

Problem 8.21

The partition function, Z_D, for N distinguishable non-interacting particles with energy levels ϵ_α is given by

$$Z_D = \sum_{\{n_\alpha\}} \left(\frac{N!}{\prod_\alpha n_\alpha!} \right) \exp\left[-\beta \sum_\alpha \epsilon_\alpha \, n_\alpha \right]$$

$$= \left(\sum_\alpha \exp[-\beta\epsilon_\alpha] \right)^N$$

Therefore, the Free-energy is found as

$$-\beta F_D = \ln Z_D$$

$$= N \, \ln \left(\sum_\alpha \exp[-\beta\epsilon_\alpha] \right)$$

The Grand-Canonical Partition function for a gas of particles obeying Quantum Statistics is given by

$$\Xi_Q = \exp[-\beta\Omega]$$

with

$$-\beta\Omega = \pm \sum_\alpha \ln(1 \pm \exp[-\beta(\epsilon_\alpha - \mu)])$$

However, the Grand-Canonical Potential is related to the Free-Energy via the Legendre Transformation

$$\Omega = F - \mu N$$

Hence, one has the relation between the Grand-Canonical and the Canonical Partition functions

$$\ln \Xi_Q = \ln Z_Q + \beta\mu N$$

In particular, one finds

$$\pm \sum_\alpha \ln(1 \pm \exp[-\beta(\epsilon_\alpha - \mu)]) = \ln Z_Q + \beta\mu N$$

This is an exact relation for particles which satisfy either Fermi-Dirac or Bose-Einstein Statistics.

Show that, in the high-temperature limit, the partition functions Z_Q of gasses which obey either Fermi-Dirac or Bose-Einstein Statistic are related to the partition function of a gas of distinguishable particles, Z_D, by

$$Z_Q \to \frac{Z_D}{N!}$$

This relation shows that the factor of $\frac{1}{N!}$ in the partition for indistinguishable particles is mandated by Quantum Statistics.

Chapter 9

Phase Transitions

In 1944 Lars Onsager [74] published the exact solution of the two-dimensional Ising Model in zero-field on a square lattice. His exact solution was a *tour de force* of mathematical physics. Onsager used the transfer matrix technique to describe a finite square lattice of size $L \times L$ and then diagonalized the matrix by finding the irreducible representations of a related matrix algebra. The exact solution demonstrated that, in thermodynamic limit, the system exhibited a phase transition which is marked by singularities in the physical properties. The form of the Hamiltonian of the Ising Model is given by

$$\hat{H} = -\sum_{i,j} J_{i,j} S_i^z S_j^z \tag{9.1}$$

where i labels each spin on the lattice and where $J_{i,j} = J$ if i and j are on neighboring lattice sites and $J_{i,j} = 0$ otherwise. The spin variables in units of $\frac{\hbar}{2}$ can only have the allowed values $S_i^z = \pm 1$. We shall assume $J > 0$ which favors states where the neighboring S^z values have the same sign. Onsager found that the exact partition function is given by the expression

$$\frac{\ln Z}{N} = \ln 2 + \frac{a^2}{2(2\pi)^2} \int_{-\frac{\pi}{a}}^{\frac{\pi}{a}} dk_x \int_{-\frac{\pi}{a}}^{\frac{\pi}{a}} dk_y$$
$$\times \ln[\cosh^2(2\beta J) - \sinh(2\beta J) (\cos k_x a + \cos k_y a)] \tag{9.2}$$

The argument of the logarithm is non-negative and the integral exists for all values of βJ. For $J > 0$, the minimum value of the argument occurs for $\underline{k} = 0$, and is given by

$$\cosh^2(2\beta J) - 2 \sinh(2\beta J) = (1 - \sinh(2\beta J))^2 \tag{9.3}$$

The non-analytic behavior of F occurs at the temperature when this minimum value vanishes. This gives rise to the identification of the critical temperature as the solution of the equation

$$\sinh(2\beta J) = 1 \qquad (9.4)$$

or equivalently

$$\tanh \beta_c J = \sqrt{2} - 1 \qquad (9.5)$$

For temperatures below the critical temperature, the system is in a ferromagnetic state. The non-analyticity originates with the long wavelength behavior of the integral, and can be found by approximating the integral by

$$\frac{\ln Z}{N} \approx \ln 2 + \frac{a^2}{2(2\pi)} \int_0^{\frac{\pi}{a}} dk\ k$$
$$\times \ln\left[(1 - \sinh(2\beta J))^2 + \frac{1}{2}\ \sinh(2\beta J)(ka)^2\right]$$
$$\approx \ln 2 + \frac{1}{(4\pi)} \int_0^{\frac{\pi^2}{2}} dx\ \ln[(1 - \sinh(2\beta J))^2 + \sinh(2\beta J)\ x] \qquad (9.6)$$

This yields the expression for the non-analytic part of the Free-Energy

$$\frac{\ln Z}{N} \sim -\frac{(1 - \sinh(2\beta J))^2}{4\pi\ \sinh(2\beta J)}\ [\ln(1 - \sinh(2\beta J))^2 - 1] \qquad (9.7)$$

Hence, the specific heat is found to diverge logarithmically at the transition temperature

$$\frac{C}{N} \sim -k_B\ \frac{8}{\pi}\ (\beta_c J)^2\ \ln|T - T_c| \qquad (9.8)$$

which is symmetrical around the transition temperature. Onsager stated without proof that, for temperatures below T_c, the zero-field limit of the magnetization defined by the average value

$$M = \sum_{i=1}^{N} \langle S_i^z \rangle \qquad (9.9)$$

varies as

$$M = N(1 - \sinh^{-4}(2\beta J))^{\frac{1}{8}} \qquad (9.10)$$

The proof of this last result was eventually published by C.N. Yang [75]. Onsager's success is of great historical importance and is without parallel, since to date, no exact solutions have been found for models of similar physical importance, such as the three-dimensional Ising model, or such as the two or three-dimensional versions of the Heisenberg model. Onsager's results provided the only rigorous treatment of a phase transition, until three decades later when the renormalization group technique was finally formulated.

9.1 Phase Transitions and Singularities

The non-analytic behavior of the Free-Energy that is seen in the vicinity of phase transition is, at first sight, quite disturbing since this implies that the partition function is also anomalous. This is unexpected since the partition function is defined as

$$Z(\beta) = \text{Trace } \exp[-\beta \hat{H}] \tag{9.11}$$

which is understood to be the sum of a finite number of positive terms and, therefore, is an analytic function of β (except possibly when $\beta \to \infty$). Furthermore, the sum of these terms exponentiates to yield a Free-Energy $F_N(\beta)$ which has an extensive part (proportional to N) and other parts which become negligible when $N \to \infty$. This would lead us to expect that the Free-Energy is also an analytic function. However, the sequence of analytic functions $F_N(\beta)$ need not be analytic in the limit $N \to \infty$.

Just to make this explicit, consider an Ising Spin Hamiltonian, with long-ranged pair-wise interactions

$$\hat{H} = -J \sum_{i>j} S_i^z S_j^z \tag{9.12}$$

where $S_i^z = \pm \frac{1}{2}$. Any spin interacts with every other spin with the same interaction strength. The Hamiltonian can be re-written as

$$\hat{H} = -\frac{J}{2} \left[\left(\sum_{i=1}^{N} S_i^z \right)^2 - \sum_{i=1}^{N} (S_i^z)^2 \right]$$

$$= -\frac{J}{2} \left[\left(\sum_{i=1}^{N} S_i^z \right)^2 - \frac{N}{4} \right] \tag{9.13}$$

Strictly speaking, it is necessary to formally define $J = J'/N$, if we were to demand that the Free-Energy be extensive. We shall ignore this simply to avoid carrying around extra factors of N. If J is positive, the ground state corresponds to a ferromagnetic state for which

$$\sum_{i=1}^{N} S_i^z = \pm \frac{N}{2} \qquad \text{for } J > 0 \tag{9.14}$$

The ferromagnetic state is two-fold degenerate. On the other hand if $J < 0$ the ground state (when N is even) corresponds to

$$\sum_{i=1}^{N} S_i^z = 0 \qquad \text{for } J < 0 \tag{9.15}$$

which is highly-degenerate. It is easy to show that the degeneracy is given by $C_{\frac{N}{2}}^N$. The partition function $Z_N(\beta)$ can be expanded in powers of a parameter z defined as

$$z = \exp\left[\frac{\beta J}{8}\right] \tag{9.16}$$

For negative J, z is reduced from 1 to 0 when T is reduced from ∞ to 0. For positive J, z increases from 1 to ∞ when T is reduced from ∞ to 0. The partition function is found to be given by

$$Z_N(\beta) = \exp\left[-N\,\frac{\beta J}{8}\right] \sum_{m=0}^{N} C_m^N\, z^{(N-2m)^2}$$

$$= z^{-N} \sum_{m=0}^{N} C_m^N\, z^{(N-2m)^2} \tag{9.17}$$

where C_m^N are the binomial coefficients. This expression contains a factor which is a very high-order polynomial in z. The polynomial has no roots on the positive real axis except, perhaps, at the point $z = 0$. However, it does have pairs of complex conjugate roots in the complex z plane. The roots may be multiple roots. For our model, it is seen that the pairs of roots are located on circles enclosing the origin $z = 0$. As z approaches a point

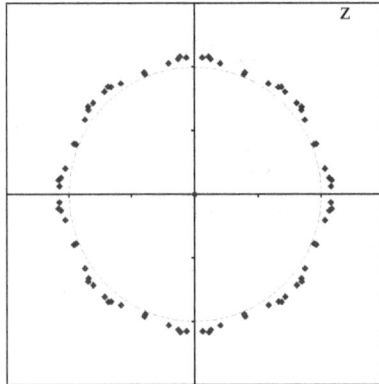

Fig. 9.1   The distribution of zeroes of the partition function $Z_9(z)$ for the Ising Model with long-ranged interactions, in the complex z-plane. The dashed blue circle has a radius of unity.

which is a root, the partition function approaches zero and the Free-Energy $F_N(z)$ diverges logarithmically.

In general, the partition function is expected to have the form

$$Z_N(z) = \exp[-NA(z)] \prod_\alpha (z - z_\alpha)(z - z_\alpha^*) \qquad (9.18)$$

where z_α and z_α^* are the pairs of complex conjugate roots in the complex z plane, and $A(z)$ is a simple function. The Free-Energy $F_N(z)$ is given by

$$F_N(z) = k_B \, TNA(z) - k_B \, T \sum_\alpha \ln(z - z_\alpha)(z - z_\alpha^*) \qquad (9.19)$$

which has singularities in the complex z plane. Lee and Yang [76,77] proved that the limit

$$\lim_{N \to \infty} \frac{1}{N} F_N(z) = -k_B \, T \lim_{N \to \infty} \frac{1}{N} \ln Z_N(z) \qquad (9.20)$$

exits for all real positive z and is a continuous monotonically increasing function of z. Also, for any region Ω which does not contain any roots of $Z_N(z)$ then $\lim_{N \to \infty} \frac{1}{N} F_N(z)$ is analytic in this region. If these conditions are satisfied for all physical values of z, the system does not exhibit a phase transition.

As the limit $N \to \infty$ is approached, the zeroes of $Z_N(z)$ may approach the real axis and pinch it off at a real value of z, z_c. The conditions of the Yang-Lee Theorem do not apply in the immediate vicinity of this point. If the zeroes approach a point z_c on the real axis continuously as N is increased, then the point z_c may be located on a branch cut of $F(z)$ which would yield non-analytic behavior at z_c or equivalently at β_c. In such a case, z_c would defines a critical temperature T_c at which the Free-Energy is singular.

Example: Yang-Lee Zeros for the Two-Dimensional Ising Model

Lars Onsager found an exact expression for the partition function Z_N for a two-dimensional Ising model on a square lattice with N-sites. The partition function can be expanded as

$$Z_N = 2^N (\cosh \beta J)^N \left[1 + \sum_{n=2} S_n \tanh^{2n} \beta J \right]$$

where S_n are numerical coefficients corresponding to closed polygons. The term in the square brackets is a polynomial in $z = \tanh \beta J$. The polynomial has a number of roots in the complex z-plane. The distribution of roots cuts the real axis of the complex z-plane in the limit $N \to \infty$. The exact

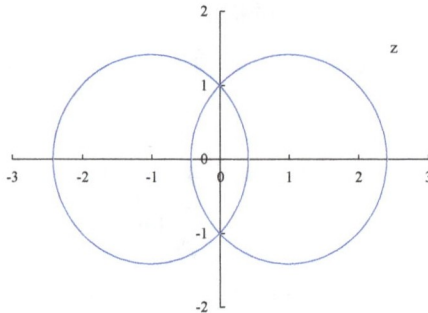

Fig. 9.2 In the limit $N \to \infty$ the zeroes of the two-dimensional Ising model on a square lattice are found for complex values of $z = \tanh \beta J$ on two circles of radius $\sqrt{2}$ centered ± 1.

expression for Z_N has a distribution of zeros which are on two circles of radius $\sqrt{2}$, centered on the points $(-1, 0)$ and $(+1, 0)$. Therefore, the zeros of the partition function cross the real z-axis at $\pm(\sqrt{2} \mp 1)$. Since $\tanh \beta J$ is real and less than unity, the only physically accessible zero, z_c, is given by

$$\tanh \beta J = \sqrt{2} - 1$$

at which temperature the Free-Energy is non-analytic. The temperature associated with this real value of z_c is the transition temperature T_c for the model.

Example: Yang-Lee Zeros of the Antiferromagnetic Triangular Lattice

The Ising Model on a triangular lattice has an exact solution [78, 79]. The antiferromagnetic Ising model on a triangular lattice is highly frustrated. It can be shown that the lowest energy, $U(0)$, is given by

$$U(0) = -2N|J|$$

and corresponds to states in which one-third of the interactions give a positive contribution to U. However, this does not correspond to a unique state since a large number of different spin configurations have the same energy. In fact, if the triangular lattice is partitioned into three sublattices, the spins on two sublattices can be energetically optimized but the spins on the third sublattice cannot. The spins on the third lattice can be assigned randomly without changing the energy, giving a contribution of $k_B N/3 \ln 2$ to the entropy. Thus, the antiferromagnetic system is disordered and has a

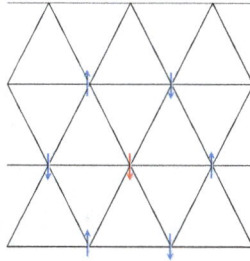

Fig. 9.3 A configuration of spins on a triangular lattice that produces a macroscopic residual $T = 0$ entropy.

finite $T = 0$ value of the entropy given by

$$S(0) = Nk_B \ \frac{3}{\pi} \ \int_0^{\frac{\pi}{6}} d\phi \ \ln(2 \ \cos \phi) \sim 0.3383 \ k_B \ N$$

This violates the third-law of thermodynamics. The system is critical in that its correlations do not decay exponentially.

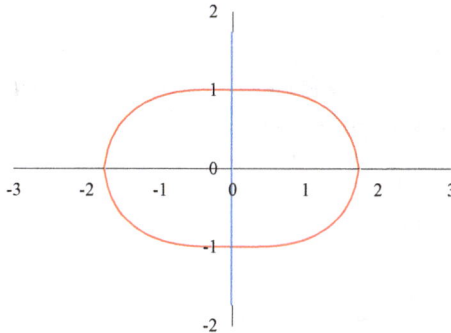

Fig. 9.4 The zeroes of the partition function for the Ising model on a triangular lattice.

The partition function $Z_N(\beta J)$ for the Ising Model on a triangular lattice has zeroes as a function of the complex parameter

$$z = \exp[2\beta J]$$

The zeroes are on two curves, one being a line segment on the imaginary z-axis between $z = \pm i\sqrt{3}$ and the second being a closed ellipse-like curve

$$\Im m \ z^2 = -(1 + \Re e \ z^2) + 2 \ \sqrt{1 + \Re e \ z^2}$$

For $J > 0$, the physical region of z is between $z = \infty$ and $z = 1$. The ellipse crosses the positive real axis at $z_c = \sqrt{3}$, signalling a ferromagnetic phase transition. For antiferromagnetic interactions $J < 0$, the physical region of z is between $1 > z > 0$. The straight-line segment crosses the real-axis at $z = 0$. Hence, the antiferromagnetic Ising Model on a triangular lattice has a $T = 0$ phase transition.

9.2 The Mean-Field Approximation: Ising Model

Consider a spin S Ising Hamiltonian in the presence of an applied magnetic field H^z

$$\hat{H} = -\sum_{i,j} J_{i,j} S_i^z S_j^z - \sum_i \frac{g\mu_B}{\hbar} S_i^z H^z \tag{9.21}$$

The operator S^z has $(2S + \hbar)/\hbar$ possible eigenvalues which are $-S$, $-S + \hbar$, ..., $S - \hbar$, S. The interaction $J_{i,j}$ couples the z-components of nearest-neighbor spins. We shall assume that the interaction J is short-ranged and takes on the same positive value between each pair of nearest-neighboring spins, so that the lowest energy configuration is ferromagnetic in which all the spins are aligned parallel to each other. Although the Hamiltonian has an extremely simple form, the only known exact expressions for the Free-Energy have been found for the special cases where the spins are arranged on one or two-dimensional lattices [71, 74]. Therefore, we shall have to describe this system approximately by using the mean-field approximation, first introduced by Weiss.

We shall define the average magnetization per spin as m and express the Hamiltonian as

$$\hat{H} = -\sum_{i,j} J_{i,j}(m + (S_i^z - m))(m + (S_j^z - m))$$
$$- \sum_i \frac{g\mu_B}{\hbar}(m + (S_i^z - m)) H^z \tag{9.22}$$

and expand in power of the fluctuations of the spins $(S_i^z - m)$ from their average value. To first-order in the fluctuations, one has

$$\hat{H}_{MF} = \sum_{i,j} J_{i,j} m^2$$
$$- \sum_{i,j} J_{i,j} m(S_i^z + S_j^z) - \sum_i \frac{g\mu_B}{\hbar} S_i^z H^z \tag{9.23}$$

where we have neglected the terms of second-order in the fluctuations. One should note that the above Hamiltonian resembles that expected for non-interacting spins in an effective magnetic field H_{eff}, given by

$$H^z_{eff} = H + \frac{\hbar}{g\mu_B} \sum_j J_{i,j} m \tag{9.24}$$

The mean-field partition function Z can be calculated as

$$Z_{MF} = \text{Trace } \exp[-\beta H_{MF}] \tag{9.25}$$

where the Trace runs over all the possible spin configurations. Thus, the Trace corresponds to the products of sums over the $(2S/\hbar + 1)$ possible configuration of each spin. Since the spins are no longer coupled, the mean-field Hamiltonian factorizes. Hence, the partition function has the form

$$Z_{MF} = \exp\left[-\beta \sum_{i,j} J_{i,j} m^2\right] \prod_{i=1}^N \left\{ \sum_{S^z_i = -S}^{+S} \exp\left[+\beta \frac{g\mu_B}{\hbar} S^z_i H^z_{eff}\right] \right\} \tag{9.26}$$

The Trace can be performed yielding the result

$$Z_{MF} = \exp\left[-\beta \sum_{i,j} J_{i,j} m^2\right] \left[\frac{\sinh \frac{(2S+\hbar)}{\hbar} \left(\frac{\beta g\mu_B H_{eff}}{2}\right)}{\sinh \left(\frac{\beta g\mu_B H_{eff}}{2}\right)} \right]^N \tag{9.27}$$

Hence, in the mean-field approximation, the Free-Energy is given by

$$F_{MF}(m) = \sum_{i,j} J_{i,j} m^2 - Nk_B T \ln\left[\sinh \frac{(2S+\hbar)}{\hbar} \left(\frac{\beta g\mu_B H_{eff}}{2}\right)\right]$$
$$+ Nk_B T \ln\left[\sinh \left(\frac{\beta g\mu_B H_{eff}}{2}\right)\right] \tag{9.28}$$

The magnetization is found from the thermodynamic relation

$$M^z = -\left(\frac{\partial F}{\partial H^z}\right) \tag{9.29}$$

which yields

$$M^z = Ng\mu_B \left[\frac{2S+\hbar}{2\hbar} \coth \frac{(2S+\hbar)}{\hbar} \left(\frac{\beta g\mu_B H_{eff}}{2}\right)\right.$$
$$\left. - \frac{1}{2} \coth \left(\frac{\beta g\mu_B H_{eff}}{2}\right)\right] \tag{9.30}$$

On recognizing that

$$M^z = \frac{g\mu_B}{\hbar} \sum_i \overline{S^z}_i \tag{9.31}$$

one finds that the average value of S^z is independent of the site and is given by

$$
\overline{S_0^z} = \left[\frac{2S + \hbar}{2} \coth \frac{(2S + \hbar)}{\hbar} \left(\frac{\beta g \mu_B H_{eff}}{2} \right) \right.
$$
$$
\left. - \frac{\hbar}{2} \coth \left(\frac{\beta g \mu_B H_{eff}}{2} \right) \right] \tag{9.32}
$$

or, equivalently, on using the definition of m as the average value of the z-component of the spin

$$
m = \left[\frac{2S + \hbar}{2} \coth \frac{(2S + \hbar)}{\hbar} \left(\frac{\beta g \mu_B H^z + \beta \sum_j J_{j,0} \hbar m}{2} \right) \right.
$$
$$
\left. - \frac{\hbar}{2} \coth \left(\frac{\beta g \mu_B H^z + \beta \sum_j J_{j,0} \hbar m}{2} \right) \right] \tag{9.33}
$$

This non-linear equation determines the value of m. This equation is known as the self-consistency equation, since the equation for m has to be solved self-consistently since m enters non-linearly in the right-hand side. The equation can be solved graphically.

For $H^z = 0$ the two spin directions are equivalent, and the equation simplifies to

$$
m = \left[\frac{2S + \hbar}{2} \coth \frac{(2S + \hbar)}{\hbar} \left(\frac{\beta \sum_j J_{j,0} \hbar m}{2} \right) \right.
$$
$$
\left. - \frac{\hbar}{2} \coth \left(\frac{\beta \sum_j J_{j,0} \hbar m}{2} \right) \right] \tag{9.34}
$$

Both the left and right hand sides are odd functions of m. This symmetry is a consequence of the symmetry under spin inversion $S_i^z \to -S_i^z$ of the Hamiltonian, when $H^z = 0$. The graphical solution is illustrated in Figs. 9.5 and 9.6. At high temperatures, the equation has only one solution $m = 0$, whereas at low temperatures, there are three solutions, one solution corresponds to $m = 0$ and the other two solutions corresponds to $m = \pm m_0(T)$ located symmetrically about $m = 0$. The magnetization serves m as the order-parameter for the transition. The value of $m_0(T)$ increases continuously from 0 and saturates at S as T is decreased towards zero. The "*Critical Temperature*", T_c at which the pair of non-zero solutions first appears can be found by expanding the right-hand side w.r.t. m, since it is expected that $m \approx 0$ just below T_c. This leads to the equation

$$
m \approx \beta \frac{2}{3} \sum_j J_{j,0} S(S + \hbar) m + O(m^3) \tag{9.35}
$$

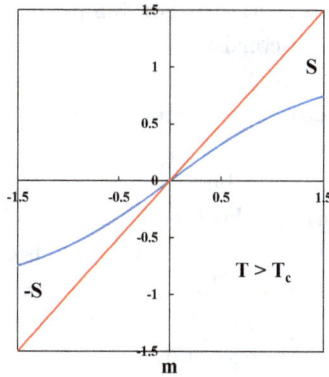

Fig. 9.5 The graphical solution of the mean-field self-consistency equation, for temperatures above T_c.

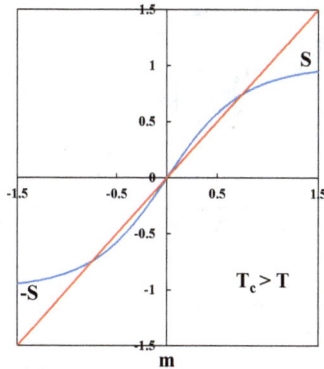

Fig. 9.6 The graphical solution of the mean-field self-consistency equation, for temperatures below T_c.

On assuming that the cubic terms are negligibly small, this has the solution $m = 0$ unless the temperature is equal to T_c which satisfies

$$k_B \ T_c = \frac{2}{3} \sum_j J_{j,0} S(S + \hbar) \tag{9.36}$$

This equation determines the mean-field approximation for the critical temperature T_c. It is a poor approximation to the true T_c since it may be significantly reduced by the effect of fluctuations which are neglected in mean-field approximation. At the critical temperature T_c, a non-zero (but still infinitesimal) value of $m_0(T)$ is first allowed. At this temperature, the graphical solution shows that the line $m = m$ is tangent to the magnetization curve at the origin. For temperatures just below T_c, the cubic terms in m in the self-consistency equation determine the magnitude of the solution m_0 to be a function of $(T_c - T)$. At the critical temperature, the order parameter is zero and the Free-Energy of the possible phases are equal.

Below T_c, the system must be described by one of the three possible values for m. In equilibrium, the Free-Energy should be minimized. The condition that $F_{MF}(m)$ is an extremum w.r.t. m is equivalent to the above self-consistency condition for m. In Fig. 9.7, it is seen that when the non-zero solutions of m exist, the Free-Energy is minimized at these values of m, and that the minima are degenerate.

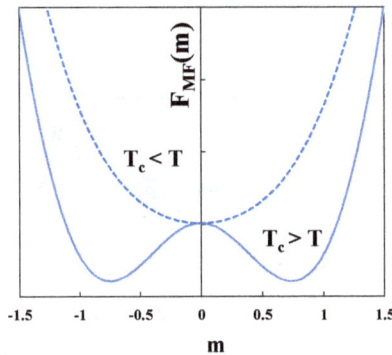

Fig. 9.7   The m-dependence of the mean-field Free-Energy, for temperatures above and below T_c.

The system will spontaneously break the spin inversion symmetry, by settling into one of the two equivalent states. In order for the system to select only one of the two states, it has to be subjected to an infinitesimal magnetic field that breaks the symmetry. In the limit $H^z \to 0$, the mean-field approximation indicates that there is a "*Second-Order Phase Transition*" at T_c. The transition occurs at the temperature at which the location of the absolute minimum value of $F_{MF}(m)$ changes from $m = 0$ for temperatures above T_c to the value of $F_{MF}(m)$ evaluated at one of the locations $m = \pm m_0(T)$ for temperatures below T_c. The magnetization $m(T)$, which is a first-order derivative of the Free-Energy, changes continuously when the temperature is reduced through T_c.

The mean-field approximation shows the existence of another type of phase transition, which occurs when the small symmetry breaking applied magnetic field H^z is varied. In this case, the applied field stabilizes one of

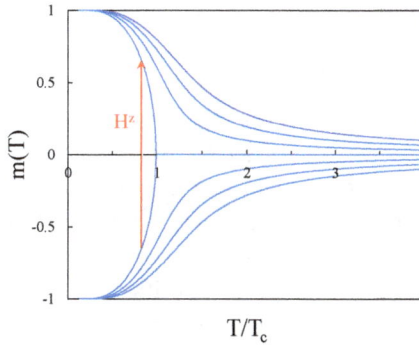

Fig. 9.8 The T-dependence of the magnetization $m(T)$ for various values of the applied field H^z. For $T < T_c$ the magnetization jumps discontinuously as H^z is increased from $-\eta$ to $+\eta$.

the minima (say m_0) w.r.t. to the minima at $-m_0$. However, on changing the applied field from a slightly positive to a slightly negative value, the equilibrium value of m jumps discontinuously from $+m_0$ to $-m_0$ [See Fig. 9.8]. This discontinuous jump between two pre-existing minima of $F(m)$ is characteristic of a *"First-Order Phase Transition"*. The defining

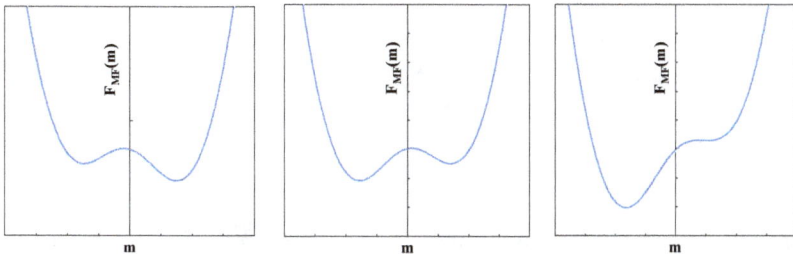

Fig. 9.9 The m-dependence of the mean-field Free-Energy, for temperatures below T_c and various values of the applied field.

characteristic of a first-order transition is the presence of discontinuous first-derivatives of the Free-Energy, such as the magnetization [81]. Ehrenfest proposed a classification of phase transitions based on the discontinuities of the derivatives of the Free-Energy. A transition was defined to be n-th order transition if the lowest-order derivative of the Free-Energy which was discontinuous across the transition was the n-th order derivative. This classification fell into disfavor when it was realized that a derivative could diverge without it being discontinuous. Since the entropy is also a first-derivative of the Free-Energy with respect to temperature, first-order transitions usually involve a latent heat. Because the two phases are distinct, and because equilibration between a part of a system and its environment is not instantaneous, first-order transitions are frequently associated with "*Mixed-Phase Regimes*" in which some parts of the system have completed the transition and others have not. A related characteristic is the presence of "*Hysteresis*". For systems where the transformation proceeds slowly, a phase may exist at fields where its Free-Energy is not a global minimum but only a local minimum. Hence, on decreasing the field, the transformation may first occur at a negative value of the field. Furthermore, if the field is subsequently increased, the reverse transformation may occur at a positive value of the field. The point at which the transformation occurs is determined by the rate at which the field is changed and the time-scale required for the system to nucleate the new phase.

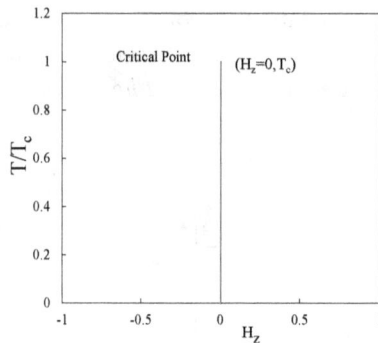

Fig. 9.10   The $H - T$ phase diagram for the mean-field description of a ferromagnet.

The phase diagram in the $H - T$ plane, shows a line of first-order transitions at low-temperatures which ends at the "*Critical Point*" ($H = 0, T = T_c$). Due to the symmetry of the magnetic system, the line of first-order transitions is vertical. On keeping $H = 0$, the system exhibits a second-order phase transition at the critical point.

Example: Stability and the Susceptibility

The mean-field approximation for the Free-Energy $F'(m, H)$ in the presence of a field is actually a Gibbs Free-Energy and $F(m, H = 0)$ is the Helmholtz

Free-Energy, since they are related by the Legendre Transform

$$F'(m, H) = F(m) - mH \tag{9.37}$$

Therefore, the magnetization is given by

$$m = -\left(\frac{\partial F'}{\partial H}\right) \tag{9.38}$$

The extrema of F' are given by

$$\frac{\partial F'}{\partial m} = \frac{\partial F}{\partial m} - H$$
$$= 0 \tag{9.39}$$

so F' is independent of m. Furthermore, since the stable states are minima of F', they satisfy

$$\frac{\partial^2 F'}{\partial m^2} > 0$$
$$\frac{\partial^2 F}{\partial m^2} > 0 \tag{9.40}$$

whereas the unstable state would be characterized by a negative second derivative.

The susceptibility can then be evaluated as

$$\chi = -\left(\frac{\partial^2 F'}{\partial H^2}\right)$$
$$= -\frac{\partial}{\partial H}\left[\left(\frac{\partial F'}{\partial H}\right) + \left(\frac{\partial F'}{\partial m}\right)\left(\frac{\partial m}{\partial H}\right)\right]$$
$$= \left(\frac{\partial m}{\partial H}\right) \tag{9.41}$$

where we have used the fact that F' is independent of H. Since, in the mean-field approximation, the magnetization is given by a solution of the equation

$$m = \left[\frac{(2S+1)}{2}\coth\frac{(2S+1)}{2}\beta(zJm + H) - \frac{1}{2}\coth\frac{1}{2}\beta(zJm + H)\right] \tag{9.42}$$

one finds that the zero-field susceptibility χ is given by

$$\chi(m) = \left(\frac{\partial m}{\partial H}\right)\Bigg|_{H=0}$$
$$= \frac{\chi_0(m)}{1 - zJ\chi_0(m)} \tag{9.43}$$

where

$$\chi_0(m) = \beta \left[\frac{(2S+1)^2}{4} \sinh^{-2} \frac{(2S+1)}{2} \beta z J m - \frac{1}{4} \sinh^{-2} \frac{1}{2} \beta z J m \right]$$

(9.44)

which is an even function of m. In the paramagnetic state where $m = 0$, $\chi_0(0)$ reduces to the Curie susceptibility

$$\chi_0(0) = \beta \frac{S(S+1)}{3}$$

(9.45)

therefore, the paramagnetic susceptibility $\chi(0)$ reduces to the Curie-Weiss susceptibility which diverges when

$$k_B T_c = z J \frac{S(S+1)}{3}$$

(9.46)

The value of T_c is the critical temperature. For temperatures below T_c, the susceptibility with $m = 0$ is negative. From consideration of the conditions for thermodynamic stability, we conclude that the paramagnetic state is unstable below T_c. The stability of the solution with magnetization $\pm m_0$ in the magnetic state, can be established by examining the value of $\chi(m_0)$

$$\chi(m_0) = \frac{\chi_0(m_0)}{1 - z J \chi_0(m_0)}$$

(9.47)

For example, at $T = 0$ the magnetization is given by $m = \pm S$, so

$$\chi_0(S) \sim \beta S^2 \exp[-2\beta z J S^2]$$

(9.48)

The susceptibility in the ferromagnetic state is found to be positive, therefore, the ferromagnetic state is stable.

For zero applied field, the states with the two non-zero solutions for m are degenerate as both solutions have the same Free-Energy. The application of an infinitesimal field δH lifts the degeneracy, which if δH is increased from a negative to a positive value causes a level crossing at which the magnetization switches discontinuously from $-m_0$ to m_0. This is a first-order transition since m is defined as a first-order derivative of F wrt H. The $m - T - H$ phase diagram is shown in Fig. 9.11. For $T < T_c$, there can be phase separation. The boundaries of the coexistence region is shown in red. In the $H = 0$ plane, the experimentally observed magnetization m is given by a linear superposition of m_0 and $-m_0$ with weights associated with the volumes of each phase. Because

$$0 = F(m_0) - F(-m_0)$$
$$= \int_{-m_0}^{m_0} dm \, H(m)$$

(9.49)

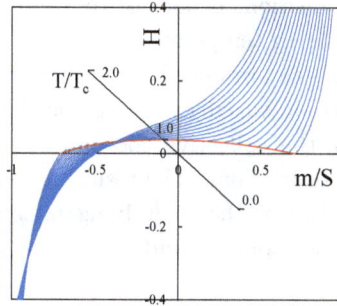

Fig. 9.11   The Mean-Field $m - T - H$ Phase Diagram. The red line in the $H = 0$ plane demarcates the boundaries of the coexistence region.

the boundary of the coexistence region is equivalent to that obtained with the Maxwell equal-area construction, which is based on the stability criterion

$$\left(\frac{\partial m}{\partial H}\right)_T > 0 \tag{9.50}$$

Phase transitions are found in many different types of system. Despite the differences between the microscopic descriptions, phase transitions can usually be described in the same manner. For example, for the liquid-gas phase transition the role of the magnetization is replaced by the density and the magnetic field is replaced by the pressure, as is shown in Fig. 9.12.

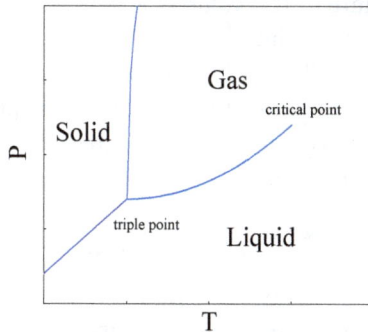

Fig. 9.12   The $P - T$ phase diagram for the liquid-gas transition. The liquid and gas are separated by a line of first-order transitions which ends at a critical point.

The line of first-order transitions is not horizontal but has a finite slope so it can be crossed by changing temperature.

The various transitions can be described in similar manners. In a unified description, the microscopic variables are replaced by *"Coarse-Grained"* variables that represent the collective coordinates of a large number of the microscopic degrees of freedom. The resulting description only retains the essential characteristics of the underlying microscopic systems, such as the symmetry, the dimensionality and the character of the ordering.

9.3   The Landau-Ginzburg Free-Energy Functional

The Landau-Ginzburg formulation of critical phenomena is based on a coarse grained version of statistical mechanics, in which the microscopic degrees of freedom, pertaining to some small region of space, have been replaced by a small set of collective variables that describe the state of the volume element. This leads to a formulation of the statistical mechanics of the system in terms of the collective variables for each small volume element. It is common to introduce an *"Order Parameter"* ϕ defined for the microscopic d-dimensional volume element $d^d\underline{r}$ surrounding the point $\underline{r}$. The order parameter characterizes the change in the system that occurs at the phase transition. For a system which changes symmetry, the order parameter is an extensive quantity (for example it could be a scalar quantity or it could be a vector quantity with a number of components that we will denote by n, etc.) which is non-zero in the state with lower-symmetry and is zero in the state with higher-symmetry. For simplicity, we shall only consider the case when one type of symmetry is broken, in which case, only one order parameter is required. The volume of the system is to be partitioned into a set of identical infinitesimal volumes (or cells) $d^d\underline{r}$ that surrounding a set of points that we shall label by the variable $\underline{r}$. The order parameter $\phi(\underline{r})$ can be expressed as the sum of terms of microscopic quantities ϕ_i that are expressed in terms of the microscopic degrees of freedom contained in the volume element $d^d\underline{r}$ surrounding the point $\underline{r}$

$$\phi(\underline{r}) = \sum_{i \in \underline{r}} \phi_i \qquad (9.51)$$

The partition function Z is expressed as the Trace over the Hamiltonian

$$Z = \text{Trace } \exp[-\beta\hat{H}] \qquad (9.52)$$

The "*Landau-Ginzburg Free-Energy Functional*" $F[\phi(\underline{r})]$ is a number that depends on the function $\phi(\underline{r})$ and is expressed as in terms of a Trace

$$\exp[-\beta F[\phi(\underline{r})]] = \text{Trace} \prod_{\underline{r}} \left\{ \delta\left(\phi(\underline{r}) - \sum_{i \in \underline{r}} \phi_i \right) \right\} \exp[-\beta \hat{H}] \quad (9.53)$$

where the product contains a delta function which constrains the microscopic variables in the volume elements around each point $\underline{r}$ to be consistent with the value of $\phi(\underline{r})$ at that point. Hence, the partition function can be expressed as an integral over the possible values of $\phi(\underline{r})$ for each cell labeled by $\underline{r}$.

$$Z = \prod_{\underline{r}} \left\{ \int d\phi(\underline{r}) \right\} \exp[-\beta F[\phi]] \quad (9.54)$$

This is recognized as a functional integral, and it should be noted that the set of possible "functions" defined by the values of $\phi(\underline{r})$ at each point of space $\underline{r}$ include many wild functions that change discontinuously from point to point and also includes functions vary smoothly over space. The functional integral over the set of all possible functions $\phi(\underline{r})$ is weighted exponentially by the Landau-Ginzburg Free-Energy Functional. The functional integral is conventionally denoted by

$$Z = \int \mathcal{D}\phi \, \exp[-\beta F[\phi]] \quad (9.55)$$

The Landau-Ginzburg Free-Energy plays the role of a Hamiltonian, which generally depends on T, that describes the physical probabilities in terms of the collective variables $\phi(\underline{r})$ described on the length scale dictated by the choice of the size of the volume elements $d^3\underline{r}$. It contains all the physics that is encoded in the Helmholtz Free-Energy F. Like the Hamiltonian, the Landau-Ginzburg Free-Energy Functional is a scalar. In principle, the Landau-Ginzburg Free-Energy Functional should be calculated from knowledge of the model and its symmetries. In practice, one can understand properties of phase transitions in a quite universal way close to a second-order phase transition or a weakly first-order transition, where the order parameter is quite small. In such cases, one can expand the Landau-Ginzburg Free-Energy Functional in powers of the order parameter, keeping only terms of low-order. The constraints imposed by stability and symmetry on the finite number of terms retained, provides severe restrictions on the form of the Landau-Ginzburg Free-Energy Functional that describes the phase transition of a system. This severe restriction causes all the different phase transitions of physical systems to fall into a small number of "*Universality*

Classes", which are determined only by the symmetry, dimensionality of the system d, the dimensionality of the order parameter n and the range of the interactions. Systems which fall into the same universality class have the same types of non-analytic temperature variations.

For example, a system residing in a d-dimensional Euclidean space which is characterized by an n-dimensional vector order parameter $\phi \equiv (\phi_1, \phi_2 \ldots \phi_n)$ and has a Hamiltonian which is symmetric under rotations of the order parameter, can be described by the expanded Landau-Ginzburg Free-Energy Functional

$$F[\phi] = \int d^d\underline{r} \left[F_0 + F_2 \, \underline{\phi}(\underline{r}) \cdot \underline{\phi}(\underline{r}) + F_4 \, (\underline{\phi}(\underline{r}) \cdot \underline{\phi}(\underline{r}))^2 - \underline{\Pi}(\underline{r}) \cdot \underline{\phi}(\underline{r}) \right.$$

$$\left. + \sum_{i=1}^{n} c(\nabla\phi_i \cdot \nabla\phi_i) \right] \tag{9.56}$$

In this expression F_0, F_2, F_4 and c are constants, that might depend on temperature and may also depend on the microscopic length scales of the system. If the above expansion is to describe stable systems that have small values of the order parameter, it is necessary to assume that $F_4 > 0$. The Free-Energy Functional has been expressed in terms of quantities that are invariant under the symmetries of space and the order parameter. The invariant quantities include the identity and the scalar product

$$\underline{\phi}(\underline{r}) \cdot \underline{\phi}(\underline{r}) = \sum_{i=1}^{n} \phi_i(\underline{r}) \, \phi_i(\underline{r}) \tag{9.57}$$

The first three terms represent the Free-Energy density for the cells, in the absence of an external field. Since the material is assumed to be homogeneous, the coefficients F_0, F_2 and F_4 are independent of $\underline{r}$. The fourth term represents the effect of a spatially varying applied external field $\underline{\Pi}(\underline{r})$ that is conjugate to $\underline{\phi}(\underline{r})$. The application of the field breaks the symmetry under rotations of the order parameter. The final term represents the interaction between neighboring cells, which tends to suppress rapid spatial variations of the order parameter and, hence, gives large weights to the functions $\underline{\phi}(\underline{r})$ which are smoothly varying. The gradient term involves two types of scalar products, one type is associated with the d-dimensional scalar product of the gradients and the other is associated with an n-dimensional scalar product of the vector order parameter. The appearance of the gradient is due to the restriction to large length scales in the Landau-Ginzburg formulation.

In this case, expressions such as

$$\kappa \sum_{\underline{\delta}} (\underline{\phi}(\underline{r} + \underline{\delta}) - \underline{\phi}(\underline{r}))^2 \tag{9.58}$$

which tend to keep the value of ϕ in the cell $\underline{r}$ close to the values of ϕ in the neighboring cells at $\underline{r} + \underline{\delta}$ can be expanded, leading to

$$\kappa \sum_{\underline{\delta}} \sum_{i=1}^{n} (\phi_i(\underline{r} + \underline{\delta}) - \phi_i(\underline{r}))^2 \approx \kappa \sum_{i=1}^{n} \sum_{\underline{\delta}} (\underline{\delta} \cdot \underline{\nabla} \, \phi_i)^2$$

$$= c \sum_{i=1}^{n} (\underline{\nabla} \, \phi_i \cdot \underline{\nabla} \, \phi_i) \tag{9.59}$$

where we have assumed that the higher-order terms in the small length scale δ are negligibly small and that the neighboring cells are distributed isotropically in space. This assumption of isotropic space and slow variations leads to the Landau-Ginzburg Functional having a form similar to the Lagrangians of continuum Field Theories. Apart from the coefficients F_0, F_2, F_4 and c, the form of the Lagrangian only depends on the symmetry and the values of n and d. However, for systems which undergo more than one type of phase transition, it may be necessary to introduce more than one order parameter, in which case the Landau-Ginzburg Free-Energy functional can have a more complicated form.

9.3.1 *Linear Response Theory*

Here, we shall calculate the response of a system described by a Landau-Ginzburg field theory to a weak static perturbation. The perturbation is described as an applied external field which couples to the order parameter.

For simplicity, we shall consider the case where the order parameter is a scalar. In general, if a system is subject to a time-independent uniform applied field with an additional small (perhaps non-uniform) component $\delta\Pi(\underline{r})$, so that

$$\Pi(\underline{r}) = \Pi_0 + \delta\Pi(\underline{r}) \tag{9.60}$$

then one expects that the additional small component of the field will induce a small additional (non-uniform) static component into the expectation value of the local order-parameter $\langle \phi(\underline{r}) \rangle$

$$\langle \phi(\underline{r}) \rangle = \phi_0 + \delta\phi(\underline{r}) \tag{9.61}$$

The average of the order-parameter is defined by

$$\langle \phi(\underline{r}) \rangle = \frac{1}{Z} \int \mathcal{D}\phi \, \phi(\underline{r}) \, \exp[-\beta F[\phi]] \tag{9.62}$$

where the Trace has been replaced by a functional integral, and the Hamiltonian H has been replaced by the Landau-Ginzburg Free-Energy Functional $F[\phi]$. The Landau-Ginzburg Free-Energy Functional includes both the uniform applied field and the small (non-uniform) component. On expanding the exponent and the partition function in powers of $\delta\Pi(\underline{r})$, one has

$$\langle\phi(\underline{r})\rangle = \frac{\int \mathcal{D}\phi \ \phi(\underline{r})[1 + \beta \int d^d\underline{r}' \ \phi(\underline{r}') \ \delta\Pi(\underline{r}')] \ \exp[-\beta F[\phi]_{\delta\Pi=0}]}{\int \mathcal{D}\phi[1 + \beta \int d^d\underline{r}' \ \phi(\underline{r}') \ \delta\Pi(\underline{r}')] \ \exp[-\beta F[\phi]_{\delta\Pi=0}]} \qquad (9.63)$$

The divisor, Z^{-1}, is expanded to lowest non-trivial order as

$$\frac{1}{\int \mathcal{D}\phi[1 + \beta \int d^d\underline{r}' \ \phi(\underline{r}') \ \delta\Pi(\underline{r}')] \ \exp[-\beta F[\phi]_{\delta\Pi=0}]}$$

$$= \frac{1}{Z_{\delta\Pi=0}}$$

$$- \frac{\int \mathcal{D}\phi[\beta \int d^d\underline{r}' \ \phi(\underline{r}') \ \delta\Pi(\underline{r}')] \exp[-\beta F[\phi]_{\delta\Pi=0}]}{Z^2_{\delta\Pi=0}} \qquad (9.64)$$

Since the uniform part of the order parameter satisfies

$$\phi_0 = \frac{\int \mathcal{D}\phi \ \phi(\underline{r}) \ \exp[-\beta F[\phi]_{\delta\Pi=0}]}{Z_{\delta\Pi=0}} \qquad (9.65)$$

then the small additional (non-uniform) component of the order parameter is given by the expression

$$\delta\phi(\underline{r}) = \beta \int d^d\underline{r}' \ \delta\Pi(\underline{r}') \left[\frac{\int \mathcal{D}\phi \ \phi(\underline{r}) \ \phi(\underline{r}') \ \exp[-\beta F[\phi]_{\delta\Pi=0}]}{Z_{\delta\Pi=0}} \right.$$

$$\left. - \frac{\int \mathcal{D}\phi \ \phi(\underline{r}) \ \exp[-\beta F[\phi]_{\delta\Pi=0}]}{Z_{\delta\Pi=0}} \frac{\int \mathcal{D}\phi \ \phi(\underline{r}') \ \exp[-\beta F[\phi]_{\delta\Pi=0}]}{Z_{\delta\Pi=0}} \right] \qquad (9.66)$$

where the integration over the additional part of the applied field $\delta\Pi(\underline{r}')$ has been taken out of the averages. The above equation can be written in a more compact form as

$$\delta\phi(\underline{r}) = \beta \int d^d\underline{r}' \ [\langle\phi(\underline{r}) \ \phi(\underline{r}')\rangle - \langle\phi(\underline{r})\rangle\langle\phi(\underline{r}')\rangle] \ \delta\Pi(\underline{r}') \qquad (9.67)$$

in which the averages are calculated with $\delta\Pi = 0$. On defining the two-point correlation function $S(\underline{r}, \underline{r}')$ as

$$S(\underline{r}, \underline{r}') = \langle\phi(\underline{r}) \ \phi(\underline{r}')\rangle - \langle\phi(\underline{r})\rangle\langle\phi(\underline{r}')\rangle \qquad (9.68)$$

then the small induced component of the order-parameter is given by

$$\delta\phi(\underline{r}) = \beta \int d^d\underline{r}' \ S(\underline{r}, \underline{r}') \ \delta\Pi(\underline{r}') \qquad (9.69)$$

which is a linear response relation which connects the small change in the order-parameter $\delta\phi$ at position $\underline{r}$ to the change in the applied field $\delta\Pi$ at position $\underline{r}'$. For translational invariant systems, the correlation function does not depend separately on $\underline{r}$ and $\underline{r}'$, but only on the relative separation $\underline{r} - \underline{r}'$. For materials which are translational invariant, one can displace the origin through a distance $\underline{r}'$ leading to the expression

$$S(\underline{r} - \underline{r}') = \langle \phi(\underline{r} - \underline{r}')\,\phi(0)\rangle - \langle \phi(\underline{r} - \underline{r}')\rangle\langle \phi(0)\rangle \tag{9.70}$$

which only depends on the difference $\underline{r} - \underline{r}'$. In the limit where $\delta\Pi$ becomes uniform one finds that, due to translational invariance, $\delta\phi$ becomes uniform and is given by

$$\delta\phi = \beta \int d^d\underline{r}'\ S(-\underline{r}')\ \delta\Pi \tag{9.71}$$

Due to the isotropy of space, the induced value of $\delta\phi$ can be expressed as

$$\delta\phi = \beta \int d^d\underline{r}'\ S(\underline{r}')\ \delta\Pi \tag{9.72}$$

Hence, the uniform differential susceptibility, χ, defined by

$$\chi = \left(\frac{\delta\phi}{\delta\Pi}\right) \tag{9.73}$$

is found to be proportional to the spatial integral of the correlation function

$$\chi = \beta \int d^d\underline{r}\ S(\underline{r}) \tag{9.74}$$

where the correlation function $S(r)$ is calculated with $\delta\Pi = 0$.

In general, linear response theory relates the response of a system to a perturbation. The system could require either a quantum mechanical or a classical description. The perturbation, and therefore the response, could be either time-dependent or time-independent. In the above description the system was described by a coarse-grained Landau-Ginzburg Free-Energy Functional that should be considered as a classical field theory, therefore, the linear response was treated classically. Also, since the perturbation was a static applied field, only a static version of linear response theory was used. A more complete description of linear response theory can found in Chapter 11.

9.4   Critical Phenomena

The simplest phase diagram exhibiting a phase transition, is a line of first-order transitions which ends in a critical point. If the line of first-order transitions is traversed along its length towards the critical point, the system is approaching a second-order phase transition. The order parameter (which is given by a first-order derivative of the Free-Energy w.r.t. to an applied field Π) is discontinuous when the system is taken on a path which crosses the line of first-order transitions. The difference in the order parameter, from the opposite side of the line, characterizes the difference between the two phases. The magnitude of the discontinuity diminishes for paths that cross the line of first-order transition closer to the critical point. This signifies that there is no discernable difference between the phases at the critical point. Above the critical point, only a unique phase exists and the order parameter is zero.

Phenomena that happen at a critical point are known as *"Critical Phenomena"*. Close to the critical point, the temperature variation in physical quantities may show non-analytic temperature variations. For example, one

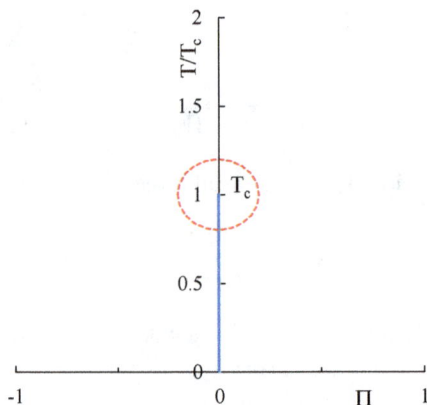

Fig. 9.13   Schematic depiction of the critical region which surrounds the critical point $(0, T_c)$ in the $\Pi - T$ phase diagram of a ferromagnet.

may consider a uniform scalar order parameter ϕ defined by the derivative of the Free-Energy w.r.t a uniform applied field Π

$$\phi = -\left(\frac{\partial F}{\partial \Pi}\right) \tag{9.75}$$

or response functions such as a zero-field susceptibility χ defined as

$$\chi = \left(\frac{\partial \phi}{\partial \Pi}\right)$$

$$= -\left(\frac{\partial^2 F}{\partial \Pi^2}\right) \tag{9.76}$$

and the heat capacity C

$$C = T\left(\frac{\partial S}{\partial T}\right)$$

$$= -T\left(\frac{\partial^2 F}{\partial T^2}\right) \tag{9.77}$$

In the vicinity of the critical point, these quantities exhibit non-analytic temperature dependencies. A dimensionless parameter t is introduced as

$$t = \left(\frac{T - T_c}{T_c}\right) \tag{9.78}$$

as a measure of the "distance" to the critical point. The temperature variation of any quantity is decomposed into a regular part and the non-analytic part. The temperature dependence of the non-analytic part can be expressed in terms of t. It is found that the leading non-analytic part follows temperature variations given by

$$\begin{aligned}
\phi &\sim |t|^\beta && \text{for} \quad T < T_c \quad \text{and} \quad \Pi = 0 \\
\chi &\sim |t|^{-\gamma} && \text{for} \quad \Pi = 0 \\
C &\sim |t|^{-\alpha} && \text{for} \quad \Pi = 0 \\
\phi &\sim \Pi^{\frac{1}{\delta}} && \text{for} \quad T = T_c
\end{aligned} \tag{9.79}$$

where the exponents α, β, γ and δ are known as the "*Critical Exponents*". Generally, the value of a critical exponent (say λ is the exponent of a quantity A) is determined by taking the limit

$$\lambda = \lim_{t \to 0}\left(\frac{\ln A}{\ln |t|}\right) \tag{9.80}$$

The critical exponent describes the leading order temperature variation. However, one expects correction terms, so that a quantity A may vary as

$$A = c|t|^\lambda (1 + D|t|^\mu + \ldots) \tag{9.81}$$

where $\mu > 0$. For a second-order transition, there should be no latent heat on passing through the transition, thus

$$L = \int_{T_c - \epsilon}^{T_c + \epsilon} dT\, C(T) = 0 \tag{9.82}$$

so, $1 > \alpha$. The value of α is actually significantly smaller than unity, and for some systems (for example the two-dimensional Ising Model) C varies logarithmically

$$C \sim \ln \left| \frac{T_c}{T - T_c} \right| \tag{9.83}$$

when T is close to T_c. Since

$$-\ln |t| = \lim_{\alpha \to 0} \frac{1}{\alpha} \left(|t|^{-\alpha} - 1 \right) \tag{9.84}$$

the logarithmic variation corresponds to $\alpha = 0$.

There are other critical exponents that are introduced to characterize the spatial correlations of the order parameter. Thus, for example, one can introduce a correlation function $S(\underline{r})$ as the an average of the product of the fluctuations of a local scalar order parameter $\Delta\phi(\underline{r})$ defined via

$$\Delta\phi(\underline{r}) = \phi(\underline{r}) - \langle \phi(\underline{r}) \rangle \tag{9.85}$$

The correlation function $S(\underline{r} - \underline{r}')$ is introduced as

$$\begin{aligned} S(\underline{r} - \underline{r}') &= \langle \Delta\phi(\underline{r}) \, \Delta\phi(\underline{r}') \rangle \\ &= \langle \phi(\underline{r}) \, \phi(\underline{r}') \rangle - \langle \phi(\underline{r}) \rangle \langle \phi(\underline{r}') \rangle \end{aligned} \tag{9.86}$$

The last term has the effect that the correlation function decays to zero at large distances for temperatures above and below T_c. Since we are assuming the system is invariant under translations, one expects that the average value is non-zero below T_c where it satisfies

$$\langle \phi(\underline{r}) \rangle = \langle \phi(\underline{r}') \rangle \tag{9.87}$$

so that it is independent of the position. Also, due to translational invariance, one has

$$\langle \phi(\underline{r}) \, \phi(\underline{r}') \rangle = \langle \phi(\underline{r} - \underline{r}') \, \phi(0) \rangle \tag{9.88}$$

so the correlation function only depends on the displacement $\underline{r} - \underline{r}$ and does not depend on the location $\underline{r}$. Furthermore, due to isotropy in space

$$S(\underline{r} - \underline{r}') = S(|\underline{r} - \underline{r}'|) \tag{9.89}$$

independent of the orientation of the vector $\underline{r}$. In the limit $|\underline{r} - \underline{r}'| \to \infty$, one expects that the value of ϕ at $\underline{r}$ will be unrelated to the value of ϕ at $\underline{r}'$. Therefore, one expects that in this limit

$$\lim_{|\underline{r} - \underline{r}'| \to \infty} \langle \phi(\underline{r}) \, \phi(\underline{r}') \rangle \to \langle \phi(\underline{r}) \rangle \langle \phi(\underline{r}') \rangle \tag{9.90}$$

Hence, one expects that correlation function decays to zero at large distances

$$\lim_{|\underline{r}-\underline{r}'|\to\infty} S(\underline{r}-\underline{r}') \to 0 \tag{9.91}$$

One can define a correlation length ξ from the form of the correlation function, valid when $\Pi = 0$ and $T \neq T_c$, which can be approximated by

$$S(\underline{r}-\underline{r}') \sim \left(\frac{1}{|\underline{r}-\underline{r}'|}\right)^{d-2+\eta} \exp\left[-\frac{|\underline{r}-\underline{r}'|}{\xi}\right] \tag{9.92}$$

where d is the number of spatial dimensions. The exponent η is the "*Anomalous Dimension*", it should only appear in $S(r)$ when $r < \xi$. The correlation length ξ describes the length scale above which the correlations die out. The correlation length is found to diverge as the critical temperature is approached, so that

$$\xi \sim |t|^{-\nu} \tag{9.93}$$

where the critical exponent is denoted by ν. One can define another exponent η which describes the spatial correlations at $T = T_c$. If one defines the Fourier components $\phi_{\underline{k}}$ of the local order parameter $\phi(\underline{r})$ via

$$\phi_{\underline{k}} = \frac{1}{\sqrt{V}} \int d^d\underline{r} \ \exp[-i\underline{k}\cdot\underline{r}] \ \phi(\underline{r}) \tag{9.94}$$

then one may define a momentum-space correlation function

$$\langle \phi_{\underline{k}} \ \phi_{-\underline{k}} \rangle = \frac{1}{V} \int d^d\underline{r}_1 \int d^d\underline{r}_2 \ \exp[-i\underline{k}\ (\underline{r}_1 - \underline{r}_2)] \langle \phi(\underline{r}_1) \ \phi(\underline{r}_2) \rangle \tag{9.95}$$

as the Fourier transformation of $S(\underline{r}-\underline{r}')$ for $\underline{k} \neq 0$. For $T = T_c$ one defines the exponent η via

$$\langle \phi_{\underline{k}} \ \phi_{-\underline{k}} \rangle \sim |\underline{k}|^{-2+\eta} \quad \text{as} \quad \underline{k} \to 0 \tag{9.96}$$

Experiments reveal that to within experimental accuracy the exponents, such as α, γ and ν, have the same value no matter in which direction the limit $t \to 0$ is taken. That is, the exponents have the same values for $T > T_c$ as for $T < T_c$. However, the coefficients of the non-analytic terms usually have different magnitudes for temperatures above and below T_c. Nevertheless, the singularities are of the same type no matter whether the transition is approached from above or below.

The same sets of values for the critical exponents are found for many different types of transitions. The same sets of values of the critical exponents are found for systems with transition temperatures that differ by many orders of magnitude, and for transitions that occur in crystalline materials,

Table 9.1   The experimentally determined values of the critical exponents for a number of three-dimensional systems with order parameters of dimension n.

	Liquid-Gas Xe	β-brass Cu-Zn	Norm-Super ^{4}He	Ferro Fe	Antiferro RbMnF$_3$
n	1	1	2	3	3
α	0.08 ± 0.02	$0.05 \pm .06$	$-0.014 \pm .016$	$-0.12 \pm .01$	$-0.139 \pm .007$
β	$0.344 \pm .003$	$0.305 \pm .005$	$0.34 \pm .01$	$0.37 \pm .01$	$0.316 \pm .008$
γ	$1.203 \pm .002$	$1.25 \pm .02$	$1.33 \pm .03$	$1.33 \pm .015$	$1.397 \pm .034$
δ	$4.4 \pm .4$	—	$3.95 \pm .15$	$4.3 \pm .1$	—
η	$0.1 \pm .1$	$0.08 \pm .017$	$0.21 \pm .05$	$0.07 \pm .04$	$0.067 \pm .01$
ν	≈ 0.57	$0.65 \pm .02$	$0.672 \pm .001$	$0.69 \pm .02$	—

are independent of the type of crystal structure. The set of transitions which share the same set of values of the critical exponents form what is known as a *"Universality Class"*.

The values of the critical exponents are not independent of one another. Historically, it was first shown that thermodynamics requires that the six exponents must satisfy (more than) four inequalities. For example, using convexity arguments, Rushbrooke showed that the heat capacity measured at constant ϕ, C_ϕ, must be positive, hence

$$C_\Pi \geq T \, \frac{\left(\frac{\partial \phi}{\partial T}\right)_\Pi^2}{\chi_T} \tag{9.97}$$

Since $\chi_T > 0$, the above inequality implies that, close to a critical point and in the limit $\Pi \to 0$, the exponents must also satisfy the inequality

$$\alpha + 2\beta + \gamma \geq 2 \tag{9.98}$$

Griffiths derived a similar inequality

$$\alpha + (\delta + 1)\,\beta \geq 2 \tag{9.99}$$

which holds on the critical isotherm $T = T_c$. Experimental evidence accumulated which showed that the inequalities were in fact equalities. The exponents of $\alpha = 0$, $\beta = \frac{1}{8}$, $\gamma = \frac{7}{4}$, $\nu = 1$ and $\eta = \frac{1}{4}$, found from Onsager's exact solution of the two-dimensional Ising Model, also satisfy the same equalities. The exponent equalities describing thermodynamic properties include the Rushbrooke relation

$$\alpha + 2\beta + \gamma = 2 \tag{9.100}$$

the Griffiths relation

$$\alpha + \beta(\delta + 1) = 2 \tag{9.101}$$

Widom argued [82] that, if one assumes that the singular part of the Helmholtz Free-Energy Density $\mathcal{F}$ does not depend separately on T and Π but, instead, is a homogeneous function which can be written as

$$\mathcal{F}(T, \Pi) = |t|^{\frac{1}{v}}\, \psi(\Pi/|t|^{\frac{x}{v}}) \tag{9.102}$$

where

$$t = \left(\frac{T - T_c}{T_c}\right) \tag{9.103}$$

then the inequalities relating the critical exponents are actually equalities. The hypothesis that the singular part of the Free-Energy is a homogenous function, is known as the *"Scaling Hypothesis"*. Leo Kadanoff introduced the idea that the equalities relating critical exponents are consequences of the existence of spatial correlations [83]. The relations which follow from the scaling hypothesis include the Fisher relation

$$(2 - \eta)\, \nu = \gamma \tag{9.104}$$

and the hyper-scaling relation or Josephson relation

$$\nu d = 2 - \alpha \tag{9.105}$$

The Josephson relation is the only relation which involves the dimensionality d. It becomes invalid for sufficiently large d, that is when d exceeds the upper critical dimension d_c. For $d > d_c$, all the critical exponents become independent of d.

The scaling that is found in the proximity of a phase transition can be understood as a consequence of the fluctuations of the order parameter that occur as the phase transition is approached. The picture is that as the temperature is decreased towards the critical temperature, the material exhibits islands of order whose spatial extent ξ increases with decreasing t. Furthermore, it is the long-ranged large-scale fluctuations that are solely responsible for the physical divergences. At the transition $t \to 0$ so $\xi \to \infty$, therefore, the system becomes scale invariant. Kadanoff's version of the scaling hypothesis assumes that the correlation length ξ is the only relevant characteristic length scale of the system close to the transition and that all other length scales must be expressible in terms of ξ. Hence, the effects of the microscopic length scale a should be expressible in terms of the ratio $\frac{a}{\xi}$ which vanishes close to the transition. The temperature dependence of static properties can then be inferred from dimensional analysis. Thus, the Free-Energy (measured in units of $k_B T$) per unit volume has dimensions L^{-d} which, on substituting ξ for L, leads to a variation as ξ^{-d}. Since the specific heat has exponent $-\alpha$, and involves the second derivative of F w.r.t. T, the Free-Energy density $\mathcal{F}$ should scale as $|t|^{2-\alpha}$. That is, we have

$$\mathcal{F} \sim \xi^{-d} \sim |t|^{2-\alpha} \qquad (9.106)$$

If the correlation function has the homogeneous form

$$S(r) \sim \left(\frac{1}{r}\right)^{d-2+\eta} \exp\left[-\frac{|r|}{\xi}\right] \qquad (9.107)$$

and is normalized to $L^{2-d-\eta}$ then, on noting that $S(r)$ is proportional to $\langle \phi(r)\,\phi(0)\rangle$, one has

$$\langle \phi(0)\rangle \sim L^{\frac{2-d-\eta}{2}} \qquad (9.108)$$

which then sets

$$\langle \phi(0)\rangle \sim \xi^{\frac{2-d-\eta}{2}} \qquad (9.109)$$

Also since, from linear response theory, the susceptibility χ can be expressed as

$$\chi \sim \int d^d\underline{r}\langle \phi(\underline{r})\,\phi(0)\rangle \qquad (9.110)$$

one has $\chi \sim L^{2-\eta}$ or

$$\chi \sim \xi^{2-\eta} \qquad (9.111)$$

On using $\xi \sim |t|^{-\nu}$ in the above scaling relations for $\mathcal{F}$, ϕ and χ, one obtains the scaling relations

$$2 - \alpha = d\nu$$

$$\beta = \frac{1}{2}\,(d - 2 + \eta)\,\nu$$

$$\gamma = (2 - \eta)\,\nu \qquad (9.112)$$

The first is recognized as the Josephson hyper-scaling relation and the last is the Fisher relation. The exponent δ can be obtained by first determining the length scale of the conjugate field Π from the definitive relation

$$\phi = -\left(\frac{\partial \mathcal{F}}{\partial \Pi}\right) \qquad (9.113)$$

which leads to the identification that

$$\Pi \sim \frac{\mathcal{F}}{\phi}$$

$$\sim L^{-d} \, L^{-\frac{2-d-\eta}{2}}$$

$$\sim L^{-\frac{2+d-\eta}{2}} \tag{9.114}$$

Since $\Pi \sim \phi^\delta$, one has

$$\phi^\delta \sim L^{-\frac{2+d-\eta}{2}}$$

$$\sim \xi^{-\frac{2+d-\eta}{2}} \tag{9.115}$$

Hence, one finds the scaling relation

$$\beta\delta = \left(\frac{2+d-\eta}{2}\right)\nu \tag{9.116}$$

The two relations consisting of

$$\beta\delta = \left(\frac{2+d-\eta}{2}\right)\nu \tag{9.117}$$

and

$$\beta = \left(\frac{d-2+\eta}{2}\right)\nu \tag{9.118}$$

can be shown to be equivalent to the Griffiths and Rushbrooke relations. The Griffith relation can be found by first adding the two relations, yielding

$$\beta(\delta+1) = d\nu \tag{9.119}$$

On using the Josephson relation, the above equation is recognized as yielding the Griffiths relation

$$\beta(\delta+1) = (2-\alpha) \tag{9.120}$$

Likewise, on subtracting the relations expressed in Eq. (9.117) and Eq. (9.118), one obtains

$$\beta(\delta-1) = (2-\eta)\nu \tag{9.121}$$

which using the Fisher relation, yields

$$\beta(\delta-1) = \gamma \tag{9.122}$$

Subtracting the above equation from the Griffiths relation leads to the Rushbrooke relation.

Scaling analysis indicates that one may consider there to be only two independent exponents, such as ν and the anomalous dimension η, but does not fix their values. As mentioned previously, the sets of the independent critical exponents, or the universality classes, are determined by the symmetry, the range of the interactions, the number of spatial dimensions and the dimensionality of the order parameter. Although the scaling analysis is ad hoc, the great success it had in describing experimental properties can be considered as providing strong evidence that critical phenomena are a result of long-ranged correlations that occur close to critical points.

9.5   Mean-Field Theory

Landau-Ginzburg Mean-Field Theory is an approximation which replaces the exact evaluation of the functional integral for the partition function

$$Z = \int \mathcal{D}\phi \ \exp[-\beta F[\phi]] \tag{9.123}$$

and, hence, approximates the Free-Energy. It can be viewed as an approximate evaluation of the functional integral analogous to the evaluation of an ordinary integral by using the method of steepest descents. In the method of steepest descents, one evaluates an integral of the form

$$I = \int_{-\infty}^{\infty} dx \ \exp[-\alpha f(x)] \tag{9.124}$$

by finding the value of x, say x_0, for which the exponent $f(x)$ is minimum and then approximating $f(x)$ by a parabola

$$f(x) = f(x_0) + \frac{1}{2} \left(\frac{d^2 f}{dx^2} \right)\bigg|_{x_0} (x - x_0)^2 + \dots \tag{9.125}$$

This approximation is based on the assumption that the value of the integral I has its largest contribution from the region around x_0. This leads to the result

$$I \sim \sqrt{\frac{2\pi}{\alpha \left(\frac{d^2 f}{dx^2} \right)\big|_{x_0}}} \ \exp[-\alpha f(x_0)] \tag{9.126}$$

The dominant contribution to the integral is given by the exponential factor. That is, the integral is approximated as a Gaussian integral for a range of x for which the exponent of the integrand is maximized. Mean-field theory approximates the functional integral by its extremal value. That is, the partition function Z is approximated by

$$Z \sim \exp[-\beta F[\phi_0]] \qquad (9.127)$$

where ϕ_0 is the ϕ which minimizes the functional $F[\phi]$.

Let us assume that there is a smooth set of $\phi_{j,0}(\underline{r})$ which minimizes the functional $F[\{\phi_j\}]$ and then consider a set of ϕ_j's with the form

$$\phi_j(\underline{r}) = \phi_{0,j}(\underline{r}) + \lambda_j \, \delta\phi_j(\underline{r}) \qquad (9.128)$$

where the $\delta\phi_j(\underline{r})$ are arbitrary functions that vanish at the boundaries of the system. The parameters λ_j can be continuously varied from zero to unity, and represents the amplitude of the deviation from ϕ_0. For fixed $\delta\phi_j$, the Free-Energy functional depends on the set of λ_j and can be written as $F(\{\lambda_j\})$. The extremum condition implies that

$$\left(\frac{\partial F(\{\lambda_j\})}{\partial \lambda_i}\right)\bigg|_{\lambda_i = 0} = 0 \qquad (9.129)$$

for all i. Therefore, the terms first-order in the λ_i's in the expansion of $F[\{\phi_{j,0} + \lambda_j \delta_j \phi\}]$ must be zero. That is, if

$$F[\{\phi_{j,0} + \lambda_j \delta_j\}\phi] = F[\{\phi_{j,0}\}] + \sum_i \lambda_i \, \delta F_i^1 + \ldots \qquad (9.130)$$

then for $\{\phi_{j,0}\}$ to be an extremum, one requires that

$$\delta F_i^1 = 0 \qquad (9.131)$$

for all i. For a d-dimensional vector order parameter, the explicit form of the above condition is

$$0 = \int d^d\underline{r} \, [2F_2(\underline{\phi}_0 \cdot \delta\underline{\phi}) + 4F_4(\underline{\phi}_0 \cdot \underline{\phi}_0)(\underline{\phi}_0 \cdot \delta\underline{\phi}) - \underline{\Pi} \cdot \delta\underline{\phi} + 2c\underline{\nabla\phi}_0 \cdot \underline{\nabla\delta\phi}] \qquad (9.132)$$

where the last term must be interpreted as involving two types of scalar product, as previously discussed. On integrating the last term by parts and recalling that the $\delta\phi_j$ vanish at the boundaries, one can write the integrand as the sum of terms that depend on the product of a factor involving $\phi_{j,0}$ and $\delta\phi_j$,

$$0 = \int d^d\underline{r} \, \delta\underline{\phi} \cdot [2F_2\underline{\phi}_0 + 4F_4\underline{\phi}_0(\underline{\phi}_0 \cdot \underline{\phi}_0) - \underline{\Pi} - 2c\nabla^2\underline{\phi}_0] \qquad (9.133)$$

This must vanish for an arbitrary $\delta\phi_j$. One can chose $\delta\phi_j$ such that

$$\delta\phi_j(\underline{r}) = \delta^d(\underline{r} - \underline{r}_0) \qquad (9.134)$$

for some arbitrary point $\underline{r}_0$. The integration over $d^d\underline{r}$ can be performed leading to the requirement that ϕ_0 must satisfy the equation

$$0 = [2F_2\underline{\phi}_0(\underline{r}_0) + 4F_4\underline{\phi}_0(\underline{r}_0)(\underline{\phi}_0(\underline{r}_0) \cdot \underline{\phi}_0(\underline{r}_0)) - \underline{\Pi}(\underline{r}_0) - 2c\nabla^2\underline{\phi}_0(\underline{r}_0)] \quad (9.135)$$

for any arbitrarily chosen point $\underline{r}_0$. The functions $\phi_0(\underline{r})$ which satisfy the above equation extremalize $F[\phi]$ for any choice of $\delta\phi(\underline{r})$. We shall write Eq. (9.135) in the form

$$[2F_2 + 4F_4(\underline{\phi}_0(\underline{r}) \cdot \underline{\phi}_0(\underline{r})) - 2c\nabla^2] \, \underline{\phi}_0(\underline{r}) = \underline{\Pi}(\underline{r}) \quad (9.136)$$

in which the spatially varying applied field acts as a source. This equation governs all the extrema of $F[\phi]$.

We shall first consider physical properties associated with the extrema for which ϕ is uniform across the system, and then consider the physical properties associated with the spatially varying solutions.

9.5.1 *Uniform Solutions*

The differential equation simplifies for spatially uniform solution and zero applied fields, $\underline{\Pi} = 0$, to

$$[2F_2 + 4F_4\underline{\phi} \cdot \underline{\phi}] \, \underline{\phi} = 0 \quad (9.137)$$

which has the solutions

$$\underline{\phi} = 0 \quad (9.138)$$

and

$$\underline{\phi} \cdot \underline{\phi} = -\frac{2F_2}{4F_4} \quad (9.139)$$

The second solution only makes sense if

$$\underline{\phi} \cdot \underline{\phi} > 0 \quad (9.140)$$

since the order parameter is assumed to be a real vector quantity. When

$$-\frac{2F_2}{4F_4} < 0 \quad (9.141)$$

the only physically acceptable solution corresponds to $\underline{\phi} = 0$. The magnitude of order parameter may take on a non-zero value if

$$-\frac{2F_2}{4F_4} > 0 \quad (9.142)$$

In this case, in addition to the solution $\underline{\phi} = 0$, there exists the possibility of continuously degenerate solutions. Due to stability considerations, one must have $F_4 > 0$. Hence, a second-order phase transition may occur when F_2 changes sign. This motivates the notation

$$F_2 = A(T - T_c) \tag{9.143}$$

with $A > 0$. For $T > T_c$, there is only one unique solution which is given by $\underline{\phi} = 0$, so the mean-field value of the Free-Energy is given by

$$F[\phi = 0] = V F_0 \tag{9.144}$$

whereas for $T < T_c$, one has the possibility of an additional solution corresponding to

$$\underline{\phi}_0 \cdot \underline{\phi}_0 = \frac{2A(T_c - T)}{4F_4} \tag{9.145}$$

which fixes the magnitude of ϕ_0 as

$$\phi_0 = \sqrt{\frac{2A(T_c - T)}{4F_4}} \tag{9.146}$$

The non-zero solution is continuously degenerate with respect to the orientation of the vector order parameter. The presence of an infinitesimal applied field allows the system to choose a direction of $\underline{\phi}$ which spontaneously breaks the rotational symmetry of the Hamiltonian. The mean-field Free-Energy corresponding the broken symmetry solution is found as

$$\begin{aligned} F[\phi_0] &= V \left[F_0 - \frac{2A^2(T_c - T)^2}{4F_4} + \frac{A^2(T_c - T)^2}{4F_4} \right] \\ &= V \left[F_0 - \frac{A^2(T_c - T)^2}{4F_4} \right] \end{aligned} \tag{9.147}$$

which is lower than the Free-Energy of the solution with $\phi = 0$. Hence, in the mean-field approximation, the system will condense into the broken symmetry state. Thus, the mean-field approximation describes a second-order transition at T_c and, for $T < T_c$, the magnitude of the order parameter will have a temperature-dependence given by

$$|\underline{\phi}_0| = \sqrt{\frac{2A(T_c - T)}{4F_4}} \tag{9.148}$$

The value of the mean-field critical exponent for the order parameter, β, is, therefore, fixed at $\beta = \frac{1}{2}$.

The order parameter exponent on the critical isotherm, δ, defined via

$$|\Pi| \sim |\phi|^\delta \tag{9.149}$$

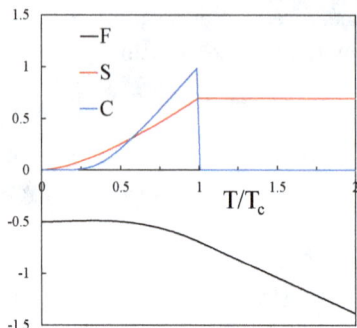

Fig. 9.14   Mean-Field Approximation for the Free-Energy F, entropy S and specific heat C, as a function of temperature.

is found from the equation

$$[2A(T - T_c) + 4F_4 \underline{\phi}_0 \cdot \underline{\phi}_0] \, \underline{\phi}_0 = \underline{\Pi} \tag{9.150}$$

which for $T = T_c$ reduces to

$$4F_4(\underline{\phi}_0 \cdot \underline{\phi}_0) \, \underline{\phi}_0 = \underline{\Pi} \tag{9.151}$$

and leads to the identification of the critical exponent $\delta = 3$.

The mean-field specific heat C is calculated from

$$C = -T \left(\frac{\partial^2 F}{\partial T^2} \right) \tag{9.152}$$

which, for $T > T_c$, yields

$$C = 0 \qquad \text{for } T > T_c \tag{9.153}$$

but, for $T < T_c$ gives

$$C = V \frac{A^2 T}{2F_4} \qquad \text{for } T > T_c \tag{9.154}$$

Hence, in the mean-field approximation, the specific heat exhibits a discontinuous jump at T_c, of magnitude $V \frac{A^2 T_c}{2F_4}$. The appropriate mean-field specific heat exponents are $\alpha = \alpha' = 0$.

The above solutions are all uniform, due to the homogeneity of space and due to the uniformity of the applied fields.

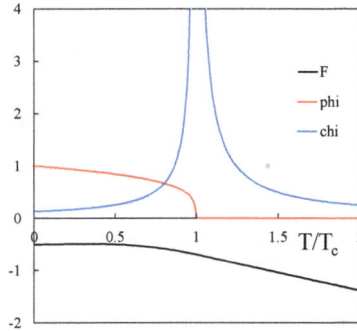

Fig. 9.15 Mean-Field Approximation for the Free-Energy F, order parameter ϕ and susceptibility χ, as a function of temperature.

9.5.2 *Spatially Varying Solutions*

It has been shown that the non-uniform configurations $\underline{\phi}(\underline{r})$ which minimize $F[\phi]$ satisfy the partial differential equation

$$[2F_2 + 4F_4\underline{\phi}(\underline{r}) \cdot \underline{\phi}(\underline{r}) - 2c\nabla^2] \, \underline{\phi}(\underline{r}) = \underline{\Pi}(\underline{r}) \qquad (9.155)$$

We shall consider spatially varying solutions, in which the spatial variations are induced by an applied field which has a small non-uniform part

$$\underline{\Pi}(\underline{r}) = \underline{\Pi}_0 + \delta\underline{\Pi}(\underline{r}) \qquad (9.156)$$

The local order-parameter $\underline{\phi}(\underline{r})$ will be expressed in terms of a uniform component $\underline{\phi}_0$ and a spatially varying part $\delta\underline{\phi}(\underline{r})$

$$\underline{\phi}(\underline{r}) = \underline{\phi}_0 + \delta\underline{\phi}(\underline{r}) \qquad (9.157)$$

The spatially varying part of the order-parameter $\delta\underline{\phi}(\underline{r})$ vanishes in the limit that $\delta\underline{\Pi}(\underline{r})$ vanishes. In this limit, $\underline{\phi}_0$ minimizes $F[\phi]$ in the presence of $\underline{\Pi}_0$. The terms of first-order in the small spatially varying components satisfy the equation

$$[2F_2 + 4F_4\underline{\phi}_0 \cdot \underline{\phi}_0 - 2c\nabla^2] \, \delta\underline{\phi}(\underline{r}) + 8F_4\underline{\phi}_0(\underline{\phi}_0 \cdot \delta\underline{\phi}(\underline{r})) = \delta\underline{\Pi}(\underline{r}) \qquad (9.158)$$

This equation indicates that, for temperatures below T_c, the mean-field response will be different depending on the relative orientation of the spatially varying field $\delta\underline{\Pi}(\underline{r})$ and the direction of the uniform order parameter $\underline{\phi}_0$. For temperatures above the critical temperature T_c, the vector order parameter vanishes and the equation simplifies to

$$[2F_2 - 2c\nabla^2] \, \delta\underline{\phi}(\underline{r}) = \delta\underline{\Pi}(\underline{r}) \qquad (9.159)$$

In this case, the magnitude of the induced vector order parameter is independent of the orientation of the spatially varying applied field. This is expected since, in the absence of the order parameter which has spontaneously broken the symmetry, the system is isotropic.

Longitudinal Response

For temperatures below the critical temperature and when the non-uniform part of the applied field $\delta\underline{\Pi}(\underline{r})$ is parallel to $\underline{\phi}_0$, the mean-field response is longitudinal and satisfies the equation

$$[2F_2 + 12F_4\underline{\phi}_0 \cdot \underline{\phi}_0 - 2c\nabla^2]\,\delta\phi_L(\underline{r}) = \delta\Pi_L(\underline{r}) \qquad (9.160)$$

This equation is also valid for temperatures above T_c, where ϕ_0 vanishes, although longitudinal and transverse is undefined. It should be noted that the equation has a different form in the two temperature regimes.

Transverse Response

The transverse response is only defined for temperatures below the critical temperature. If $\delta\underline{\Pi}(\underline{r})$ is transverse to $\underline{\phi}_0$, the mean-field response is determined from the partial differential equation

$$[2F_2 + 4F_4\underline{\phi}_0 \cdot \underline{\phi}_0 - 2c\nabla^2]\,\delta\phi_T(\underline{r}) = \delta\Pi_T(\underline{r}) \qquad (9.161)$$

In the limit $\underline{\Pi}_0 \to 0$, the uniform order parameter $\underline{\phi}_0$ satisfies the equation

$$[2F_2 + 4F_4\underline{\phi}_0 \cdot \underline{\phi}_0] = 0 \qquad (9.162)$$

Hence, the partial differential equation for the transverse response in the mean-field approximation simplifies to

$$-2c\nabla^2\delta\phi_T(\underline{r}) = \delta\Pi_T(\underline{r}) \qquad (9.163)$$

The solution of this equation determines the order parameter for which the Landau-Ginzburg Free-Energy Functional is extremal.

9.5.3   *The Mean-Field Correlation Functions*

The differences in the response show that the correlation function $S(\underline{r})$ involved in the linear response theory must be considered as a tensor quantity. The mean-field equations for the order parameter allows one to calculate the (mean-field) tensor correlation function. Linear response theory describes how $\delta\underline{\phi}(\underline{r})$ is related to $\delta\underline{\Pi}(\underline{r})$. In particular, if a tensor correlation function $S_{i,j}(\underline{r})$ is defined via

$$S_{i,j}(\underline{r} - \underline{r}') = \langle \phi_i(\underline{r})\,\phi_j(\underline{r}')\rangle - \langle \phi_i(\underline{r})\rangle\langle \phi_j(\underline{r}')\rangle \qquad (9.164)$$

one finds that components satisfy the linear response relations

$$\delta\phi_i(\underline{r}) = \beta \sum_j \int d^d\underline{r}'\, S_{i,j}(\underline{r} - \underline{r}')\, \delta\Pi_j(\underline{r}') \qquad (9.165)$$

Substitution of the linear response relation into Eq. (9.158) results in an integral equation which is linear in the components of $\delta\Pi$. The left-hand side involves an integral over $\underline{r}'$ of the components of $\delta\Pi(\underline{r}')$ weighted by a matrix function involving the derivatives of $S_{i,j}(\underline{r} - \underline{r}')$. The right-hand side only depends on a component of $\delta\Pi(r)$. The integral equation has to be satisfied for an arbitrary $\delta\Pi(r)$. The equation is satisfied if the matrix function is proportional to a d-dimensional delta function $\delta^d(r - r')$. The integration over $\underline{r}'$ can then be trivially performed. This results in a matrix equation which, after eliminating the arbitrary spatially varying field, determines the tensor response function.

We shall first examine the longitudinal response function S_L which satisfies the linear response relation

$$\delta\phi_L(\underline{r}) = \beta \int d^d\underline{r}'\, S_L(\underline{r} - \underline{r}')\, \delta\Pi_L(\underline{r}') \qquad (9.166)$$

On substituting the longitudinal linear response relation into Eq. (9.160) which describes the mean-field order parameter, one finds

$$[2F_2 + 12F_4\underline{\phi}_0 \cdot \underline{\phi}_0 - 2c\nabla_{\underline{r}}^2] \int d^d\underline{r}'\, S_L(\underline{r} - \underline{r}')\, \delta\Pi_L(\underline{r}') = k_B T\, \delta\Pi_L(\underline{r})$$
$$(9.167)$$

On changing the order of integration and differentiation, one obtains

$$\int d^d\underline{r}' [2F_2 + 12F_4\underline{\phi}_0 \cdot \underline{\phi}_0 - 2c\nabla_{\underline{r}}^2]\, S_L(\underline{r} - \underline{r}')\, \delta\Pi_L(\underline{r}') = k_B T\, \delta\Pi_L(\underline{r}) \quad (9.168)$$

where $\delta\Pi_L$ is to be regarded as an arbitrary function. The equation is satisfied for any $\delta\Pi_L(\underline{r})$, if the (mean-field) longitudinal correlation function satisfies

$$[2F_2 + 12F_4\underline{\phi}_0 \cdot \underline{\phi}_0 - 2c\nabla_{\underline{r}}^2]\, S_L(\underline{r} - \underline{r}') = k_B T\, \delta^d(\underline{r} - \underline{r}') \qquad (9.169)$$

as can be verified by substitution. Hence, we have determined a partial differential equation which determines the longitudinal correlation function below T_c and also determines the correlation function above T_c.

Likewise, one can determine the transverse correlation function $S_T(\underline{r} - \underline{r}')$ which satisfies the linear response relation

$$\delta\phi_T(\underline{r}) = \beta \int d^d\underline{r}'\, S_T(\underline{r} - \underline{r}')\, \delta\Pi_T(\underline{r}') \qquad (9.170)$$

On substituting the transverse linear response relation into Eq. (9.163), which describes the mean-field order parameter, one finds

$$- 2c\nabla_r^2 \int d^d\underline{r}' \, S_T(\underline{r} - \underline{r}') \, \delta\Pi_T(\underline{r}') = k_B T \, \delta\Pi_T(\underline{r}) \qquad (9.171)$$

On changing the order of integration and differentiation, one obtains

$$\int d^d\underline{r}' \, [-2c\nabla_r^2 \, S_T(\underline{r} - \underline{r}')] \, \delta\Pi_T(\underline{r}') = k_B T \, \delta\Pi_T(\underline{r}) \qquad (9.172)$$

where $\delta\Pi_T$ is to be regarded as an arbitrary function. The equation is satisfied for any $\delta\Pi_T(\underline{r})$, if the (mean-field) transverse correlation function satisfies

$$- 2c\nabla_r^2 \, S_T(\underline{r} - \underline{r}') = k_B T \, \delta^d(\underline{r} - \underline{r}') \qquad (9.173)$$

The above equation can be used to determine the transverse correlation function in the mean-field approximation.

For $T > T_c$ the mean-field order parameter is zero, so the correlation function satisfies

$$(2A(T - T_c) - 2c\nabla^2) \, S(\underline{r} - \underline{r}') = k_B T \, \delta^d(\underline{r} - \underline{r}') \qquad (9.174)$$

In the low temperature phase, $T < T_c$, the order parameter is given by

$$|\underline{\phi}_0|^2 = -\frac{A(T - T_c)}{2F_4} \qquad (9.175)$$

therefore, the longitudinal correlation function satisfies the partial differential equation

$$(-4A(T - T_c) - 2c\nabla_r^2) \, S_L(\underline{r} - \underline{r}') = k_B T \, \delta^d(\underline{r} - \underline{r}') \qquad (9.176)$$

Dimensional analysis of the above two equations indicates that the correlation length ξ should be given by

$$\xi = \sqrt{\frac{c}{A(T - T_c)}} \qquad (9.177)$$

for $T > T_c$, and

$$\xi = \sqrt{\frac{c}{2A(T_c - T)}} \qquad (9.178)$$

for $T < T_c$. Since the correlation length is defined to scale as

$$\xi \sim |T - T_c|^{-\nu} \qquad (9.179)$$

one finds that the critical exponents are the same above and below T_c and are given by $\nu = \nu' = \frac{1}{2}$.

The above two equations for S_L and S can be written as a single equation when expressed in terms of the correlation length ξ. The equations both have the form

$$(\xi^{-2} - \nabla_{\underline{r}}^2) \, S(\underline{r} - \underline{r}') = \frac{k_B T}{2c} \, \delta^d(\underline{r} - \underline{r}') \tag{9.180}$$

This equation can be solved by Fourier transforming, leading to

$$(\xi^{-2} + \underline{k}^2) \, S(\underline{k}) = \frac{k_B T}{2c} \tag{9.181}$$

Thus,

$$S(\underline{k}) = \left(\frac{k_B T}{2c} \right) \left(\frac{1}{\xi^{-2} + \underline{k}^2} \right) \tag{9.182}$$

The correlation function $S(\underline{r})$ is given by the inverse Fourier Transformation

$$\begin{aligned}
S(\underline{r}) &= \int \frac{d^d k}{(2\pi)^d} \, \exp[i\underline{k} \cdot \underline{r}] \, S(\underline{k}) \\
&= \left(\frac{k_B T}{2c} \right) \int \frac{d^d k}{(2\pi)^d} \frac{\exp[i\underline{k} \cdot \underline{r}]}{\xi^{-2} + \underline{k}^2}
\end{aligned} \tag{9.183}$$

For three dimensions, the integral is evaluated as

$$\begin{aligned}
S(\underline{r}) &= \left(\frac{k_B T}{2c} \right) \int_0^{2\pi} d\varphi \int_0^\pi d\theta \, \sin\theta \int_0^\infty \frac{d^3 k}{(2\pi)^3} \frac{\exp[ikr\cos\theta]}{\xi^{-2} + k^2} \\
&= \left(\frac{k_B T}{2c} \right) \int_0^\pi d\theta \, \sin\theta \int_0^\infty \frac{dk}{(2\pi)^2} k^2 \frac{\exp[ikr\cos\theta]}{\xi^2 + k^2} \\
&= \left(\frac{k_B T}{2c} \right) \int_0^\infty \frac{dk}{(2\pi)^2} \frac{k^2}{ikr} \left(\frac{\exp[+ikr] - \exp[-ikr]}{(\xi^{-2} + k^2)} \right) \\
&= \left(\frac{k_B T}{4\pi c} \right) \frac{1}{r} \int_{-\infty}^\infty \frac{dk}{(2\pi i)} \exp[+ikr] \frac{k}{(\xi^{-2} + k^2)}
\end{aligned} \tag{9.184}$$

where two terms have been combined extending the integration over k to the range $-\infty$ to $+\infty$. The remaining integration can be performed using Cauchy's Theorem by completing the contour with a semi-circle in the upper-half complex plane. The contribution of the semi-circular contour at infinity vanishes due to Jordan's lemma. The integral is dominated by the residue at the pole $k = i\xi^{-1}$, leading to

$$S(r) = \left(\frac{k_B T}{8\pi c} \right) \frac{\exp\left[-\frac{r}{\xi} \right]}{r} \tag{9.185}$$

which leads to the identification of the (three-dimensional)[1] mean-field value of the critical exponent η is $\eta = 0$. The quantity η is also known as the anomalous dimension.

The susceptibility χ in the high temperature phase and the longitudinal susceptibility χ_L in the low temperature ordered phase can be obtained from linear response theory. The susceptibility is given by

$$\chi = \beta \int d^d\underline{r}\, S(\underline{r}) \tag{9.186}$$

which, in three-dimensions leads to the mean-field susceptibility given by

$$\chi = \left(\frac{1}{8\pi c}\right) \int d^3\underline{r}\, \frac{\exp\left[-\frac{r}{\xi}\right]}{r}$$

$$= \left(\frac{1}{2c}\right) \int_0^\infty dr\, r \exp\left[-\frac{r}{\xi}\right]$$

$$= \left(\frac{\xi^2}{2c}\right) \int_0^\infty dx\, x \exp[-x] \tag{9.187}$$

where the dimensionless parameter x

$$x = \left(\frac{r}{\xi}\right) \tag{9.188}$$

has been introduced in the last line. The integral can be evaluated using integration by parts, leading to

$$\chi = \frac{\xi^2}{2c} \tag{9.189}$$

The same result also holds in d-dimensions, where the susceptibility is given by the integral

$$\chi = \left(\frac{1}{8\pi c}\right) \int d^d\underline{r}\, F\left(\frac{r}{\xi}\right) \frac{\exp\left[-\frac{r}{\xi}\right]}{r^{d-2}}$$

$$= \left(\frac{\xi^2}{2c}\right) \int_0^\infty dx\, x\, F(x) \exp[-x] \tag{9.190}$$

---

[1]The three-dimensional result is quite special. A more general form of the mean-field form of $S(\underline{r})$ in d-dimensions is given by

$$S(r) = \left(\frac{k_B T}{2cS_d(d-2)}\right) \frac{\exp\left[-\frac{r}{\xi}\right]}{r^{(d-2)}} F\left(\frac{r}{\xi}\right)$$

where $F(x)$ has the properties that $F(0) = 1$ and for large r

$$F\left(\frac{r}{\xi}\right) \sim \left(\frac{r}{\xi}\right)^{\frac{(d-3)}{2}}$$

For $d = 3$ the function reduces to a constant, which is unity.

and where $F(x)$ is a dimensionless function. Hence, the susceptibilities are given by

$$\chi = \frac{1}{2A(T - T_c)} \tag{9.191}$$

for $T > T_c$, whereas, for $T < T_c$, the longitudinal susceptibility is given by

$$\chi_L = \frac{1}{4A(T_c - T)} \tag{9.192}$$

Both susceptibilities diverge as the transition temperature is approached. Thus, in mean-field theory the susceptibility critical exponent γ is the same in the low and high temperature phases and is given by $\gamma = \gamma' = 1$. However, the amplitude of the divergent term is a factor of two greater for the high temperature phase.

The transverse susceptibility χ_T can also be calculated from the transverse correlation function $S_T(\underline{r})$. The correlation length for the transverse correlation function diverges for all temperatures below T_c. The divergence of transverse correlation length is connected with Goldstone's Theorem (See Section 9.8). It has the effect that, $S_T(\underline{r})$ satisfies the equation

$$- 2c\nabla^2 \, S_T(\underline{r}) = k_B T \delta^d(\underline{r}) \tag{9.193}$$

One can integrate the equation for S_T over $\underline{r}$, where the integration runs over a volume which contains the origin, and then use Gauss's Theorem to express the remaining volume integral as an integral over the surface of the volume

$$-2c \int d^d\underline{r} \, \nabla^2 \, S_T(\underline{r}) = k_B T \int d^d\underline{r} \, \delta^d(\underline{r})$$

$$-2c \int d^d\underline{r} \, \nabla^2 \, S_T(\underline{r}) = k_B T$$

$$-2c \int d\underline{S}^{d-1} \cdot \underline{\nabla} \, S_T(\underline{r}) = k_B T \tag{9.194}$$

where the direction of the $(d-1)$ dimensional surface element, $d\underline{S}^{d-1}$, is defined to be normal to the surface of integration. Since $S_T(\underline{r})$ is spherically symmetric, the integration is easily performed over the surface of a hypersphere of radius r, leading to

$$- 2c \, S_d \, r^{d-1} \, \hat{e}_r \cdot \underline{\nabla} \, S_T(\underline{r}) = k_B T \tag{9.195}$$

where S_d is the surface area of a d-dimensional unit sphere. Thus,

$$\left(\frac{\partial S_T}{\partial r} \right) = - \left(\frac{k_B T}{2cS_d} \right) \frac{1}{r^{d-1}} \tag{9.196}$$

For $d > 2$, the expression can be integrated leading to the transverse correlation function $S_T(\underline{r})$ being given by

$$S_T(r) = \left(\frac{k_B T}{2c(d-2)\ S_d} \right) \frac{1}{r^{d-2}} \qquad (9.197)$$

The transverse susceptibility, χ_T, can then be found from S_T via

$$\chi_T = \beta \int d^d \underline{r}\ S_T(\underline{r})$$

$$\sim \int_0^L dr\ r^{d-1}\ \frac{1}{r^{d-2}}$$

$$\sim \int_0^L dr\ r$$

$$\sim L^2 \qquad (9.198)$$

which diverges in the thermodynamic limit where the linear dimension of the system, L, is sent to infinity. The transverse susceptibility χ_T is infinite since the application of a small transverse field can cause the vector order parameter to re-orient.

The results for the longitudinal and the high temperature transverse susceptibilities could have been determined directly from the equation

$$[2F_2 + 4F_4\ \underline{\phi}_0 \cdot \underline{\phi}_0 - 2c\nabla^2]\ \underline{\delta\phi}(\underline{r}) + 8F_4\ \underline{\phi}_0\ (\underline{\phi}_0 \cdot \underline{\delta\phi}(\underline{r})) = \underline{\delta\Pi}(\underline{r}) \quad (9.199)$$

by considering the limit in which $\delta\phi$ and $\delta\Pi$ become independent of $\underline{r}$. In this limit, one obtains

$$[2F_2 + 4F_4\ \underline{\phi}_0 \cdot \underline{\phi}_0]\ \underline{\delta\phi} + 8F_4\ \underline{\phi}_0\ (\underline{\phi}_0 \cdot \underline{\delta\phi}) = \underline{\delta\Pi} \qquad (9.200)$$

The differential longitudinal susceptibility χ_L is defined as

$$\chi_L = \frac{\delta\phi_L}{\delta\Pi_L} \qquad (9.201)$$

and is found to be given by

$$\chi_L = \frac{1}{[2F_2 + 12F_4\ \underline{\phi}_0 \cdot \underline{\phi}_0]} \qquad (9.202)$$

which is evaluated as

$$\chi_L = \frac{1}{4A(T_c - T)} \qquad (9.203)$$

The susceptibility in the disordered phase, where $\phi_0 = 0$, is given by

$$\chi = \frac{1}{2F_2} \qquad (9.204)$$

which reduces to

$$\chi = \frac{1}{2A(T - T_c)} \tag{9.205}$$

as found previously.

9.6 The Gaussian Approximation

The "*Gaussian Approximation*" is an approximation in which one evaluates the corrections to mean-field theory due to the small amplitude fluctuations about the mean-field ground state. The Gaussian approximation is analogous to the steepest descents evaluation of an integral of the form

$$I = \int_{-\infty}^{\infty} dx \ \exp[-\alpha f(x)] \tag{9.206}$$

The mean-field approximation corresponds to approximating the integral by the exponential of the function $f(x_0)$ at the value x_0 which minimizes $f(x)$. The Gaussian approximation includes the corrections due to integrating the lowest-order (quadratic) terms in the Taylor expansion of $f(x)$. This leads to the result

$$I \sim \sqrt{\frac{2\pi}{\alpha \left(\frac{d^2 f}{dx^2}\right)\Big|_{x_0}}} \ \exp[-\alpha f(x_0)]$$

$$= \exp\left[-\alpha f(x_0) - \frac{1}{2} \ln \left(\frac{\alpha \left(\frac{d^2 f}{dx^2}\right)\Big|_{x_0}}{2\pi}\right)\right] \tag{9.207}$$

which gives a correction to the exponent which is a factor of α^{-1} smaller than the leading contribution.

Example: Gaussian Integration

As an example of how to evaluate a Functional Integral in the Gaussian Approximation, consider an expression for a partition function Z which is of the form

$$Z = \prod_{i=1}^{N} \left\{ \int_{-\infty}^{\infty} d\phi_i \right\} \ \exp\left[-\beta \sum_{i,j} \phi_j \ F_{j,i} \ \phi_j\right] \tag{9.208}$$

where the matrix $F_{j,i}$ is assumed to be a positive real symmetric $N \times N$ matrix and the ϕ_i are dimensionless variables representing the i-th coarse-grained unit cell. The exponent in the expression represents a discretized version of the Landau Free-Energy Functional for a system with N coarse-grained unit cells. The Gaussian approximation neglects all terms in the exponent that are higher-order than quadratic in ϕ. Since the matrix $F_{j,i}$ is real and symmetric, the eigenvalues $\tilde{f}_k$ and eigenvectors $R_{j,k}$ exist

$$\sum_{i=1}^{N} F_{j,i} \, R_{i,k} = \tilde{f}_k \, R_{j,k} \tag{9.209}$$

The eigenvalues of the matrix, $\tilde{f}_k$, are positive and real. The eigenvectors R_k can be used to form an orthonormal basis. Hence, one can decompose the components, ϕ_i, of an arbitrary vector ϕ in terms of the components of the eigenvector basis with expansion coefficients $\tilde{\phi}_k$

$$\phi_i = \sum_{k=1}^{N} R_{i,k} \, \tilde{\phi}_k \tag{9.210}$$

The $\tilde{\phi}_k$ play the role of the amplitudes of "normal modes". Then the matrix F can be diagonalized by transforming with the matrix of eigenvectors R

$$R^{-1} \, FR = \tilde{F} \tag{9.211}$$

where, due to the orthonormality of the eigenvectors,

$$\tilde{F}_{k',k} = \tilde{f}_k \, \delta_{k',k} \tag{9.212}$$

The diagonalization of the matrix F leads to the exponent being expressed as a sum over the normal modes

$$\sum_{i,j} \phi_j \, F_{j,i} \, \phi_i = \sum_{k=1}^{N} \tilde{\phi}_k \, \tilde{f}_k \, \tilde{\phi}_k \tag{9.213}$$

Hence, when expressed in terms of the transformed variables, the Gaussian integration has the form

$$Z = \prod_{i=1}^{N} \left\{ \int_{-\infty}^{\infty} d\phi_i \right\} \exp\left[-\beta \sum_{i,j} \phi_j \, F_{j,i} \, \phi_j \right]$$

$$= |\det(R)| \prod_{i=1}^{N} \left\{ \int_{-\infty}^{\infty} d\tilde{\phi}_i \right\} \exp\left[-\beta \sum_{k} \tilde{\phi}_k \, \tilde{f}_k \, \tilde{\phi}_k \right] \tag{9.214}$$

Due to the orthogonality of R, the Jacobian is given by

$$|\det(R)| = 1 \tag{9.215}$$

Thus, the Gaussian integration has been reduced to a product of integrations over the amplitudes of the "normal modes" and is evaluated as

$$
\begin{aligned}
Z &= \prod_{i=1}^{N} \left\{ \int_{-\infty}^{\infty} d\tilde{\phi}_i \right\} \exp\left[-\beta \sum_k \tilde{\phi}_k \tilde{f}_k \tilde{\phi}_k \right] \\
&= \prod_{k=1}^{N} \sqrt{\frac{\pi}{\beta \tilde{f}_k}} \\
&= (\pi k_B T)^{\frac{N}{2}} \det(F)^{-\frac{1}{2}}
\end{aligned}
\tag{9.216}
$$

where we have used $\det(R) = 1$ in

$$
\begin{aligned}
\prod_{k=1} \tilde{f}_k &= \det(\tilde{F}) = \det(R^{-1} F R) \\
&= \det(R^{-1}) \det(F) \det(R) = \det(F)
\end{aligned}
\tag{9.217}
$$

to obtain the last line. This example can be extended to Hermitean matrices and complex fields ϕ.

The Gaussian approximation goes one step beyond the mean-field approximation. The mean-field approximation finds the most probable configuration ϕ_0 and takes into account the small amplitude fluctuations $\underline{\phi}(\underline{r})$. The spatially varying order parameter can be represented as

$$
\underline{\phi}(\underline{r}) = \underline{\phi}_0 + \delta\underline{\phi}(\underline{r})
\tag{9.218}
$$

in which the fluctuations $\delta\underline{\phi}(\underline{r})$ are assumed to have small amplitudes

$$
|\delta\underline{\phi}(\underline{r})| \ll |\underline{\phi}_0|
\tag{9.219}
$$

The functional integral expression for the partition function can be expressed in terms of a product of integrations over the variables for the N coarse-grained unit cells

$$
\begin{aligned}
Z &= \int \mathcal{D}\underline{\phi} \, \exp[-\beta F[\underline{\phi}(\underline{r})]] \\
&= \prod_{\underline{r}=1}^{N} \{d\delta\underline{\phi}(\underline{r})\} \, \exp[-\beta F[\underline{\phi}_0 + \delta\underline{\phi}(\underline{r})]]
\end{aligned}
\tag{9.220}
$$

where the coarse-grained unit cells each have volume $\Omega_c = a^d$. The spatial integration in the Free-Energy Functional $F[\underline{\phi}]$ ranges over a d-dimensional volume V, where

$$
V = L^d = N\Omega_c
\tag{9.221}
$$

and we shall impose periodic boundary conditions on $\delta\underline{\phi}(\underline{r})$. Since the Landau-Ginzburg Free-Energy Functional is translationally invariant, so momentum is conserved, it is advantageous to Fourier Transform the spatially varying fields. The small amplitude fluctuations are expressed in terms of their Fourier components $\underline{\phi}_{\underline{k}}$ via

$$\delta\underline{\phi}(\underline{r}) = \frac{1}{\sqrt{V}} \sum_{\underline{k}} \exp[+i\underline{k}\cdot\underline{r}] \, \underline{\phi}_{\underline{k}} \qquad (9.222)$$

If periodic boundary conditions are imposed on $\delta\underline{\phi}(\underline{r})$, then, due to the boundary conditions, the vector $\underline{k}$ has discrete components k_i which are specified by

$$k_i = n_i \frac{\pi}{L} \qquad (9.223)$$

where the n_i are integers. Furthermore, since the continuous field $\delta\underline{\phi}(\underline{r})$ originated from the coarse-grained discrete lattice, the k_i must have an upper cut-off Λ which is related to the lattice parameter a by

$$\Lambda \sim \frac{\pi}{a} \qquad (9.224)$$

Therefore, $\underline{\phi}_{\underline{k}} = 0$ if $|\underline{k}| > \Lambda$. The inverse Transformation is given by

$$\underline{\phi}_{\underline{k}} = \frac{1}{\sqrt{V}} \int d^d\underline{r} \, \exp[-i\underline{k}\cdot\underline{r}] \, \delta\underline{\phi}(\underline{r}) \qquad (9.225)$$

It should be noted that since the $\delta\underline{\phi}(\underline{r})$ are real functions, then

$$\underline{\phi}_{\underline{k}}^* = \underline{\phi}_{-\underline{k}} \qquad (9.226)$$

and so $\underline{\phi}_{\underline{k}}$ are complex functions with real and imaginary parts. The $\underline{k}$-space representation can be used to simplify the expression for the Landau-Ginzburg Free-Energy Functional. The transformation can be expressed in terms of the discrete lattice of coarse-grained unit cells

$$\delta\underline{\phi}_{\underline{r}} = \frac{1}{\sqrt{N}} \sum_{\underline{k}} \exp[+i\underline{k}\cdot\underline{r}] \frac{\underline{\phi}_{\underline{k}}}{\sqrt{\Omega_c}} \qquad (9.227)$$

in which the factor involving the unit-cell volume is being incorporated into the independent variables. The inverse Fourier Transform can then be expressed as a summation

$$\frac{\underline{\phi}_{\underline{k}}}{\sqrt{\Omega_c}} = \frac{1}{\sqrt{N}} \sum_{\underline{r}=1}^{N} \exp[-i\underline{k}\cdot\underline{r}] \, \delta\underline{\phi}_{\underline{r}} \qquad (9.228)$$

From this discrete finite-dimensional formulation of the transformation, one finds that the Jacobian J_N is unity. This follows from applying the transformation in Eq. (9.228) twice, which, using periodic boundary conditions, leads to

$$\widetilde{\delta\phi}_{\underline{r}'} = \frac{1}{N} \sum_{\underline{r},\underline{k}} \exp[-i\underline{k} \cdot (\underline{r} + \underline{r}')] \, \underline{\delta\phi}_{\underline{r}}$$

$$= \pm \underline{\delta\phi}_{\underline{r}'} \tag{9.229}$$

Hence, the Jacobian of the doubled transformation is found as

$$\det \frac{\widetilde{\phi}_{i,r}}{\phi_{i,r}} = \det \left(\frac{\phi_{i,k}}{\sqrt{\Omega_c} \, \phi_r} \right)^2 = \pm 1 \tag{9.230}$$

which leads to the identification

$$J_N = \left| \det \left(\frac{\phi_{i,k}}{\sqrt{\Omega_c} \, \phi_r} \right) \right| = 1 \tag{9.231}$$

Thus, the functional integral can be re-expressed in terms of integrals over the Fourier components $\phi_{i,\underline{k}}$

$$Z = \exp[-\beta F]$$

$$= \prod_{i,\underline{k}} \left\{ \frac{\int d\phi_{i,k}}{\sqrt{\Omega_c}} \right\} \exp[-\beta F[\phi]] \tag{9.232}$$

where we are formally assuming that the Fourier components $\phi_{i,\underline{k}}$ are independent fields. The Gaussian approximation only retains terms in $F[\phi]$ up to quadratic order in the $\phi_{i,\underline{k}}$ and is diagonal in this Fourier Transformed basis. This truncation allows the functional integral of the resulting approximate integrand to be evaluated exactly.

Like the mean-field approximation, the Gaussian approximation takes on different forms in the ordered and disordered phases.

9.6.1   *The Gaussian Approximation for $T > T_c$*

For $T > T_c$ the mean-field order parameter is given by $\underline{\phi}_0 = 0$, and the non-trivial part of the Free-Energy Functional can be expressed as

$$\int d^d\underline{r}[F_2\underline{\delta\phi}^2(\underline{r}) + c(\nabla \, \underline{\delta\phi})^2 + F_4(\underline{\delta\phi}^2(\underline{r}))^2] \tag{9.233}$$

which can be written in terms of the Fourier components. The quadratic terms

$$\int d^d\underline{r}(F_2\underline{\delta\phi}^2(\underline{r}) + c(\nabla \, \underline{\delta\phi})^2) \tag{9.234}$$

can be expressed as

$$\frac{1}{V} \sum_{\underline{k},\underline{k}'} \int d^d\underline{r} (F_2 \underline{\phi}_{\underline{k}} \cdot \underline{\phi}_{\underline{k}'} - c\underline{k} \cdot \underline{k}' \underline{\phi}_{\underline{k}} \cdot \underline{\phi}_{\underline{k}'}) \, \exp[i(\underline{k} + \underline{k}') \cdot \underline{r}] \quad (9.235)$$

The integration over $\underline{r}$ can be evaluated by using the identity

$$\frac{1}{V} \int d^d\underline{r} \, \exp[i(\underline{k} + \underline{k}') \cdot \underline{r}] = \delta_{\underline{k}+\underline{k}'} \quad (9.236)$$

where δ is the Kronecker delta function. Hence, the quadratic terms in the Free-Energy Functional reduce to

$$\sum_{\underline{k}} (F_2 \underline{\phi}_{\underline{k}} \cdot \underline{\phi}_{-\underline{k}} + c\underline{k} \cdot \underline{k} \underline{\phi}_{\underline{k}} \cdot \underline{\phi}_{-\underline{k}}) \quad (9.237)$$

The quartic terms can be evaluated in a similar manner, by using the identity

$$\int d^d\underline{r} \, \exp[i(\underline{k}_1 + \underline{k}_2 + \underline{k}_3 + \underline{k}_4) \cdot \underline{r}] = V\delta_{\underline{k}_1+\underline{k}_2+\underline{k}_3+\underline{k}_4} \quad (9.238)$$

which expresses conservation of momentum. This procedure leads to the Free-Energy Functional being expressed in terms of the Fourier components by

$$\sum_{\underline{k}} (F_2 + c\underline{k}^2) \, \underline{\phi}_{\underline{k}} \cdot \underline{\phi}_{-\underline{k}} + \frac{1}{V} \sum_{\underline{k}_1, \underline{k}_2, \underline{k}_3} F_4 (\underline{\phi}_{\underline{k}_1} \cdot \underline{\phi}_{\underline{k}_2})(\underline{\phi}_{\underline{k}_3} \cdot \underline{\phi}_{-\underline{k}_1-\underline{k}_2-\underline{k}_3}) \quad (9.239)$$

In the Gaussian approximation, where $\phi_{\underline{k}}$ is assumed to be small, one neglects the fourth-order term proportional to F_4. Hence, in this approximation one has

$$F[\phi] \approx F_0 V + \sum_{\underline{k}} (F_2 + c\underline{k}^2) \, \underline{\phi}_{\underline{k}} \cdot \underline{\phi}_{-\underline{k}}$$

$$\approx F_0 V + \sum_{\underline{k}} \sum_{i=1}^{n} (A(T - T_c) + c\underline{k}^2) \, \phi_{i,\underline{k}} \, \phi_{i,-\underline{k}} \quad (9.240)$$

Thus, the Gaussian Free-Energy Functional is diagonal in the momentum space representation as can be seen by using the condition

$$\phi_{\underline{k}}^* = \phi_{-\underline{k}} \quad (9.241)$$

which relates the fields at points $\underline{k}$ and $-\underline{k}$. Since the two fields are not independent, it is convenient to partition $\underline{k}$-space into two disjoint regions; one region denoted by $'$ which contains the set of points $\underline{k}$ and a second region that contains all the points $-\underline{k}$ obtained by inversion of the points in the region $'$. The primed region, $'$, is chosen such that all points of k-space are contained in either the region $'$ or its inversion partner. The Gaussian functional integral is evaluated by first re-writing it as

$$Z = \exp[-\beta F]$$

$$= \prod_{i,\underline{k}} \left\{ \frac{\int d\phi_{i,\underline{k}}}{\sqrt{\Omega_c}} \right\} \exp[-\beta F[\phi]]$$

$$= \prod_{i,\underline{k}}^{\prime} \left\{ \frac{\int d\phi_{i,\underline{k}} \int d\phi_{i,-\underline{k}}}{\Omega_c} \right\} \exp[-\beta F[\phi]]$$

$$= \prod_{i,\underline{k}}^{\prime} \left\{ \frac{\int d\phi_{i,\underline{k}} \int d\phi_{i,\underline{k}}^*}{\Omega_c} \right\} \exp[-\beta F[\phi]] \tag{9.242}$$

where the values of $\underline{k}$ in the primed products are restricted to the region $'$. The variable of integration is changed from $\phi_{i,\underline{k}}$ and $\phi_{i,\underline{k}}^*$ to the real and imaginary parts of the components of the field, $\Re e\ \phi_{i,\underline{k}}$ and $\Im m\ \phi_{i,\underline{k}}$

$$\phi_{i,\underline{k}} = \Re e\ \phi_{i,\underline{k}} + i\ \Im m\ \phi_{i,\underline{k}} \tag{9.243}$$

Hence, the partition function can be written as

$$Z = \prod_{i,\underline{k}}^{\prime} \left\{ 2 \frac{\int d\Re e\ \phi_{i,\underline{k}} \int d\Im m\ \phi_{i,\underline{k}}}{\Omega_c} \right\} \exp[-\beta F[\phi]] \tag{9.244}$$

where the Jacobian of the transformation is 2. The approximate Free-Energy functional is also re-written as a summation over $\underline{k}$ where the $\underline{k}$-values in the summation are restricted to the primed region.

$$F[\phi] \approx F_0 V + 2 \sum_{\underline{k}}^{\prime} \sum_{i=1}^{n} \left(A(T - T_c) + c\underline{k}^2 \right) \phi_{i,\underline{k}}\ \phi_{i,-\underline{k}}$$

$$\approx F_0 V + 2 \sum_{\underline{k}}^{\prime} \sum_{i=1}^{n} \left(A(T - T_c) + c\underline{k}^2 \right) \phi_{i,\underline{k}}\ \phi_{i,\underline{k}}^*$$

$$\approx F_0 V + 2 \sum_{\underline{k}}^{\prime} \sum_{i=1}^{n} \left[(A(T - T_c) + c\underline{k}^2) \right.$$

$$\left. \times ((\Re e\ \phi_{i,\underline{k}})^2 + (\Im m\ \phi_{i,\underline{k}})^2) \right] \tag{9.245}$$

On performing the Gaussian integrals, one finds that the Partition Function Z is approximated by

$$Z \approx \exp[-\beta F_0 V] \prod_{i=1}^{n} \prod_{\underline{k}}{}' \left[\frac{\Omega_c [A(T - T_c) + c\underline{k}^2]}{\pi k_B T} \right]^{-1}$$

$$\approx \exp[-\beta F_0 V] \prod_{i=1}^{n} \prod_{\underline{k}} \left[\frac{\Omega_c [A(T - T_c) + c\underline{k}^2]}{\pi k_B T} \right]^{-\frac{1}{2}} \tag{9.246}$$

where in the last line we have restored the product to run over the entire range of $\underline{k}$. In this expression, each of the n components of the order parameter yields an identical factor. Thus, since

$$Z = \exp[-\beta F] \tag{9.247}$$

the Gaussian approximation to the Free-Energy, for $T > T_c$, is given by the expression

$$F \approx F_0 V + \frac{k_B T n}{2} \sum_{\underline{k}} \ln \left[\frac{\Omega_c [A(T - T_c) + c\underline{k}^2]}{\pi k_B T} \right] \tag{9.248}$$

where the summation runs over the full range of $\underline{k}$. The specific heat can be obtained from the expression

$$C = -T \left(\frac{\partial^2 F}{\partial T^2} \right) \tag{9.249}$$

Hence, C is found from

$$\frac{2C}{n k_B} = - \left(T^2 \frac{\partial^2}{\partial T^2} + 2T \frac{\partial}{\partial T} \right) \sum_{\underline{k}} \ln \left[\frac{\Omega_c (A(T - T_c) + c\underline{k}^2)}{\pi k_B T} \right] \tag{9.250}$$

The least singular term comes from the temperature derivative of the denominator of the argument of the logarithm. It yields the contribution

$$\frac{n}{2} k_B \tag{9.251}$$

which is $k_B/2$ per normal mode, as is expected from the Equipartition Theorem. The next least singular term converges for $d > 2$. The most divergent term in C is recognized to be

$$C \sim \frac{T^2 n k_B}{2} V \int \frac{d^d \underline{k}}{(2\pi)^d} \frac{A^2}{(A(T - T_c) + c\underline{k}^2)^2} \tag{9.252}$$

As we shall see, this exhibits different types of behavior depending on whether $d > 4$ or $d < 4$. On setting

$$x = k\xi \tag{9.253}$$

where the correlation length ξ is defined by

$$\xi^{-2} = \frac{A(T - T_c)}{c} \tag{9.254}$$

one finds that the leading divergence of the specific heat is given by

$$C \sim \frac{nk_B \, A^2 T^2}{2c^2} \, V\xi^{4-d} \int \frac{d^d x}{(2\pi)^d} \left(\frac{1}{1 + x^2} \right)^2 \tag{9.255}$$

The integral is convergent at large k for $4 > d$, in which case it is independent of the cut-off for k. For larger dimensions, d, the integral may exhibit an ultra-violet divergence if the cut-off is ignored. However, the upper cut-off for x does depend on the lattice spacing a and is given by $x_c \sim \xi\Lambda \sim \pi\xi/a$, so the expression for the part of C displayed becomes independent of ξ when $d > 4$, leading to

$$C \sim \frac{nk_B \, A^2 T^2}{2c^2} \, Va^{4-d} \tag{9.256}$$

Thus, we have found that for $d > 4$ the critical exponent is given by $\alpha = 0$. In general, the only divergences of interest are those which occur at $k \sim 0$ when $\xi \to \infty$. Divergences that occur due to the behavior of the integrand in the region $k \sim 0$ are known as infra-red divergences. For $4 > d$, the integral in Eq. (9.255) is convergent when ξ is finite, and one obtains

$$C \sim \xi^{4-d}$$
$$\sim (T - T_c)^{-\nu(4-d)} \tag{9.257}$$

This expression reflects the infra-red divergence which occurs at $T = T_c$. Thus, in the Gaussian approximation and with $4 > d$, the critical exponent α has been calculated as

$$\alpha = \nu(4 - d)$$
$$= 2 - \frac{d}{2} \tag{9.258}$$

which differs from the value $\alpha = 0$ found from the discontinuous specific heat as calculated in the mean-field approximation. For $d > 4$, the Gaussian approximation was found to yield the specific heat exponent $\alpha = 0$, just like the value of α found in the mean-field approximation. The exponents of the Gaussian and the mean-field approximation first coincide when $d = 4$. The difference in the α the exponents found for $d < 4$ indicates that the fluctuations of the order parameter must be included for dimensions less than four.

9.6.2   *The Gaussian Approximation for $T_c > T$*

For temperatures below T_c, the mean-field order parameter is non-zero and will be parallel to any uniform applied field, Π_L, no matter how small. We shall orient our coordinate system so that the applied field lies along one axis, the longitudinal axis. The longitudinal component of the order parameter, $\phi_L(\underline{r})$, will be written as the sum of the mean-field order parameter ϕ_L and the spatially varying fluctuations

$$\phi_L(\underline{r}) = \phi_L + \delta\phi_L(\underline{r}) \tag{9.259}$$

The remaining $(n-1)$ components are transverse components which represent truly spatially varying fluctuations, i.e. they have no uniform components. The transverse components will be denoted as

$$\delta\phi_j(\underline{r}) \tag{9.260}$$

for $j = 1, 2, \ldots, n-1$. On substituting these expressions into the Ginzburg-Landau Functional, one obtains

$$
\begin{aligned}
F[\phi] = {}& V[F_0 + F_2\phi_L^2 + F_4\phi_L^4 - \Pi_L\phi_L] \\
&+ \int d^d\underline{r} \; \Big[F_2\delta\phi_L^2(\underline{r}) + 6F_4\phi_L^2\delta\phi_L^2(\underline{r}) + c(\nabla\delta\phi_L)^2 \\
&+ F_2 \sum_{j=1}^{n-1} \delta\phi_j^2(\underline{r}) + 2F_4\,\phi_L^2 \sum_{j=1}^{n-1} \delta\phi_j^2(\underline{r}) + c \sum_{j=1}^{n-1} (\nabla\delta\phi_j)^2 \\
&+ 4F_4\phi_L\delta\phi_L^3(\underline{r}) + 4F_4\phi_L\delta\phi_L(\underline{r}) \sum_{j=1}^{n-1} \delta\phi_j^2(\underline{r}) \\
&+ F_4\delta\phi_L^4(\underline{r}) + 2F_4\delta\phi_L^2(\underline{r}) \sum_{j=1}^{n-1} \delta\phi_j^2(\underline{r}) + F_4 \sum_{i,j}^{n-1} \delta\phi_i^2(\underline{r})\,\delta\phi_j^2(\underline{r}) \Big]
\end{aligned}
\tag{9.261}
$$

The first line represents the Landau-Ginzburg Free-Energy for a uniform longitudinal order parameter. The Gaussian approximation consists of minimizing the first line, as in mean-field theory, and retains the terms in the second and third lines as they are of quadratic order in the fluctuations. The terms in the last two lines are neglected, since they are of cubic and quartic order in the fluctuations. The fluctuating parts of the fields are expressed in terms of their Fourier components

$$\delta\phi_L(\underline{r}) = \frac{1}{\sqrt{V}} \sum_{\underline{k}} \exp[+i\underline{k}\cdot\underline{r}]\; \phi_{L,\underline{k}} \tag{9.262}$$

and

$$\delta\phi_j(\underline{r}) = \frac{1}{\sqrt{V}} \sum_{\underline{k}} \exp[+i\underline{k} \cdot \underline{r}] \; \phi_{j,\underline{k}} \tag{9.263}$$

On substituting into the Gaussian approximation for the Free-Energy Functional, one obtains

$$F[\phi] \approx V[F_0 + F_2\phi_L^2 + F_4\phi_L^4 - \Pi_L\phi_L]$$
$$+ \sum_{\underline{k}} [F_2 + 6F_4\phi_L^2 + c\underline{k}^2] \; \phi_{L,\underline{k}} \; \phi_{L,-\underline{k}}$$
$$+ \sum_{j=1}^{n-1} \sum_{\underline{k}} [F_2 + 2F_4\phi_L^2 + c\underline{k}^2] \; \phi_{j,\underline{k}} \; \phi_{j,-\underline{k}} \tag{9.264}$$

Since ϕ_L minimizes the first term in the approximate Free-Energy Functional it satisfies

$$[2F_2 + 4F_4\phi_L^2] \; \phi_L = \Pi_L \tag{9.265}$$

or

$$F_2 + 2F_4\phi_L^2 = \frac{\Pi_L}{2\phi_L} \tag{9.266}$$

On utilizing the expression for ϕ_L, one can write the approximate Free-Energy Functional as

$$F[\phi] \approx V \left[F_0 - F_4\phi_L^4 - \frac{\Pi_L\phi_L}{2} \right]$$
$$+ \sum_{\underline{k}} \left[4F_4\phi_L^2 + \frac{\Pi_L}{2\phi_L} + c\underline{k}^2 \right] \phi_{L,\underline{k}} \; \phi_{L,-\underline{k}}$$
$$+ \sum_{j=1}^{n-1} \sum_{\underline{k}} \left[\frac{\Pi_L}{2\phi_L} + c\underline{k}^2 \right] \phi_{j,\underline{k}} \; \phi_{j,-\underline{k}} \tag{9.267}$$

The longitudinal and transverse fluctuations have different behaviors. It is seen that the longitudinal fluctuations are primarily stabilized by the non-zero order parameter, whereas only the applied field stabilizes the transverse fluctuations. The Gaussian functional integral can be evaluated leading to the Gaussian approximation to the Free-Energy

$$F \approx F[\phi_L] + \frac{k_BT}{2} \sum_{\underline{k}} \ln \left[\frac{\Omega_c \left[4F_4\phi_L^2 + \frac{\Pi_L}{2\phi_L} + c\underline{k}^2 \right]}{\pi k_BT} \right]$$
$$+ (n-1) \frac{k_BT}{2} \sum_{\underline{k}} \ln \left[\frac{\Omega_c \left[\frac{\Pi_L}{2\phi_L} + c\underline{k}^2 \right]}{\pi k_BT} \right] \tag{9.268}$$

The Free-Energy can be used to calculate the divergent part of the specific heat and its critical exponent α. Below T_c, the specific heat is given by the sum of the contributions the mean-field theory and the longitudinal Gaussian fluctuations. Note that the amplitude of the singular parts of the specific heat are different above and below T_c. Above T_c there is a factor of n in the singular part of specific heat since all fluctuations are equivalent, whereas at low temperatures only the longitudinal fluctuations contribute to the singular part of the specific heat as the transverse fluctuations do not depend on $(T_c - T)$. Also, due to the factor of $2A$ in the expression for the inverse squared correlation length ξ^{-2}, the singular part of the specific heat has been increased by a factor of $2^{\frac{d}{2}}$.

The Gaussian approximation usually becomes unreliable for temperatures close to T_c, due to the neglect of the higher-order terms in the Landau-Ginzburg Free-Energy Functional. However, the Gaussian approximation does yield reasonable results for a range of temperatures. This can be ascertained by comparing the contribution of the mean-field Free-Energy to physical properties with the contributions due to fluctuations.

9.7   Gaussian Correlations

Here we shall examine Correlation Functions, based on the Gaussian Approximation. First we shall define the functional derivative.

Functional Derivatives:

Consider a functional of the form

$$F[\phi] = \int d^d\underline{r} \, [\mathcal{F}(\underline{r}) \, P(\phi(\underline{r})) + c(\underline{\nabla}\phi(\underline{r}))^2] \tag{9.269}$$

where $P(\phi(\underline{r}))$ is a polynomial. The functional derivative is denoted by

$$\frac{\delta F[\phi]}{\delta \phi(\underline{r}_0)} \tag{9.270}$$

The functional derivative is found by setting

$$\phi(r) \to \phi(\underline{r}) + \delta\phi(\underline{r}) \tag{9.271}$$

where $\delta\phi$ is a small peak located at $\underline{r}_0$, so

$$\delta\phi(\underline{r}) = \lambda\delta^d(\underline{r} - \underline{r}_0) \tag{9.272}$$

The functional derivative is found by taking the derivative w.r.t. λ and then setting $\lambda = 0$. Thus, after integrating by parts and setting the boundary term to zero, the functional derivative is calculated to be

$$\frac{\delta F[\phi]}{\delta \phi(\underline{r}_0)} = \left(\mathcal{F}(\underline{r}_0) \frac{\partial P(\phi(\underline{r}_0))}{\partial \phi(\underline{r}_0)} - 2c \nabla^2 \phi(\underline{r}_0) \right) \tag{9.273}$$

In the presence of a spatially varying applied field $\underline{\Pi}(\underline{r})$, the Free-Energy Functional is given by

$$F[\underline{\phi}, \underline{\Pi}] = F[\underline{\phi}] - \int d^d \underline{r}\ \underline{\Pi}(\underline{r}) \cdot \underline{\phi}(\underline{r}) \tag{9.274}$$

which can be regarded as Legendre Transformation of $F[\phi]$. The partition function can be expressed as

$$Z[\underline{\Pi}] = \int \mathcal{D}\phi\ \exp[-\beta F[\underline{\phi}, \underline{\Pi}]] \tag{9.275}$$

The partition function can be regarded as a functional of the applied field $\underline{\Pi}(\underline{r})$. The partition function $Z[\underline{\Pi}]$ is also a generating function for cumulant correlation functions. The functional derivative of $\ln Z[\underline{\Pi}]$ is evaluated as

$$\frac{\delta \ln Z[\underline{\Pi}]}{\delta \underline{\Pi}(\underline{r}_0)} = \frac{\beta}{Z} \int \mathcal{D}\phi\ \underline{\phi}(\underline{r}_0)\ \exp[-\beta F[\underline{\phi}, \underline{\Pi}]]$$

$$= \beta \langle \underline{\phi}(\underline{r}_0) \rangle \tag{9.276}$$

Likewise, the second-order functional derivative is given by

$$\frac{\delta^2 \ln Z[\underline{\Pi}]}{\delta \underline{\Pi}(\underline{r}_0)\, \delta \underline{\Pi}(\underline{r}_1)} = \frac{\beta^2}{Z} \int \mathcal{D}\phi\ \underline{\phi}(\underline{r}_0)\ \underline{\phi}(\underline{r}_1)\ \exp[-\beta F[\underline{\phi}, \underline{\Pi}]]$$

$$- \frac{\beta}{Z} \int \mathcal{D}\phi\ \underline{\phi}(\underline{r}_0)\ \exp[-\beta F[\underline{\phi}, \underline{\Pi}]]$$

$$\times \frac{\beta}{Z} \int \mathcal{D}\phi\ \underline{\phi}(\underline{r}_1)\ \exp[-\beta F[\underline{\phi}, \underline{\Pi}]]$$

$$= \beta^2 \left(\langle \underline{\phi}(\underline{r}_0)\ \underline{\phi}(\underline{r}_1) \rangle - \langle \underline{\phi}(\underline{r}_0) \rangle \langle \underline{\phi}(\underline{r}_1) \rangle \right) \tag{9.277}$$

which, apart from factors of β, is clearly related to the derivation of the correlation function $S(\underline{r} - \underline{r}')$ that appeared in the Linear Response Theory of Section 9.3.

In the Gaussian approximation, for $T > T_c$, the Free-Energy Functional is given by

$$F[\delta\underline{\phi}, \underline{\Pi}] = \int d^d\underline{r} \; [F_2 \; \delta\underline{\phi}^2(\underline{r}) + c(\underline{\nabla} \; \delta\underline{\phi})^2 - \delta\underline{\phi}(\underline{r}) \cdot \underline{\Pi}(\underline{r})] \qquad (9.278)$$

The partition function can be evaluated by expressing the Free-Energy functional in terms of the Fourier components

$$F[\delta\underline{\phi}, \underline{\Pi}] = \sum_{\underline{k}} \left[(F_2 + c\underline{k}^2) \; \underline{\phi}_{\underline{k}} \; \underline{\phi}_{-\underline{k}} - \frac{1}{2} (\underline{\phi}_{\underline{k}} \cdot \underline{\Pi}_{-\underline{k}} + \underline{\phi}_{-\underline{k}} \cdot \underline{\Pi}_{\underline{k}}) \right] \qquad (9.279)$$

which can be written as a complete square

$$\begin{aligned}
F[\delta\underline{\phi}, \underline{\Pi}] = \sum_{\underline{k}} & \left[(F_2 + c\underline{k}^2) \right. \\
& \left. \times \left(\underline{\phi}_{\underline{k}} - \frac{\underline{\Pi}_{\underline{k}}}{2(F_2 + c\underline{k}^2)} \right) \cdot \left(\underline{\phi}_{-\underline{k}} - \frac{\underline{\Pi}_{-\underline{k}}}{2(F_2 + c\underline{k}^2)} \right) \right] \\
& - \sum_{\underline{k}} \frac{\underline{\Pi}_{\underline{k}} \cdot \underline{\Pi}_{-\underline{k}}}{4(F_2 + c\underline{k}^2)} \qquad (9.280)
\end{aligned}$$

Therefore, the application of the field $\underline{\Pi}(\underline{r})$ results in a multiplicative factor in the expression for the partition function

$$Z[\underline{\Pi}] = Z[\Pi = 0] \; \exp\left[\sum_{\underline{k}} \frac{\beta}{4} \frac{\underline{\Pi}_{\underline{k}} \cdot \underline{\Pi}_{-\underline{k}}}{(F_2 + c\underline{k}^2)} \right] \qquad (9.281)$$

On transforming the applied fields back to real space, one obtains

$$\begin{aligned}
Z[\underline{\Pi}] &= Z[\Pi = 0] \; \exp\left[\frac{\beta}{4V} \int d^d\underline{r} \int d^d\underline{r}' \; \underline{\Pi}(\underline{r}') \cdot \underline{\Pi}(\underline{r}) \right. \\
& \left. \times \sum_{\underline{k}} \frac{\exp[i\underline{k} \cdot (\underline{r} - \underline{r}')]}{(F_2 + c\underline{k}^2)} \right] \\
&= Z[\Pi = 0] \; \exp\left[\frac{\beta^2}{2} \int d^d\underline{r} \int d^d\underline{r}' \; \underline{\Pi}(\underline{r}') \cdot \underline{\Pi}(\underline{r}) \; S(\underline{r} - \underline{r}') \right] \qquad (9.282)
\end{aligned}$$

where $S(\underline{r})$ is the correlation function

$$S(\underline{r}) = \left(\frac{k_B T}{2c} \right) \int \frac{d^d\underline{k}}{(2\pi)^d} \frac{\exp[i\underline{k} \cdot \underline{r}]}{\xi^{-2} + \underline{k}^2} \qquad (9.283)$$

Hence, the second-order Functional Derivative of $\ln Z[\underline{\Pi}]$ wrt $\underline{\Pi}$, yields

$$\frac{\delta^2 \ln Z[\underline{\Pi}]}{\delta \underline{\Pi}(\underline{r}_0)\, \delta \underline{\Pi}(\underline{r}_1)} = \beta^2\, S(\underline{r}_0 - \underline{r}_1)$$

$$= \beta^2 (\langle \underline{\phi}(\underline{r}_0)\, \underline{\phi}(\underline{r}_1)\rangle - \langle \underline{\phi}(\underline{r}_0)\rangle\langle \underline{\phi}(\underline{r}_1)\rangle) \qquad (9.284)$$

in accord with our previous results. As could be anticipated, Eq. (9.283) shows that fluctuations which have larger amplitudes correspond to the fluctuations which have a smaller cost in the Free-Energy.

The d-dimensional "*Ornstein-Zernike*" correlation function $S(\underline{r})$ can be evaluated from the inverse Fourier Transformation

$$S(\underline{r}) = \left(\frac{k_B T}{2c}\right) \int \frac{d^d \underline{k}}{(2\pi)^d}\, \exp[i\underline{k}\cdot\underline{r}]\, S(\underline{k})$$

$$= \left(\frac{k_B T}{2c}\right) \int \frac{d^d \underline{k}}{(2\pi)^d}\, \frac{\exp[i\underline{k}\cdot\underline{r}]}{\xi^{-2} + \underline{k}^2} \qquad (9.285)$$

The evaluation proceeds by first raising the denominator to an exponent by using the expression

$$\frac{1}{\xi^{-2} + \underline{k}^2} = \int_0^\infty dt\, \exp[-t(\xi^{-2} + \underline{k}^2)] \qquad (9.286)$$

The inverse Fourier Transform can then be evaluated in a d-dimensional Cartesian coordinate system by using the identity

$$\int_{-\infty}^\infty dk_i\, \exp[ik_i r_i]\, \exp[-tk_i^2] = \sqrt{\frac{\pi}{t}}\, \exp\left[-\frac{r_i^2}{4t}\right] \qquad (9.287)$$

d times. This results in the expression

$$S(r) = \left(\frac{k_B T}{2c}\right) \frac{\pi^{\frac{d}{2}}}{(2\pi)^d} \int_0^\infty dt\, t^{-\frac{d}{2}}\, \exp\left[-\frac{t}{\xi^2} - \frac{r^2}{4t}\right] \qquad (9.288)$$

which, on changing variables from t to s where $t = \xi^2 s$, leads to

$$S(r) = \left(\frac{k_B T}{2c}\right) \frac{1}{(2\pi)^{\frac{d}{2}}} \left(\frac{\xi^{2-d}}{2^{\frac{d}{2}}}\right) \int_0^\infty ds\, s^{-\frac{d}{2}}\, \exp\left[-s - \frac{1}{4s}\frac{r^2}{\xi^2}\right] \qquad (9.289)$$

Since the modified Bessel function has the integral representation

$$K_\nu(x) = \frac{1}{2} \left(\frac{x}{2}\right)^\nu \int_1^\infty ds\, s^{-(\nu+1)}\, \exp\left[-s - \frac{x^2}{4s}\right] \qquad (9.290)$$

it can be substituted into the expression for the correlation function, which yields the final result

$$S(r) = \left(\frac{k_B T}{2c}\right) \frac{1}{(2\pi)^{\frac{d}{2}} r^{d-2}} \left(\frac{r}{\xi}\right)^{\frac{d-2}{2}} K_{\frac{d-2}{2}}(r/\xi) \qquad (9.291)$$

The above expression for $S(r)$ is the "*Ornstein-Zernike*" correlation function.

The integral representation for the modified Bessel function can be expressed in an alternate form

$$K_\nu(x) = \frac{\pi^{\frac{1}{2}}}{(\nu - \frac{1}{2})!} \left(\frac{x}{2}\right)^\nu \int_1^\infty dt \ \exp[-xt](t^2 - 1)^{\nu - \frac{1}{2}} \qquad (9.292)$$

for $\nu > -\frac{1}{2}$. On changing variable

$$t = 1 + \frac{s}{x} \qquad (9.293)$$

one obtains the expression

$$K_\nu(x) = \sqrt{\frac{\pi}{2x}} \frac{\exp[-x]}{(\nu - \frac{1}{2})!} \int_0^\infty ds \left[\exp[-s] \ s^{\nu - \frac{1}{2}} \left(1 + \frac{s}{2x}\right)^{\nu - \frac{1}{2}}\right] \qquad (9.294)$$

By extracting a factor of $\frac{s}{2x}$ from the last factor, one can expand in powers of x to show that

$$\lim_{x \to 0} K_\nu(x) \to \frac{(\nu - 1)! \ 2^{\nu - 1}}{x^\nu} \qquad (9.295)$$

for $\nu > 0$. An asymptotic expansion can be found by expanding the last factor in powers of x^{-1} via the Binomial Theorem and integrating term by term. This process yields the asymptotic expansion

$$K_\nu(x) \sim \sqrt{\frac{\pi}{2x}} \ \exp[-x] \left[1 + \frac{(4\nu^2 - 1)}{1! \ (8x)}\right.$$
$$\left. + \frac{(4\nu^2 - 1)(4\nu^2 - 9)}{2! \ (8x)^2} + \ldots \right] \qquad (9.296)$$

The above results can be used to show that the transverse and all the correlation functions at the critical point where $\xi \to \infty$ behave very differently above and below $d = 2$. In particular, the results show that

$$\lim_{r \to \infty} S(r) \sim \frac{2a^{2-d}}{(d - 2) \ S_d} \qquad (9.297)$$

for $d > 2$,

$$\lim_{r \to \infty} S(r) \sim \frac{r^{2-d}}{(d-2) \, S_d} \qquad (9.298)$$

for $2 > d$ and

$$\lim_{r \to \infty} S(r) \sim -\frac{\ln r}{2\pi} \qquad (9.299)$$

for $d = 2$. This variation is related to the existence of Goldstone modes, which via the Mermin-Wagner Theorem leads to the identification of the "*Lower Critical Dimension*" $d_c^l = 2$, below which a phase with spontaneously broken continuous symmetry is unstable. The correlation function may also be used to indicate the value of the "*Upper-Critical Dimension*", above which the fluctuations may be treated in a Gaussian approximation.

9.7.1   *The Ginzburg Criterion*

The "*Ginzburg Criterion*" provides an estimate of the temperature range in which the results of mean-field theory may be reasonable. Mean-field theory (or the Gaussian approximation) may be considered reasonable whenever the fluctuations in the order parameter are smaller than the average value of the order parameter. The size of the mean-squared fluctuations can be estimated by $S(r)$ evaluated at a length scale given by the correlation length ξ. Hence, the results of mean-field theory may be reasonable when

$$1 > \frac{S(\xi)}{\phi_0^2} \qquad (9.300)$$

or, equivalently

$$\phi_0^2 > \left(\frac{k_B T}{8\pi c}\right) \frac{\exp[-1]}{\xi^{d-2}} \qquad (9.301)$$

which leads to

$$\frac{A(T_c - T)}{2F_4} > \left(\frac{k_B T}{8\pi c}\right) \left(\frac{2A(T_c - T)}{c}\right)^{\frac{d-2}{2}} \qquad (9.302)$$

or

$$\left(\frac{2\pi c^2}{k_B T F_4}\right) > \left(\frac{2A(T_c - T)}{c}\right)^{\frac{d-4}{2}} \qquad (9.303)$$

This suggests that, generally for $4 > d$, mean-field theory might be reasonable for temperatures outside the critical region which is a narrow temperature window around T_c. The fluctuations dominate in the critical region.

The Ginzburg criterion also indicates that mean-field theory, or the Gaussian approximation, might also be reasonable for all temperatures in four or higher dimensions. The upper critical dimension d_c^u is the dimension above which the critical exponents are those found in the mean-field approximation, and for an ordinary second-order transition with short-ranged interactions $d_c^u = 4$. Although, for dimensions above the upper critical dimension, the critical exponents are given by the mean-field values, fluctuations do change the amplitudes and give rise to non-singular contributions which may be large.

There is also a lower critical dimension d_c^l. Mermin and Wagner (See Appendix 10.4.1) have shown that a phase with spontaneously broken continuous symmetry is unstable for dimensions less than two, since long wavelength transverse fluctuations of the order parameter are divergent. In this case, the lower critical dimension d_c^l, below which a phase transition cannot occur, is $d_c^l = 2$. The divergence of the fluctuations for $2 > d$ found in systems with a continuously broken symmetry is related to the presence of Goldstone modes (See section 9.8). Due to the divergence of the fluctuations, the average value of the order parameter is not well-defined and, therefore, the fluctuations dynamically restore the broken symmetry. The suppression of ordering can be seen in a different way, by examining how T_c is reduced in the self-consistent Gaussian approximation.

9.7.2 *The Self-Consistent Gaussian Approximation*

To motivate the self-consistent Gaussian approximation, we first consider the Free-Energy obtained within the Gaussian approximation which has the form

$$F[\phi] \approx F[\phi_L] + \frac{k_B T}{2} \sum_{\underline{k}} \ln \left[\frac{\Omega_c (F_2 + 6F_4 \phi_L^2 + c\underline{k}^2)}{\pi k_B T} \right]$$

$$+ (n-1) \frac{k_B T}{2} \sum_{\underline{k}} \ln \left[\frac{\Omega_c (F_2 + 2F_4 \phi_L^2 + c\underline{k}^2)}{\pi k_B T} \right] \qquad (9.304)$$

This expression holds true for both T greater and T smaller than T_c. For temperatures above T_c, ϕ_L will be zero and the two logarithmic terms can be combined since there is no physical distinction between the longitudinal and transverse directions, if $\Pi = 0$. Minimization w.r.t. ϕ_L leads to the solutions of either

$$\phi_L = 0 \qquad (9.305)$$

or

$$0 = V[F_2 + 2F_4\phi_L^2] + \frac{k_B T}{2} \sum_{\underline{k}} \frac{6F_4}{F_2 + 6F_4\ \phi_L^2 + c\underline{k}^2}$$

$$+ (n-1)\ \frac{k_B T}{2} \sum_{\underline{k}} \frac{2F_4}{F_2 + 2F_4\ \phi_L^2 + c\underline{k}^2}$$

$$= V[F_2 + 2F_4\phi_L^2] + 6F_4 \sum_{\underline{k}} \langle \phi_{L,\underline{k}}\ \phi_{L,-\underline{k}} \rangle$$

$$+ (n-1)\ 2F_4 \sum_{\underline{k}} \langle \phi_{T,\underline{k}}\ \phi_{T,-\underline{k}} \rangle \tag{9.306}$$

where the last two terms have been recognized as involving the fluctuations of the order parameter, as evaluated in the Gaussian approximation. The critical temperature T_c is the temperature at which two infinitesimal but real solutions for ϕ_L first occur. This is to be contrasted with the mean-field critical temperature, $T_c^{(0)}$, defined by

$$F_2 = A(T - T_c^{(0)}) \tag{9.307}$$

A new critical temperature T_c is determined from the equation

$$0 = VF_2 + 2(n+2)\ F_4 \left(\frac{k_B T_c}{2}\right) \sum_{\underline{k}} \frac{1}{F_2 + c\underline{k}^2} \tag{9.308}$$

The last term is recognized as the Gaussian approximation expression for the correlation function, $S(0)$, evaluated at temperatures above T_c. The equation for the new critical temperature can then be expressed in the compact form

$$0 = F_2 + 2(n+2)\ F_4 S(0) \tag{9.309}$$

The term proportional to $S(0)$ is positive, since $S(0)$ is a measure of the local fluctuations of the order parameter

$$S(0) = \langle \phi(\underline{r})^2 \rangle \tag{9.310}$$

Therefore, the interaction F_4 and the fluctuations $S(0)$ reduce the value of T_c to below $T_c^{(0)}$. However, in this approximation, the susceptibility still diverges at the mean-field transition temperature $T_c^{(0)}$.

To cure this inconsistency, the correlation function $S(0)$ for the high-temperature phase ought to be calculated self-consistently using the "*Self-Consistent Gaussian Approximation*". In the self-consistent Gaussian approximation, one sets

$$\underline{\phi}_{\underline{k}} = \underline{\varphi}_{\underline{k}} + \delta\underline{\phi}_{\underline{k}} \tag{9.311}$$

and expands the Free-Energy functional to quadratic order in $\delta\phi_{\underline{k}}$ and replaces expressions involving $\varphi_{\underline{k}}$ by their average values which are to be calculated by a Gaussian integral over $\delta\phi_{\underline{k}}$. In the case under consideration, the only terms in question originate from the quartic F_4 terms and the approximation involves the replacement

$$\varphi_{i,\underline{k}}\,\varphi_{j,\underline{k}'} = \delta_{\underline{k}+\underline{k}'}\,\delta_{i,j}\langle\phi_{i,\underline{k}}\,\phi_{i,-\underline{k}}\rangle \tag{9.312}$$

Above T_c, the self-consistent Gaussian approximation for the Free-Energy is given by

$$F[\phi] = VF_0 + \sum_{\underline{k}} \left[F_2 + 2(n+2)\,F_4S(0) + c\underline{k}^2\right]\delta\phi_{\underline{k}}\cdot\delta\phi_{-\underline{k}} \tag{9.313}$$

where $S(0)$ is the correlation function which must be evaluated self-consistently from the above expression for the Free-Energy. Using the methods developed previously, a self-consistent expression for the correlation function $S(r)$ can be obtained from the self-consistent Free-Energy Functional. In the high-temperature phase, the result is

$$S(r) = \langle\phi(\underline{r})\,\phi(0)\rangle$$

$$= \frac{k_BT}{2V}\sum_{\underline{k}}\left(\frac{\exp[i\underline{k}\cdot\underline{r}]}{F_2 + 2(n+2)\,F_4S(0) + c\underline{k}^2}\right) \tag{9.314}$$

By taking the $r \to 0$ limit, one finds that $S(0)$ must satisfy the self-consistency equation

$$S(0) = \langle\phi(0)^2\rangle$$

$$= \frac{k_BT}{2V}\sum_{\underline{k}}\left(\frac{1}{F_2 + 2(n+2)\,F_4S(0) + c\underline{k}^2}\right) \tag{9.315}$$

Furthermore, the correlation function can be used to derive the uniform susceptibility via the relation

$$\chi = \beta\int d^d\underline{r}\,S(r)$$

$$= \frac{1}{2[F_2 + 2(n+2)\,F_4S(0)]} \tag{9.316}$$

Hence, one finds that the susceptibility diverges at the critical temperature T_c, which is determined by

$$F_2 + 2(n+2)\,F_4S(0) = 0 \tag{9.317}$$

thereby confirming the self-consistent nature of the approximation. At T_c, $S(0)$ reduces to the integral

$$S(0) = \left(\frac{k_B T_c}{2c}\right) \frac{S_d}{(2\pi)^d} \int dk \; k^{d-3} \qquad (9.318)$$

For $d > 3$, the integral is finite and of the order $\frac{a^{2-d}}{(d-2)}$, hence, one expects that the shift of T_c will be reasonably moderate. On the other-hand, for $3 > d$ the integral representing the order parameter fluctuations is divergent due to the behavior at $k \sim 0$, thereby suppressing T_c to much lower temperatures. The logarithmic divergence of the correction to T_c that occurs for $d = 2$ is consistent with the value of the lower critical dimension $d_c^l = 2$ that is inferred from the Mermin-Wagner Theorem (See Section 10.4.1).

To summarize, we have shown that the fluctuations of the order parameter, $S(0)$, are finite above T_c and their effect is to reduce T_c below the mean-field value of the critical temperature $T_c^{(0)}$. At the reduced critical temperature T_c the susceptibility diverges and the order parameter becomes finite.

9.8 Collective Modes and Symmetry Breaking

"Goldstone's Theorem" [84] pertains to phase transition where the Hamiltonian has a continuous symmetry that is spontaneously broken. That is, the ground state is infinitely degenerate and does not have the full symmetry of the Hamiltonian. The theorem states that in the broken symmetry state, the system will have branch of collective boson excitations that has a dispersion relation which reaches $\omega = 0$ at $k = 0$.

Furthermore, in the long wavelength limit, the bosons are non-interacting and a coherent superpositions of these bosons connects the broken symmetry state to the continuum of degenerate states. This is easily understood, since if the system was to be physically transformed from one broken symmetry state to another, no energy would have to be supplied to the system. Thus, the bosons dynamically restore the broken symmetry.

Goldstone bosons in the form of spin waves were already known to exist in ferromagnets and antiferromagnets [85], where the continuous spin rotational symmetry is spontaneously broken at low temperatures. For ferromagnets, the ground state and the spin-wave dispersion relations can be calculated exactly. Ironically, P.W. Anderson had already investigated the dynamic modes associated with a superconductor [86, 87], prior to Goldstone's work. Anderson had found, contrary to Goldstone's Theorem, that the bosons in a superconductor had a finite excitation energy similar to the plasmon energy of the metal. A posteriori, this is obvious since metals neither become transparent nor change colour when they start to superconduct. Anderson's idea was subsequently picked up by Peter Higgs [88] and by Tom Kibble and co-workers [89] and also by François Englert and Robert Brout [90] who noted that, if long-ranged interactions were present, the modes would acquire a mass. The massive modes, associated with the breaking of a continuous symmetry in the presence of long-ranged interactions, are known as Kibble-Higgs modes.

Here we shall examine the Goldstone bosons of a Heisenberg ferromagnet, which is a slightly unusual case since the order parameter of a ferromagnet is a conserved quantity.

The Ferromagnetic State

The fully polarized ferromagnetic state $|\Phi_0\rangle$ has all the spins aligned and is an exact eigenstate of the Heisenberg Hamiltonian. The Hamiltonian can be written as a scalar product

$$
\begin{aligned}
\hat{H} &= -\sum_{i,j} J_{i,j}\, \hat{\underline{S}}_i \cdot \hat{\underline{S}}_j \\
&= -\sum_{i,j} J_{i,j} \left[\hat{S}_i^z \hat{S}_j^z + \frac{1}{2}\, (\hat{S}_i^+ \hat{S}_j^- + \hat{S}_i^- \hat{S}_j^+) \right]
\end{aligned}
\tag{9.319}
$$

where the sum runs over pairs of sites. We shall assume that the spontaneous magnetization is parallel to the z-axis. Therefore, the ground state $|\Phi_0\rangle$ satisfies

$$
\hat{S}_i^z |\Phi_0\rangle = S |\Phi_0\rangle
\tag{9.320}
$$

for all lattice sites i. Then

$$
\begin{aligned}
\hat{H}\,|\Phi_0\rangle &= -\sum_{i,j} J_{i,j} \left[\hat{S}_i^z \hat{S}_j^z + \frac{1}{2}\, (\hat{S}_i^+ \hat{S}_j^- + \hat{S}_i^- \hat{S}_j^+) \right] |\Phi_0\rangle \\
&= -\sum_{i,j} J_{i,j} \left[SS + \frac{1}{2}\, (\hat{S}_i^+ \hat{S}_j^- + \hat{S}_i^- \hat{S}_j^+) \right] |\Phi_0\rangle \\
&= -\sum_{i,j} J_{i,j} S^2 |\Phi_0\rangle
\end{aligned}
\tag{9.321}
$$

The second line follows since all the spins are aligned with the z-axis and are eigenstates of $\hat{S}_i^z$ with eigenvalue S

$$\hat{S}_i^z|S_i\rangle = S|S_i\rangle \tag{9.322}$$

The third line occurs since the spin-flip terms vanish as they all involve the spin-raising operator at a site and

$$\hat{S}_i^+|S_i\rangle = 0 \tag{9.323}$$

as the spin cannot be raised further. Hence, the fully-polarized ferromagnetic state is an exact eigenstate with eigenvalue $E_0 = -\sum_{i,j} J_{i,j}S^2$. This state is infinitely degenerate since any rotation of the total magnetization leads to an equivalent symmetry broken ground state.

Conservation of Magnetization

The z-component of the total magnetization M is defined as

$$\hat{M}^z = \sum_{i=1}^{N} \hat{S}_i^z \tag{9.324}$$

The magnetization commutes with the Hamiltonian and so is conserved. The commutator

$$[\hat{M}^z, \hat{H}] \tag{9.325}$$

can be evaluated with the aid of the commutation relations

$$[\hat{S}_i^z, \hat{S}_j^+] = +\delta_{i,j}\hbar\hat{S}_i^+$$
$$[\hat{S}_i^z, \hat{S}_j^-] = -\delta_{i,j}\hbar\hat{S}_i^- \tag{9.326}$$

From which one finds that

$$[\hat{M}^z, \hat{H}] = \frac{\hbar}{2} \sum_{i,j} J_{i,j}(\hat{S}_i^+ \hat{S}_j^- - \hat{S}_i^- \hat{S}_j^+)$$
$$- \frac{\hbar}{2} \sum_{i,j} J_{i,j}(\hat{S}_i^+ \hat{S}_j^- - \hat{S}_i^- \hat{S}_j^+)$$
$$= 0 \tag{9.327}$$

independent of any choice for the exchange interaction. Hence, the total magnetization is conserved.

The Spin Wave Dispersion Relation

The spin wave state $|\Phi_{\underline{q}}\rangle$ is a linear superpositions of ferromagnetic states with a single flipped spin. The spin wave state can be expressed as

$$|\Phi_{\underline{q}}\rangle = \hat{S}_{\underline{q}}^- |\Phi_0\rangle$$

$$= \frac{1}{\sqrt{N}} \sum_j \exp[i\underline{q} \cdot \underline{R}_j] \, S_j^- |\Phi_0\rangle \tag{9.328}$$

which satisfies an energy-eigenvalue equation

$$\hat{H}|\Phi_{\underline{q}}\rangle = E_{\underline{q}}|\Phi_{\underline{q}}\rangle \tag{9.329}$$

The energy eigenvalue can be expressed in terms of the ground state energy E_0 and the spin wave excitation energy $\hbar\omega_{\underline{q}}$

$$E_{\underline{q}} = E_0 + \hbar\omega_{\underline{q}} \tag{9.330}$$

The spin wave dispersion relation can be found from

$$[\hat{H}, \hat{S}_{\underline{q}}^-] \, |\Phi_0\rangle = \hbar\omega_{\underline{q}} \hat{S}_{\underline{q}}^- |\Phi_0\rangle \tag{9.331}$$

by using the commutation relations

$$[\hat{S}_j^z, \hat{S}_i^-] = -\delta_{i,j} \, \hbar\hat{S}_i^- \tag{9.332}$$

and

$$[\hat{S}_j^+, \hat{S}_i^-] = 2\delta_{i,j} \, \hbar\hat{S}_i^z \tag{9.333}$$

On using the above two commutation relations, one finds that the commutation relation between $S_{\underline{q}}^-$ and the Hamiltonian produces

$$[\hat{H}, \hat{S}_{\underline{q}}^-] = \frac{1}{\sqrt{N}} \sum_{i,j} J_{i,j} \hbar[\hat{S}_i^z \hat{S}_j^- - \hat{S}_i^- \hat{S}_j^z] \, \exp[i\underline{q} \cdot \underline{R}_j]$$

$$+ \frac{1}{\sqrt{N}} \sum_{i,j} J_{i,j} \hbar[\hat{S}_i^- \hat{S}_j^z - \hat{S}_i^z \hat{S}_j^-] \, \exp[i\underline{q} \cdot \underline{R}_i] \tag{9.334}$$

Hence, when acting on the ferromagnetic state, the commutation relation reduces to

$$[\hat{H}, \hat{S}_{\underline{q}}^-]|\Phi_0\rangle = \frac{1}{\sqrt{N}} \sum_{i,j} J_{i,j} \hbar S[\hat{S}_j^- - \hat{S}_i^-] \exp[i\underline{q} \cdot \underline{R}_j] \, |\Phi_0\rangle$$

$$+ \frac{1}{\sqrt{N}} \sum_{i,j} J_{i,j} \hbar S[\hat{S}_i^- - \hat{S}_j^-] \, \exp[i\underline{q} \cdot \underline{R}_i] \, |\Phi_0\rangle \tag{9.335}$$

which can be further reduced by noting that since the pairwise interaction only depends on the nearest-neighbor separation and not the absolute location in the lattice. Therefore, on writing the above expression in terms of $\hat{S}_{\underline{q}}^-$ and a sum over the nearest-neighbors sites, one has

$$[\hat{H}, \hat{S}_{\underline{q}}^-]\,|\Phi_0\rangle = \sum_i J_{i,0}\hbar S(1 - \cos[\underline{q} \cdot \underline{R}_{i,0}])\,\hat{S}_{\underline{q}}^-|\Phi_0\rangle \qquad (9.336)$$

Thus, the dispersion relation is evaluated as a sum over nearest-neighbor sites

$$\hbar\omega_{\underline{q}} = \sum_i J_{0,i}\hbar S(1 - \cos[\underline{q} \cdot \underline{R}_{0,i}]) \qquad (9.337)$$

Hence, the spin-wave state is an exact eigenstate of the fully-polarized ferromagnet.

The dispersion relation vanishes quadratically as $q \to 0$

$$\hbar\omega_q \sim Dq^2 \qquad (9.338)$$

where D is the spin-wave stiffness constant. Usually, Goldstone modes have linear dispersion relations, however, due to the conserved nature of the order parameter, ferromagnetic spin waves have quadratic dispersion relations. In the limit of $q \to 0$, one recognizes that

$$\lim_{q \to 0} \hat{S}_{\underline{q}}^- = \frac{1}{\sqrt{N}}\sum_i \hat{S}_i^- \qquad (9.339)$$

so the operator reduces the z-component of the uniform magnetization by $\hbar$. Since the state with one excited spin wave state with $q = 0$ and $\omega_q = 0$ corresponds to a state in which the z-component of the ground state magnetization is uniformly reduced by $\hbar$, it corresponds to a new ground state in which the direction of the magnetization has been rotated through an infinitesimal angle. Since a uniform rotation of the magnetization should not affect the system's excitations, the Goldstone modes are non-interacting. Therefore, a coherent state made of multiple spin wave excitations of Goldstone modes with $q \to 0$ and $\omega \to 0$ can dynamically restore the spin-rotational symmetry that is broken by the ferromagnetic transition.

9.9   Problems

Problem 9.1

Consider the following model partition function developed by Uhlenbeck and Ford to describe a gas with a hard-core repulsive potential

$$Z_N(V) = \sum_{n=0}^{N} \frac{V!}{(V - N + n)! \, (N - n)!}$$

where z is the fugacity and the volume V is represented by a positive integer.
(i) Show that the Grand-Canonical Partition function is given by

$$\Xi(V, z) = \sum_{N=0}^{V} z^N Z_N(V)$$

$$= \left(\frac{z^{V+1} - 1}{z - 1} \right) (1 + z)^V$$

(ii) Determine the zeroes of the partition function.
(iii) Determine the forms of the Grand-Canonical Potential $\Omega(V, z)$ in the thermodynamic limit

$$-\beta \lim_{V \to \infty} \Omega(V, z) \to V \, \ln(1 + z) \quad \text{if} \quad |z| < 1$$

$$-\beta \lim_{V \to \infty} \Omega(V, z) \to V \, \ln z(1 + z) \quad \text{if} \quad |z| > 1$$

(iv) Using thermodynamics, find the equations of state and show that the model exhibits a first-order transition at $z = 1$. Determine the jump in volume.

Problem 9.2

By using a duality relationship between the high temperatures and low temperature physics of the two-dimensional Ising Model, Kramers and Wannier [91] showed that a phase transition, if it exists and is unique, would occur at a specific temperature T_c. The high temperature and low temperature expansions for the partition function $Z(\beta J)$ can be used to give the following duality relation for the partition function at a high temperature T and a low temperature T^*

$$Z_N(\beta^* J) = 2(\sinh(2\beta J))^{-N} Z_N(\beta J)$$

where

$$\exp[-2\beta^* J] = \tanh \beta J$$

(i) Find the relationship between the Free-Energies at T^* and T, in the thermodynamic limit $N \to \infty$.

(ii) Show that if both the expressions for the Free-Energies diverge at a common temperature T_c, then T_c is given by

$$\sinh(2\beta_c J) = 1$$

(iii) Discuss the possibility that the Free-Energy diverges at pairs of temperatures T_c^* and T_c.

Problem 9.3

Consider an Ising model with ferromagnetic interactions

$$H = -J \sum_{i,j} S_i^z S_j^z$$

where $J > 0$ and the sum i, j runs over z nearest-neighbor spins.

(i) Determine the mean-field equations for the average spin per site $\langle S_i^z \rangle = m$.

(ii) Determine the mean-field approximation for the transition temperature T_c.

Problem 9.4

The long-ranged Ising model has a Hamiltonian given by

$$H = -\frac{J}{2N} \sum_{i=1}^{N} S_i \sum_{j=1}^{N} S_j - B \sum_{i=1}^{N} S_i$$

where the factor N is introduced in the interaction to keep the energy extensive.

(i) Prove the identity

$$\exp\left[\frac{\beta J}{2N} \sum_{i=1}^{N} S_i \sum_{j=1}^{N} S_j \right] = \frac{1}{\sqrt{\pi}} \int_{-\infty}^{\infty} dx \, \exp\left[-x^2 + 2x\sqrt{\frac{\beta J}{2N}} \sum_{i=1}^{N} S_i \right]$$

(ii) Hence, show that the partition function can be written in the form of

$$Z_N = \frac{1}{\sqrt{\pi}} \int_{-\infty}^{\infty} dx \, \exp[F(x)]$$

where

$$F(x) = -x^2 + N \ln\left(2 \cosh\left[\beta B + \sqrt{\frac{2\beta J}{N}} \, x \right] \right)$$

(iii) Therefore, show that Z_N, with long-ranged interactions, can be evaluated exactly in the thermodynamic limit and compare your result with the mean-field result with $J \to \frac{J}{N}$ and $z \to N$.

Problem 9.5

Consider an Ising antiferromagnet $(J > 0)$

$$H = +J \sum_{i,j} S_i^z S_j^z - B \sum_i S_i^z$$

on a bi-partite lattice. Let the average moment on the A sub-lattice be m_A and the average spin on the B sub-lattice be m_B,
(i) Show that the mean-field equation for m_A and m_B in the presence of a uniform applied magnetic field are given by

$$m_A = \tanh \beta (B - z J m_B)$$
$$m_B = \tanh \beta (B - z J m_A)$$

(ii) Determine the critical temperature T_N for $H = 0$ at which one first finds a non-zero solution for $m_A = -m_B$ when $B \to 0$.
(iii) Show that the sub-lattice susceptibilities in the zero-field limit are given by

$$\chi_A = (1 - z J \chi_B) \beta \operatorname{sech}^2 \beta z J m$$
$$\chi_B = (1 - z J \chi_A) \beta \operatorname{sech}^2 \beta z J m$$

where $m_A = -m_B = 0$ when $B \to 0$.
(iv) Hence, show that the uniform magnetic susceptibility, considered as a function of T, is given by

$$\chi = (1 - z J \chi) \beta \operatorname{sech}^2 \beta z J m$$
$$= \frac{(1 - m^2)}{k_B T + z J (1 - m^2)}$$

and has a cusp at $T = T_N$.
(v) Express χ for $T > T_N$ in terms of T and T_N.

Problem 9.6

A ferrimagnet is a magnet with two interpenetrating sublattices, in which spins of magnitude S_A reside on the A sub-lattice and spins of magnitude S_B on the B sublattice. The magnitudes of the spin are unequal, $S_A > S_B$. Each site has the coordination number z. The Hamiltonian is given by

$$H = -J \sum_{i,j} \underline{S}_i \cdot \underline{S}_j - g_A \mu \sum_i \underline{S}_i \cdot \underline{B} - g_B \mu \sum_j \underline{S}_j \cdot \underline{B}$$

(i) Determine the mean-field equations for m_A and m_B.
(ii) Determine the susceptibility χ

$$\chi = \mu \frac{N}{2} \frac{\partial}{\partial B} (g_A m_A + g_B m_B)$$

and determine the temperature T_N at which the susceptibility diverges.

Problem 9.7

Consider the classical Heisenberg model in which a spin is represented by a unit vector in three-dimensions

$$\underline{S}_i \equiv (\sin \theta_i \cos \varphi_i, \sin \theta_i \sin \varphi_i, \cos \theta_i)$$

and the Heisenberg Hamiltonian is expressed as

$$H = -J \sum_{i,j} \underline{S}_i \cdot \underline{S}_j$$

where i and j are nearest-neighbor lattice sites of a cubic lattice.
(i) Show that the mean-field equation is given by

$$\langle \underline{S}_i \rangle = \frac{\int_0^{2\pi} d\varphi \int_0^{\pi} d\theta \ \sin \theta \ \underline{S}_i \ \exp[6\beta J \underline{m} \cdot \underline{S}_i]}{\int_0^{2\pi} d\varphi \int_0^{\pi} d\theta \ \sin \theta \ \exp[6\beta J \underline{m} \cdot \underline{S}_i]}$$

(ii) Hence, show that $\langle S_i^z \rangle$ is given by

$$\langle S_i^z \rangle \equiv m$$
$$= \coth 6\beta J m - \frac{1}{6\beta J m}$$

(iii) Show that if the system has a non-trivial solution and if m is small, then

$$m^2 \sim \frac{5}{4} \left(\frac{2\beta J - 1}{\beta^3 J^3} \right)$$

(iv) Determine the partition function and the first few terms in the expansion of the Free-Energy as a power series in m. Show that when the non-trivial solution is real, it is stable.

(v) Show that the susceptibility can be written as

$$\chi = \frac{T}{12J(T_c - T)}$$

Problem 9.8

Consider a spin one-half Ising system in which the interactions involve the four spins surrounding each unit cell in the lattice. The interaction can be written as

$$H = -J \sum_{i,j,k,l} S_i S_j S_k S_l$$

where the sum is over the four spins on the smallest squares of the lattice.
(i) Expand the interaction in terms of the average magnetization

$$m = \frac{1}{N} \sum_i \langle S_i \rangle$$

per site and the fluctuations $\Delta S_i = S_i - m$, and neglect all terms of higher order than ΔS. Hence find the mean-field Hamiltonian.
(ii) Find an expression for the mean-field Helmholtz Free-Energy and the mean-field equation for m

$$m = \tanh 4\beta J m^3$$

(iii) Show that the transition is first-order.

Problem 9.9

The Helmholtz Free-Energy of mixing of a binary alloy may be approximated by

$$F = N\left[\frac{z}{2}\left(2E_{AB} - E_{AA} - E_{BB}\right)x(1 - x)\right.$$

$$\left. + k_B T(x \ln x + (1 - x)\ln(1 - x))\right]$$

(i) Under what circumstances is the equilibrium state determined by minimizing F?
(ii) Determine what sign of $(2E_{AB} - E_{AA} - E_{BB})$ will lead to phase separation.
(iii) Show that the phase separation temperature is proportional to

$$T_c \propto (1 - 2x) \ln\left(\frac{1 - x}{x}\right)$$

(iv) Expand the Free-Energy in powers of $(x - \frac{1}{2})$ and determine the critical exponent β.

Problem 9.10

Consider an asymmetric Landau-Ginzburg Free-Energy function

$$F(\varphi) = F_0 + F_2 \phi^2 + F_3(T) \ \phi^3 + F_4 \phi^4 + c(\nabla \phi)^2$$

where for stability, $F_4 > 0$.
(i) Find the position of the extrema.
(ii) Show that for positive F_2 there can be a discontinuous jump in ϕ signaling a first-order transition.
(iii) Determine the magnitude of the jump in ϕ. Hence, show that the transition occurs when

$$F_3(T)^2 = 4F_2 F_4$$

(iv) Determine the curvature of Free-Energy functional about the minima. Therefore, determine mean-field behavior of the correlation length, ξ, in the vicinity of the transition. Show that the correlation length does not diverge.

Problem 9.11

Consider a Landau-Ginzburg Free-Energy function with a symmetric form

$$F(\phi) = F_0 + F_2(T) \ \phi^2 + F_4(T) \ \phi^4 + F_6 \phi^6$$

where $F_4(T) < 0$ and, for stability, $F_6 > 0$.

(i) Determine the extremal values of ϕ.
(ii) Show that the value of $F_2(T)$ at which the system undergoes a first-order transition is given by

$$F_2(T) = \frac{F_4(T)^2}{4F_6}$$

(iii) Show that the discontinuity in ϕ is given by

$$\Delta \phi = \sqrt{-\frac{F_4(T)}{2F_6}}$$

(iv) Determine an expression for the change in the entropy and, hence, the latent heat.

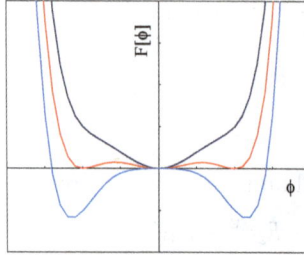

Fig. 9.16   The form of the Landau-Ginzburg Free-Energy function near a tricritical point.

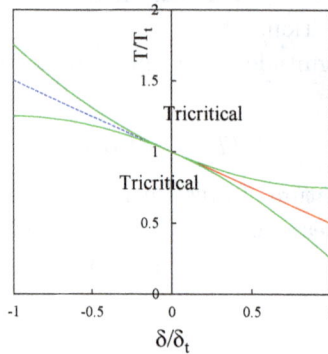

Fig. 9.17   The tricritical regime is external to the critical regime but impinges on the tricritical point, whereas the critical regime contains the second-order line and its first-order extension, but narrows to zero just at the tricritical point.

Problem 9.12

Consider a Landau-Ginzburg Free-Energy function with a symmetric form

$$F(\varphi) = F_0 + F_2(T, \delta) \; \phi^2 + F_4(T, \delta) \; \phi^4 + F_6 \phi^6 - \Pi \; \phi$$

where, for stability, $F_6 > 0$.

(i) Show that for $F_4(T) > 0$ and $\Pi = 0$, there is a line of second-order transitions which meets a line of first-order transitions at a tricritical point $(F_2 = 0, F_4 = 0)$.

(ii) Show that the slope of the line of continuous phase transitions in (δ, T) space is given by

$$\left(\frac{\partial \delta}{\partial T}\right) = -\left(\frac{\partial F_2}{\partial T}\right) \bigg/ \left(\frac{\partial F_2}{\partial \delta}\right)$$

the slope at the tricritical point is continuous.

In the tricritical regime

$$\chi^{-1} = \left(\frac{\partial \Pi}{\partial \phi}\right)\Bigg|_{\Pi=0}$$

$$= 2F_2(T,\delta) + 12F_4(T,\delta)\,\phi^2 + 30F_6\phi^4$$

where $F_2 \sim (T - T_t)$ and $F_4 \sim (T - T_t)$.

(iii) Show that near the tricritical point

$$\phi^2 = \frac{|F_4| + \sqrt{F_4^2 - 3F_6F_2}}{3F_6}$$

and the temperature dependence of the order parameter is dominated by the square root term. Hence, determine that $\beta_t = \frac{1}{4}$, $\gamma_t = 1$ and $\alpha_t = \frac{1}{2}$ near the tricritical point. At the tricritical point, there are two competing transitions which produces an upper critical dimension of three, therefore, the mean-field tricritical exponents are exact [92]. This can be seen from the Ginzburg criterion, which indicates that mean-field theories are valid for $\nu(d_c^u - 2) > 2\beta$, which for $\beta_t = \frac{1}{4}$ and $\nu_t = \frac{1}{2}$, leads to $d_c^u = 3$ for the tricritical point.

Problem 9.13

Consider a ferromagnetic Ising model in which the variables S_i^z can each take on one of three values $S_i^z = \pm 1$ and 0. The Hamiltonian is given by

$$H = -J \sum_{i,j} S_i^z S_j^z - \sum_i (\mu_B B^z S_i^z + D S_i^z S_i^z)$$

where $J > 0$. Each spin has z nearest-neighbor spins.

(i) Show that the mean-field Free-Energy F is given in terms of $m = \langle S_i^z \rangle$ by

$$F(m) = JzN \frac{m^2}{2} - Nk_B T \ \ln[2 \ \cosh \beta(zJm + \mu_B B^z) \exp[\beta D] + 1]$$

(ii) Minimize F wrt m to determine an equation for m.

(iii) Use the equation found in (ii) to find an equation for the critical temperature for $D > 0$.

(iv) Show that the fourth-order term in the expansion of the expression for F with $B^z = 0$ found in (i) changes sign for

$$\exp[\beta D] = \frac{1}{4}$$

while the sixth-order term is positive.

Problem 9.14

A three-dimensional system that exhibits a phase transition is described by a Landau-Ginzburg Free-Energy Functional $F[\phi]$

$$F[\phi] = \int d^3\underline{r} [F_0 + A(T - T_c) \phi^2 + F_6 \phi^6 + c\underline{\nabla}\phi \cdot \underline{\nabla}\phi - \Pi\phi]$$

where ϕ is a one-component order parameter and Π is a uniform applied field.

(i) Determine the mean-field critical exponent β that governs ϕ.

(ii) Calculate the critical exponent of the zero-field specific heat C, in the mean-field approximation.

(iii) Determine the critical exponent of the zero-field susceptibility

$$\chi = \left(\frac{\partial \phi}{\partial \Pi}\right)\Big|_{\Pi=0}$$

Problem 9.15

Consider a Landau-Ginzburg Free-Energy function for two real order parameters which is of the form

$$F(\phi_1, \phi_2) = F_0 + A_1 \phi_1^2 + A_2 \phi_2^2 + F_1 \phi_1^4 + F_2 \phi_2^4 + 2F_{12}\phi_1^2\phi_2^2$$

where stability requires $F_i > 0$. The symmetries of ϕ_1 and ϕ_2 are such that bi-linear coupling terms are forbidden. Assume the mean-field approximation is valid.

(i) Show that for weak repulsion $F_{12}^2 < F_1 F_2$, the phase diagram may contain four phases, consisting of the high-temperature phase, two pure phases and a mixed phase. Show that the order parameters for the mixed phase are given by

$$\phi_1^2 = -\frac{F_2 A_1 - F_{12} A_2}{2(F_1 F_2 - F_{12}^2)} > 0$$

$$\phi_2^2 = -\frac{F_1 A_2 - F_{12} A_1}{2(F_1 F_2 - F_{12}^2)} > 0$$

and the Free-Energy of the mixed phase is given by

$$F(\phi_1, \phi_2) = F_0 - \frac{F_2 A_1^2 + F_1 A_2^2 - 2F_{12} A_1 A_2}{4(F_1 F_2 - F_{12}^2)}$$

(ii) Hence, show that the pure and mixed phases are separated by lines of second-order transitions which meet at a tetracritical point.

(iii) Show that if $F_{12}^2 > F_1F_2$ corresponding to a strong repulsion of the two order parameters, the phase diagram has only three phases, where the high-temperature phase is connected to the ordered phases by two lines of second-order transitions, and the two ordered phases may be separated by a first-order transition that ends in a bicritical point.

Problem 9.16

Consider a Landau-Ginzburg Free-Energy function for two real order parameters in which ϕ_1 undergoes a phase transition but the applied field acts on a secondary order parameter ϕ_2 which is coupled to, but has different symmetry from, ϕ_1. The Free-Energy function is of the form

$$F(\phi_1, \phi_2) = F_0 + A(T - T_1)\,\phi_1^2 + F_1\phi_1^4 + A_2\phi_2^2 - \Pi\phi_2 + F_{12}\phi_1^2\phi_2^2$$

where stability requires $F_i > 0$ and $A_2 > 0$ is independent of temperature.
(i) By eliminating ϕ_2, determine the effective one-component Free-Energy function

$$F(\phi_1) = F_0 + A_1\phi_1^2 + F_1\phi_1^4 - \frac{\Pi^2}{4(A_2 + F_{12}\phi_1^2)}$$

(ii) Show that the model has a tricritical point at the tricritical temperature given by

$$T_t = T_1 - \frac{A_2 F_1}{A F_{12}}$$

and the tricritical field given by

$$\Pi_t^2 = \frac{4A_2^3 F_1}{F_{12}^2}$$

Problem 9.17

For systems in which the energy is a function of the entropy S and a macroscopic order parameter ϕ

$$E = E(S, \phi)$$

one can define a thermodynamic potential $A(T, \phi)$ by performing a Legendre Transformation. Therefore

$$S = -\left(\frac{\partial A(T, \phi)}{\partial T}\right)_\phi$$

The field Π canonically conjugate to ϕ is defined by

$$\Pi = \left(\frac{\partial A(T, \phi)}{\partial \phi} \right)_T$$

By considering convexity properties of A, Griffiths derived the thermodynamic inequality

$$A(T_c, \phi) - A(T_c, 0) \leq (T_c - T)(S(T_c) - S(T))$$

where ϕ is the order parameter at temperature T. The equation holds for $T < T_c$. Derive the Griffiths inequality by taking the limit $T \to T_c$.

Problem 9.18

Widom's hypothesis is that the non-analytic part of the Free-Energy F_s is a homogeneous function and can be written as

$$\lambda^d \, F_s(t, \Pi) = F_s(\lambda^{dy}|t|, \lambda^{dx}\Pi)$$

(i) Determine α, β, γ and δ.
(ii) Use the linear response expression for the susceptibility to relate γ to the correlation length exponent ν.
(iii) Show that the Josephson and Fisher relations are satisfied.

Problem 9.19

The correlation function $S(r)$ is assumed to have the form

$$S(r) = \langle \phi(\underline{r}) \, \phi(0) \rangle$$
$$= Cr^{-(d-2+\eta)} \, \exp[-r/\xi]$$

where the correlation length ξ scales as

$$\xi \sim t^{-\nu}$$

near T_c.
(i) Calculate the susceptibility χ

$$\chi = \beta \int d^d\underline{r} \, S(\underline{r})$$
$$= \beta \, \frac{2\pi^{\frac{d}{2}}}{\Gamma\left(\frac{d}{2}\right)} \int_0^\infty dr \, r^{d-1} \, S(r)$$

The susceptibility is assumed to diverge with a critical exponent γ

$$\chi \sim t^{-\gamma}$$

(ii) Express γ in terms of ν and η.

Problem 9.20

The spontaneous magnetization $m(T)$ for the two-dimensional Ising model is given by

$$m(T)^8 = [1 - \sinh^{-4} 2\beta J]$$

The critical behavior is given by

$$m(T) = At^{-\beta}[1 + B(-t)^\Delta + \ldots]$$

where A and B are constants and $t = (T - T_c)/T_c$.
(i) Determine the critical exponents β, and Δ the correction to scaling.
(ii) Determine the coefficients A and B in terms of $\beta_c J$.

Problem 9.21

A three-dimensional system that exhibits a phase transition is described by a Landau-Ginzburg Free-Energy Functional $F[\phi]$

$$F[\phi] = \int d^3\underline{r}[F_0 + A(T - T_c)\,\phi^2 + F_6\phi^6 + c\underline{\nabla}\phi \cdot \underline{\nabla}\phi - \Pi\phi]$$

where ϕ is a one-component order parameter and $\Pi(\underline{r})$ is a small spatially varying applied field.
(i) Determine the mean-field equation that governs the spatially varying part of the order parameter $\delta\phi(\underline{r})$.
The correlation function $S(\underline{r})$ is defined via

$$S(\underline{r} - \underline{r}') = \langle \phi(\underline{r})\,\phi(\underline{r}')\rangle - \langle\phi(\underline{r})\rangle\langle\phi(\underline{r}')\rangle$$

(ii) In the mean-field approximation, find the form of the correlation function and find expressions for the correlation length ξ, for temperatures above and below T_c.
Linear response theory provides the relation

$$\delta\phi(\underline{r}) = \beta \int d^d\underline{r}'\ S(\underline{r} - \underline{r}')\ \delta\Pi(\underline{r}')$$

(iii) Using the above relation and the mean-field form of the correlation function, express the uniform zero-field susceptibility χ in terms of the correlation length and hence determine the critical exponents.

(iv) Apply the Ginzburg criterion to find the region in which the mean-field approximation is reasonable by comparing

$$\phi_0^2 \, \frac{S_d}{d} \, \xi^d > S_d \int_0^\infty dr \; r^{d-1} \, S(r)$$

Hence, show that the upper-critical dimension d_c^u is given by

$$(d_c^u - 2) \, \nu = 2\beta$$

Problem 9.22

The correlation function $S(r)$ in d-dimensions can be found by Fourier-Transforming the Euler-Lagrange equation for the fluctuations induced by a delta function source. The result is

$$S(k) = \left(\frac{k_B T}{2c}\right) \left(\frac{1}{k^2 + \xi^{-2}}\right)$$

The inverse Fourier Transform is given by[2]

$$S(\underline{r}) = \int \frac{d^d \underline{k}}{(2\pi)^d} \; \exp[i\underline{k} \cdot \underline{r}] \, S(k)$$

The infinitesimal volume element can be written in d-dimensional hyper-spherical polar coordinates as

$$d^d \underline{k} = dk \; k^{d-1} \, d\theta_{d-2} \, \sin^{d-2}\theta_{d-2} \, d\theta_{d-3} \, \sin^{d-3}\theta_{d-3} \ldots d\theta_1 \, \sin\theta_1 \, d\varphi$$

where the θ_i range form 0 to π and φ ranges over 2π.
(i) Find a recursion relation between the surface area of a d-dimensional hypersphere of radius unity, S_d, and the surface area of a $d-1$ hypersphere of with the same unit radius, S_{d-1}.
(ii) Evaluate the correlation function $S(r)$ for $d = 3$. Show that the correlation function is consistent with

$$S(r) = \left(\frac{k_B T}{2c S_d(d-2)}\right) \frac{\exp[-r\xi^{-1}]}{r^{d-2}} \, F_d(r)$$

---

[2]The closed form expression for $S(r)$ is given by

$$S(r) = \left(\frac{k_B T}{2c}\right) \frac{1}{(2\pi)^{\frac{d}{2}} \, r^{d-2}} \left(\frac{r}{\xi}\right)^{\frac{d-2}{2}} K_{\frac{d-2}{2}}(r/\xi)$$

where $K_\nu(x)$ is a modified Bessel function.

where $F_d(0) = 1$ and

$$F_d(r) \sim \sqrt{\frac{\pi}{2}} \left(\frac{r}{\xi}\right)^{\frac{(d-3)}{2}} \left[1 + \frac{d^2 - 4d + 3}{8} \frac{\xi}{r} + \dots\right]$$

for $r \gg \xi$.

(iii) The Ginzburg criterion provides an estimate of the temperature range over which the mean-field approximation is reasonable. The criterion compares the average mean-squared fluctuations in a finite region of space with the squared order parameter, ϕ_0^2. Typically the comparison is made over a volume given by

$$\frac{S_d}{d} \xi^d$$

in which case, the mean-field approximation should be valid when

$$\frac{S_d}{d} \xi^d \phi^2 > \int_\xi d^d\underline{r}\, S(r)$$

What is the error made if the upper limit of range of integration on the right-hand side is extended from ξ to infinity for $d = 3$?

(iv) Consider the Landau-Ginzburg Free-Energy functional $F[\phi]$,

$$F[\phi] = \int d^3\underline{r}[F_0 + A(T - T_c)\, \phi^2 + F_{2m}\phi^{2m} + c\nabla\phi \cdot \nabla\phi]$$

where m is an integer. Determine β and ν.

(v) Determine the m-dependence for the upper critical dimension, d_c^u, above which the mean-field approximation is reasonable.

Problem 9.23

Consider an expression for a partition function Z in the form of a Gaussian integral

$$Z = \prod_{i=1}^N \left\{ \int_{-\infty}^\infty d\phi_i \int_{-\infty}^\infty d\phi_i^* \right\} \exp\left[-\beta \sum_{i,j} \phi_j^* F_{j,i}\phi_j\right]$$

where the matrix $F_{j,i}$ is assumed to be a Hermitean $N \times N$ matrix and the ϕ_i are complex variables.

(i) Show that

$$Z = (\pi k_B T)^N \det(F)^{-1}$$

Problem 9.24

Consider the d-dimensional correlation function $S(r)$ which satisfies Green's equation with a delta function at the origin.
(i) Provide an argument which shows that the correlation function is isotropic and, therefore, satisfies the differential equation

$$c\left[-\frac{1}{r^{d-1}}\frac{\partial}{\partial r}\left(r^{d-1}\frac{\partial}{\partial r}\right)+\xi^{-2}\right]S(r) = k_B T\delta^d(\underline{r})$$

An ansatz for $S(r)$ is given by

$$S(r) = \left(\frac{k_B T}{2c}\right)\frac{1}{(2\pi)^{\frac{d}{2}}r^{d-2}}\left(\frac{r}{\xi}\right)^{\frac{d-2}{2}}K_{\frac{d-2}{2}}(r/\xi)$$

where $K_\nu(x)$ is an unknown function.
(ii) Substitute the ansatz for $S(r)$ into the differential equation, with $r \neq 0$, and show that $K_{\frac{d-2}{2}}(r/\xi)$ must satisfy the modified Bessel equation

$$\left[-x^2\frac{\partial^2}{\partial x^2}-x\frac{\partial}{\partial x}+\left(\frac{d-2}{2}\right)^2+x^2\right]K_{\frac{d-2}{2}}(x) = 0$$

where $x = \frac{r}{\xi}$.
The modified Bessel function can be represented by the integral

$$K_\nu(x) = \frac{\pi^{\frac{1}{2}}}{(\nu-\frac{1}{2})!}\left(\frac{x}{2}\right)^\nu\int_1^\infty dt\ \exp[-xt](t^2-1)^{\nu-\frac{1}{2}}$$

for $\nu > -\frac{1}{2}$.
(iii) Show that the integral representation of the modified Bessel function satisfies the modified Bessel equation since

$$x^\nu\int_1^\infty dt\ \frac{\partial}{\partial t}\ [\exp[-xt](t^2-1)^{\nu+\frac{1}{2}}] = 0$$

and since the resulting expression is an integral of a perfect differential for which the argument vanishes at both ends of the range of integration.
(iv) By changing variable

$$t = 1+\frac{s}{x}$$

obtain the expression

$$K_\nu(x) = \sqrt{\frac{\pi}{2x}}\frac{\exp[-x]}{(\nu-\frac{1}{2})!}\int_0^\infty ds\ \exp[-s]\ s^{\nu-\frac{1}{2}}\left(1+\frac{s}{2x}\right)^{\nu-\frac{1}{2}}$$

Hence, by extracting a factor of $\frac{s}{2x}$ from the last factor, show that

$$\lim_{x \to 0} K_\nu(x) \to \frac{(\nu - 1)! \, 2^{\nu-1}}{x^\nu}$$

for $\nu > 0$.

(v) Find the asymptotic expansion by expanding in powers of x^{-1} and integrating term by term. Show that this process yields the asymptotic expansion

$$K_\nu(x) \sim \sqrt{\frac{\pi}{2x}} \, \exp[-x] \left[1 + \frac{(4\nu^2 - 1)}{1! \, (8x)} \right.$$
$$\left. + \frac{(4\nu^2 - 1)(4\nu^2 - 9)}{2! \, (8x)^2} + \cdots \right]$$

Problem 9.25

(i) Show that for a ferromagnetic solid, the spin waves with the dispersion relation

$$\hbar\omega = JS\hbar q^2$$

contribute a $T^{\frac{3}{2}}$ term to the low temperature specific heat.

(ii) Determine the low-T temperature dependent reduction of the magnetization of a ferromagnet due to thermally excited spin waves.

9.10 Appendix: The Yang-Lee Theorem for the $d = 1$ Ising Model

The partition function Z_N for the one-dimensional Ising Model for N spins in a magnetic field can be determined by the Transfer Matrix approach (See Appendix 4.18). Motivated by the Yang-Lee Theorem that establishes a relation between the zeroes of the partition function and non-analytic behavior, we shall examine the distribution of the zeroes of the partition function. The partition function is given by

$$Z_N = \lambda_1^N + \lambda_2^N \tag{9.340}$$

where the eigenvalues λ_1 and λ_2 are given by

$$\lambda = \exp[\beta J] \, \cosh \beta B \pm \sqrt{\exp[+2\beta J] \, \sinh^2 \beta B + \exp[-2\beta J]} \tag{9.341}$$

which are the solutions of a quadratic equation. We shall define the parameter z by

$$z = \exp[-4\beta J + 2\beta B] = \exp[-4\beta J + i\theta] \tag{9.342}$$

corresponding to $\beta B = i\frac{\theta}{2}$. On defining a parameter r via

$$r = \exp[-2\beta J] \tag{9.343}$$

then

$$z = r^2 \, \exp[i\theta] \tag{9.344}$$

The values of θ for which $Z_N(z) = 0$ are determined from the equation

$$\left[\cos \frac{\theta}{2} + \left(r^2 - \sin^2 \frac{\theta}{2} \right)^{\frac{1}{2}} \right]^N + \left[\cos \frac{\theta}{2} - \left(r^2 - \sin^2 \frac{\theta}{2} \right)^{\frac{1}{2}} \right]^N = 0 \tag{9.345}$$

On setting

$$\cos \frac{\theta}{2} = (1 - r^2) \, \cos \varphi \tag{9.346}$$

the equation for θ can be simplified to

$$(1 - r^2)^{\frac{N}{2}} [(\cos \varphi + i \sin \varphi)^N + (\cos \varphi - i \sin \varphi)^N] = 0 \tag{9.347}$$

or, equivalently

$$(1 - r^2)^{\frac{N}{2}} \, 2 \, \cos N\varphi = 0 \tag{9.348}$$

which has solutions given by

$$\varphi = \pm \left(k - \frac{1}{2} \right) \frac{\pi}{N} \tag{9.349}$$

for $k = 1, 2, \ldots, N$. Since

$$\cos\theta = 2\,\cos^2\frac{\theta}{2} - 1 \qquad (9.350)$$

one has the solution

$$\cos\theta = -r^2 + (1 - r^2)\,\cos 2\varphi$$

$$= -r^2 + (1 - r^2)\,\cos\left(\frac{(2k-1)\pi}{N}\right) \qquad (9.351)$$

for $k = 1, 2, \ldots, N$. This determines the distribution of zeroes of $Z_N(z)$. It is seen that they reside on a circle of radius r^2 in the complex z plane. The zeroes lie on an arc of the circle starting with the angle $\theta_i \approx \cos^{-1}(1 - 2r^2) = 2\,\sin^{-1}r$ and ending with an angle $\theta_f = 2\pi - \theta_i$. It has a gap of angular width $4\,\sin^{-1}r$ centered on the real z-axis. It only pinches of the real z-axis when $r \to 0$ or, equivalently, when $T \to 0$. Thus, the model can be considered as exhibiting a phase transition at $T = 0$. In this case, the critical exponents are identified as $\alpha = 1$, $\beta = 0$, $\gamma = 1$, $\nu = 1$ and $\delta = \infty$. The exponents satisfy the scaling relations. The existence of a phase transition in one-dimension is not inconsistent with the Mermin-Wagner Theorem (Section 10.4.1), since the broken symmetry is discrete.

The distribution of zeroes on the circle, $\rho(\theta)$, is defined as

$$\rho(\theta) = \frac{1}{N}\left(\frac{dk}{d\theta}\right) \qquad (9.352)$$

and is evaluated in the two regions as

$$\rho(\theta) = \begin{cases} \dfrac{1}{2\pi}\,\dfrac{\sin\frac{\theta}{2}}{\sqrt{\sin^2\frac{\theta}{2} - r^2}} & \text{if } \theta_f > \theta > \theta_i \\[2ex] 0 & \text{otherwise} \end{cases} \qquad (9.353)$$

respectively. The distribution is normalized to unity

$$\int_0^{2\pi} d\theta\,\rho(\theta) = 1 \qquad (9.354)$$

Chapter 10

The Renormalization Group

Scaling behavior at a critical point shows that there exists a single relevant length scale, the correlation length ξ, which describes the large amplitude, long-ranged, fluctuations that dominate the singular parts of the Free-Energy. The scaling theory and the formulation of the Landau-Ginzburg Free-Energy Functional assume that the microscopic length scales in the Hamiltonian are irrelevant. The "*Scaling Hypothesis*" describes the change in the fluctuations as the correlation length is changed by, for example, changing the temperature. Furthermore, at the critical temperature, the system appears to exhibit the same behavior at all length scales. The renormalization group technique [83,93] supplements the scaling hypothesis by incorporating the effect of the short scale physics. It shows that if the length scale is changed, then the effective interactions controlling the large scale fluctuations also change. The interactions between the long-ranged fluctuations are re-scaled, when the short-ranged fluctuations are removed by integrating them out. The method involves the following three steps:

(1) Integrating out the short scale fluctuations of the system, thereby increasing the effective short distance cut-off for the system.
(2) Re-defining all length scales, so that the new cut-off appears indistinguishable from the old cut-off.
(3) Re-definition or renormalization of the interaction strengths which govern the fluctuations of the order parameter.

The above procedure introduces the idea of an operation that can be compounded, resulting a semi-group rather than a group, since the operations are not uniquely invertible. The operations result in flows of both the form of the Landau-Ginzburg Functional and its parameters as the length scale is changed by successive infinitesimal increments. The set of parameters $\{F_2, F_4, \ldots, c\}$ that describe the most general form of the Landau-Ginzburg Free-Energy Functional describe a point in parameter space. A change in scale by a factor of b, where $b > 1$, results in a flow between two different points in parameter space

$$\{F_2', F_4', \ldots, c'\} = \mathcal{R}(b)\{F_2, F_4, \ldots, c\} \tag{10.1}$$

At the critical point, the above operations should leave the renormalized Landau-Ginzburg Functional invariant, reflecting the scale-invariance that occurs at the critical point,

$$\{F_2^*, F_4^*, \ldots, c^*\} = \mathcal{R}(b)\{F_2^*, F_4^*, \ldots, c^*\} \tag{10.2}$$

The corresponding invariant point of parameter space $\{F_2^*, F_4^*, \ldots, c^*\}$ is known as a "*Fixed Point*". Sometimes the properties of a system which is close to a fixed point can be inferred from the flow of the parameters under the renormalization group operations by linearizing the flow around the fixed point. In this case, the procedure results in the recovery of the phenomena described by the scaling hypothesis together with the actual values of the critical exponents.

The renormalization group transformation may have other fixed-points, such as a fixed-point that represents the high-temperature limit $T \to \infty$ at which the correlation length ξ is zero. In this limit, $F_2 \to \infty$, so the other parameters $F_4, \ldots, c$ may be neglected. Therefore, one expects that there is a fixed-point

$$\{F_2^*, 0, \ldots, 0'\} = \mathcal{R}(b)\{F_2^*, 0, \ldots, 0\} \tag{10.3}$$

at which there are no correlations. To be sure, the high-temperature state is characterized as having completely uncorrelated probabilities. Any renormalization group transformation should maintain the probabilities and, therefore, does not change the Landau-Ginzburg Functional. Hence, the disordered state at $T \to \infty$ is a fixed point. Similarly, if there is an ordered state at $T = 0$, one expects that the $T = 0$ state should be completely ordered and, since the transformation maintains the perfect correlations, the Free-Energy Functional is fixed.

10.1   Real Space Renormalization for Ising Models

Consider a Hamiltonian $\hat{H}_{A,B}$ which depends on the set of N variables located at the sites (i, j) of a two-dimensional lattice. Specifically, consider the nearest-neighbor Ising model with spins on the sites of a two-dimensional square lattice. The spin variables can be broken into two sets of site variables $S_{i,j}^A$ and $S_{i,j}^B$ on inter-penetrating lattices depending on whether $(i+j)$ is either even or odd. The Partition Function Z is obtained by performing a Trace over both sets of sublattice spins. If one only performs the Trace over the spins on the B lattice sites, one obtains an expression for the Partition Function which involves a new effective Hamiltonian $\hat{H}_A$ with a new form that only depends on the spin variables of the A sub-lattice

$$Z = \text{Trace}_{\{S_{i,j}^A\}} \ \text{Trace}_{\{S_{i',j'}^B\}} \ \exp[-\beta \hat{H}_{A,B}] \tag{10.4}$$

$$= \text{Trace}_{\{S_{i,j}^A\}} \ \exp[-\beta \hat{H}_A] \tag{10.5}$$

The new effective Hamiltonian $\hat{H}_A$ is defined by

$$\hat{H}_A = -k_B T \ \ln(\text{Trace}_{\{S_{i',j'}^B\}} \ \exp[-\beta \hat{H}_{A,B}]) \tag{10.6}$$

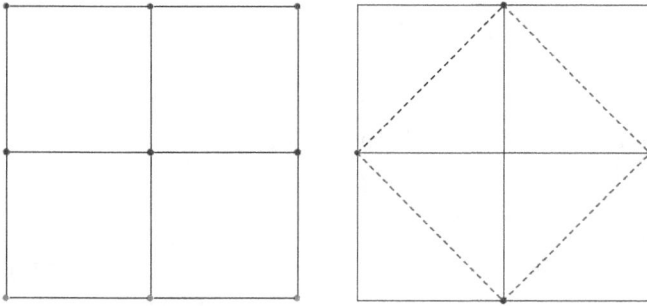

Fig. 10.1 A square lattice with an Ising spin at each site (i,j). The lattice of remaining spins in which the Trace over the spins $S_{i,j}$ at odd values of $i+j$ has been performed.

After the partial Trace of the B site variables has been performed, the number of degrees of freedom have been reduced from N to $N/2$. Likewise, the length scale of the remaining sublattice has been increased from a of the original lattice to $\sqrt{2}\,a$. If the length scale is re-scaled by a factor of $\sqrt{2}$, the original lattice and the lattice describing the remaining spin variables are similar (apart from the rotation through $\pi/4$) but the shortest-ranged fluctuations have been eliminated. If this process is iterated n times, the length scale r' is given by

$$r' = \frac{r}{(\sqrt{2})^n} \tag{10.7}$$

and the number of degrees remaining degrees of freedom N' are given by

$$N' = \frac{N}{(\sqrt{2})^{nd}} \tag{10.8}$$

where d is two, the dimension of the lattice. More generally, one may replace the factor of $\sqrt{2}$ by a scaling parameter $b > 1$, which allows discussion of more general methods of performing partial traces. The series of effective Hamiltonians describes fluctuations at successively longer length scales. The Free-Energy per degree of freedom or Free-Energy density $\mathcal{F}$ is defined as

$$-\beta\mathcal{F} = \lim_{N\to\infty} N^{-1} \ln Z_N[\hat{H}] \tag{10.9}$$

We shall subsume the factor of β into $\mathcal{F}$. Since the partition function is invariant under the renormalization group transformation, the Free-Energy density clearly transforms as

$$\mathcal{F}' = \frac{N}{N'} \mathcal{F} \tag{10.10}$$

Since

$$N' = b^{-nd} N \tag{10.11}$$

the above relation may be written as

$$\mathcal{F}' = b^{nd} \mathcal{F} \tag{10.12}$$

The above relation is intimately connected with the re-scaling of the length scale ξ that characterizes the system. The correlations length of the system is defined by the spatial correlation function $S(\underline{r}_i - \underline{r}_j)$, such as

$$S(\underline{r}_i - \underline{r}_j) = \langle S_i S_j \rangle \tag{10.13}$$

which, for translationally invariant and isotropic systems, may have the form

$$S(\underline{r}_i - \underline{r}_j) \sim \exp\left[-\frac{|\underline{r}_i - \underline{r}_j|}{\xi}\right] \tag{10.14}$$

where ξ is governed by $\beta\hat{H}$. After the scale transformation, the correlation function becomes

$$S'\left(\frac{|\underline{r}_i - \underline{r}_j|}{b^n}\right) \sim S(|\underline{r}_i - \underline{r}_j|) \tag{10.15}$$

The above relation does involve an over all scaling of the correlations which is independent of the distance since it is due to a change in the magnitude of the spin variables. This uniform scale factor is of no importance in what follows. Since the spatial correlations are to be preserved, the transformation defines the new correlation length ξ' as

$$b^n \xi' = \xi \tag{10.16}$$

The critical behavior of the system is revealed by infinite repetition of the above two processes. The transformation of the length scale together with the transformation of the Free-Energy per degree of freedom determines the critical exponents. The sequence of effective Hamiltonians are not simply related but, nevertheless, are related by a complicated transformation

$$\hat{H}'_n = \mathcal{R}_n \hat{H} \tag{10.17}$$

This transformation establishes a map between the parameters of a series of effective Hamiltonians and the sequence of transformations forms a semigroup.

If a system is at the critical point, the system should have fluctuations at all length scales and remain unchanged after arbitrarily many transformations. In this case, the Hamiltonian should be invariant at the critical point. In addition to the critical point, there should be other points (fixed points) at which the Hamiltonian remains invariant under the transformation. Denoting the form of the Hamiltonian at a fixed point by $\hat{H}^*$, one has

$$\hat{H}^* = \mathcal{R}_n \hat{H}^* \tag{10.18}$$

The correlation length ξ is determined by $\hat{H}$ and ξ'_n is determined by $\hat{H}'_n$. The correlation lengths are related via

$$\xi'_n = \frac{\xi}{b^n} \tag{10.19}$$

At the fixed point, $\hat{H}$ and $\hat{H}'_n$ are both equal to $\hat{H}^*$. Then, since at the fixed point both the correlation lengths ξ'_n and ξ are determined by the fixed-point Hamiltonian $\hat{H}^*$, both correlation lengths are equal to ξ^*. This leads to the equation

$$\xi^* = \frac{\xi^*}{b^n} \tag{10.20}$$

Hence, at a fixed point ξ^* can only take on two values, either 0 or ∞. The correlation length is infinity at a critical fixed point and is zero at a trivial fixed point.

Each Hamiltonian in the series can be written as a product of a set of parameters, denoted by a vector $\underline{u}$, and a set of well-defined operators, denoted by the vector $\hat{\underline{O}}$, so that

$$\hat{H} = \underline{u} \cdot \hat{\underline{O}} \tag{10.21}$$

and

$$\hat{H}' = \underline{u}' \cdot \hat{O} \tag{10.22}$$

Therefore, the parameter sets are related by the non-linear transformation

$$\underline{u}' = \mathcal{R}_n \underline{u} \tag{10.23}$$

involving a matrix $\mathcal{R}_n$. At the fixed point, the transformation reduces to

$$\underline{u}^* = \mathcal{R}_n \underline{u}^* \tag{10.24}$$

Close to a fixed point, the parameter vectors may be written as

$$\underline{u}' = \underline{u}^* + \delta \underline{u}'$$
$$\underline{u} = \underline{u}^* + \delta \underline{u} \tag{10.25}$$

where $\delta \underline{u}$ and $\delta \underline{u}'$ are small. The transformation between $\underline{u}'$ and $\underline{u}$ is non-linear, but the small deviations from the fixed point are linearly related by

$$\delta \underline{u}' = \delta \underline{u} \cdot \nabla_{\underline{u}}(\mathcal{R}_n \underline{u}^*) \tag{10.26}$$

Near the fixed point, $T(\underline{u}^*) = \nabla_{\underline{u}}(\mathcal{R}_n \underline{u}^*)$ is a real matrix which has components that are given by

$$T(\underline{u}^*)_{i,j} = \frac{\delta u'_i}{\delta u_j}\bigg|_{u^*} \tag{10.27}$$

The scaling flows allow the properties of the system to be calculated at any point in parameter space $\underline{u}$ in terms of the values at a fixed point. It is expected that the system will have fixed points which corresponding to the high-temperature and low-temperature fixed points. It is expected that all the scaling fields will be irrelevant in the vicinity of these fixed points, so these fixed points will be stable. It is easy to calculate properties near trivial or high-temperature fixed points at which $\xi = 0$. Critical properties are determined by the eigenvalues and eigenvectors of $T(\underline{u}^*)$ of the linearized transformation near the critical points where $\xi \to \infty$.

10.1.1   Critical Phenomenon

There is no reason to assume that the matrix $T(\underline{u}^*)$ is symmetric, so in general the left eigenvalues will be different form the right eigenvalues. The left eigenvalue equation for the transformation matrix $T_b(\underline{u}^*)$ is

$$\underline{\phi}_i T_b(\underline{u}^*) = \lambda_i(b) \, \underline{\phi}_i \tag{10.28}$$

where the eigenfunctions are denoted by ϕ_i. The left eigenvalues of the matrix $T_b(\underline{u}^*)$ are functions of the re-scaling factor $b = \sqrt{2}$. The i-th eigenvalue can be expressed as

$$\lambda_i(b) = b^{\Lambda_i} \qquad (10.29)$$

The eigenvalue must be a power of b since, if one performs two successive transformations by b, the transformation is represented either by $T_{b^2}(\underline{u}^*)$ or by the product $T_b(\underline{u}^*) \, T_b(\underline{u}^*)$. Hence, the eigenvalues must satisfy

$$\lambda_i(b^2) = \lambda_i(b) \, \lambda_i(b) \qquad (10.30)$$

etc. The above relation is only satisfied if $\lambda_i(b)$ is a power of b.

Close to a fixed point, the parameters of the Hamiltonian can be expressed in terms of the eigenfunctions of the linearized transformation

$$\underline{u} = \underline{u}^* + \sum_i g_i \, \underline{\phi}_i$$

$$\underline{u}' = \underline{u}^* + \sum_i g_i' \, \underline{\phi}_i \qquad (10.31)$$

Since $\underline{u}$ and $\underline{u}'$ are related by $T_b(\underline{u}^*)$, one finds that

$$g_i' = g_i \, b^{\Lambda_i} \qquad (10.32)$$

The parameters g_i are known as *"Scaling Fields"*. The scaling fields determine the deviation from a fixed point.

10.1.2 *Renormalization Group Flow*

The exponents Λ_i in the eigenvalues of the linearized transformation determine the flow of the Hamiltonian parameters under successive transformations. If an exponent is greater than zero $\Lambda_i > 0$ the scaling field g_i becomes amplified by the transformation. A scaling field, g_i is classified as a *"Relevant"* parameter if the renormalization group flow moves them away from the fixed point. On the other hand, if $\Lambda_i < 0$, the scaling field is classified as *"Irrelevant"* since the deviation from the fixed point decreases under the transformation. A scaling field which that has an exponent $\Lambda_i = 0$ is called *"Marginal"*. Marginal variables give rise to logarithmic corrections to scaling. To re-iterate:

- If $\Lambda_i > 0$, g_i is a relevant scaling field.
- If $\Lambda_i = 0$, g_i is a marginal scaling field.
- If $\Lambda_i < 0$, g_i is an irrelevant scaling field.

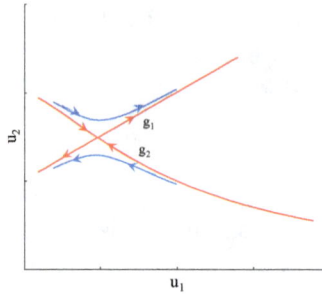

Fig. 10.2 Schematic renormalization group flows of the parameters. The scaling field g_1 is relevant since $\Lambda_1 > 0$ and the scaling field g_2 is irrelevant since $\Lambda_2 < 0$. The scaling fields are zero at the fixed point, $g_1 = g_2 = 0$.

Since scaling fields are only defined through the linearized transformation at a specific fixed point $T_b(\underline{u}^*)$, their classification as relevant, irrelevant or marginal only pertains to the vicinity of the specific fixed point $\underline{u}^*$.

For a critical point that has N scaling variables and n irrelevant variables, in the vicinity of the critical point, one may define a *"Critical Surface"* of dimension n consisting of the points $\{g_i\}$ which flow towards the critical point. That is, the critical surface is the subspace spanned by the irrelevant eigenvectors. Only a finite number of scale transformations are required to bring a point on the critical surface to the critical point, since the transformations only involve the irrelevant scaling fields. All points on the critical surface have the same long-range physics as the critical point. The correlation length ξ diverges on the critical surface,[1] so it defines the phase transition for all values of the irrelevant parameters. The critical surface acts as a separatrix which divides the space into points which flow way from the surface in opposite senses, either to the high-temperature fixed point or to the $T = 0$ fixed point. If the concept of a critical surface is extended to a very general form of the Hamiltonian with an extended parameter space, one may define a universality class as the set of all systems which have Hamiltonians with parameters that flow into points on the same critical surface. Systems in other universality classes have flows that flow into points on other critical surfaces.

---

[1]Since the correlation length ξ is determined by the effective Hamiltonian and, therefore, the scaling parameters $\{g_1, g_2, \ldots\}$, one has

$$\xi(g_1, g_2, \ldots) = b^n \, \xi(b^{n\Lambda_1} g_1, b^{n\Lambda_2} g_2, \ldots)$$

Hence, for sufficiently large b^n, the irrelevant scaling fields go to zero, and the divergence of ξ is only controlled by the relevant parameters.

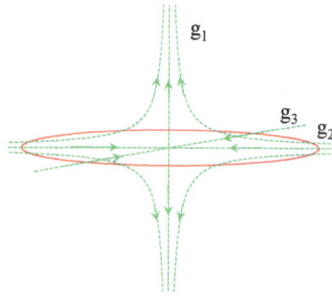

Fig. 10.3 Schematic renormalization group flows of the parameters near the critical surface (red). The scaling field g_1 is relevant whereas g_2 and g_3 are irrelevant.

Since the transformation $T_b(\underline{u}^*)$ is analytic, the scaling fields should be analytic functions of the parameters of the Hamiltonian. In particular, the scaling field that represents temperature should have a positive exponent and can be expanded in powers of t, where

$$t = \left(\frac{T - T_c}{T_c} \right) \tag{10.33}$$

For a ferromagnetic transition, the magnetic field B is also a relevant variable. Since most of the scaling fields are irrelevant, the critical exponents only depend on a few parameters. This limited dependence enables phase transitions to be classified into a small set of universality classes.

10.1.3 *Scaling Laws and Critical Exponents*

Under the renormalization group transformation, the non-analytic portion of the Free-Energy density, $\mathcal{F}$, transforms as

$$\mathcal{F}(g_1, g_2, \ldots) = b^{-nd} \; \mathcal{F}(b^{n\Lambda_1} g_1, b^{n\Lambda_2} g_2, \ldots) \tag{10.34}$$

Hence, the Free-Energy density is a homogeneous function of the scaling fields. The renormalization group has provided the basis for Widom's assumption. By selecting a value of n such that

$$b^{n\Lambda_1} g_1 = 1 \tag{10.35}$$

one finds

$$\mathcal{F}(g_1, g_2, \ldots) = g_1^{d/\Lambda_1} \; \mathcal{F}(1, g_1^{-\Lambda_2/\Lambda_1} g_2, \ldots) \tag{10.36}$$

If g_1 is a relevant variable then, to be in close proximity to the critical point, g_1 will be small and must vanish at the critical point. Therefore, close the critical point, the dependence of the Free-Energy density $\mathcal{F}$ on a irrelevant scaling fields g_i can be expressed as $g_i g_1^{-\Lambda_i/\Lambda_1}$ which vanishes as g_1 approaches zero. Hence, the critical properties don't depend on the irrelevant scaling fields.

For a ferromagnetic transition, the temperature t and the magnetic field are relevant variables. Therefore, one may set $g_1 = t$ and $g_2 = B$, so that

$$\mathcal{F}(t, B, \ldots) = t^{d/\Lambda_1} \, \mathcal{F}(1, t^{-\Lambda_2/\Lambda_1} B, \ldots) \tag{10.37}$$

The critical exponents can be obtained from the non-analytic part of the Free-Energy density $\mathcal{F}$. For example, since the non-analytic part of the specific heat C is given by

$$C \sim \frac{\partial \mathcal{F}}{\partial t^2} \sim t^{\frac{d}{\Lambda_1} - 2} \tag{10.38}$$

then using the definition of its critical exponent $C \sim t^{-\alpha}$, one identifies α as

$$\alpha = 2 - \frac{d}{\Lambda_1} \tag{10.39}$$

Likewise, since the magnetization is given by

$$M \sim -\frac{\partial \mathcal{F}}{\partial B}\bigg|_{B=0} \sim t^{\frac{d-\Lambda_2}{\Lambda_1}} \tag{10.40}$$

and since $m \sim t^\beta$, one finds that β is given by

$$\beta = \frac{d - \Lambda_2}{\Lambda_1} \tag{10.41}$$

Since susceptibility χ is given by

$$\chi \sim -\frac{\partial^2 \mathcal{F}}{\partial B^2}\bigg|_{B=0} \sim t^{\frac{d-2\Lambda_2}{\Lambda_1}} \tag{10.42}$$

and its critical exponent γ, defined by $\chi \sim t^{-\gamma}$, is found to be

$$\gamma = \frac{2\Lambda_2 - d}{\Lambda_1} \tag{10.43}$$

The re-scaling parameter b played a prominent role in the above analysis. However, since all length scales are reduced by a factor of b under a renormalization group transformation, b is simply related to the n-th re-scaling of the correlation length ξ by

$$\xi'_n = \xi b^{-n} \tag{10.44}$$

Furthermore, the $i = 1$ scaling field t scales as

$$t'_n = tb^{n\Lambda_1} \tag{10.45}$$

Therefore, we may eliminate b to express the correlation function in terms of t

$$\xi'_n = \xi \left(\frac{t}{t'_n} \right)^{\Lambda_1^{-1}} \tag{10.46}$$

From the definition of the critical exponent for the correlation length ν

$$\xi \sim t^{-\nu} \tag{10.47}$$

one finds the critical exponent is given by

$$\nu = \frac{1}{\Lambda_1} \tag{10.48}$$

To summarize, we found that the critical exponents only depend on two quantities and the dimension d as

$$\alpha = 2 - \frac{d}{\Lambda_1}$$

$$\beta = \frac{d - \Lambda_2}{\Lambda_1}$$

$$\gamma = \frac{2\Lambda_2 - d}{\Lambda_1}$$

$$\nu = \frac{1}{\Lambda_1} \tag{10.49}$$

One of the quantities, Λ_1, is uniquely determined by the critical exponent for the correlation length ν. In fact, as argued by Kadanoff, the values of the critical exponents can be evaluated directly using the above reasoning in which b has been eliminated in favour of the physically measurable quantities, ξ. The above analysis shows that the values of J and T_c neither play crucial roles nor affect the values of the critical exponents. The critical exponents are only determined by quantities such as d the dimension of the lattice, n the dimension of the spin vector (For the Ising model $n = 1$, since the spins are quantized along the z-axis and have neither x nor y components.) and the range of the interaction. It is because of the restricted dependence on fundamental quantities that phase transitions fall into a few universality classes which are characterized by having the same sets of values for the critical exponents.

All long-ranged behavior is preserved by the renormalization group transformation, and that includes spatial correlations. This statement can be used to obtain the scaling behavior of the correlation function and can be used to determine the anomalous dimension. The spatial correlations can be found by considering the presence of a spatially varying external field which contributes a term

$$-\int d^d\underline{r} \; S(\underline{r}) \; B(\underline{r}) \tag{10.50}$$

to the Landau-Ginzburg Free-Energy Functional. After the renormalization group transformation, the interaction term becomes

$$-\int d^d\underline{r}' \; S'(\underline{r}') \; B'(\underline{r}') \tag{10.51}$$

where $B' = b^{n\Lambda_2}B$. Since the correlation function can be expressed as a second-order functional derivative of the partition function with respect to the applied field

$$S(\underline{r}_1 - \underline{r}_2) = \frac{\delta^2 Z}{\delta B(\underline{r}_1)\delta B(\underline{r}_2)} \tag{10.52}$$

and since Z is invariant under the renormalization group transformation, one has

$$\frac{\delta^2 Z(B')}{\delta B'(\underline{r}_1)\delta B'(\underline{r}_2)} = \frac{\delta^2 Z(B)}{\delta B'(\underline{r}_1)\delta B'(\underline{r}_2)} \tag{10.53}$$

The partition function $Z(B)$ includes the field $B = b^{-n\Lambda_2}B'$ which acts the spins in a volume element ΔV. However, in the derivative of $Z(B')$ wrt to B', the field B' acts on the spins which reside in a volume $b^{nd}\Delta V$. Therefore, the second-order derivative yields

$$S(r/b^n) = b^{n2(d-\Lambda_2)} \; S(r) \tag{10.54}$$

Since we have set

$$b^{n\Lambda_1} \, t = 1 \tag{10.55}$$

one finds

$$\begin{aligned}
S(r) &= b^{2n(\Lambda_2-1)} \; S(r/b^n) \\
&= t^{\frac{2(d-\Lambda_2)}{\Lambda_1}} S(rt^{\frac{1}{\Lambda_1}}) \\
&= \xi^{2(\Lambda_2-d)} \; S(r/\xi) \tag{10.56}
\end{aligned}$$

which shows that the correlation function is a homogenous function of r/ξ. On examining the region where $r < \xi$ for which the ansatz

$$S(r) \sim \frac{1}{r^{d-2+\eta}} \; \exp[-r/\xi] \tag{10.57}$$

holds, one finds that

$$2(\Lambda_2 - d) = (2 - d - \eta) \tag{10.58}$$

The above equation yields identification of the anomalous dimension as

$$\eta = d + 2 - 2\Lambda_2 \tag{10.59}$$

The quantity η is known as the anomalous dimension since it measures the difference between the actual dimension of the order parameter and that calculated within the mean-field approximation.

In the mean-field approximation, the quadratic term in the energy functional $\xi^{-2} \phi^2(r)$ determines the non-analytic temperature dependence of the Free-Energy density ξ^{-d}. This identification leads to the relation

$$\xi^{-2} \langle \phi(0) \ \phi(0) \rangle \sim \xi^{-d} \tag{10.60}$$

which suggests that the order parameter varies as $\xi^{\frac{2-d}{2}}$ and has dimension $\frac{2-d}{2}$. Furthermore, if one assumes that if the mean-field correlation function $S(r)$ is a homogeneous function of r/ξ then, close to the critical temperature where $\xi > r$, the mean-field correlation should vary as $S(r) \sim \frac{1}{r^{d-2}}$. By contrast, on utilizing the expression for the correlation function in terms of the anomalous dimension then, close to the critical point, one has

$$S(0) \sim \xi^{-(d+\eta-2)} \tag{10.61}$$

It is seen that the anomalous dimension represents a re-scaling of the dimension of the order parameter which is not present in mean-field theory. Furthermore, the homogeneity of $S(r)$ implies, that close to the critical point where $\xi \to \infty$, the anomalous dimension will appear in the susceptibility

$$\chi \sim \xi^{2-\eta} = t^{-\nu(2-\eta)} \tag{10.62}$$

Hence, the anomalous dimension appears in γ, the susceptibility critical exponent

$$\gamma = \nu(2 - \eta) \tag{10.63}$$

which contrasts with the mean-field result $\gamma = 2\nu$.

10.1.4 *Application: One-Dimensional Ising Model*

The Partition Function is given by

$$Z = \text{Trace}_{S_i} \ \exp\left[\beta J \sum_i S_i S_{i+1} + \beta B \sum_i S_i\right] \tag{10.64}$$

On performing the Trace over the odd indexed spins, one has

$$Z = \text{Trace}_{S_{2i}} \prod_i \exp[\beta B S_{2i}] \left[\exp[\beta B] \exp[\beta J(S_{2i} + S_{2i+2})]\right.$$
$$\left. + \exp[-\beta B] \exp[-\beta J(S_{2i} + S_{2i+2})]\right] \tag{10.65}$$

since only the even spins variables remain, the new Trace only runs over the even spins. The new Hamiltonian is given by

$$\hat{H}' = -\sum_{2i} [B S_{2i} + k_B T \ln(\exp[\beta B] \exp[\beta J(S_{2i} + S_{2i+2})]$$
$$+ \exp[-\beta B] \exp[-\beta J(S_{2i} + S_{2i+2})])] \tag{10.66}$$

The second term, which depends on $(S_{2i} + S_{2i+2})$, describes a constant energy shift due to the field and a pairwise interaction. Although there are four possible states of S_{2i} and S_{2i+2}, there are only three possible values of $(S_{2i} + S_{2i+2})$. The three projection operators that project onto the latter sets of states are

$$P_2 = \frac{(1 + S_{2i}S_{2i+2})}{4} + \frac{S_{2i} + S_{2i+2}}{4}$$

$$P_0 = \frac{(1 - S_{2i}S_{2i+2})}{2}$$

$$P_{\bar{2}} = \frac{(1 + S_{2i}S_{2i+2})}{4} - \frac{S_{2i} + S_{2i+2}}{4} \tag{10.67}$$

Let

$$K(S_{2i} + S_{2i+2}) = -k_B T \ln (\exp[\beta B] \exp[\beta J(S_{2i} + S_{2i+2})]$$
$$+ \exp[-\beta B] \exp[-\beta J(S_{2i} + S_{2i+2})]) \tag{10.68}$$

Then, the effective interaction can be written as

$$K(S_{2i} + S_{2i+2}) = \frac{K(2) + 2K(0) + K(\bar{2})}{4} + \frac{K(2) - K(\bar{2})}{4} (S_{2i} + S_{2i+2})$$
$$+ \frac{K(2) - 2K(0) + K(\bar{2})}{4} S_{2i}S_{2i+2} \tag{10.69}$$

The first term which is independent of the spin variables, represents an additive contribution to the Free-Energy which is analytic. In what follows, we shall focus on the non-analytic terms of the Free-Energy and shall neglect the analytic terms, since they are additive and do not become singular after multiply repeated transformations. In this example, the remaining terms represent an effective magnetic field and a nearest (remaining) neighbor interaction. That is, the transformed Ising interaction is expressed in terms of the same types of operators products of S_i and $S_i S_{i+1}$ as in the original Hamiltonian, but where the position indices i have been replaced by $2i$. This reflects that the effective length scale has been increased by two. For more general models, new interactions will be generated such as-next nearest-neighbor interactions and four-spin interactions. That is, more realistic models will require the addition of new operators to the set $\hat{O}$. The new Hamiltonian is expressed in terms of new parameters. The new field parameter is given by

$$B' = B - \frac{K(2) - K(\bar{2})}{2} \tag{10.70}$$

and the new exchange interaction parameter J' is given by

$$J' = -\frac{K(2) - 2K(0) + K(\bar{2})}{4} \tag{10.71}$$

These are non-linear transformations. Furthermore, the scaling parameters are identified as βB and βJ. The transformation can be written as

$$\beta J' = \frac{1}{4} \ln \left(\frac{\cosh \beta(2J + B) \cosh \beta(2J - B)}{\cosh^2 \beta B} \right)$$

$$\beta B' = \beta B + \frac{1}{2} \ln \left(\frac{\cosh \beta(2J + B)}{\cosh \beta(2J - B)} \right) \tag{10.72}$$

The fixed points of the Hamiltonian are given by

$$\beta B^* = 0$$

$$\beta J^* \to \infty \tag{10.73}$$

or

$$\beta J^* = 0 \tag{10.74}$$

The first fixed point is the critical point (where $T_c = 0$) and the second is the trivial non-interacting fixed point.

Critical Phenomenon

The Partition Function Z is invariant under the renormalization transformation, so the Free-Energy density $\mathcal{F}$ or Free-Energy per degree of freedom must be scaled as

$$\mathcal{F}[\hat{H}'] = 2^d \; \mathcal{F}[\hat{H}] \tag{10.75}$$

where $d = 1$, corresponding to the reduction in the spin degrees of freedom. Similarly, the spatial correlations are preserved, so the correlation lengths

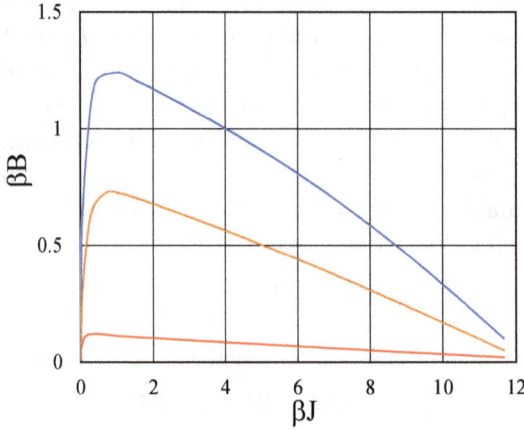

Fig. 10.4 Scaling flows of the Hamiltonian parameters of the One-dimensional Ising Model. The flows are away from the critical point $\beta J \to \infty$, $\beta B = 0$ and towards the trivial fixed point $\beta J = 0$, $\beta B = 0$.

must be re-scaled as

$$\xi[\hat{H}'] = 2^{-1}\,\xi[\hat{H}] \tag{10.76}$$

since the lattice has shrunk. At the fixed points

$$\hat{H}^* = \mathcal{R}_n\,\hat{H}^* \tag{10.77}$$

It is convenient to introduce the variables (x, y) that lead to the parameters having finite values at the fixed points. The new variables are denoted by x and y and are defined by

$$x = \exp[-4\beta J] \tag{10.78}$$

and

$$y = \exp[2\beta B] \tag{10.79}$$

The renormalization group transformation is expressed in terms of the new variables as

$$x' = x\left(\frac{(1+y)^2}{(1+xy)(y+x)}\right)$$

$$y' = y\left(\frac{y+x}{1+xy}\right) \tag{10.80}$$

The quantity x vanishes at the critical point $(x^*, y^*) = (0, 1)$. From this, we may infer that δx behaves like $t = \frac{T - T_c}{T_c}$ for a finite temperature phase transition since they both vanish at the critical point and because both describe the result of changing temperature. We shall identity δx with t. Furthermore, since the expression for the partition function is invariant under the transformation $\{S_i\} \to \{-S_i\}$ with $B \to -B$, the renormalization group transformation is seen to be invariant under the replacement $y \to y^{-1}$. At $\beta B = 0$ or equivalently $y = 1$, the renormalization group flow for x reduces to

$$x' = \frac{4x}{(1+x)^2} \tag{10.81}$$

so (at $\beta B = 0$) the critical points are located at $x^* = 0$ and $x^* = 1$. The renormalization flow is linearized about the fixed points

$$x = x^* + \delta x$$
$$y = y^* + \delta y \tag{10.82}$$

which, close to the critical point $x^* = 0, y^* = 1$, leads to the linear transformation

$$\begin{pmatrix} \delta x' \\ \delta y' \end{pmatrix} = \begin{pmatrix} 4 & 0 \\ 0 & 2 \end{pmatrix} \begin{pmatrix} \delta x \\ \delta y \end{pmatrix} \tag{10.83}$$

The linearized transformation matrix has eigenvalues of $\Lambda_x = 4$ and $\Lambda_B = 2$. Iterating the transformation n times with $\delta y = 0$, leads to

$$\delta x'_n = 4^n \delta x \tag{10.84}$$

where Λ_x is an eigenvalue of the linearized recursion relations around the critical point. Since the deviation of the variable y from its value at the $(0,1)$ fixed point is identified as $\delta y = 2\beta B$, then the eigenvalue of the transformation matrix $\Lambda_B = 2$ can be found directly by expanding the relation

$$\beta B' = \beta B + \frac{1}{2} \ln \left(\frac{x \exp[-\beta B] + \exp[\beta B]}{x \exp[\beta B] + \exp[-\beta B]} \right) \tag{10.85}$$

about the fixed point $x^* = 0$. Using the linearized flow transformation and iterating n times shows that the field scales as

$$\beta B'_n = 2^n \beta B \tag{10.86}$$

Hence, we have

$$\delta x'_n = \Lambda_x^n \delta x$$
$$\delta \beta B'_n = \Lambda_B^n \delta \beta B \tag{10.87}$$

which involves powers of the eigenvalues of the linearized transformation operator $\Lambda_x = 2^2$ and $\Lambda_B = 2^1$. Since the parameters flow away from the critical point under the transformation, the critical point is unstable. Furthermore, since δx and δy flow away from the critical point, δx and δy are defined as relevant variables. Variables that don't flow away from the fixed point are defined to be irrelevant variables. Since the Partition Function is invariant under the transformation, with $y^* = 1$, one has

$$\mathcal{F}(x^* + \delta x'_n) = 2^{nd}\ \mathcal{F}(x^* + \delta x) \tag{10.88}$$

or, since $x^* = 0$ and $d = 1$

$$\mathcal{F}(\delta x'_n) = 2^n\ \mathcal{F}(\delta x) \tag{10.89}$$

For values of n which leads to sufficiently small values of $\delta x'_n$ such that the linear approximation Eq. (10.84) is valid, one may eliminate n by setting

$$n = \frac{\ln(\delta x'_n / \delta x)}{\ln \Lambda_x} \quad \text{with } \Lambda_x = 4 \tag{10.90}$$

On eliminating n from the re-scaling relation for the Free-Energy density, one finds that the non-analytic part of the Free-Energy density scales as

$$\mathcal{F}(\delta x'_n) = 2^{\frac{\ln(\delta x'_n / \delta x)}{\ln \Lambda_x}}\ \mathcal{F}(\delta x) \tag{10.91}$$

On taking the quantities involving $\delta x'_n$ to one side of the equation and those involving δx to the other side, one finds

$$\delta x'_n{}^{-\frac{\ln 2}{\ln \Lambda_x}}\ \mathcal{F}(\delta x'_n) = \delta x^{-\frac{\ln 2}{\ln \Lambda_x}}\ \mathcal{F}(\delta x) \tag{10.92}$$

which is clearly a constant that we shall denote by F_0. Hence, one finds the x-dependence of the Free-Energy density in terms of the constant F_0. That is, for small δx

$$\mathcal{F}(\delta x) = F_0 \delta x^{\frac{\ln 2}{\ln \Lambda_x}} = F_0 \delta x^{\frac{1}{2}} \tag{10.93}$$

The specific heat is the second derivative of the Free-Energy wrt to temperature and since temperature is related to x, one may express the non-analytic part of the Free-Energy density as

$$\mathcal{F}(\delta x) \sim \delta x^{2-\alpha} \tag{10.94}$$

which leads to the identification of the specific heat exponent α as

$$\alpha = \frac{3}{2} \tag{10.95}$$

which reflects a non-trivial power law dependence on $t = \frac{T - T_c}{T_c}$. Likewise, one finds the correlation length ξ satisfies

$$\xi'_n = 2^{-n}\ \xi = \exp[-n\ \ln 2]\ \xi \tag{10.96}$$

On substituting the equality

$$n = \frac{\ln(\delta x'_n / \delta x)}{\ln \Lambda_x} \qquad (10.97)$$

in the above equation, one finds

$$\xi'_n \, \delta x'^{\frac{\ln 2}{\ln \Lambda_x}}_n = \xi \, \delta x^{\frac{\ln 2}{\ln \Lambda_x}} \qquad (10.98)$$

which is a constant. Thus, we find that

$$\xi \sim \delta x^{-\frac{1}{2}} \qquad (10.99)$$

Hence, the correlation length ξ diverges at the critical point $\delta x = 0$. The critical exponent for the correlation length, ν is defined as

$$\xi \sim \delta x^{-\nu} \qquad (10.100)$$

therefore $\nu = \frac{1}{2}$. We can compare this exponent with the exact solution for the correlation function, where

$$\xi = \frac{1}{|\ln(\tanh \beta J)|} \qquad (10.101)$$

by noting that close to the $T = 0$ critical point

$$\tanh \beta J \sim 1 - 2 \, \exp[-2\beta J] \qquad (10.102)$$

Hence, with $x = \exp[-4\beta J]$ and $x^* = 0$, one finds

$$\xi \sim \exp[2\beta J] \sim \delta x^{-\frac{1}{2}} \qquad (10.103)$$

The critical exponents for the specific heat and the correlation length are related by the hyper-scaling relation

$$2 - \alpha = d\nu \qquad (10.104)$$

which is satisfied since $d = 1$.

In the presence of the magnetic field, the Free-Energy density $\mathcal{F}(\delta x, \delta y)$ can be considered to be a function of δx and δy close to the fixed point $(0, 1)$. The scaling parameters, which are the deviations from the fixed are transformed according to

$$\delta x'_n = 4^n \delta x$$
$$\delta y'_n = 2^n \delta y \qquad (10.105)$$

and the Free-Energy per degree of freedom transforms according to

$$\mathcal{F}(\delta x'_n, \delta y'_n) = 2^n \, \mathcal{F}(\delta x, \delta y) \qquad (10.106)$$

A value of n is chosen such that the linear approximation is valid

$$n = \frac{\ln \frac{\delta x'_n}{\delta x}}{\ln 4} \tag{10.107}$$

The Free-Energy density can be written as

$$\mathcal{F}(\delta x'_n, \delta y'_n) = \left(\frac{\delta x'_n}{\delta x}\right)^{\frac{1}{2}} \mathcal{F}(\delta x, \delta y) \tag{10.108}$$

but, since

$$\delta y'_n = \left(\frac{\delta x'_n}{\delta x}\right)^{\frac{1}{2}} \delta y \tag{10.109}$$

one finds that

$$\mathcal{F}(\delta x, \delta y) = \left(\frac{\delta x}{\delta x'_n}\right)^{\frac{1}{2}} \mathcal{F}\left(\delta x'_n, \left(\frac{\delta x'_n}{\delta x}\right)^{\frac{1}{2}} \delta y\right) \tag{10.110}$$

Setting $\delta x'_n = 1$, the above relation reduces to

$$\mathcal{F}(\delta x, \delta y) = \delta x^{\frac{1}{2}} \, \mathcal{F}(1, \delta x^{-\frac{1}{2}} \delta y) \tag{10.111}$$

Thus, one finds

$$2 - \alpha = \frac{1}{2} \tag{10.112}$$

Since the magnetic susceptibility χ is related to the second derivative of F wrt B evaluated at $B = 0$, one finds that the singular part of the susceptibility is given by

$$\chi \sim \delta x^{-\gamma} \tag{10.113}$$

which leads to the identification of the susceptibility critical exponent as $\gamma = \frac{1}{2}$. Since $x = \exp[-4\beta J]$, the above expression reduces to

$$\chi \sim \exp[2\beta J] \tag{10.114}$$

which is in agreement with the exact solution for the magnetic susceptibility.

10.2   Momentum Space Renormalization

We shall introduce momentum space renormalization through the example of the Gaussian model which neglects the interaction F_4 and then examine the renormalization group for the Landau-Ginzburg model.

10.2.1   *The Renormalization Group for the Gaussian Model*

The Gaussian approximation is an approximation that neglects the interaction F_4. If one neglects F_4, the Landau-Ginzburg Free-Energy for an
n-dimensional vector order parameter $\underline{\phi}$, for $T > T_c$ is given by

$$F[\underline{\phi}] = F_0 V + \int d^d\underline{r}[F_2\underline{\phi}^2(\underline{r}) + c\underline{\nabla}\phi \cdot \underline{\nabla}\phi] \tag{10.115}$$

which, in momentum space, can be expressed as

$$F[\underline{\phi}] = F_0 V + \sum_{\underline{k}} [F_2 + c\underline{k}^2]\,\underline{\phi}_{\underline{k}} \cdot \underline{\phi}_{-\underline{k}} \tag{10.116}$$

The partition function is given by the discretized functional integral

$$Z = \prod_{\Lambda > k} \int d^3\underline{\phi}_{\underline{k}}\ \exp[-\beta F[\underline{\phi}]] \tag{10.117}$$

where $\Lambda = \frac{\pi}{a}$ is momentum cut-off. The first step of the renormalization
group consists of integrating out momentum in the momentum shell $\Lambda >
k > \frac{\Lambda}{b}$ where b is the change in the length scale. Denoting $\phi_{\underline{k}}$ by $\delta\phi_k$ if
$\underline{k}$ is in the momentum shell, one notices that the Gaussian Free-Energy
Functional separates into two parts

$$F[\underline{\phi}] = F[\underline{\phi}_{\underline{k}}] + F[\delta\underline{\phi}_{\underline{k}}] \tag{10.118}$$

On integrating over the momentum shell just below the cut-off, one finds

$$Z = Z_s \prod_{\Lambda/b > k} \int d^n\underline{\phi}_{\underline{k}}\ \exp[-\beta F[\underline{\phi}_{\underline{k}}]] \tag{10.119}$$

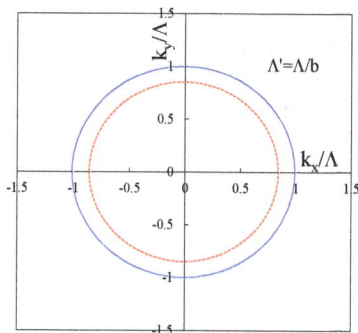

Fig. 10.5   The original momenta k have values smaller than the cut-off Λ, the degrees
of freedom with momenta between the cut-off and Λ/b are integrated out in the renormalization group transformation.

where the factor Z_s represents the Gaussian functional integration over the variables $\delta\phi_{\underline{k}}$ and is evaluated as

$$Z_s = \prod_{\Lambda>k>\Lambda/b} \int d^n\delta\phi_{\underline{k}} \, \exp[-\beta F[\delta\phi_{\underline{k}}]]$$

$$= \prod_{\Lambda>k>\Lambda/b} \left[\frac{\Omega_c(F_2 + c\underline{k}^2)}{\pi k_B T}\right]^{-\frac{n}{2}}$$

$$= \exp[-\beta V \Delta F_0] \tag{10.120}$$

where ΔF_0 is a renormalization of the constant F_0. The quantity ΔF_0 is not expected to be singular because it is evaluated for large momenta $k \sim \Lambda$. The second step of the renormalization group transformation is re-scaling the remaining momenta via

$$k' = bk \tag{10.121}$$

in order that the momentum cut-off Λ/b is restored to Λ. In the presence of an applied field, the re-scaled Free-Energy density $F'[\phi_k]$ is given by

$$F'[\underline{\phi}] = (F_0 + \Delta F_0)\,V + b^{-d} \sum_{\underline{k}'} [F_2 + cb^{-2}\,\underline{k}'^2]\,\underline{\phi}_{\underline{k}'} \cdot \underline{\phi}_{-\underline{k}'}$$

$$- \underline{\phi}_{\underline{k}=0} \cdot \underline{\Pi} \tag{10.122}$$

where $\Lambda > k'$. The next step corresponds to redefining the order parameter, to preserve the coefficient of c, this is achieved by setting

$$\underline{\phi}'_{\underline{k}'} = b^{-\frac{d+2}{2}}\,\underline{\phi}_{\underline{k}'} \tag{10.123}$$

The transformed Free-Energy Functional takes the form

$$F'[\underline{\phi}] = (F_0 + \Delta F_0)\,V + \sum_{\underline{k}} [b^2 F_2 + c\underline{k}^2]\,\underline{\phi}_{\underline{k}} \cdot \underline{\phi}_{-\underline{k}}$$

$$- b^{\frac{d+2}{2}}\,\underline{\phi}_{\underline{k}=0} \cdot \underline{\Pi} \tag{10.124}$$

Hence, in the presence of a uniform applied field $\underline{\Pi}$, the renormalization group transformation yields the renormalized parameters

$$F'_2 = b^2 F_2$$

$$\underline{\Pi}' = b^{\frac{d+2}{2}}\,\underline{\Pi}$$

$$c' = c \tag{10.125}$$

These relations can be converted to a set of differential equations in a variable $\tau = \ln b$ which is a semi-continuous variable when $b \sim 1$, so

$$b = 1 + d\tau \tag{10.126}$$

Using the relation

$$b^n = 1 + nd\tau + O(d\tau^2) \tag{10.127}$$

one finds that the scaling relations become

$$\frac{\partial F_2}{\partial \tau} = 2F_2$$

$$\frac{\partial \Pi_i}{\partial \tau} = \frac{d+2}{2} \Pi_i \tag{10.128}$$

which have the solutions

$$F_2 = F_2(0)\, b^2$$

$$\Pi_i = \Pi_i(0)\, b^{\frac{d+2}{2}} \tag{10.129}$$

Therefore, the exponents for the scaling fields g_1 and g_2 are given by

$$\Lambda_1 = 2$$

$$\Lambda_2 = \frac{2+d}{2} \tag{10.130}$$

Since $F_2 \sim t$, the correlation length exponent ν is given by

$$\nu = \frac{1}{\Lambda_1}$$

$$= \frac{1}{2} \tag{10.131}$$

which is the mean-field value. The scaling exponent for the applied field, Λ_2, is related to the susceptibility exponent γ via

$$\gamma = \frac{2\Lambda_2 - d}{2} \tag{10.132}$$

Therefore, one finds the susceptibility exponent has the mean-field value

$$\gamma = 1 \tag{10.133}$$

Likewise, using the scaling relation and the values of γ and ν, one finds that since

$$\gamma = \nu(2 - \eta)$$

$$= \frac{2 + \eta}{2} \tag{10.134}$$

the anomalous dimension is found to be zero, $\eta = 0$.

The Gaussian approximation for the thermodynamic Free-Energy density $\mathcal{F}$ has the scaling properties

$$\mathcal{F}(t, \Pi) = b^{-d} \, \mathcal{F}(b^2 t, b^{\frac{2+d}{2}} \Pi) \tag{10.135}$$

then with $b^2 t = 1$, one finds the scaling property

$$\mathcal{F}(t, \Pi) = t^{\frac{d}{2}} \, \mathcal{F}(1, t^{-\frac{2+d}{4}} \Pi) \tag{10.136}$$

The specific heat exponent is found from the second order derivative of $\mathcal{F}$ wrt t, thus

$$\alpha = \frac{4 - d}{2} \tag{10.137}$$

Finally, since

$$\phi \sim -\frac{\partial \mathcal{F}}{\partial \Pi}$$

$$\sim t^{\frac{d}{2}} \, t^{-\frac{2+d}{4}} \tag{10.138}$$

one finds that the order parameter has the critical exponent

$$\beta = \frac{(d - 2)}{4} \tag{10.139}$$

The second-derivative of $\mathcal{F}$ wrt Π leads to the susceptibility

$$\chi \sim -\frac{\partial^2 \mathcal{F}}{\partial \Pi^2}$$

$$\sim t^{\frac{d}{2}} \, t^{-\frac{2+d}{2}} \tag{10.140}$$

Hence, the susceptibilities critical exponent is given by

$$\gamma = 1 \tag{10.141}$$

The exponents α, γ, and ν agree with the mean-field results which should be valid for $d > 4$, but β is incorrect.

This disagreement between the Gaussian and mean-field approximation is not surprising since F_4 is neglected in the Gaussian approximation and, in mean-field theory

$$\phi = -\frac{F_2^2}{2F_4} \tag{10.142}$$

We may expect that F_4 has a scaling exponent $(4 - d)$ because of the Ginzburg criterion which also suggest that mean-field exponents are correct for $4 > d$. Therefore, for $4 > d$, F_4 is expected to be an irrelevant variable, but nevertheless it is a dangerous irrelevant variable since its presence is required stabilize a finite value of ϕ.

10.2.2 *Landau-Ginzburg Renormalization Group*

We shall consider the case where ϕ is a scalar field. The Landau-Ginzburg Free-Energy Functional can be written as

$$F[\phi] = \int d^d\underline{r}[F_2\phi^2(\underline{r}) + F_4\phi^4(\underline{r}) + c(\nabla\phi)^2] \tag{10.143}$$

The first step of the renormalization group process, is to transform to the momentum representation and integrate out k values in the range $\Lambda > k > \Lambda - \delta\Lambda$ where $\Lambda = \frac{\pi}{a}$ and $\Lambda - \delta\Lambda = \frac{\pi}{ab}$ and b is close to 1. Therefore

$$\frac{\delta\Lambda}{\Lambda} = \delta\ln\Lambda$$

$$\approx \frac{b-1}{b} \sim \ln b \tag{10.144}$$

The volume of the momentum shell being integrated out contains ΔN states

$$\Delta N = \frac{1}{2} V S_d \Lambda^{d-1} \delta\Lambda \tag{10.145}$$

The values of ϕ in the shell are denoted by $\delta\phi_k$. The integrations over $\delta\phi_k$ are performed after making a number of approximations, such as neglecting terms of fourth-order in $\delta\phi_k$. This is equivalent to a Gaussian integral and results in

$$F'[\phi] = F[\varphi]$$

$$+ \frac{1}{2} \sum_{r=1}^{\Delta N} \ln\left[\left(\frac{2\Lambda^2}{\pi}\right)\left(c + \frac{2F_2}{\Lambda^2} + \frac{12F_4}{\Lambda^2}\varphi_r^2\right)\right] \tag{10.146}$$

where φ_r are the order parameters in discretized real-space. The coefficient of c remains unscaled. Expanding the logarithm and replacing the sum over the ΔN states (the states in the shell that are being traced out) by an integral

$$\sum_k = \frac{\Lambda^d \, 2\pi^{\frac{d}{2}}}{\Gamma\left(\frac{d}{2}+1\right)} \, \ln b \int_0^{\delta\Lambda^{-1}} dr \qquad (10.147)$$

one finds

$$F'[\phi] = F[\varphi]$$

$$+ \frac{\Lambda^d \, 2\pi^{\frac{d}{2}}}{\Gamma\left(\frac{d}{2}+1\right)} \, \ln b \int_0^{\delta\Lambda^{-1}} dr \left[6 \left(\frac{F_4}{c\Lambda^2} - \frac{2F_2 F_4}{c^2\Lambda^4} \right) \varphi_r^2 - \frac{36 F_4^2}{c^2\Lambda^4} \, \varphi_r^4 \right]$$

$$(10.148)$$

The changes in the Landau Free-Energy Functional can be absorbed into new parameters

$$F_2' = F_2 + 6 \, \frac{2\pi^{\frac{d}{2}}}{\Gamma\left(\frac{d}{2}+1\right)} \, \ln b \left(\Lambda^{d-3} \frac{F_4}{c} - \Lambda^{d-5} \frac{2F_2 F_4}{c^2} \right)$$

$$F_4' = F_4 - 18 \, \frac{2\pi^{\frac{d}{2}}}{\Gamma\left(\frac{d}{2}+1\right)} \, \ln b \, \Lambda^{d-5} \frac{F_4^2}{c^2} \qquad (10.149)$$

The next step is change the spatial variable r to $r' = r/b$. The result is

$$F_2'' = b^2 F_2'$$

$$F_4'' = b^{4-d} F_4' \qquad (10.150)$$

Since $b \sim 1$, $b^n \approx 1 + n \ln b$, one can express the result in terms of differences

$$F_2'' - F_2 = \left[2F_2 + 6 \, \frac{2\pi^{\frac{d}{2}}}{\Gamma\left(\frac{d}{2}+1\right)} \left(\Lambda^{d-3} \frac{F_4}{c} - \Lambda^{d-3} \frac{2F_2 F_4}{\Lambda^2} \right) \right] \ln b$$

$$F_4'' - F_4 = \left[(4-d) \, F_4 - 18 \, \frac{2\pi^{\frac{d}{2}}}{\Gamma\left(\frac{d}{2}+1\right)} \, \Lambda^{d-5} \frac{F_4^2}{c} \right] \ln b \qquad (10.151)$$

The first terms are those found in the Gaussian analysis. The second terms are non-Gaussian terms. The d-dimensional surface areas can be absorbed by changing the definitions of Λ and F_2.

The renormalization group flow equations can be obtained by changing from b to a continuous variable τ defined by

$$\tau = \ln b \qquad (10.152)$$

and set

$$F_2(\tau) = F_2''$$

$$F_4(\tau) = F_4''$$

(10.153)

The difference equations reduce to the differential equations

$$\frac{\partial F_2}{\partial \tau} = 2F_2 + 6\Lambda^{d-2} \frac{F_2}{c} - 6\Lambda^{d-4} \frac{F_2 F_4}{c^2}$$

$$\frac{\partial F_4}{\partial \tau} = (4 - d) \, F_4 - 18\Lambda^{d-4} \frac{F_4^2}{c^2}$$

(10.154)

These are the renormalization group flow equations.

The Flow Equations and Fixed Points

On changing variables to the dimensionless quantities x and y defined by

$$x = \frac{F_2}{c\Lambda^2}$$

$$y = \frac{F_4}{c\Lambda^{4-d}}$$

(10.155)

the renormalization group equations take on a simple form

$$\frac{\partial x}{\partial \tau} = 2x - 6xy + 6y$$

$$\frac{\partial y}{\partial \tau} = (4 - d) \, y - 18y^2$$

(10.156)

The flow equations have a trivial Gaussian fixed point

$$x^* = 0$$

$$y^* = 0$$

(10.157)

and a non-trivial fixed point

$$x^* = -\frac{4-d}{6} \left[1 - \frac{4-d}{6} \right]^{-1}$$

$$y^* = \frac{4-d}{18}$$

(10.158)

The non-trivial fixed point is known as the Wilson-Fisher fixed point, and this fixed point governs the physics for $4 > d$ and is unphysical for $d > 4$.

The flow equations can be linearized about the fixed points.

$$x = x^* + \delta x$$
$$y = y^* + \delta y \tag{10.159}$$

The linearized flow equations can be written as

$$\frac{\partial \delta x}{\partial \tau} = (2 - 6y^*)\,\delta x + (6 - 6x^*)\,\delta y$$

$$\frac{\partial \delta y}{\partial \tau} = ((d-4) - 36y^*)\,\delta y \tag{10.160}$$

which can be written in matrix form

$$\frac{\partial}{\partial \tau}\begin{pmatrix} \delta x \\ \delta y \end{pmatrix} = \begin{pmatrix} 2 - 6y^* & (6 - 6x^*) \\ 0 & (4-d) - 36y^* \end{pmatrix}\begin{pmatrix} \delta x \\ \delta y \end{pmatrix} \tag{10.161}$$

Gaussian Fixed Point

At the Gaussian fixed point, the linearized matrix $\hat{T}(\underline{u}^*)$ simplifies

$$\begin{pmatrix} 2 - 6y^* & (6 - 6x^*) \\ 0 & (4-d) - 36y^* \end{pmatrix} = \begin{pmatrix} 2 & 6 \\ 0 & (4-d) \end{pmatrix} \tag{10.162}$$

and has two left eigenvalues and eigenvectors: One eigenvalue is given by

$$\Lambda_1 = 2 \tag{10.163}$$

and its corresponding eigenvector is

$$\underline{\phi}_1 = \left(\tfrac{2-(4-d)}{6} \quad 1 \right) \tag{10.164}$$

The other eigenvalue is given by

$$\Lambda_2 = (4 - d) \tag{10.165}$$

and its eigenvector is

$$\underline{\phi}_2 = (0 \quad 1) \tag{10.166}$$

Hence, the scaling fields g_1 and g_2 are given by

$$g_i(\tau) = g_i\,\exp[\Lambda_i\,\tau]$$
$$= g_i\,b^{\Lambda_i} \tag{10.167}$$

The critical exponents for the scaling fields are the eigenvalues $\Lambda_1 = 2$ and $\Lambda_2 = (d-4)$.

The Gaussian critical exponents are found to be

$$\alpha = 2 - \frac{d}{2}$$

$$\gamma = 1$$

$$\nu = \frac{1}{2}$$

$$\eta = 0 \tag{10.168}$$

Wilson-Fisher Fixed Point

In the neighborhood of the non-trivial fixed point, one may further approximate the linearized transformation matrix by expanding it in powers of $\epsilon = (4 - d)$

$$\begin{pmatrix} 2 - 6y^* & 6(1 - x^*) \\ 0 & (d-4) - 36y^* \end{pmatrix} \approx \begin{pmatrix} 2(1 - \frac{\epsilon}{6}) & 6(1 + \frac{\epsilon}{6}) \\ 0 & -\epsilon \end{pmatrix} \tag{10.169}$$

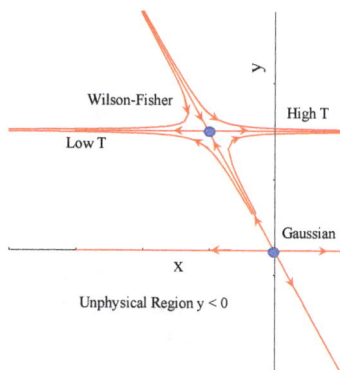

Fig. 10.6   The fixed points and scaling flows for the Landau-Ginzburg/Wilson-Fisher Model for $4 > d$.

The 2×2 matrix has a two left eigenvalues and eigenvectors. One eigenvalue is given by

$$\Lambda_1 = 2 - \frac{\epsilon}{3} \tag{10.170}$$

and its corresponding eigenvector is

$$\underline{\phi}_1 = (1 + \tfrac{\epsilon}{3} \quad 3 + \tfrac{\epsilon}{2}) \tag{10.171}$$

The other eigenvalue and eigenvector is given by

$$\Lambda_2 = -\epsilon \tag{10.172}$$

and

$$\underline{\phi}_2 = (0 \quad 1) \tag{10.173}$$

For $d > 4$, the non-trivial fixed point for lies in the unphysical region since $F_4 < 0$, but is in the physical region for $4 > d$. Hence, for $d > 4$, the critical behavior is governed by the Gaussian fixed point, but for $4 > d$, the physics is governed by the non-trivial fixed point.

The scaling fields have exponents which are given by

$$\Lambda_x = 2 - \frac{\epsilon}{3}$$

$$\Lambda_y = -\epsilon \tag{10.174}$$

Hence, to linear order in ϵ, the critical exponents are given by

$$\alpha = \frac{\epsilon}{6}$$

$$\beta = \frac{1}{2} - \frac{\epsilon}{6}$$

$$\gamma = 1 + \frac{\epsilon}{6}$$

$$\nu = \frac{1}{2} + \frac{\epsilon}{12}$$

$$\eta = 0 \tag{10.175}$$

The exponents have only been calculated to first-order in ϵ. The ϵ expansion has been carried out to higher-orders by Wilson and Fisher [94].

10.3　Problems

Problem 10.1

Consider the nearest-neighbour Ising Model on a square lattice, with ferro-magnetic interactions J in the absence of a magnetic field.

(i) Show that after performing a Trace over the spins on every other lattice site, the transformed Hamiltonian has nearest-neighbor interactions J'_{nn}, next-nearest-neighbour interactions J'_{nnn} and four spin interactions K'. The consecutive spins on the boundaries of the new unit cell are labelled as S_1, S_2, S_3 and S_4. The transformed Hamiltonian is given by

$$\hat{H}' = \beta J'_{nn}(S_1S_2 + S_2S_3 + S_3S_4 + S_4S_1$$
$$+ \beta J'_{nnn}(S_1S_3 + S_2S_4) + \beta K' \, S_1S_2S_3S_4$$

where

$$\beta J'_{nn} = \frac{1}{4} \, \ln \cosh 4\beta J$$

$$\beta J'_{nnn} = \frac{1}{8} \, \ln \cosh 4\beta J$$

$$\beta K' = \frac{1}{8} \, \ln \cosh 4\beta J - \frac{1}{2} \ln \cosh 2\beta J$$

(ii) Why is the nearest-neighbor interaction larger than the next-nearest-neighbor interaction by a factor of 2?

Problem 10.2

Consider a real-space renormalization group transformation on the triangular lattice [95]. The spins are grouped into threes and the triangles are replaced by block spins. Since the resulting lattice must have the same hexagonal symmetry as the original lattice, the new lattice has one block spin per original hexagon. The block spin has three internal bonds and each internal spin of the block is connected to the vertices of four neighboring block spins. However, two neighboring blocks are connected by two bonds which produce the renormalized interaction.

The block spin M of a triangle is decided by majority rule. That is

$$M = \text{sgn}(S_1 + S_2 + S_3)$$

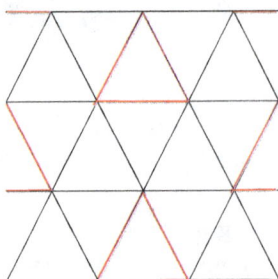

Fig. 10.7   A decimation scheme for the two-dimensional triangular lattice.

Fig. 10.8   The interactions between a pair of block spins (dashed lines), and the internal interactions (red lines).

where S_1, S_2 and S_3 are the spins on the vertex of the triangle. Since there are an odd number of sites m is unambiguously ± 1

Since three spins are to be collapsed onto one block spin, the density of spins will be reduced so

$$\frac{\rho'}{\rho} = \frac{1}{3}$$

Therefore, since $\rho \sim 1/a^2$, the length scale should be re-scaled by a factor of $b = \sqrt{3}$ to restore the original lattice.

The interactions are written as an internal interaction within the block spin

$$H_i = -J(S_1^i S_2^i + S_2^i S_3^i + S_3^i S_1^i)$$

and pairwise interactions between spins in the neighboring blocks i, j is

$$H_{i,j} = -JS_1^i(S_2^j + S_3^j)$$

The partition function is written as

$$Z = \text{Trace}_{M^i} \, \text{Trace}_{\{S_k^i\}} \left[\prod_i \left(\delta(M^i - \text{sgn}(S_1^i - S_2^i - S_3^i)) \exp[-\beta H_i] \right) \right.$$

$$\left. \times \exp\left[-\beta \sum_{i,j} H_{i,j} \right] \right]$$

The interaction between blocks can be expressed in terms of the cumulants of $H_{i,j}$. Here we shall only consider the lowest-order cumulant. The averaging over a spin in the i-th block is assumed to be only dependent on M^i and not on the other block spins. Hence, the averaging can be factorized so the average is only to be taken within the block spin labeled by i. Within a given configuration with the given block spin M^i, the average over S_1^i can be evaluated as

$$\langle S_1^i \rangle_{M^i} = \text{Trace}_{S_1^i, S_2^i, S_3^i} \frac{\exp[-\beta H_i]}{Z_{M^i}} S_1^i \, \delta(M^i - \text{sgn}(S_1^i - S_2^i - S_3^i))$$

where

$$Z_{M^i} = \text{Trace}_{S_1^i, S_2^i, S_3^i} \exp[-\beta H_i] \, \delta(M^i - \text{sgn}(S_1^i - S_2^i - S_3^i))$$

is just the partition function of the i-th triangle given that its block spin is M^i.

(i) Determine Z_{M^I}.

(ii) Find $\langle S_1^i \rangle$ for each value of M^i. Show that the average of S_1^i is in the direction of the block spin M^i.

(iii) Therefore, show that the effective interaction between neighboring block spins is given by

$$J_{i,j} = 2J \left(\frac{\exp[3\beta J] + \exp[-\beta J]}{\exp[3\beta J] + 3\exp[-\beta J]} \right)^2$$

Problem 10.3

An approximate real-space renormalization transformation for the ferromagnetic Ising model on a triangular lattice increases the length scale of the lattice by a factor of $\sqrt{3}$ and produces the following mapping of the coupling constant

$$J' = 2J \left(\frac{\exp[4\beta J] + 1}{\exp[4\beta J] + 3} \right)^2$$

(i) Show that the transformation has the following three fixed points:
(a) A high-temperature fixed point $\beta J^* = 0$.
(b) A low-temperature fixed point $\beta J^* \to \infty$.
(c) An intermediate fixed point

$$\beta J^* = \frac{1}{4} \ln \left(\frac{3 - \sqrt{2}}{\sqrt{2} - 1} \right) \approx 0.3356$$

[The exact value is $\beta J^* = \ln(3)/4$.]
(ii) By linearizing the transformation around the fixed points show that (a)
and (b) are stable and that (c) satisfies

$$\left. \frac{\partial \beta J'}{\partial \beta J} \right|_{\beta J^*} = 2 \left(\frac{\exp[4\beta J^*] + 1}{\exp[4\beta J^*] + 3} \right)^2$$

$$+ 32\beta J^* \, \exp[4\beta J^*] \, \frac{(\exp[4\beta J^*] + 1)}{(\exp[4\beta J^*] + 3)^3}$$

$$\approx 1.624$$

(iii) Identify the eigenvalue of the linearized transformation

$$1.624 = \sqrt{3}^{\Lambda_1}$$

with the thermodynamic scaling field. By considering the dimensions of
the Free-Energy density and the re-scaling of the correlation length, ξ, determine the exponents α and ν. [The exact values are $\alpha = 0$ and $\nu = 1$].

10.4 Appendix: Topological Transitions

Systems with short-ranged interactions that couple discrete variables behave differently from systems where the short-ranged interactions couple continuous variables. In fact, systems in which the degrees of freedom are designated as components of a vector, behave differently according to the dimensionality of the vector. The dimensionality of the vector degrees of freedom, n, and the dimensionality of the lattice d are important quantities that (along with other factors, such as the range of the interactions) determine the properties of a system near a phase transition.

The Ising Model is an example of a model with discrete variables. The variables in the Ising model represent the z-components of a set of N spins of magnitude $\pm\frac{\hbar}{2}$. In units with $\hbar = 2$, the spins take on values ± 1. The spin variables can be consider as a one-dimensional vector ($n = 1$). The Hamiltonian is represented by

$$H = -J \sum_{R,R'} S_R^z S_{R'}^z \tag{10.176}$$

where S_R^z is the spin at lattice site R and the sum runs over pairs of nearest-neighbor spins. We shall assume $J > 0$.

The classical Heisenberg model describes a set of vector spins ($n = 3$) that have nearest-neighbor interactions. The Hamiltonian can be written in terms of the sums of scalar product

$$H = -J \sum_{R,R'} \underline{S}_R \cdot \underline{S}_{R'} \tag{10.177}$$

where $\underline{S}_R$ is the vector spin at lattice site R and the sum runs over pairs of nearest-neighbor spins. The classical vector of a spin $S = 1$ at lattice site R can be represented in terms of the polar coordinates (φ_R, θ_R).

$$\underline{S}_R = (\cos\varphi_R \ \sin\theta_R, \sin\varphi_R \ \sin\theta_R, \cos\theta_R) \tag{10.178}$$

The $x - y$ model corresponds to the Heisenberg model in which the spins are confined within a plane ($n = 2$), i.e. $\theta_R = \frac{\pi}{2}$ for all R. Thus, the Hamiltonian for the $x - y$ model is given by

$$H = -J \sum_{R,R'} \cos(\varphi_R - \varphi_{R'}) \tag{10.179}$$

where the interaction runs over nearest-neighbor pairs of spins and the angle variable φ_R is only defined modulo 2π.

$$\varphi_R + 2\pi = \varphi_R \tag{10.180}$$

This identification of the phase allows for non-trivial topological configurations. For a standard real valued field (such as $\cos \varphi$) the field must return to its initial value when one traverses a closed loop. For a phase field, the field need only return to its initial value modulo 2π. It should be noted that the $x - y$ model has two important symmetries. First, since the interaction only depends on the difference $(\varphi_i - \varphi_j)$, the Hamiltonian is invariant under a global (gauge-like) transformation in which $\varphi_i \rightarrow \varphi_i + C$, for all i. Secondly, the interaction is invariant under the global transformation $\varphi_i \rightarrow -\varphi_i$ for all i.

10.4.1 *The Lower Critical Dimension*

We shall define the lower critical dimension and show that the lower critical dimension of the Ising Model (with discrete variables) is different from that of the $x - y$ model which has continuous variables. The lower critical dimension is the dimension below which a model does not exhibit a phase transition.

The Ising Model

In particular models with scalar variables, such as the Ising Model, do not exhibit finite-temperature phase transitions for dimensions less than two. For one-dimension, the lowest-energy state is doubly-degenerate and corresponds to either $S_R^z = +1$ or to $S^z(R) = -1$ for all R. The magnetization per spin is defined by

$$\frac{M^z}{N} = \frac{1}{N} \sum_R S_R^z \tag{10.181}$$

The magnetization per spin is the order parameter. For the lowest energy states, the value of the order parameter is ± 1. The energy of this state is given by

$$E = -J(N - 1) \tag{10.182}$$

since there are $(N - 1)$ pairs of parallel spins and each pair of parallel spins contribute an energy $-2J$. The entropy S of this energy level is given by

$$S = k_B \ln 2 \tag{10.183}$$

This model does not exhibit a phase transition at any finite temperature. This can be seen by examining the next-highest energy states in which the interaction between one pair of spins does not have its lowest value. The energy of these spin configurations is

$$E = -J(N-1) + 2J \tag{10.184}$$

since the interaction between the pair of antiparallel spins is $+J$ which replaces the energy $-J$. The excitation with energy $+2J$ separates the one-dimensional chain of spins into two segments. The spins on each of the segments have parallel spins but are antiparallel with the spins of the other segment. If the unsatisfied (non-minimized) interaction is located within $L/2$ lattice sites from the center of the chain and $N \gg L$, the magnetization per spin is infinitesimal. In this case, approximately half of the spins point upwards and approximately half the spins point down. The entropy of these lowest excited energy configurations is given by

$$S = k_B \ln 2L \tag{10.185}$$

since the unsatisfied interaction could sit on any of the L sites. The difference in Free-Energy between the excited configurations for which the order parameter is zero and the perfectly magnetically ordered configuration ΔF is given by

$$\Delta F = 2J - k_B T \ln L \tag{10.186}$$

For

$$k_B T > \frac{2J}{\ln L} \tag{10.187}$$

the perfectly magnetically ordered configurations have higher Free-Energy and, therefore, are unstable to the configurations in which the magnetization per spin or order parameter is zero. The perfectly ordered state where all the spins are parallel is destroyed by any small temperature. Hence, in the thermodynamic limit $N \to \infty$ where one can also take $L \to \infty$, one sees the one-dimensional Ising model does not exhibit a magnetically ordered phase (except at zero temperature).

A similar argument can be applied to the two-dimensional Ising Model, but the argument leads to a different conclusion. The lowest energy configuration corresponds to states in which all the spins are parallel. These states are two-fold degenerate. For a square lattice with N sites, the extensive part of the energy is $-2JN$. This value is a consequence of each site having four nearest-neighbors and each of the four interactions or bonds contributing an energy $-J$ to the total energy. However, since each interaction is shared between a pair of sites, when summing over sites, each interaction should be counted with a factor of one-half. We shall consider a state in which an island of perimeter length L confines a region in which all the spins are flipped. The perimeter bisects the bonds between antiparallel spins and forms a closed polygon. The perimeter of the island is known as a domain wall. The energy of this domain wall is higher than the energy of the lowest-energy configuration by an approximate amount $2LJ$, since the interactions between the spins on the perimeter are between antiparallel spins and are unsatisfied.

$$\Delta E \approx 2JL \tag{10.188}$$

The entropy associated with the island of reversed spins corresponds to the logarithm of the number of different islands that have the same perimeter. Consider the construction of a perimeter which bisects the bonds between nearest-neighbor spins and which returns to its origin, but does not pass through any other bond more than once. Given one link on the perimeter, the choice of the next link on the perimeter can only be made in three different ways since, although there are four nearest-neighbors bonds, the perimeter cannot double back on itself. Since there are L links, the perimeter can be constructed in approximately $N \times 3^L$ different ways, since the first link on the perimeter can be chosen N different ways. This construction does neglect the constraint that the perimeters must close, but, for large perimeters the contours do almost close. This leads to an approximate increase of the entropy ΔS given by

$$\Delta S \approx L \ln 3 + \ln N \tag{10.189}$$

Thus, the Free-Energy associated with the island of flipped spins is approximately given by

$$\Delta F \approx L(2J - k_B T \ln 3) + \ln N \tag{10.190}$$

For $k_B T \ln 3 > 2J$, it is favorable to create islands of reversed spins, so there should be no long ranged order. For $k_B T \ln 3 < 2J$, the creation of islands of reversed spins is unfavorable so all spins should be parallel. This argument, due to Rudolf Peierls, predicts that the two-dimensional Ising model has a phase transition at the critical temperature T_c given by

$$k_B T_c \approx \frac{2}{\ln 3} J \tag{10.191}$$

If one enforces the constraint that the perimeter must close, one finds that

$$k_B T_c \approx \frac{2}{\ln 2.639} J \tag{10.192}$$

The approximate value of $1.8J$ found by Peierls should be compared with the exact value of $2.2J$ for $k_B T_c$ previously identified in our discussion of Lars Onsager's exact solution as

$$k_B T_c = \frac{2}{\ln(1 + \sqrt{2})} J \qquad (10.193)$$

These values of the critical temperature are in fair agreement.

The above argument suggests that systems with short-ranged interactions and scalar variables have a lower critical dimension $d_{lc} = 1$, below which the order parameter must be zero and long-ranged order is not possible.

The x-y Model

The lower critical dimension of the $x - y$ model in d-dimensions can be inferred from the following arguments. First, we shall consider slowly varying spin configurations, so that the Hamiltonian can be expanded

$$H = -J \sum_{R,R'} \cos(\varphi_R - \varphi_{R'})$$

$$= J \sum_{R,R'} \left[-1 + \frac{1}{2} (\varphi_R - \varphi_{R'})^2 \right]$$

$$= \text{Const.} + \frac{J}{2} \sum_{R,R'} (\varphi_R^2 + \varphi_{R'}^2 - 2\varphi_R \varphi_{R'}) \qquad (10.194)$$

The first term is a constant which is the energy of the uniform configurations of spins which does not have a physical consequence and can be discarded. In the continuum approximation and assuming the phases can be defined unambiguously, the phase differences can be expressed as

$$\varphi_R - \varphi_{R'} = (\underline{R} - \underline{R'}) \cdot \nabla \varphi \qquad (10.195)$$

so, one finds

$$H = \text{Const.} + \frac{Ja^2}{2} \sum_{R,R'} (\nabla \varphi_{R,R'})^2 \qquad (10.196)$$

where the sum runs over the pairs of bonds between R and R'. The second term can be expressed in terms of the Fourier transform of φ_R, which is defined by

$$\varphi_q = \frac{1}{\sqrt{N}} \sum_R \exp[i\underline{q} \cdot \underline{R}] \, \varphi_R \qquad (10.197)$$

Hence,

$$H = \frac{Ja^2}{2} \sum_q q^2 \varphi_q \varphi_{-q} \qquad (10.198)$$

apart from a constant. The $T = 0$ configurations that minimize the continuum approximation for the energy satisfy Laplace's equation

$$\nabla^2 \varphi = 0 \qquad (10.199)$$

The perfectly ordered configuration of parallel spins $\varphi = C$ obviously minimizes the spin-wave Hamiltonian. Small amplitude plane waves are exact eigenstates of this Hamiltonian and correspond to spin waves. In addition, to the small-amplitude spin waves there may be large amplitude excitations.

For a system of linear spatial dimensions L with a fixed phase φ_0 on one boundary and a phase φ_L, on the other boundary, one can consider the large amplitude solution of Laplace's equation

$$\varphi = \varphi_0 + \frac{x}{L} \, (\varphi_L - \varphi_0) \qquad (10.200)$$

Hence,

$$E = \text{Const.} + \frac{Ja^2}{2} \left(\frac{L}{a}\right)^d \left(\frac{\varphi_L - \varphi_0}{L}\right)^2 \qquad (10.201)$$

where the volume of the system is given by $V = L^d$. The energy of this spin configuration is proportional to L^{d-2} and increases indefinitely as $L \to \infty$ for $d > 2$. Since a very large energy is required to twist the spins on the two opposite boundaries of a large system, the system is robust against twists for $d > 2$. Hence, at $T = 0$ and $d > 2$, the phase on opposite sides of the system are likely to be the same, therefore, the system has long-ranged order. On the other hand, for $2 > d$, the phases on the boundaries can be considered as being independent, and so the system does not exhibit long-ranged order at $T = 0$.

At finite temperatures, one can evaluate the real-space correlation function

$$\langle (\varphi_R - \varphi_0)^2 \rangle = \frac{1}{N} \sum_{q,q'} (\exp[i\underline{q}.\underline{R}] - 1)(\exp[i\underline{q}'.\underline{R}] - 1)\langle \varphi_q \varphi_{q'} \rangle \qquad (10.202)$$

However, at low temperatures, the correlation function $\langle \varphi_q \varphi_{q'} \rangle$ can be evaluated as

$$\langle \varphi_q \varphi_{q'} \rangle = \frac{1}{Z} \prod_{q_1} \left\{ \int d\varphi_{q_1} \int d\varphi_{-q_1} \right\}$$

$$\times \varphi_q \varphi_{q'} \exp\left[-\frac{\beta J a^2}{2} \sum_{q_2} q_2^2 \, \varphi_{q_2} \varphi_{-q_2} \right]$$

$$= \left(\frac{k_B T}{J q^2 a^2} \right) \delta_{q,-q'} \tag{10.203}$$

which represents the effect of the small-amplitude spin-waves. The presence of the Kronecker delta function is expected from consideration of translational invariance. Thus, the real-space correlation function becomes

$$\langle (\varphi_R - \varphi_0)^2 \rangle = \frac{2 k_B T}{J} a^{d-2} \int \frac{d^d q}{(2\pi)^d} \frac{(1 - \cos qR)}{q^2}$$

$$\sim \frac{k_B T}{J} \frac{S_d}{(2\pi)^d} a^{d-2} \int_{R^{-1}}^{a^{-1}} dq \; q^{d-3} \tag{10.204}$$

where S_d is the surface area of a sphere in d-dimensions. The upper limit of integration corresponds to the magnitude of the primitive reciprocal lattice vector and the lower limit corresponds to the q value at which the numerator becomes negligibly small, ie. $\cos qR \sim 1$. For $d > 2$, the integral is evaluated as

$$\frac{k_B T}{J(d-2)} \left[1 - \left(\frac{R}{a} \right)^{2-d} \right] \tag{10.205}$$

This converges to a finite value for large R, since $d > 2$. This result suggests that the phase difference remains finite and long-ranged order is maintained for $d > 2$. For $d = 2$, the integral results in terms proportional to $\ln R$, which diverges as $R \to \infty$. Hence, for $d = 2$, the fluctuations grow indefinitely for large R at any finite temperature. Therefore, long-ranged order is destroyed at finite temperatures for $d = 2$. Thus, we expect that the lower critical dimension is $d_{lc} = 2$ for the $x - y$ model.

The Mermin-Wagner Theorem

Mermin and Wagner [96] have given the above argument for the lower critical dimension of models with continuous variables a rigorous basis, starting from the Schwartz Inequality.

Renormalization Group

Polyakov [97] performed a renormalization group treatment of n-component classical spin models with short-ranged interactions in d-dimensions. Following Phil Anderson's renormalization group method, as introduced by Leo Kadanoff and popularized by Ken Wilson, Polyakov considered the flow of the ratio $\kappa = J/T$ with the length scale L. The dimensionless interaction was found to scale according to

$$\frac{d\kappa^{-1}}{dL} = -(d-2)\,\kappa^{-1} + \left(\frac{n-2}{2\pi}\right)\kappa^{-2} + \dots \tag{10.206}$$

The dimensionless interaction strength remains invariant under re-scaling when the fixed point κ_c

$$(d-2)\,\kappa_c^{-1} = \left(\frac{n-2}{2\pi}\right)\kappa_c^{-2} \tag{10.207}$$

is reached. At the fixed point, the physics looks similar for all length scales L. For $d = 2$, and $n > 2$ the effective temperature increases with increasing L, suggesting that the fixed point of the model represents the disordered phase. For $n = 1$, the effective temperature decreases to zero, confirming that the $d = 2$ Ising model exhibits long-ranged order at low temperatures as found by Lars Onsager. On the other hand for $n = 2$ and $d = 2$, the model is marginal as

$$\frac{d\kappa^{-1}}{dL} = 0 \tag{10.208}$$

to all orders in κ^{-1}.

10.4.2   *The Kosterlitz-Thouless Transition*

At low temperatures, the two-dimensional $x-y$ model has correlations that don't fall off exponentially with distance, as expected at high temperatures, but instead the correlations fall of as a power law. This suggests that the two-dimensional $x-y$ model appears to exhibit some type of critical behavior at low temperatures, but does not exhibit long-ranged order. This peculiar behavior is a result of the two-dimensional $x-y$ model being at the lower critical dimension.

The Spatial Correlations

The correlation function, evaluated in the spin-wave approximation with the aid of Bloch's Identity, can be expressed as

$$\langle \cos(\varphi_R - \varphi_0) \rangle = \exp[i\langle(\varphi_R - \varphi_0)\rangle]$$

$$= \exp\left[-\frac{1}{2}\langle(\varphi_R - \varphi_0)^2\rangle\right] \tag{10.209}$$

The expression in the exponent can be evaluated with $d = 2$ in the spin-wave approximation and is found to be given by

$$\langle(\varphi_R - \varphi_0)^2\rangle = \frac{2k_B T}{J(2\pi)^2} \int_0^{2\pi} d\theta \int_{R^{-1}}^{a^{-1}} \frac{dq}{q}$$

$$= \left(\frac{k_B T}{\pi J}\right) \ln\left(\frac{R}{a}\right) \tag{10.210}$$

The correlation function, evaluated by only considering spin waves, is given by

$$\langle \cos(\varphi_R - \varphi_0) \rangle = \left(\frac{a}{R}\right)^{\left(\frac{k_B T}{2\pi J}\right)} \tag{10.211}$$

This correlation function falls of as a power law, in contrast to the Ising model where the correlation functions usually fall of exponentially in the paramagnetic phase. The power-law variation found for the $x - y$ model at low temperatures is characteristic of a system at the critical point for which $\xi \to \infty$. Specifically, the scale invariance of the correlations indicates that each value of T/J where this occurs is a fixed point. That is there is a line of fixed points at low temperatures. As T approaches zero, the correlations decay more slowly but the system does not exhibit long-ranged magnetic order. However, at high temperatures, one expects that the correlations should fall off more rapidly than inferred from the above spin-wave analysis. This could be expected if the spin configurations vary quite rapidly in space such as those which have large amplitude variations.

A Single Vortex Configuration

The solutions of the Laplace equation

$$\nabla^2 \varphi = 0 \tag{10.212}$$

are supplemented by large amplitude topological solutions which satisfy
Laplace's equation everywhere except for a finite set of points. Topological
solutions are inhomogeneous configurations that cannot be continuously
deformed into a uniform configuration. Topological vortex states exist in
the $x - y$ model and they affect the state of the system and are responsible
for the destruction of the power-law correlations at high temperatures and
result in a phase transition. However, the system does not show long-ranged
order in the low-temperature phase.

A family of vortex configurations can be described by

$$\varphi(\underline{r}) = \text{Const.} + n\theta \tag{10.213}$$

where the strength or winding number n is a positive integer $n = m$ for a
vortex and a negative integer $n = -m$ for an anti-vortex. The vortices are
solutions of Laplace's equation. We shall show this using the representation

$$\varphi = n \ \tan^{-1}\left(\frac{y}{x}\right) + C \tag{10.214}$$

which shows that φ is only defined up to multiples of 2π and is not defined
at the origin. From the representation of the family of vortices, one finds
the Cartesian components of the gradient are given by

$$\left(\frac{\partial\varphi}{\partial x}\right)_y = -\frac{ny}{x^2 + y^2}$$

$$\left(\frac{\partial\varphi}{\partial y}\right)_x = +\frac{nx}{x^2 + y^2} \tag{10.215}$$

so one finds that φ is non-analytic as it does not satisfy the Cauchy condition as

$$\left(\frac{\partial^2\varphi}{\partial y\partial x}\right) = -n\frac{(x^2 - y^2)}{(x^2 + y^2)^2}$$

$$\left(\frac{\partial^2\varphi}{\partial x\partial y}\right) = +n\frac{(x^2 - y^2)}{(x^2 + y^2)^2} \tag{10.216}$$

This is a consequence of the phase not being defined at the origin. The
second derivatives of the phase field are given by

$$\left(\frac{\partial^2\varphi}{\partial x^2}\right)_y = +\frac{2nxy}{(x^2 + y^2)^2}$$

$$\left(\frac{\partial^2\varphi}{\partial y^2}\right)_x = -\frac{2nxy}{(x^2 + y^2)^2} \tag{10.217}$$

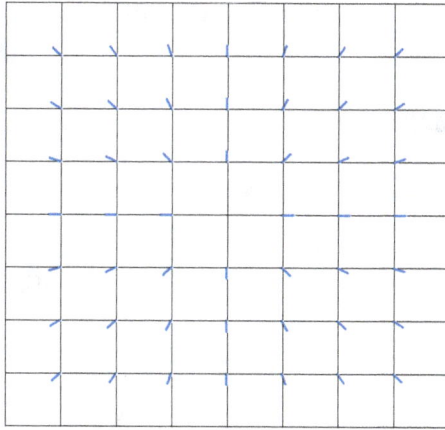

(a)

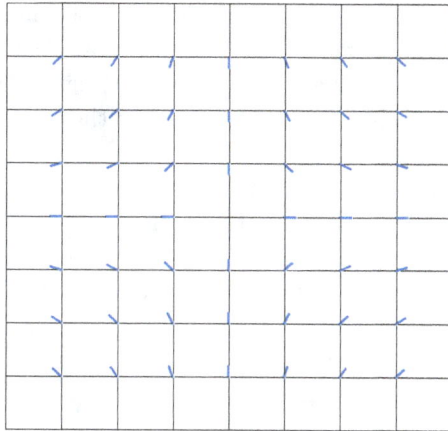

(b)

Fig. 10.9 (a) Single vortices with $n = +1$ and (b) $n = -1$ centered on the origin. The vortex configuration is specified by the spin angles $\varphi = n \tan^{-1} \frac{y}{x} + C$ where C is an arbitrary constant. For convenience we have set $C = 0$. The value of the constant of integration C does not affect the physics of the $x - y$ model since it merely corresponds to a sort of global gauge-like transformation. We note that, no matter what value C has, the spins of the positive n configurations rotate counter-clockwise when one traverses a counterclockwise path encircling the origin. In the anti-vortex configurations with negative n, the spins rotate clockwise as one encircles the origin on a counter-clockwise path.

so the vortex solutions satisfy Laplace's equation

$$\nabla^2 \varphi = 0 \tag{10.218}$$

except at the origin. The circulation integral of the phase field near a single vortex is given by

$$\oint d\underline{r} \cdot \underline{\nabla}\varphi = 2\pi n \tag{10.219}$$

when the contour encloses the vortex at the origin. Otherwise

$$\oint d\underline{r} \cdot \underline{\nabla}\varphi = 0 \tag{10.220}$$

Thus, the vortex can be considered as a topological singularity in the field. The topological excitations are topologically stable, in the sense they cannot be deformed into a uniform configuration by a sequence of continuous rotations of the phase. We note that a vortex and anti-vortex configurations corresponding to the same $|n|$ value are degenerate, since they are merely related by the global symmetry $\varphi \to -\varphi$. In what follows we may, at our convenience, focus on singularities with $n = \pm 1$, since the energies of the configurations with higher values of n (such as those shown in Fig. 10.10) have energies which grow as n^2.

Since we recognize that the phase is only defined within 2π and, therefore, φ may have a part with a finite rotation or curl, we denote the generalized derivative of the phase field as $\underline{v}$, and in addition to a longitudinal or gradient term we include a transverse or rotational part

$$\underline{v} = \underline{v}_L + \underline{v}_T$$
$$= \underline{\nabla}\varphi + \underline{\nabla} \wedge (\hat{e}_z\psi) \tag{10.221}$$

where

$$\underline{\nabla} \cdot \underline{v}_T = 0$$
$$\underline{\nabla} \wedge \underline{v}_L = 0 \tag{10.222}$$

The choice of $\hat{e}_z$ keeps the rotational part in-plane. The rotational part has components

$$v_{T,x} = \frac{\partial \psi}{\partial y}$$

$$v_{T,y} = -\frac{\partial \psi}{\partial x} \tag{10.223}$$

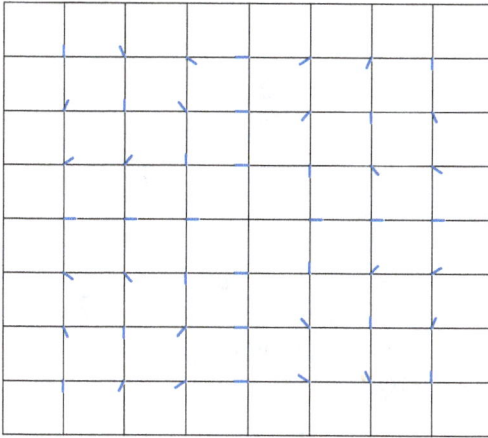

(a)

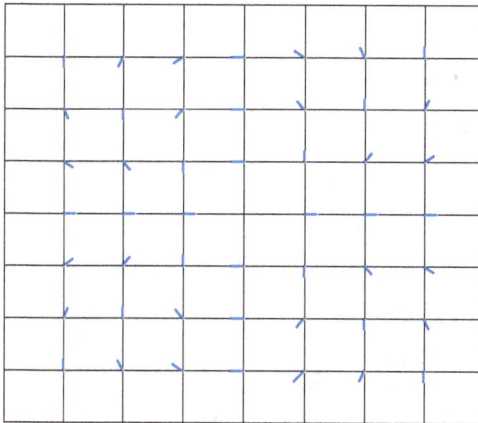

(b)

Fig. 10.10 (a) Single vortices with $n = +2$ and (b) $n = -2$ centered on the origin. We note that the spins of the positive n configurations rotate counter-clockwise when one traverses a counterclockwise path encircling the origin. In the anti-vortex configurations with negative n, the spins rotate clockwise as one encircles the origin on a counter-clockwise path.

The circulation integral is evaluated as

$$\oint d\underline{r} \cdot \underline{v} = \int d^2 \underline{S} \cdot (\underline{\nabla} \wedge \underline{v})$$

$$= 2\pi n \qquad (10.224)$$

when the contour encloses the vortex at the origin and is zero otherwise. Thus, we can identify

$$\underline{\nabla} \wedge \underline{v} = 2\pi n \delta^2(\underline{r}) \hat{e}_z \qquad (10.225)$$

Since the curl of the longitudinal term vanishes, the above equation determines the transverse part of $\underline{v}$. The relation between ψ and the strength and position of the vortex is given by

$$\underline{\nabla} \wedge (\underline{\nabla} \wedge (\hat{e}_z \psi)) = n 2\pi \delta^2(\underline{r}) \hat{e}_z \qquad (10.226)$$

Thus, ψ satisfies Poisson's equation

$$-\nabla^2 \psi = n 2\pi \delta^2(\underline{r}) \qquad (10.227)$$

The solution of the above equation for ψ is

$$\psi = -n \ln r \qquad (10.228)$$

This solution for ψ determines the transverse part of the phase field

$$\underline{v}_T = \underline{\nabla} \wedge (\hat{e}_z \psi) \qquad (10.229)$$

The energy of the field can be expressed as

$$H = \text{Const.} + \frac{J}{2} \int d^2\underline{r} \, \underline{v}^2$$

$$= \text{Const.} + \frac{J}{2} \int d^2\underline{r} (\underline{v}_L^2 + \underline{v}_T^2 + 2 \, \underline{v}_L \cdot \underline{v}_T) \qquad (10.230)$$

The interaction between the transverse and longitudinal fields vanish, since one can integrating by parts

$$\int d^2\underline{r}\underline{v}_L \cdot \underline{v}_T = \int d^2\underline{r}\nabla\varphi \cdot \underline{v}_T$$

$$= -\int d^2\underline{r}\varphi\underline{\nabla} \cdot \underline{v}_T$$

$$= 0 \qquad (10.231)$$

For a lattice, this argument can be made rigorous by positioning the singularities on the sites of a dual lattice, i.e. a dual lattice where the sites are located at the center of each plaquette. Not only do the transverse and longitudinal excitations decouple in the energy but also their contributions to correlation functions factorize. Since the magnitude of the purely rotational vector, $\underline{v}_T$, can be found from

$$v_T^2 = (\underline{\nabla}\psi)^2 \tag{10.232}$$

the energy of the vortex can also be expressed in terms of the scalar field ψ as

$$H = \text{Const.} + \frac{J}{2} \int d^2\underline{r}(\underline{\nabla}\psi)^2 \tag{10.233}$$

We shall now examine the change in Free-Energy due to a single vortex excitation.

First, we shall examine the change in energy due to the presence of a single vortex. The energy is expressed in terms of the gradient of the field. The radial and azimuthal components of the gradient of φ is given by

$$\hat{e}_r \cdot \underline{\nabla}\varphi = 0 \tag{10.234}$$

and is given by

$$\hat{e}_\theta \cdot \underline{\nabla}\varphi = \frac{1}{r} \frac{\partial \varphi}{\partial \theta}$$

$$= \frac{n}{r} \tag{10.235}$$

Hence, the change in energy ΔE due to the presence of a single vortex is given by

$$\Delta E = \frac{J}{2} \int d^2\underline{r}\, (\underline{\nabla}\varphi)^2$$

$$= \frac{J}{2} \int dr\, r \int_0^{2\pi} d\theta \, \frac{n^2}{r^2}$$

$$= n^2 \pi J \, \ln\frac{L}{a} + E_c \tag{10.236}$$

where L is the linear dimension of the system and a is the radius of the vortex core and E_c is the energy associated with the core spins.

The change in entropy ΔS associated with a single vortex is given by

$$\Delta S = k_B \, \ln\left[\pi \left(\frac{L}{a}\right)^2\right] \tag{10.237}$$

since this is the number of ways of locating the center of the vortex in an area πL^2.

Hence, the change in the Free-Energy associated with a single vortex is given by

$$\Delta F = (\pi J - 2k_B T) \ \ln\left(\frac{L}{a}\right) + E_c \qquad (10.238)$$

The Free-Energy ΔF changes sign at the temperature T_{KT} given by

$$k_B T_{KT} = \frac{\pi}{2} \ J \qquad (10.239)$$

For temperatures above T_{KT}, the system lowers its Free-Energy by creating vortices. In the presence of a vortex, the spin variable changes direction rapidly, invalidating the continuum or spin wave approximation. Thus, above T_{KT}, the power-law variation of the correlation function is destroyed. For temperatures below T_{KT}, the creation of a single vortex increases the Free-Energy of the system and is thermodynamically unstable [98]. However, the creation of vortex/anti-vortex pairs does not increase the Free-Energy. In the low-temperature phase which has bound vortex/anti-vortex pairs, the system exhibits the power-law spatial variation. In the high-temperature phase, the vortex/anti-vortex pairs become unbound. The unbinding is a collective effect. The vortex pairs, that are created as the temperature is raised, screen the interactions between the vortices so the interactions becomes zero at large vortex separations.

Vortex/Anti-Vortex Pairs

Since Laplace's equation is a linear equation, a more general solution is given by the linear superposition

$$\varphi = \sum_{\underline{R_0}} n_{R_0} \ \tan^{-1}\left(\frac{y - Y_0}{x - X_0}\right) \qquad (10.240)$$

where $\underline{R_0}$ label the positions of the vortices (X_0, Y_0) which are on the sites of a dual lattice and n_{R_0} is the strength of the vortex. The sites of the dual lattice are located at the positions $X_0 = a(\frac{1}{2} + m_x)$ and $Y_0 = a(\frac{1}{2} + m_y)$, where m_x and m_y are integers.

Consider a number of vortices or anti-vortices that are located close to the origin, then the gradient field of the spin configuration at large distances from the origin can be expressed as

$$\hat{e}_r \cdot \underline{\nabla}\varphi \approx 0 \qquad (10.241)$$

and

$$\hat{e}_\theta \cdot \underline{\nabla}\varphi \approx \frac{\sum_{R_0} n_{R_0}}{r} \qquad (10.242)$$

Therefore, the energy associated with this configuration is given by

$$\Delta E = \frac{J}{2} \int d^2\underline{r}(\nabla\varphi)^2$$

$$\approx \frac{J}{2} \int dr \, r \int_0^{2\pi} d\theta \, \frac{\left(\sum_{R_0} n_{R_0}\right)^2}{r^2}$$

$$\approx \left(\sum_{R_0} n_{R_0}\right)^2 \pi J \, \ln \frac{L}{a} + \sum_{R_0} E_c \qquad (10.243)$$

From this, it is seen that spin configurations that have

$$\sum_{R_0} n_{R_0} = 0 \qquad (10.244)$$

have no divergent energy associated with them and so can exist in the low-temperature phase. This result is a consequence that the far-field configuration in the presence of vortex/anti-vortex pairs continuously deforms into a uniform configuration as $r \to \infty$. This can be seen by examining the

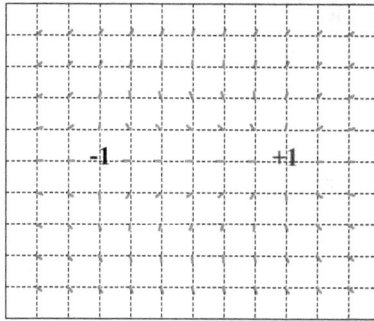

Fig. 10.11   The spin configuration with an $n = -1$ anti-vortex separated from a $n = +1$ vortex by a distance along the x-axis. The spin configuration reduces to a uniform configuration as $r \to \infty$.

spin configuration of a vortex and anti-vortex separated by a distance $2d$ along the x axis

$$\varphi = \tan^{-1}\left(\frac{2dy}{r^2 - d^2}\right) \qquad (10.245)$$

This reduces to $\varphi \to 0$ as $r \to \infty$. Hence, we conclude that any configuration in which the vortices are bound in pairs is likely to exist at low temperatures. Since the bound vortices do not effect the long-ranged variation of the spin configurations, the vortex factor in the correlation function is a constant. Thus, the correlation function is dominated by the spin-wave factor, so the correlations will decay like a power law. The Kosterlitz-Thouless transition [99] is associated with the un-binding of the vortex pairs at temperatures above T_{KT} which introduces the exponentially decaying correlations.

Interactions between Vortices

If we define the vortex density $\rho(\underline{r})$ via

$$\rho(\underline{r}) = \sum_{\underline{R_0}} n_{R_0} \, \delta^2(\underline{r} - \underline{R_0}) \tag{10.246}$$

where $\underline{R_0}$ are located on the sites of the dual lattice, then

$$\nabla^2 \psi = -2\pi\rho \tag{10.247}$$

On Fourier transforming Poisson's equation, one obtains

$$q^2 \psi_q = 2\pi\rho_q \tag{10.248}$$

The rotational part of the energy

$$E_V = \frac{J}{2} \int d^2\underline{r} (\nabla\psi)^2 \tag{10.249}$$

can also be expressed as

$$E_V = \frac{Ja^2}{2} \sum_{\underline{q}} q^2 \, \psi_q \, \psi_{-q} \tag{10.250}$$

Hence, the interaction between vortices can be expressed as

$$E_V = \frac{Ja^2}{2} (2\pi)^2 \sum_{\underline{q}} \frac{\rho_q \rho_{-q}}{q^2} \tag{10.251}$$

which can then be expressed in terms of real space by introducing the Green's function $G(\underline{R} - \underline{R'})$ defined by

$$G(\underline{R} - \underline{R'}) = \sum_{\underline{q}} \frac{\exp[i\underline{q} \cdot (\underline{R} - \underline{R'})]}{q^2 a^2} \tag{10.252}$$

so the energy can be written as

$$\begin{aligned} E_V &= \frac{Ja^4}{2} (2\pi)^2 \sum_{R,R'} \rho(\underline{R}) \, G(\underline{R} - \underline{R'}) \, \rho(\underline{R'}) \\ &= \frac{J}{2} \, 4\pi^2 \int d^2\underline{r} \int d^2\underline{r'} \, \rho(\underline{r}) \, G(\underline{r} - \underline{r'}) \, \rho(\underline{r'}) \end{aligned} \tag{10.253}$$

The Green's function can be evaluated as

$$G(r) = \frac{1}{(2\pi)^2} \int dq \, q \int d\theta \, \frac{\exp[iqr \cos\theta]}{q^2}$$

$$= \frac{1}{(2\pi)^2} \int_{\frac{\pi}{L}}^{\frac{\pi}{a}} dq \int_0^{2\pi} d\theta \, \frac{\cos[qr \cos\theta]}{q} \tag{10.254}$$

Setting $u = qr$, one finds

$$G(r) = \frac{1}{2\pi} \int_{\frac{\pi r}{L}}^{\frac{\pi r}{a}} du \, \frac{J_0(u)}{u} \tag{10.255}$$

This can be separated into two pieces

$$G(r) = \frac{1}{2\pi} \int_{\frac{\pi r}{L}}^{\frac{\pi r}{a}} du \left[\frac{1}{u} - \frac{1 - J_0(u)}{u} \right]$$

$$= \frac{1}{2\pi} \ln \frac{L}{a} - \frac{1}{2\pi} \int_{\frac{\pi r}{L}}^{\frac{\pi r}{a}} du \left(\frac{1 - J_0(u)}{u} \right) \tag{10.256}$$

where the first term diverges when $L \to \infty$. The integrand in the last line varies as u when $u \to 0$

$$\lim_{u \to 0} \left(\frac{1 - J_0(u)}{u} \right) \to O(u) \tag{10.257}$$

and varies as

$$\lim_{u \to \infty} \left(\frac{J_0(u)}{u} \right) \to \frac{1}{u^{\frac{3}{2}}} \tag{10.258}$$

for asymptotically large u. Thus, when $r > a$, one arrives at the approximation

$$G(r) \approx \frac{1}{2\pi} \ln \frac{L}{a} - \frac{1}{2\pi} \ln \left(\frac{\pi r}{a} \right) + \text{Const.} \tag{10.259}$$

Thus, the energy of the vortices can be expressed as

$$E_V = \left(\int d^2\underline{r} \, \rho(\underline{r}) \right)^2 \pi J \ln \frac{L}{a}$$

$$- \pi J \int_{|r-r'|>a} d^2\underline{r} \int d^2\underline{r}' \, \rho(\underline{r}) \ln \left(\frac{|r-r'|}{a} \right) \rho(\underline{r}')$$

$$+ \sum_R n_R^2 E_c \tag{10.260}$$

The first term diverges when $L \to \infty$ so, minimizing E_v requires that the total vorticity must vanish. The second term represents the interaction energy between pairs of vortices that are separated by a distances greater than a. The last term represents the energies of the vortex cores. The energy $2E_c$ is the energy of a vortex/anti-vortex pair [with strengths (± 1)] separated by the distance a.

$$E_c \sim \pi J \left[\int_0^\pi du \left(\frac{1 - J_0(u)}{u} \right) - \int_\pi^\infty du \left(\frac{J_0(u)}{u} \right) \right] \qquad (10.261)$$

The precise value of E_c also depends on the cutoff for the lattice.[2]

Screening by Vortex/Anti-vortex Pairs

At finite temperatures, one expects that the thermally activated vortex pairs will screen the interaction between the vortex/anti-vortex pairs. This can be seen by examining the response of the longitudinal degrees of freedom by a twist force $\underline{f}$, which produces an interaction energy given by

$$H_{int}^L = - \int d^2 \underline{r} \, \underline{f} \cdot \underline{v}_L \qquad (10.262)$$

which amounts to a twist energy of $-fL(\varphi(L) - \varphi(0))$. Minimizing the total energy of the longitudinal field

$$H^L = \int d^2 \underline{r} \left(\frac{J}{2} \, \underline{v}_L^2 - \underline{f} \cdot \underline{v}_L \right) \qquad (10.263)$$

one obtains

$$\underline{f} = J \underline{v}_L \qquad (10.264)$$

This twist produces an increase in the longitudinal (spin-wave) energy of

$$\Delta E_L = - \int d^2 \underline{r} \left(\frac{f^2}{2J} \right) \qquad (10.265)$$

At finite temperatures, the coupling of the twist field to the transverse excitations

$$H_{int}^T = - \int d^2 \underline{r} \, \underline{f} \cdot \underline{v}_T \qquad (10.266)$$

produces a contribution to the Free-Energy which can be found from the high temperature expansion

$$\Delta F_T = \text{Trace} \left[H_{int}^T \, \frac{\exp[-\beta H]}{Z} \right] - \frac{\beta}{2} \, \text{Trace} \left[H_{int}^{T\,2} \, \frac{\exp[-\beta H]}{Z} \right] \qquad (10.267)$$

---

[2] For the square lattice $E_c = 2\pi J(\gamma + \frac{3}{2} \ln 2)$, where γ is Euler's constant.

The first term (proportional to $\underline{f}$) can be expressed as

$$-\int d^2\underline{r}(\underline{f} \cdot \langle \underline{v}_T \rangle) \tag{10.268}$$

and its contribution is found to vanish since the contributions vortices and anti-vortices come in pairs and cancel. The second term is non-zero and is proportional to f^2.

$$-\frac{\beta}{2}\int d^2\underline{r}\int d^2\underline{r}' \langle (\underline{f}\cdot\underline{v}_T(\underline{r}))(\underline{f}\cdot\underline{v}_T(\underline{r}')) \rangle \tag{10.269}$$

and is evaluated as

$$-\frac{\beta f^2}{4}\int d^2\underline{r}\int d^2\underline{r}' \langle \underline{v}_T(\underline{r})\cdot\underline{v}_T(\underline{r}')\rangle \tag{10.270}$$

where we have used the isotropy of the unconstrained equilibrium vortex configuration. Since

$$\underline{v}_T(\underline{r}) = \underline{\nabla}\wedge(\hat{e}_z\psi(\underline{r}))$$
$$-\nabla^2\psi(\underline{r}) = 2\pi\rho(\underline{r}) \tag{10.271}$$

and the vortex correlation is translational invariant and isotropic, so it only depends on the magnitude $|\underline{r}-\underline{r}'|$. Thus, one finds that the second-order contribution of the transverse component of the Free-Energy is given by

$$-\frac{\beta f^2}{4}\int d^2\underline{r}\int d^2\underline{r}'\int d^2\underline{q}\,\frac{\langle \rho_{\underline{q}}\,\rho_{-\underline{q}}\rangle}{q^2}\,\exp[i\underline{q}\cdot(\underline{r}-\underline{r}')] \tag{10.272}$$

The large distance part of the integral can be evaluated from the $q\to 0$ limit of the correlation function

$$\langle \rho_{\underline{q}}\,\rho_{-\underline{q}}\rangle = \frac{1}{L^4}\int d^2\underline{r}\int d^2\underline{r}' \langle \rho(\underline{r})\,\rho(\underline{r}')\rangle \exp[-i\underline{q}\cdot(\underline{r}-\underline{r}')]$$

$$= \frac{1}{L^4}\int d^2\underline{r}\int d^2\underline{r}' \langle \rho(\underline{r})\rho(\underline{r}')\rangle\left[1-i\underline{q}\cdot(\underline{r}-\underline{r}')\right.$$

$$\left.-\frac{1}{2}(\underline{q}\cdot(\underline{r}-\underline{r}'))^2+\dots\right] \tag{10.273}$$

The first term in the expansion vanishes, since there are as many vortices as anti-vortices in the area of integration $\underline{r}$ and likewise for the area of integration $\underline{r}'$. The second term vanishes as it is odd in $\underline{r}-\underline{r}'$. The remaining term can be expressed as

$$\langle \rho_{\underline{q}}\,\rho_{-\underline{q}}\rangle \approx -\frac{(2\pi)^2}{4L^4}\int d^2\underline{r}\int d^2\underline{r}' \langle \rho(\underline{r})\,\rho(\underline{r}')\rangle q^2|\underline{r}-\underline{r}'|^2 \tag{10.274}$$

after performing the angular averages. Therefore, the Free-Energy from the vortex pairs can be expressed as

$$\Delta F_T = -\frac{\beta f^2}{16L^2} (2\pi)^2 \int d^2r \int d^2r' |r - r'|^2 \langle \rho(r) \, \rho(r') \rangle \qquad (10.275)$$

For large core energies E_c, the dominant contribution to the correlation function comes from the vortex/anti-vortex correlations and can be expressed as

$$\langle \rho(r) \, \rho(r') \rangle \approx -\frac{2}{a^4} \exp[-2\beta E_c] \, \exp\left[-2\pi\beta J \, \ln\left(\frac{|r - r'|}{a} \right) \right] \qquad (10.276)$$

The effective interaction energy is given by the sum of the longitudinal and transverse energies. The inverse effective interaction J_{eff} has a contribution from the screening by the vortex/anti-vortex pairs

$$\frac{f^2}{2J_{eff}} = \frac{f^2}{2J} - \frac{\beta f^2}{16L^2} (2\pi)^2 \int d^2r \int d^2r' |r - r'|^2 \langle \rho(r) \, \rho(r') \rangle \qquad (10.277)$$

An analysis of this formula shows that increasing the temperature to T_{KT} causes the effective interaction between pairs of vortices vanishes since the polarizability term proportional to r^2 diverges. The vanishing of J_{eff} occurs because the integral

$$I = \int_a^d d|r - r'| \, |r - r'|^{3 - 2\pi\beta J}$$
$$\sim \frac{d^{4 - 2\pi\beta J} - a^{4 - 2\pi\beta J}}{4 - 2\pi\beta J} \qquad (10.278)$$

diverges when $k_B T_{KT} = \frac{\pi}{2} J$. Thus, the vanishing of the effective interaction leads to the unbinding of vortex/anti-vortex pairs at the Kosterlitz-Thouless transition temperature.

Scaling Flows

One can introduce a self-consistent approximation, in which the interaction between pairs of vortex and anti-vortex are screened by vortex/antivortex pairs. The energy of a vortex/antivortex pair separated by a distance r is given by

$$2\beta E(r) = \int_a^r dr' \, \frac{K(r')}{r'} \qquad (10.279)$$

where the effective force (divided by temperature) between the pair of vortices is given by

$$\frac{2\pi K(r)}{r} \qquad (10.280)$$

The dimensionless effective interaction $K(r) = \beta J_{eff}(r)$ is distance-dependent since the dimensionless pair interaction parameter $K_0 = \beta J$ is screened by the dielectric function $\epsilon(r)$

$$\epsilon(r) = \frac{K_0}{K(r)} \qquad (10.281)$$

Then, one has

$$K^{-1}(r) = K_0^{-1} + 2\pi^3 \int_a^r \frac{dr'}{a} \frac{r'}{a} \left(\frac{r'}{a}\right)^2 \exp[-2\beta E(r)] \qquad (10.282)$$

The above equation motivates the introduction of a distance dependent (single-vortex) fugacity $y(r)$ via

$$y(r) = \left(\frac{r}{a}\right)^2 \exp[-\beta E(r)] \qquad (10.283)$$

or

$$y(r) = y_0 \left(\frac{r}{a}\right)^2 \exp\left[-\pi \int_a^r dr' \frac{K(r')}{r'}\right] \qquad (10.284)$$

Thus,

$$K^{-1}(r) = K_0^{-1} + 4\pi^3 \int_a^r \frac{dr'}{r'} y(r')^2 \qquad (10.285)$$

Differentiating with respect to the logarithmic distance variable

$$l = \ln \frac{r}{a} \qquad (10.286)$$

one obtains

$$\left(\frac{\partial y}{\partial l}\right) = (2 - \pi K) y$$

$$\left(\frac{\partial K^{-1}}{\partial l}\right) = 4\pi^3 y^2 \qquad (10.287)$$

Multiplying the second equation by a factor of $(2 - \pi K)$ and then using the first equation to eliminate a factor of y on the right-hand side, one finds

$$(2 - \pi K) \left(\frac{\partial K^{-1}}{\partial l}\right) = 4\pi^3 y \left(\frac{\partial y}{\partial l}\right) \qquad (10.288)$$

This can be integrated, leading to the scaling trajectories being determined from

$$\ln\left(\frac{\pi K(l)}{2}\right) + \frac{2}{\pi K(l)} - 2\pi^2 y^2(l) = 1 + C \qquad (10.289)$$

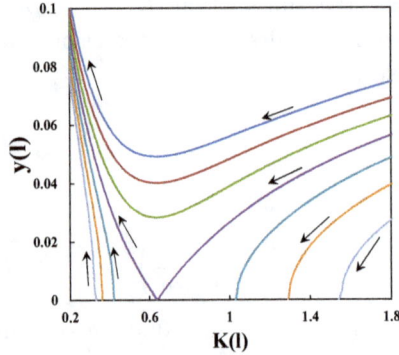

Fig. 10.12   The Scaling Flow for the two-dimensional $x - y$ model. For $C < 0$, the flow always scales to $y \to \infty$ and $K \to 0$, but for $C > 0$ it scales to either $y \to 0$ with finite K or $y \to \infty$ and $K \to 0$, depending on whether $\frac{\pi K}{2}$ either is > 1 or < 1.

The critical line is $C = 0$, on which the fugacity $y(l)$ obtains both the limit $y(l) \to 0$ and $y(l) \to \infty$ so the vortex/anti-vortex pairs can bind or unbind. The critical point between the two-phases is determined by the line $C = 0$ and

$$\frac{\pi K(l)}{2} = 1 \tag{10.290}$$

Criticality of the low-T phase

At low temperatures, the vortices are bound and so there contribution to the correlation function

$$\langle \underline{S}_r \cdot \underline{S}_0 \rangle = \langle \cos(\varphi_r - \varphi_0) \rangle \tag{10.291}$$

is negligible and so the correlation function is mainly determined by the spin-fluctuations.

$$\langle \cos(\varphi_r - \varphi_0) \rangle = \left(\frac{r}{a} \right)^{-\left(\frac{k_B T}{2\pi J} \right)} \tag{10.292}$$

At $T = T_{KT}$, one has

$$k_B T_{KT} = \frac{\pi}{2} J \tag{10.293}$$

Hence at the transition temperature, the correlation function takes the form

$$\langle \cos(\varphi_r - \varphi_0) \rangle \sim \left(\frac{r}{a} \right)^{-\frac{1}{4}} \tag{10.294}$$

Below, the transition temperature, the correlation function falls off more slowly. The susceptibility χ is given as the average of the product of fluctuations at different sites

$$\Delta \underline{S}_R = \underline{S}_R - \langle \underline{S}_R \rangle \tag{10.295}$$

by

$$k_B T \chi = \sum_R \langle \Delta \underline{S}_R \cdot \Delta \underline{S}_0 \rangle$$

$$\sim 2\pi \int_0^L dR \, R \, R^{-\eta} \to \infty \tag{10.296}$$

as $L \to \infty$ since $\eta < \frac{1}{4}$. Therefore, the susceptibility is divergent in the low-temperature phase. In this sense, the low-temperature phase is critical because of the algebraic decay of the correlation function.

Chapter 11

Non-Equilibrium Phenomena

The study of non-equilibrium properties is difficult. The only systematic analytic methods that are available are those that treat systems which are close to equilibrium, such as systems in steady states or systems subjected to weak perturbations.

Interlude: Equilibration and Poincaré Recurrence

Statistical Mechanics is based on the assumption that systems evolve towards an equilibrium state. It is often assumed that equilibration occurs due to coupling between a system and an environment which results in a gradual loss of coherence. The *"Jaynes-Cummings Model"* shows that the coupling can yield unexpected phenomena.

The Jaynes-Cummings model [100] consists of a two-level system which is coupled to a bosonic system. The two-level system has an excitation energy ω_0 and is occupied by a fermion. The two-level system is coupled to bosons that are described by a Harmonic Oscillator of frequency ω. The Hamiltonian, in the absence of the coupling, $\hat{H}_0$, is given by

$$\hat{H}_0 = \hbar\omega_0 \begin{pmatrix} 1 & 0 \\ 0 & -1 \end{pmatrix} + \hbar\omega \left(\hat{a}^\dagger \, \hat{a} + \frac{1}{2} \right) \tag{11.1}$$

The coupling between the two-level system and the Harmonic Oscillator is given by the interaction Hamiltonian

$$\hat{H}_{int} = \frac{\hbar}{2} \begin{pmatrix} 0 & g\hat{a} \\ g\hat{a}^\dagger & 0 \end{pmatrix} \tag{11.2}$$

The coupling is such that the absorption of a boson is accompanied by an excitation of the fermion and the emission of a boson is accompanied by the de-excitation of the fermion. Hence, the number of excited particles,

$\hat{N}_{exc} = \hat{a}^\dagger \, \hat{a} + \hat{c}_+^\dagger \, \hat{c}_+$, is a conserved quantity. Due to this conservation law, the Hilbert space of the model decouples into sectors spanned by the states

$$\begin{pmatrix} 1 \\ 0 \end{pmatrix} |n-1\rangle \tag{11.3}$$

and

$$\begin{pmatrix} 0 \\ 1 \end{pmatrix} |n\rangle \tag{11.4}$$

The n-th sector is described by the reduced Hamiltonian

$$\hat{H}_n = n\hbar\omega \begin{pmatrix} 1 & 0 \\ 0 & 1 \end{pmatrix} + \frac{\hbar}{2}(\omega_0 - \omega) \begin{pmatrix} 1 & 0 \\ 0 & -1 \end{pmatrix} + \frac{\hbar}{2}g\sqrt{n} \begin{pmatrix} 0 & 1 \\ 1 & 0 \end{pmatrix} \tag{11.5}$$

The reduced Hamiltonian can be diagonalized to yield the eigenvalues

$$E_{n,\pm} = n\hbar\omega \pm \frac{\hbar}{2} \sqrt{g^2 n + (\omega_0 - \omega)^2} \tag{11.6}$$

and eigenstates. The eigenstates are found as

$$|n_+\rangle = \sin\Theta \begin{pmatrix} 1 \\ 0 \end{pmatrix} |n-1\rangle - \cos\Theta \begin{pmatrix} 0 \\ 1 \end{pmatrix} \tag{11.7}$$

and

$$|n_-\rangle = \cos\Theta \begin{pmatrix} 1 \\ 0 \end{pmatrix} |n-1\rangle + \sin\Theta \begin{pmatrix} 0 \\ 1 \end{pmatrix} \tag{11.8}$$

where the mixing angle Θ is given by

$$\tan 2\Theta = \frac{g\sqrt{n}}{\omega_0 - \omega} \tag{11.9}$$

If the system initially is in the lowest level $|\Psi(0)\rangle$, and is in the n-th sector, it can be decomposed into the energy eigenstates

$$|\Psi(0)\rangle = \begin{pmatrix} 0 \\ 1 \end{pmatrix} |n\rangle$$
$$= \sin\Theta \, |n_-\rangle - \cos\Theta \, |n_+\rangle \tag{11.10}$$

Therefore, the initial state evolves with time as

$$|\Psi(t)\rangle = \sin\Theta \, \exp\left[-\frac{i}{\hbar}E_{n_-}t\right]|n_-\rangle - \cos\Theta \, \exp\left[-\frac{i}{\hbar}E_{n_+}t\right]|n_+\rangle \tag{11.11}$$

The probability $P_+(t)$ that the electron is found in the excited state at time t is given by

$$P_+(t) = \frac{g^2 n}{g^2 n + (\omega_0 - \omega)^2} \sin^2 \frac{\sqrt{g^2 n + (\omega_0 - \omega)^2} \; t}{2} \tag{11.12}$$

The probability that excited electron level is occupied oscillates with a frequency given by

$$\left(\frac{2\pi}{T_0}\right) = \sqrt{g^2 n + (\omega_0 - \omega)^2} \tag{11.13}$$

which is the Rabi frequency. The time-average of $P_\pm(t)$ over a long time approaches $\frac{1}{2}$.

Now consider a Jaynes-Cummings model which is resonant, $\omega_0 = \omega$, so the two uncoupled states become degenerate. The assumption of equal a priori probabilities leads one to expect that, due to the coupling, the system will evolve into an equilibrium state with $\lim_{t \to \infty} P_\pm(t) \to \frac{1}{2}$. In the $t = 0$ initial state the electron occupies the lower energy level. At $t = 0$, the fermionic system is placed in contact with the Harmonic Oscillator which is in coherent state with a large average number of bosons

$$|\Psi(0)\rangle = \exp\left[-\frac{1}{2}|\alpha|^2\right] \begin{pmatrix} 0 \\ 1 \end{pmatrix} \exp[\alpha \hat{a}^\dagger] |0\rangle$$

$$= \exp\left[-\frac{1}{2}|\alpha|^2\right] \begin{pmatrix} 0 \\ 1 \end{pmatrix} \sum_{n=0}^{\infty} \frac{\alpha^n}{\sqrt{n!}} |n\rangle \tag{11.14}$$

Since $\sin\Theta = \cos\Theta$ for the resonant model, the initial state evolves into

$$|\Psi(t)\rangle = \exp\left[-\frac{|\alpha|^2 - i\omega t}{2}\right] \sum_{n=0}^{\infty} \frac{(\alpha \exp[-i\omega t])^n}{\sqrt{n!}}$$

$$\times \left[\cos\left(\frac{gt\sqrt{n}}{2}\right) \begin{pmatrix} 0 \\ 1 \end{pmatrix} |n\rangle + i\sin\left(\frac{gt\sqrt{n}}{2}\right) \begin{pmatrix} 1 \\ 0 \end{pmatrix} |n-1\rangle\right] \tag{11.15}$$

The probability to find the electron in the excited state is given by

$$P_+(t) = \exp[-|\alpha|^2] \sum_{n=0}^{\infty} \frac{|\alpha|^{2n}}{n!} \sin^2 \frac{gt\sqrt{n}}{2}$$

$$= \frac{1}{2} \exp[-|\alpha|^2] \sum_{n=0}^{\infty} \frac{|\alpha|^{2n}}{n!} (1 - \cos gt\sqrt{n})$$

$$= \frac{1}{2}\left[1 - \exp[-|\alpha|^2] \sum_{n=0}^{\infty} \frac{|\alpha|^{2n}}{n!} \cos gt\sqrt{n}\right] \tag{11.16}$$

The various components of the state produce oscillations with frequencies $g\sqrt{n}$. The amplitudes of the oscillatory terms are sharply peaked function of n and have a maximum at $n = |\alpha|^2$. This can be seen by using Stirling's formula

$$\frac{|\alpha|^{2n}}{n!} = \frac{1}{\sqrt{2\pi n}} \exp[n \ln |\alpha^2| - n \ln n + n] \tag{11.17}$$

and minimizing with respect to n shows an extremum at

$$\ln |\alpha|^2 - \ln n = 0 \tag{11.18}$$

The width of the distribution is given by $|\alpha|$, so the amplitude of the oscillatory terms can be approximated by a Gaussian

$$\exp[-|\alpha|^2] \sum_{n=0}^{\infty} \frac{|\alpha|^{2n}}{n!} \approx \frac{1}{\sqrt{2\pi |\alpha|^2}} \exp\left[-\frac{(n - |\alpha|^2)^2}{2|\alpha|^2}\right] \tag{11.19}$$

valid for large $|\alpha|$. Setting $m = n - |\alpha|^2$, one finds

$$P_+(t) \approx \frac{1}{2}\left[1 - \frac{1}{\sqrt{2\pi |\alpha|^2}} \sum_{m=-\infty}^{\infty} \exp\left[-\frac{m^2}{2|\alpha|^2}\right] \cos gt \sqrt{m + |\alpha|^2}\right]$$

$$\tag{11.20}$$

For short times, the oscillations have the period T_0

$$T_0 = \left(\frac{2\pi}{g|\alpha|}\right) \tag{11.21}$$

which is the on-resonance period of the Rabi oscillations.

For longer times, the various oscillating components start to dephase and start to interfere destructively. The dephasing time can be estimated

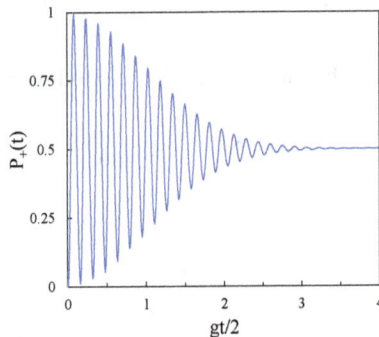

Fig. 11.1   The initial time-dependence of the probability, $P_+(t)$, for the on-resonance Jaynes-Cummings model in which the boson is in a coherent state. The probability oscillates with the Rabi frequency and decays over a dephasing time, τ_d.

as the time T_d at which the oscillating terms at the shoulders of the distribution $m = \pm|\alpha|$ are out of phase with the oscillations from the peak.

$$gT_d[\sqrt{|\alpha|^2 + |\alpha|} - |\alpha|] = \pi \qquad (11.22)$$

or

$$T_d \approx \frac{\pi}{g\sqrt{|\alpha|^2 + |\alpha|} - |\alpha|}$$

$$\approx \left(\frac{2\pi}{g}\right) \qquad (11.23)$$

The decay is exponential as can be found by expanding the argument of the cosine in terms of m

$$\cos gt \left(|\alpha| + \frac{m}{2|\alpha|} + \dots\right) \approx \frac{1}{2} \Re \exp[ig|\alpha|t] \exp\left[i \frac{gt}{2|\alpha|} m\right] \qquad (11.24)$$

valid for short time scales. Hence, for short times, the decaying term in $P_+(t)$ can be approximated by

$$\approx \frac{1}{\sqrt{8\pi|\alpha|^2}} \Re \exp[ig|\alpha|t] \sum_{m=-\infty}^{\infty} \exp\left[-\frac{m^2}{2|\alpha|^2}\right] \exp\left[i \frac{gt}{2|\alpha|} m\right]$$

$$\approx \cos(g|\alpha|t) \exp\left[-\frac{g^2t^2}{8}\right] \qquad (11.25)$$

Therefore, we have arrived at the approximation

$$P_+(t) \approx \frac{1}{2}\left[1 - \cos(g|\alpha|t) \exp\left[-\frac{g^2t^2}{8}\right]\right] \qquad (11.26)$$

which shows that the amplitude of the Rabi oscillations decays exponentially. The amplitude of the oscillations decay over a long de-phasing time-scale

$$\tau_d = \left(\frac{2\sqrt{2}}{g}\right) \qquad (11.27)$$

and the probability approaches the expected equilibrium value. This type of dephasing phenomena is a classical phenomenon and was first predicted by Lord Rayleigh [101].

The model shows a surprise at even longer time scales when the oscillations start to reappear. The Poincaré recurrence time T_R is the time-scale where the phases from the peak frequency and its neighboring frequencies become in phase again. That is, the recurrence occurs when the neighboring

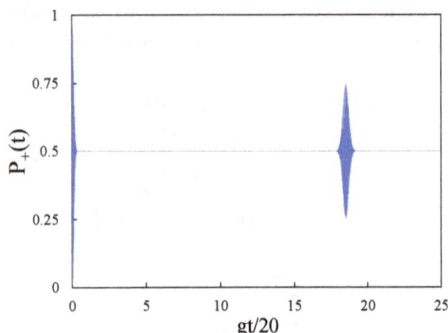

Fig. 11.2   The long time-scale time-dependence of the probability, $P_+(t)$, for the on-resonance Jaynes-Cummings model in which the boson is in a coherent state. The probability oscillations recur on the times scale T_R. (Schematic)

terms have phase differences which approach 2π

$$gT_R[\sqrt{|\alpha|^2 + 1} - |\alpha|] = 2\pi \qquad (11.28)$$

or

$$T_R = \left(\frac{4\pi|\alpha|}{g}\right) \qquad (11.29)$$

The recurrence is not complete as the amplitude of the probability remains less than unity. The partial recurrence forms and decays over a decoherence time-scale. Further partial recurrences occur and, eventually, the probability randomly fluctuates about $\frac{1}{2}$. The phenomena of recurrence is a quantum interference effect that crucially depends on the frequency spectrum of the bosons and the initial probabilities for observing n bosons. For a thermalized boson distribution [102], the distribution is so broad that the Rabi oscillations do not survive and there is no recurrence. Here we have assumed a discrete spectrum, usually one expects that the boson spectra should be quasi-continuous. For a quasi-continuous spectrum, the Poincaré recurrence time would correspond to the minuscule energy separations between the discrete energy levels.

11.1   Linear Response Theory

Many experiments are aimed at analyzing how a system responds to a small external perturbation, such as the measurements of the specific heat, the magnetic susceptibility or the electrical conductivity. If the perturbations are small enough compared with an appropriate energy scale of the unperturbed system, the response is linear in the perturbation. In such cases, the response can be calculated via perturbation theory and the state of the system is expected to be close to thermal equilibrium. This constitutes linear response theory [103].

Consider an experiment on a quantum mechanical system. In the experiment, a quantity $\hat{A}$ is measured at time t while the system is perturbed by a time-dependent field $F(t)$ produced by a signal generator. Since the experimental apparatus can be considered as classical, the time-dependent field $F(t)$ can be considered as a classical field. The system is governed by the Hamiltonian $\hat{H}(t)$ where

$$\hat{H}(t) = \hat{H}_0 + \hat{H}_{int}(t) \tag{11.30}$$

where $\hat{H}_0$ is a time-independent Hamiltonian that describes the closed system in the absence of the field and $\hat{H}_{int}(t)$ represents the interaction of the system with the field produced by an external source. The Hamiltonian $\hat{H}_0$ is the full Hamiltonian of the closed system including all the many-body interactions. The field couples to the system via an interaction $\hat{H}_{int}$ that has the form

$$\hat{H}_{int}(t) = -\hat{B}(t)\, F(t) \tag{11.31}$$

where $\hat{B}$ is some Hermitean operator conjugate to F which acts on the quantum system. We assume that neither $\hat{A}$ nor $\hat{B}$ are constants of motion and that, in the experiment, the external field $F(t)$ is turned on adiabatically at $t \to -\infty$. The adiabatic switching on is done to avoid the observation of transient effects which are dependent on the details of how the apparatus is turned on. The adiabatic switching on of the interaction may be described by letting

$$F(t) = F_0(t)\, \exp[\eta t] \tag{11.32}$$

where η is a positive constant such that $\eta \to 0$.

The quantity being measured is the thermal average of $\hat{A}$ at time t. The thermal average is a real number and is denoted by $A(t)$ and is expected to be time-dependent since the system is being perturbed by a time-dependent field. The thermal average at time t can be expressed as

$$\langle \hat{A}(t) \rangle = \text{Trace } \hat{A}\, \hat{\rho}(t) \tag{11.33}$$

where $\hat{A}$ is a time-independent Hermitean operator and where $\hat{\rho}(t)$ is the time-dependent density operator. The time-dependence of the density operator can be evaluated within the Dirac or interaction representation.

11.1.1   *The Dirac Representation*

If a state of a system is in state $|\psi\rangle$ when $t \to -\infty$, its value at time t is denoted by $|\psi(t)\rangle$. The time-dependence of the state can be expressed in terms of the unitary time-evolution $\hat{U}(t)$ operator which is defined via

$$|\psi(t)\rangle = \hat{U}(t)\,|\psi\rangle \tag{11.34}$$

Since the state satisfies the time-dependent Schrödinger equation, the time-evolution operator satisfies

$$i\hbar\,\frac{\partial}{\partial t}\,\hat{U}(t) = \hat{H}(t)\,\hat{U}(t) \tag{11.35}$$

In the absence of the interaction, the time-evolution operator $\hat{U}_0(t)$ satisfies

$$i\hbar\,\frac{\partial}{\partial t}\,\hat{U}_0(t) = \hat{H}_0(t)\,\hat{U}_0(t) \tag{11.36}$$

The Dirac time-evolution operator $\hat{U}_D(t)$ is defined by

$$\hat{U}(t) = \hat{U}_0(t)\,\hat{U}_D(t) \tag{11.37}$$

or, equivalently by

$$\hat{U}_D(t) = \hat{U}_0^\dagger(t)\,\hat{U}(t) \tag{11.38}$$

Hence, the time-evolution operator in the Dirac representation satisfies

$$i\hbar\,\frac{\partial}{\partial t}\,\hat{U}_D(t) = \hat{U}_0^\dagger(t)\,\hat{H}_{int}(t)\,\hat{U}_0(t)\,\hat{U}_D(t) \tag{11.39}$$

If the effective interaction Hamiltonian is expressed as

$$\hat{H}_{int}^D(t) = \hat{U}_0^\dagger(t)\,\hat{H}_{int}(t)\,\hat{U}_0(t) \tag{11.40}$$

then the time-evolution operator, in the Dirac representation, satisfies the equation of motion

$$i\hbar\,\frac{\partial}{\partial t}\,\hat{U}_D(t) = \hat{H}_{int}^D(t)\,\hat{U}_D(t) \tag{11.41}$$

The expectation of an operator $\hat{A}$ in a state $|\psi\rangle$ is given by

$$\begin{aligned}\langle\psi|\,\hat{U}^\dagger(t)\,\hat{A}\hat{U}(t)|\psi\rangle &= \langle\psi|\,\hat{U}_D^\dagger(t)\,\hat{U}_0^\dagger(t)\,\hat{A}\hat{U}_0(t)\,\hat{U}_D(t)|\psi\rangle \\ &= \langle\psi|\,\hat{U}_D^\dagger(t)\,\hat{A}_D(t)\hat{U}_D(t)|\psi\rangle \\ &= \langle\psi_D(t)|\,\hat{A}_D(t)|\psi_D(t)\rangle \end{aligned} \tag{11.42}$$

where the operators in the Dirac representation evolve under the unperturbed Hamiltonian

$$\hat{A}_D(t) = \hat{U}_0^\dagger(t)\,\hat{A}\hat{U}_0(t) \tag{11.43}$$

and the states $|\,\psi_D(t)\rangle$ evolve with the Dirac time evolution operator $\hat{U}_D(t)$. Thus, in the Dirac representation the states evolve via the Dirac interaction

$$i\hbar\,\frac{\partial}{\partial t}\,|\psi_D(t)\rangle = \hat{H}_{int}^D(t)\,|\psi_D(t)\rangle \tag{11.44}$$

11.1.2 *Thermal Averages and the Density Matrix*

Previously, in the Schrödinger representation, the time-dependent density operator was shown to satisfy

$$i\hbar \frac{\partial}{\partial t} \hat{\rho}(t) + [\hat{\rho}(t), \hat{H}(t)] = 0 \tag{11.45}$$

which has the solution

$$\hat{\rho}(t) = \hat{U}(t) \hat{\rho} \hat{U}^\dagger(t) \tag{11.46}$$

where $\hat{\rho}$ is the density operator of the initial state. The time-evolution of the density operator can be expressed in terms of the Dirac time-evolution operators via

$$\hat{\rho}(t) = \hat{U}_0(t) \hat{U}_D(t) \hat{\rho} \hat{U}_D^\dagger(t) \hat{U}_0^\dagger(t) \tag{11.47}$$

The time-dependent density operator in the Dirac representation $\hat{\rho}_D(t)$ is defined by

$$\hat{\rho}_D(t) = \hat{U}_D(t) \hat{\rho} \hat{U}_D^\dagger(t) \tag{11.48}$$

The time-evolution of the thermal average can be expressed in terms of the Trace of the Dirac density operator and the operator $\hat{A}_D(t)$. This can be shown by substituting the expression for the density operator in Eq. (11.47) into the thermal average

$$\begin{aligned}
\langle \hat{A}(t) \rangle &= \text{Trace } \hat{A}\hat{\rho}(t) \\
&= \text{Trace } \hat{A}\hat{U}_0(t) \hat{\rho}_D(t) \hat{U}_0^\dagger(t) \\
&= \text{Trace } \hat{U}_0^\dagger(t) \hat{A}\hat{U}_0(t) \hat{\rho}_D(t) \\
&= \text{Trace } \hat{A}_D(t) \hat{\rho}_D(t)
\end{aligned} \tag{11.49}$$

where the cyclic invariance has been used. Hence, the thermal average takes a similar form in the Dirac representation, except that the operator $\hat{A}$ evolves with time due to $\hat{H}_0$ and the Dirac density operator evolves due to the perturbation $\hat{H}_{int}^D(t)$.

The relation between the Dirac density operator and the density operator in the Schrödinger representation is

$$\hat{\rho}_D(t) = \hat{U}_D(t) \hat{\rho} \hat{U}_D^\dagger(t) \tag{11.50}$$

Thus, on taking the time-derivative of the above equation and using the expressions for the time derivatives of $\hat{U}_D(t)$ and the Hermitean conjugate expression, one finds that the Dirac density operator satisfies the equation of motion

$$i\hbar \frac{\partial}{\partial t} \hat{\rho}_D(t) + [\hat{\rho}_D(t), \hat{H}_{int}^D(t)] = 0 \tag{11.51}$$

Formally, the time-dependence of density operator can be found by integration

$$\hat{\rho}_D(t) = \hat{\rho}_D(-\infty) + \frac{i}{\hbar} \int_{-\infty}^{t} dt' [\hat{\rho}_D(t'), \hat{H}_{int}^D(t')] \qquad (11.52)$$

After solving for the density operator in the Dirac representation, the time-dependent thermal average can be evaluated from

$$\langle \hat{A}(t) \rangle = \text{Trace } \hat{A}_D(t) \, \hat{\rho}_D(t) \qquad (11.53)$$

where $\hat{A}_D(t)$ evolves with $\hat{H}_0$ but the time-evolution of $\hat{\rho}_D$ is governed by $\hat{H}_{int}^D(t)$.

11.1.3   Derivation of the Linear Response Relation

To first-order in the interaction, the density operator is given by

$$\hat{\rho}_D(t) = \hat{\rho}_D(-\infty) + \frac{i}{\hbar} \int_{-\infty}^{t} dt' [\hat{\rho}_D(-\infty), \hat{H}_{int}^D(t')] \qquad (11.54)$$

As $t \to -\infty$, $\hat{H}_{int}^D$ vanishes so $\hat{U}_D(t) \to 1$ and the system is in state of thermal equilibrium that is governed by $\hat{H}_0$. This leads to

$$\lim_{t \to -\infty} \hat{\rho}_D(t) \to \lim_{t \to -\infty} \hat{U}_0(t) \, \hat{\rho} \hat{U}_0^\dagger(t) \qquad (11.55)$$

Therefore, since $\hat{\rho}$ is expected to be a function of $\hat{H}_0$, it should commute with $\hat{U}_0(t)$. The expression for the density operator reduces to

$$\hat{\rho}_D(t) = \hat{\rho} - \frac{i}{\hbar} \int_{-\infty}^{t} dt' [\hat{\rho}, \hat{B}_D(t')] \, F(t') \qquad (11.56)$$

The thermal average of $\hat{A}$ at time t is then given by

$$\langle \hat{A}(t) \rangle = \text{Trace } \hat{A}_D(t) \, \hat{\rho}$$
$$- \frac{i}{\hbar} \text{Trace } \int_{-\infty}^{t} dt' \hat{A}_D(t) \, [\hat{\rho}, \hat{B}_D(t')] \, F(t') \qquad (11.57)$$

On using the cyclic invariance of the Trace, the expectation value simplifies to

$$\langle \hat{A}(t) \rangle = \text{Trace } \hat{A}_D(t) \, \hat{\rho}$$
$$+ \frac{i}{\hbar} \text{Trace } \int_{-\infty}^{t} dt' \hat{\rho} [\hat{A}_D(t), \hat{B}_D(t')] \, F(t') \qquad (11.58)$$

However, the first term is recognized to be the value of A in the thermal equilibrium state

$$
\begin{aligned}
\text{Trace } \hat{A}_D(t) \, \hat{\rho} &= \text{Trace } \hat{U}_0^\dagger(t) \, \hat{A}\hat{U}_0(t) \, \hat{\rho} \\
&= \text{Trace } \hat{A}\hat{U}_0(t) \, \hat{\rho}\hat{U}_0^\dagger(t) \\
&= \text{Trace } \hat{A}\hat{\rho} \\
&= \langle \hat{A}(-\infty) \rangle \qquad (11.59)
\end{aligned}
$$

where $\hat{\rho}$ is the time-independent density operator of the initial equilibrium state. Hence, the response to the external perturbation $\Delta A(t)$ defined by

$$
\Delta A(t) = \langle \hat{A}(t) \rangle - \langle \hat{A}(-\infty) \rangle \qquad (11.60)
$$

is given by

$$
\Delta A(t) = \int_{-\infty}^{t} dt' \chi_{A,B}(t,t') \, F(t') \qquad (11.61)
$$

where the response function $\chi_{A,B}(t,t')$ is given by

$$
\chi_{A,B}(t,t') = \frac{i}{\hbar} \, \text{Trace } \hat{\rho}[\hat{A}_D(t), \hat{B}_D(t')] \qquad (11.62)
$$

The linear response relation is causal. The response that occurs at time t is related to the perturbation at the earlier times t'. Therefore, one can define the response function as

$$
\chi_{A,B}(t,t') = \frac{i}{\hbar} \, \Theta(t-t') \, \text{Trace } \hat{\rho}[\hat{A}_D(t), \hat{B}_D(t')] \qquad (11.63)
$$

which is non-zero when $t > t'$. Thus

$$
\begin{aligned}
\chi_{A,B}(t-t') &\neq 0 \quad t-t' > 0 \\
\chi_{A,B}(t-t') &= 0 \quad t-t' < 0
\end{aligned} \qquad (11.64)
$$

The response function involves operators that evolve under the unperturbed Hamiltonian $\hat{H}_0$ and depends on the density operator $\hat{\rho}$ evaluated in the absence of the interaction and, thus, is a property of the isolated system and not on the external perturbation. Furthermore, since $\hat{U}_0(t)$ commutes with $\hat{\rho}$, the response function only depends on the time difference $t - t'$

$$
\chi_{A,B}(t-t') = \frac{i}{\hbar} \, \Theta(t-t') \, \text{Trace } \hat{\rho}[\hat{A}_D(t-t'), \hat{B}_D(0)] \qquad (11.65)
$$

One might have expected separate dependencies on t and on the time between cause and effect, $t-t'$, but since there is no intrinsic time-dependence of thermal equilibrium states, the susceptibility only depends on $(t - t')$.

The above formula is sometimes known as the *"Kubo Formula"*. The upper limit of the integration over t' in the linear response relation can be extended from t to ∞ because of the introduction of the Heaviside step function

$$\Delta A(t) = \int_{-\infty}^{\infty} dt' \chi_{A,B}(t - t') \, F(t') \tag{11.66}$$

With this change, the linear response relation is recognized to be in the form of a convolution. Furthermore, since both $\hat{A}$ and $\hat{B}$ are Hermitean operators and since the classical field $F(t')$ is real, the response function $\chi_{A,B}(t - t')$ is also a real quantity.

11.2 The Response Function

In the time-domain, the relation between cause and effect has the form of a convolution. The expression simplifies if one takes the Fourier transform of the relation with respect to time. The Fourier transform of $\Delta A(t)$ is defined as

$$\Delta A(\omega) = \int_{-\infty}^{\infty} dt \, \exp[-i\omega t] \, \Delta A(t) \tag{11.67}$$

To ensure the convergence of the integral, in the limit $t \to \infty$, one may introduce a convergence factor of $\exp[-\eta t]$, where η is a positive infinitesimal quantity. The inverse Fourier transform is given by

$$\Delta A(t) = \int_{-\infty}^{\infty} \frac{d\omega}{2\pi} \, \exp[+i\omega t] \, \Delta A(\omega) \tag{11.68}$$

The Fourier transform of a convolution is just the product of the Fourier transforms. Therefore, in the frequency domain, the linear response relation takes on the algebraic form

$$\Delta A(\omega) = \chi_{A,B}(\omega) \, F(\omega) \tag{11.69}$$

where

$$\chi_{A,B}(\omega) = \int_{-\infty}^{\infty} dt \, \exp[-i\omega t] \, \chi_{A,B}(t) \tag{11.70}$$

and where the Fourier transform of the classical field is given by

$$F(\omega) = \int_{-\infty}^{\infty} dt \, \exp[-i\omega t] \, F(t) \tag{11.71}$$

For a static field $F(t) \equiv F_s$, one finds

$$F(\omega = 0) = 2\pi F_s \, \delta(\omega) \tag{11.72}$$

which elicits a zero frequency response $\Delta A(\omega = 0)$ since

$$\Delta A(\omega) = \chi_{AB}(\omega)(2\pi) \, F_s \delta(\omega) \qquad (11.73)$$

In the time domain, the response $\Delta A(t)$ is time-independent and is given by

$$\Delta A = \int \frac{d\omega}{2\pi} \, \exp[i\omega t] \, \chi_{A,B}(\omega)(2\pi) \, F\delta(\omega)$$

$$= \chi_{A,B}(\omega = 0) \, F_s \qquad (11.74)$$

Hence, the zero frequency limit of the susceptibility $\chi_{AB}(\omega = 0)$ is identified with the static susceptibility. On the other-hand, an oscillatory field $F(t)$ given by

$$F(t) = F_0 \, \cos(\omega_0 t + \varphi) \qquad (11.75)$$

gives rise to

$$F(\omega) = \pi F_0(\exp[+i\varphi] \, \delta(\omega - \omega_0) + \exp[-i\varphi] \, \delta(\omega + \omega_0)) \qquad (11.76)$$

which is complex. Since the frequency domain response is given by

$$\Delta A(\omega) = \chi_{A,B}(\omega) \, F(\omega) \qquad (11.77)$$

the response only has two non-zero frequency components; one at $+\omega_0$ and the second at $-\omega_0$.

Causality and Analytic Properties of $\chi_{A,B}$

The Fourier-Transform of the response function is defined as

$$\chi_{A,B}(\omega) = \int_{-\infty}^{\infty} dt \, \exp[-i\omega t] \, \chi_{A,B}(t) \qquad (11.78)$$

The causality condition puts restrictions on the type of analytic properties of $\chi_{A,B}(\omega)$. In particular, "*a necessary and sufficient condition for causality is that $\chi_{A,B}(\omega)$ is analytic in the lower-half complex ω plane*".

For sufficiency, it shall be assumed that $\chi_{A,B}(\omega)$ is analytic for $\Im m \, \omega < 0$ to show that $\chi_{A,B}(t)$ is causal. If $\chi_{A,B}(\omega)$ is analytic in the lower-half complex ω plane, the singularities must be located in the upper-half complex ω plane. The time-domain response function is given by the inverse Fourier transform

$$\chi_{A,B}(t) = \int_{-\infty}^{\infty} \frac{d\omega}{2\pi} \, \exp[+i\omega t] \, \chi_{A,B}(\omega) \qquad (11.79)$$

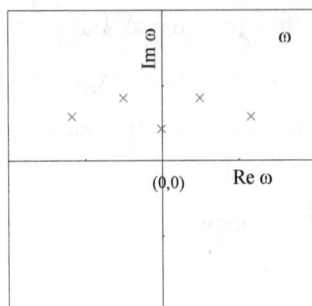

Fig. 11.3 Causality Requires the poles of the response function are confined to the upper-half complex ω plane.

where the integral runs over the real ω axis. For $t > 0$, the integration can be closed in the upper-half complex plane since the semi-circular contour at infinity $\mathcal{C}$ vanishes, since

$$\lim_{\Im m\, \omega \to \infty} \int_{\mathcal{C}} d\omega \, \exp[+i\, \Re e\, \omega t - \Im m\, \omega t] \, \chi_{A,B}(\omega) \to 0 \qquad (11.80)$$

as $\Im m\, \omega t \to \infty$. Hence, the integration around the closed contour, consisting of the real axis and the semicircle at infinity, encloses all the singularities in the upper-half complex plane. By Cauchy's Theorem, the contour integration is given by $2\pi i$ times the sum of the residues. This is expected to be non-zero, so for $t > 0$ the susceptibility satisfies

$$\chi_{A,B}(t) \neq 0 \qquad \text{for } t > 0 \qquad (11.81)$$

However, for $t < 0$, the contour must be closed in the lower-half complex plane where $\Im m\, \omega t > 0$. Since there are no singularities in the lower-half complex plane

$$\chi_{A,B}(t) = 0 \qquad \text{for } t < 0 \qquad (11.82)$$

The requirement of the analyticity of the frequency-dependent susceptibility $\chi_{A,B}(\omega)$ in the lower-half complex plane can be related to the stability of the system. Heuristically, this is illustrated by examining the time-dependence of the response $A(t)$ which occurs when the susceptibility has a simple pole. If the pole is in the upper-half complex plane at

$$\omega = \omega_0 + i\Gamma \qquad (11.83)$$

then, by Jordan's Lemma, for $t > 0$ the integral

$$\chi_{A,B}(t) = \int_{-\infty}^{\infty} d\omega \, \exp[+i\omega t] \, \chi_{A,B}(\omega) \qquad (11.84)$$

can be completed in the upper-half complex plane by the semi-circle at infinity. Since the contour encloses the pole, from Cauchy's Theorem, the time-dependence of the susceptibility is given by

$$\chi_{A,B}(t) \sim \exp[i\omega_0 t] \ \exp[-\Gamma t] \tag{11.85}$$

Therefore, for a delta function pulse $F(t) = F_0 \ \tau \ \delta(t)$, one finds that for $t > 0$,

$$\Delta A(t) \sim \exp[i\omega_0 t] \ \exp[-\Gamma t] \tag{11.86}$$

which indicates that the response relaxes to its equilibrium value. If on the other hand, the pole is in the lower-half complex plane at

$$\omega = \omega_0 - i\Gamma \tag{11.87}$$

then for $t < 0$, the integral over the real ω axis can be completed in the lower-half complex plane, enclosing the pole. Hence, for $t < 0$, one finds that

$$\Delta A(t) \sim \exp[i\omega_0 t] \ \exp[+\Gamma t] \tag{11.88}$$

which illustrates that the system spontaneously generates exponentially growing distortions, even before the delta function pulse has been applied. Since causality is a fundamental principle which should not be lightly discarded, one concludes that the initial system is unstable as ΔA will grow arbitrarily large even if there is no external perturbation. That is, the system is unstable if there are poles in the lower-half complex ω plane.

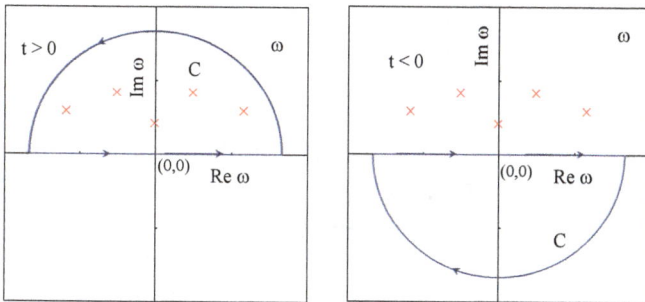

Fig. 11.4 For $t > 0$ the inverse Fourier transform can be complemented by an integral over the semi-circle at infinity in the upper-half complex ω plane. This encloses the poles of $\chi_{AB}(\omega)$ and yields a finite response. For $t < 0$, the contour must be complemented by a contour in the lower-half complex plane and results in a zero response, as expected from causality.

The real-time susceptibility $\chi_{A,B}(t)$ is real. This will be explicitly demonstrated later, but can be motivated by noting that the response $A(t)$ is real, the field $F(t)$ is real and the time t is also real, which suggests that the quantity that relates them is also real even though it appears under an integral.

The real-time susceptibility $\chi_{A,B}(t)$ is real, but since the Fourier transform is defined by

$$\chi_{A,B}(\omega) = \int_{-\infty}^{\infty} dt \, \exp[-i\omega t] \, \chi_{A,B}(t) \tag{11.89}$$

the frequency domain susceptibility $\chi_{A,B}(\omega)$ is complex

$$\chi_{A,B}(\omega) = \Re \, \chi_{A,B}(\omega) + i \, \Im \, \chi_{A,B}(\omega) \tag{11.90}$$

On taking the complex conjugate of the Fourier transform and using the fact that $\chi_{A,B}(t)$ is real, one finds

$$\chi_{A,B}^*(\omega) = \chi_{A,B}(-\omega) \tag{11.91}$$

Separating the real and imaginary parts of the above relation, one has

$$\Re \, \chi_{A,B}(\omega) = \Re \, \chi_{A,B}(-\omega)$$
$$\Im \, \chi_{A,B}(\omega) = -\Im \, \chi_{A,B}(-\omega) \tag{11.92}$$

Thus, the real part of the susceptibility is an even function of ω and the imaginary part is an odd function of ω.

11.3 Explicit Expressions for $\chi_{A,B}$

An expression for the time domain response function $\chi_{A,B}(t)$ can be evaluated in the basis of eigenstates of $\hat{H}_0$

$$\hat{H}_0|n\rangle = E_n|n\rangle \tag{11.93}$$

Since all operators in $\chi_{A,B}$ evolve under the Hamiltonian $\hat{H}_0$, they can be regarded as operators in the Heisenberg representation of the closed system. Henceforth, the index D shall be dropped. Thus the susceptibility is given by

$$\chi_{A,B}(t) = \frac{i}{\hbar} \, \Theta(t) \, \mathrm{Trace} \, \hat{\rho}[\hat{A}(t), \hat{B}(0)] \tag{11.94}$$

In the basis of the energy eigenstates, the density operator can be expressed as

$$\hat{\rho} = \sum_n p_n |n\rangle\langle n| \tag{11.95}$$

So on expressing the commutator in terms of its two terms and evaluating the Trace in the basis of eigenstates of $\hat{H}_0$, one finds

$$\chi_{A,B}(t) = \frac{i}{\hbar} \,\Theta(t) \sum_n p_n [\langle n|\hat{A}(t)\hat{B}(0)|n\rangle$$
$$- \langle n|\hat{B}(0)\hat{A}(t)|n\rangle] \tag{11.96}$$

On inserting the completeness relation between the pair of operators $\hat{A}$ and $\hat{B}$, the susceptibility reduces to

$$\chi_{A,B}(t) = \frac{i}{\hbar} \,\Theta(t) \sum_{n,m} p_n [\langle n|\hat{A}(t)|m\rangle\langle m|\hat{B}(0)|n\rangle$$
$$- \langle n|\hat{B}(0)|m\rangle\langle m|\hat{A}(t)|n\rangle]$$
$$= \frac{i}{\hbar} \,\Theta(t) \sum_{n,m} p_n$$
$$\times \left[\exp\left[+i\,\frac{(E_n - E_m)\,t}{\hbar} \right] \langle n|\hat{A}|m\rangle\langle m|\hat{B}|n\rangle \right.$$
$$\left. - \exp\left[-i\frac{(E_n - E_m)\,t}{\hbar} \right] \langle n|\hat{B}|m\rangle\langle m|\hat{A}|n\rangle \right] \tag{11.97}$$

Since the second term in the big parenthesis is the complex conjugate of the first, one finds that the susceptibility $\chi_{A,B}(t)$ is real despite the presence of the explicit multiplicative factor of i. That is, one has

$$\chi_{A,B}(t) = -\frac{2}{\hbar} \,\Theta(t) \sum_{n,m} p_n \,\Im m \left[\exp\left[+i\,\frac{(E_n - E_m)\,t}{\hbar} \right] \right.$$
$$\left. \times \langle n|\hat{A}|m\rangle\langle m|\hat{B}|n\rangle \right] \tag{11.98}$$

which is purely real.

The frequency domain response function is evaluated as

$$\chi_{A,B}(\omega) = \int_{-\infty}^{\infty} dt \, \exp[-i\omega t]\, \chi_{A,B}(t) \tag{11.99}$$

or, on taking the causality condition into account

$$\chi_{A,B}(\omega) = \int_{0}^{\infty} dt \, \exp[-i\omega t]\, \chi_{A,B}(t) \tag{11.100}$$

To ensure that the integral converges one may introduce a convergence factor

$$\exp[-\eta t] \tag{11.101}$$

where $\eta \to 0$. This is equivalent to adding an infinitesimal imaginary part to the frequency

$$z = \omega - i\eta \tag{11.102}$$

Therefore, the frequency-dependent susceptibility is given by

$$\chi_{A,B}(\omega) = \int_0^\infty dt \, \exp[-izt] \, \chi_{A,B}(t) \tag{11.103}$$

On substituting $\chi_{A,B}(t)$ expressed in terms of the basis of energy eigenstates and integrating over t, one finds

$$\chi_{A,B}(\omega) = \sum_{n,m} p_n \left[\frac{\langle n|\hat{A}|m\rangle\langle m|\hat{B}|n\rangle}{\hbar(\omega - i\eta) - E_n + E_m} \right.$$
$$\left. - \frac{\langle n|\hat{B}|m\rangle\langle m|\hat{A}|n\rangle}{\hbar(\omega - i\eta) - E_m + E_n} \right] \tag{11.104}$$

The small imaginary part η has the effect of suppressing singularities of $\chi_{A,B}(\omega)$ on the real ω-axis by moving them into the upper-half of the complex ω plane. Note that on setting $z \to -z$, one has

$$\chi_{A,B}(\omega)_\eta = \chi_{B,A}(-\omega)_{-\eta} \tag{11.105}$$

which proves to be of use in discussing Lars Onsager's "*Reciprocity Relations*" [104]. On interchanging the summation indices m and n in the second term, one obtains

$$\chi_{A,B}(\omega) = \sum_{n,m} \left[\frac{p_n - p_m}{\hbar(\omega - i\eta) - E_n + E_m} \right]$$
$$\times \langle n|\hat{A}|m\rangle\langle m|\hat{B}|n\rangle \tag{11.106}$$

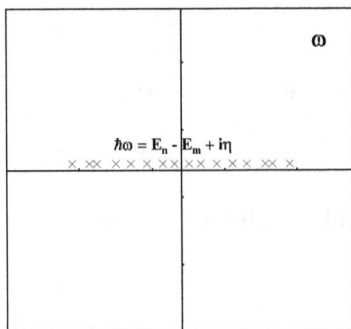

Fig. 11.5   The small imaginary part η causes the poles of the susceptibility to sit just above the real ω-axis. For a dense spectrum, this can be considered as forming a branch cut.

The term outside the parenthesis is expected to be complex, unless $\hat{A} = \hat{B}$, in which case it is real. The term in the square parenthesis is complex. It can be expressed in terms of a real and imaginary part via the identity

$$\frac{1}{\hbar(\omega - i\eta) - E_n + E_m} = \frac{(\hbar\omega - E_n + E_m)}{(\hbar\omega - E_n + E_m)^2 + \hbar^2\eta^2}$$

$$+ i \, \frac{\hbar\eta}{(\hbar\omega - E_n + E_m)^2 + \hbar^2\eta^2} \qquad (11.107)$$

In the limit, $\eta \to 0$, the imaginary part reduces to a Dirac delta function.

$$\lim_{\eta \to 0} \frac{\hbar\eta}{(\hbar\omega - E_n + E_m)^2 + \hbar^2\eta^2} = \pi \, \delta(\hbar\omega - E_n + E_m) \qquad (11.108)$$

The response function $\chi_{A,B}(\omega)$ has a frequency-dependent phase. Frequency-dependent phase shifts between a perturbing potential and the response usually signal a dissipative process.

The symmetries of the susceptibility can be demonstrated most readily if the measured quantity $\hat{A}$ is the same as $\hat{B}$. The susceptibility is given by

$$\chi_{A,A}(\omega) = \sum_{n,m} p_n |\langle n|\hat{A}|m\rangle|^2$$

$$\times \left[\frac{1}{\hbar(\omega - i\eta) - E_n + E_m} + \frac{1}{-\hbar(\omega - i\eta) - E_n + E_m} \right]$$

$$(11.109)$$

and the imaginary part is given by

$$\Im m \, \chi_{A,A}(\omega) = \pi \sum_{n,m} p_n |\langle n|\hat{A}|m\rangle|^2$$

$$\times [\delta(\hbar\omega - E_n + E_m) - \delta(-\hbar\omega - E_n + E_m)]$$

$$(11.110)$$

which is an odd function of ω. The imaginary part involves delta functions which are non-zero for ω values that match the excitation energies of the systems

$$\hbar\omega = \pm(E_n - E_m) \qquad (11.111)$$

Thus, the imaginary part of the response function measures the spectrum of excitations caused by the absorption or emission of quanta of the field. The intensity of the excitations is given by the squared modulus of the matrix elements, $|\langle n|\hat{A}|m\rangle|^2$. The real part of the susceptibility can be expressed as

$$\Re \chi_{A,A}(\omega) = \sum_{n,m} p_n |\langle n|\hat{A}|m\rangle|^2$$

$$\times \left[\frac{\hbar\omega - E_n + E_m}{(\hbar\omega - E_n + E_m)^2 + \hbar^2\eta^2} + \frac{-\hbar\omega - E_n + E_m}{(-\hbar\omega - E_n + E_m)^2 + \hbar^2\eta^2} \right] \tag{11.112}$$

which is an even function of ω. Since $\Re \chi_{A,A}(\omega)$ is an even function of ω and $\Im \chi_{A,A}(\omega)$ is odd, one has

$$\chi_{A,A}^*(\omega) = \chi_{A,A}(-\omega) \tag{11.113}$$

Example: The Damped Quantum Mechanical Harmonic Oscillator

We shall consider a Hamiltonian describing a Harmonic Oscillator coupled to a heat bath by a bilinear interaction. The Hamiltonian is written as

$$\hat{H} = \left(\frac{\hat{P}^2}{2M} + \frac{M\omega_0^2}{2} \hat{Q}^2 \right)$$

$$+ \sum_{i=1}^{N} \left(\frac{\hat{p}_i^2}{2m_i} + \frac{m_i\omega_i^2}{2} \hat{q}_i^2 \right)$$

$$+ \sum_{i=1}^{N} \lambda_i \hat{Q} \hat{q}_i \tag{11.114}$$

where the first line describes the Harmonic Oscillator with frequency ω_0 and mass M, the second line describes a heat bath composed of N harmonic oscillators with frequencies ω_i. The last term represents the coupling between the harmonic oscillator and the heat bath degrees of freedom.

We shall define the causal response function as

$$\chi_{A,B}(t) = \frac{i}{\hbar} \Theta(t) \, \text{Trace} \, \hat{\rho}[\hat{A}(t), \hat{B}(0)] \tag{11.115}$$

The equation of motion is found as

$$i\hbar \frac{\partial}{\partial t} \chi_{A,B}(t) = -\delta(t) \, \text{Trace} \, \hat{\rho}[\hat{A}(0), \hat{B}(0)]$$

$$+ \frac{i}{\hbar} \Theta(t) \, \text{Trace} \, \hat{\rho}[[\hat{A}(t), \hat{H}(t)], \hat{B}(0)] \tag{11.116}$$

which leads to the closed set of coupled first-order differential equations

$$\frac{\partial}{\partial t}\chi_{Q,Q}(t) = \frac{1}{M}\ \chi_{P,Q}(t)$$

$$\frac{\partial}{\partial t}\chi_{P,Q}(t) = \delta(t) - M\omega_0^2\ \chi_{Q,Q}(t) - \sum_{i=1}^{N}\lambda_i\ \chi_{q_i,Q}(t)$$

$$\frac{\partial}{\partial t}\chi_{q_i,Q}(t) = \frac{1}{m_i}\ \chi_{p_i,Q}(t)$$

$$\frac{\partial}{\partial t}\chi_{p_i,Q}(t) = -\ m_i\omega_i^2\ \chi_{q_i,Q}(t) - \lambda_i\ \chi_{Q,Q}(t) \tag{11.117}$$

in which $\hbar$ has dropped out. On introducing the Fourier Transformation as

$$\tilde{\chi}_{A,B}(\omega) = \int_{-\infty}^{\infty} dt\ \exp[-i\omega t]\ \chi_{A,B}(t) \tag{11.118}$$

and Fourier Transforming the equations of motion, one obtains the closed set of coupled algebraic equations

$$i\omega\ \tilde{\chi}_{Q,Q}(\omega) = \frac{1}{M}\ \tilde{\chi}_{P,Q}(\omega)$$

$$i\omega\ \tilde{\chi}_{P,Q}(\omega) = 1 - M\omega_0^2\ \tilde{\chi}_{Q,Q}(\omega) - \sum_{i=1}^{N}\lambda_i\ \tilde{\chi}_{q_i,Q}(\omega)$$

$$i\omega\ \tilde{\chi}_{q_i,Q}(\omega) = \frac{1}{m_i}\ \tilde{\chi}_{p_i,Q}(\omega)$$

$$i\omega\ \tilde{\chi}_{p_i,Q}(\omega) = -m_i\omega_i^2\ \tilde{\chi}_{q_i,Q}(\omega) - \lambda_i\ \tilde{\chi}_{Q,Q}(\omega) \tag{11.119}$$

The equations can be solved to yield

$$\tilde{\chi}_{Q,Q}(\omega) = -\frac{1}{M}\ [\omega^2 - \omega_0^2 - 2\omega_0\Pi(\omega)]^{-1} \tag{11.120}$$

where the polarization part $\Pi(\omega)$ is defined by

$$2\omega_0\Pi(\omega) = \sum_{i=1}^{N}\frac{\lambda_i^2}{Mm_i}\ (\omega^2 - \omega_i^2)^{-1} \tag{11.121}$$

Due to the dense set of poles on the real axis, $\Pi(\omega)$ has a branch cut on the real ω axis. Furthermore, one finds

$$2\omega_0\ \Im m\ \Pi(\omega \pm i\eta) = \mp\sum_{i=1}^{N}\frac{\pi\lambda_i^2}{2Mm_i\omega_i}\ [\delta(\omega - \omega_i) - \delta(\omega + \omega_i)] \tag{11.122}$$

which shows the response function neither has poles in the upper-half nor the lower-half of the complex ω-plane. Furthermore, it shows that $\Im m\,\Pi(\omega)$ is an odd function of ω, while $\Re e\,\Pi(\omega)$ is an even function of ω.

The above behavior suggests an approximate method can be introduced to evaluate integrations involving $\tilde{\chi}_{Q,Q}(\omega)$ on lines just below the real ω-axis, in which the branch cut is replaced by poles in the upper-half complex plane. In this phenomenology, $\Re e\,\Pi(\omega)$ is approximated by a constant and the imaginary part, $2\omega_0\Im m\,\Pi(\omega)$, by $i\gamma\omega$ which has the effect of producing poles of $\tilde{\chi}_{Q,Q}(\omega)$ in the upper-half complex plane.

11.4   The Dissipation of Energy and $\Im m\,\chi_{A,A}(\omega)$

A change in the state of the system as a result of the applied external field is accompanied by the absorption of energy. The source of the energy is the external field and this energy is dissipated into heat.

The power dissipated by the external field is

$$\frac{d}{dt}\langle \hat{H}(t)\rangle \geq 0 \tag{11.123}$$

so energy flows into the system. Here another the notation for the thermal average has been introduced

$$\langle \hat{H}(t)\rangle = \text{Trace}\,\hat{H}(t)\,\hat{\rho}(t) \tag{11.124}$$

where $\hat{H}(t)$ is the Hamiltonian in the Schrödinger representation and $\hat{\rho}(t)$ is the corresponding density operator. The derivative of the thermal averaged energy is

$$\frac{d}{dt}\langle \hat{H}(t)\rangle = \text{Trace}\,\frac{\partial \hat{H}(t)}{\partial t}\,\hat{\rho}(t) + \text{Trace}\,\hat{H}(t)\,\frac{\partial \hat{\rho}(t)}{\partial t} \tag{11.125}$$

The term proportional to the derivative of the density operator vanishes since it is assumed that $\rho(t)$ evolves under $\hat{H}(t)$ via

$$i\hbar\,\frac{\partial \hat{\rho}(t)}{\partial t} + [\hat{\rho}(t),\hat{H}(t)] = 0 \tag{11.126}$$

so

$$\text{Trace}\,\hat{H}(t)\,\frac{\partial \hat{\rho}(t)}{\partial t} = \frac{i}{\hbar}\,\text{Trace}\,\hat{H}(t)[\hat{\rho}(t),\hat{H}(t)]$$

$$= 0 \tag{11.127}$$

where the cyclic invariance of the Trace has been used. Hence, the power dissipated by the signal generator is

$$\frac{d}{dt}\langle \hat{H}(t)\rangle = -\text{Trace } \hat{A}\hat{\rho}(t)\, \frac{\partial F}{\partial t}$$

$$= -\langle \hat{A}(t)\rangle\, \frac{\partial F}{\partial t} \tag{11.128}$$

The average power dissipated by an oscillatory external field of the form

$$F(t) = F_0\, \cos(\omega t + \varphi)$$

$$= \frac{F_0}{2}\, [\exp[i(\omega t + \varphi)] + \exp[-i(\omega t + \varphi)]] \tag{11.129}$$

shall be calculated to lowest order in the field. The expectation value of $\hat{A}$ is given by

$$\langle \hat{A}(t)\rangle = \int_{-\infty}^{\infty} dt'\, \chi_{A,A}(t-t')\frac{F_0}{2}[\exp[i(\omega t' + \varphi)] + \exp[-i(\omega t' + \varphi)]]$$

$$= \frac{F_0}{2}\, [\chi_{A,A}(\omega)\, \exp[i(\omega t + \varphi)] + \chi_{A,A}(-\omega)\, \exp[-i(\omega t + \varphi)]] \tag{11.130}$$

but since

$$\chi_{A,A}^*(\omega) = \chi_{A,A}(-\omega) \tag{11.131}$$

one finds

$$\langle \hat{A}(t)\rangle = \frac{F_0}{2}\, [\chi_{A,A}(\omega)\, \exp[i(\omega t+\varphi)]+\chi_{A,A}^*(\omega)\, \exp[-i(\omega t+\varphi)]] \tag{11.132}$$

which is real. To lowest order in the field, the power dissipated at time t is given by

$$\left\langle \frac{d\hat{H}}{dt}\right\rangle = -\langle \hat{A}(t)\rangle\, \frac{\partial F}{\partial t}$$

$$= \frac{F_0^2\omega}{4i}\, [\exp[i(\omega t + \varphi)] - \exp[-i(\omega t + \varphi)]]$$

$$\times\, [\chi_{A,A}(\omega)\, \exp[i(\omega t + \varphi)] + \chi_{A,A}^*(\omega)\, \exp[-i(\omega t + \varphi)]] \tag{11.133}$$

On averaging over time t, where the averaging is performed over a time interval much longer than the period, one finds

$$\overline{\left\langle \frac{d\hat{H}}{dt}\right\rangle} = -\frac{F_0^2\omega}{4i}\, [\chi_{A,A}(\omega) - \chi_{A,A}^*(\omega)]$$

$$= -\frac{F_0^2}{2}\, \omega\, \Im m\, \chi_{A,A}(\omega) \tag{11.134}$$

The power dissipated is positive. Hence, one has the inequality

$$\omega\, \Im m\, \chi_{A,A}(\omega) \le 0 \tag{11.135}$$

The imaginary part of the susceptibility is connected with the dissipation of energy.

Example: A Damped Classical Harmonic Oscillator

The equation of motion for a driven damped classical Harmonic Oscillator can be written as

$$m(\ddot{x} + \gamma\dot{x} + \omega_0^2 x) = F(t) \qquad (11.136)$$

where γ represents the dissipation. The solution of the inhomogeneous linear differential equation is composed of a linear superposition of a particular solution of the homogeneous equation, which represents the transients, together with a solution of the inhomogeneous equation which represents the response to the driving force $F(t)$. The solution of the inhomogeneous equation can be found by considering the response, $\chi(t, t')$, to a delta function force

$$F(t) = \delta(t - t') \qquad (11.137)$$

The response function $\chi(t, t')$ is defined to be the solution of the equation

$$m\left(\frac{\partial^2}{\partial t^2} + \gamma\frac{\partial}{\partial t} + \omega_0^2\right)\chi(t, t') = \delta(t - t') \qquad (11.138)$$

Since the equation of motion for the simple Harmonic Oscillator is homogeneous in time, χ is a function of $t - t'$. The solution of the driven Harmonic equation with any arbitrary driving force $F(t)$ can be expressed in terms of the response function as a convolution

$$x(t) = \int_{-\infty}^{\infty} dt' \chi(t - t')\, F(t') \qquad (11.139)$$

That the above expression represents the solution for an arbitrary driving force can be verified simply by substituting it into the equation of motion and noting that the differentials only act on the t dependence of χ and the integral over t' reproduces the driving function on the r.h.s.

The Frequency Domain Response

The response function can be found from $\chi(t - t')$ by its inverse Fourier Transform $\tilde{\chi}(\omega)$

$$\chi(t - t') = \int_{-\infty}^{\infty} \frac{d\omega}{2\pi}\, \exp[i\omega(t - t')]\, \tilde{\chi}(\omega) \qquad (11.140)$$

which leads to

$$m \int \frac{d\omega}{2\pi} \, [-\omega^2 + i\omega\gamma + \omega_0^2] \, \exp[i\omega(t - t')] \, \tilde{\chi}(\omega) = \delta(t - t')$$

(11.141)

The Fourier Transform $\tilde{\chi}(\omega)$ can be found by noting that the delta function can be expressed as

$$\int \frac{d\omega}{2\pi} \, \exp[i\omega(t - t')] = \delta(t - t')$$

(11.142)

Hence, one finds that $\tilde{\chi}(\omega)$ is given by

$$\tilde{\chi}(\omega) = \frac{1}{m} \left(\frac{1}{-\omega^2 + i\omega\gamma + \omega_0^2} \right)$$

(11.143)

which has real and imaginary parts that are given by

$$\Re \, \tilde{\chi}(\omega) = \frac{1}{m} \left(\frac{\omega_0^2 - \omega^2}{(\omega^2 - \omega_0^2)^2 + \gamma^2\omega^2} \right)$$

$$\Im \, \tilde{\chi}(\omega) = \frac{1}{m} \left(\frac{-i\gamma\omega}{(\omega^2 - \omega_0^2)^2 + \gamma^2\omega^2} \right)$$

(11.144)

The real part of the response function is an even function of ω and the imaginary part is odd.

However, if the function $\chi(z)$ is analytically continued from the lower-half to the upper-half of the complex plane, the response function has poles located at

$$\omega_\pm = +i \, \frac{\gamma}{2} \pm \sqrt{\omega_0^2 - \frac{\gamma^2}{4}}$$

(11.145)

Therefore, the response differs for $\frac{\gamma}{2} > \omega_0$ and $\frac{\gamma}{2} < \omega_0$.

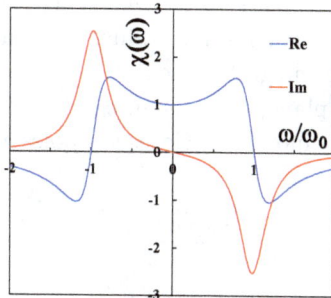

Fig. 11.6   The real and imaginary parts of the response function $\tilde{\chi}(\omega)$ as a function of ω for a Classical Harmonic Oscillator with frequency ω_0 and damping strength $\gamma = 0.4\omega_0$.

For $\frac{\gamma}{2} > \omega_0$, the oscillator is "*Overdamped*" and has poles on the upper part of the imaginary axis at

$$\omega_\pm = +i\,\frac{\gamma}{2} \pm i\sqrt{\frac{\gamma^2}{4} - \omega_0^2} \qquad (11.146)$$

For $\frac{\gamma}{2} < \omega_0$, the oscillator is "*Underdamped*" and has poles in the upper-half complex plane at at

$$\omega_\pm = +i\,\frac{\gamma}{2} \pm \sqrt{\omega_0^2 - \frac{\gamma^2}{4}} \qquad (11.147)$$

Causality

The response function in the time domain, $\chi(t)$ is causal. The time domain response function $\chi(t)$ is obtained by the Inverse Fourier Transform

$$\chi(t) = \int_{-\infty}^{\infty} \frac{d\omega}{2\pi} \, \exp[i\omega t] \, \tilde{\chi}(\omega)$$

$$= \frac{1}{m} \int_{-\infty}^{\infty} \frac{d\omega}{2\pi} \left(\frac{\exp[i\omega t]}{-\omega^2 + i\omega\gamma + \omega_0^2} \right) \qquad (11.148)$$

The integral can be evaluated using Cauchy's Theorem. The integrand can be written as

$$\chi(t) = -\frac{1}{m} \int_{-\infty}^{\infty} \frac{d\omega}{2\pi} \frac{\exp[i\omega t]}{(\omega - \omega_+)(\omega - \omega_-)}$$

$$= \frac{1}{m} \int_{-\infty}^{\infty} \frac{d\omega}{2\pi} \frac{\exp[i\omega t]}{\omega_+ - \omega_-} \left(\frac{1}{\omega - \omega_-} - \frac{1}{\omega - \omega_+} \right) \qquad (11.149)$$

Causality requires that the response can only occur after the driving pulse has occurred, this corresponds to $t > 0$. For negative values of t, the response should be zero. For positive values of t, then the integral can be obtained by closing the integration contour by a semi-circle at infinity in the upper-half complex plane, since the integral over the semicircle vanishes by Jordan's lemma. The integral encloses the poles at $\omega = \omega_+$ and $\omega = \omega_-$, yielding the result

$$\chi(t) = \frac{1}{im} \left[\frac{\exp[i\omega_+ t] - \exp[i\omega_- t]}{\omega_+ - \omega_-} \right] \qquad (11.150)$$

valid for $t > 0$. For negative values of t, the contour must be completed in the lower-half complex plane. Since there are no poles in the lower-half complex plane, the integral is zero and so the time domain response function is

$$\chi(t) = 0 \tag{11.151}$$

for $t < 0$. Since there is no response prior to the impulse, which occurs at $t = 0$, the response function is causal.

In the overdamped regime with $t > 0$, the response is given by

$$\chi(t) = \frac{1}{m} \exp\left[-\frac{\gamma}{2} t\right] \left(\frac{\sinh\left[\sqrt{\frac{\gamma^2}{4} - \omega_0^2}\, t\right]}{\sqrt{\frac{\gamma^2}{4} - \omega_0^2}}\right) \tag{11.152}$$

which is purely decaying for large times.

In the underdamped region for $t > 0$

$$\chi(t) = \frac{1}{m} \exp\left[-\frac{\gamma}{2} t\right] \left(\frac{\sin\left[\sqrt{\omega_0^2 - \frac{\gamma^2}{4}}\, t\right]}{\sqrt{\omega_0^2 - \frac{\gamma^2}{4}}}\right) \tag{11.153}$$

which shows oscillations but with an exponentially decaying envelope.

Dissipation

The rate at which energy is absorbed by the system is equivalent to the rate at which work is done on the system. Therefore

$$\frac{\partial E}{\partial t} = F(t)\, \dot{x}$$

$$= F(t)\, \frac{\partial}{\partial t} \int_{-\infty}^{\infty} dt'\, \chi(t - t')\, F(t')$$

$$= F(t) \int_{-\infty}^{\infty} dt'\, \frac{\partial}{\partial t} \chi(t - t')\, F(t') \tag{11.154}$$

On expressing $\chi(t - t')$ in term of its Fourier Transform

$$\chi(t - t') = \int_{-\infty}^{\infty} \frac{d\omega}{2\pi}\, \exp[i\omega(t - t')]\, \tilde{\chi}(\omega) \tag{11.155}$$

one obtains

$$\frac{\partial E}{\partial t} = F(t) \int_{-\infty}^{\infty} dt' \int_{-\infty}^{\infty} \frac{d\omega}{2\pi}\, i\omega\, \exp[i\omega(t - t')]\, \tilde{\chi}(\omega)\, F(t')$$

$$= F(t) \int_{-\infty}^{\infty} \frac{d\omega}{2\pi}\, i\omega\, \exp[i\omega t]\, \tilde{\chi}(\omega)\, F(\omega) \tag{11.156}$$

The periodic driving force

$$F(t) = F_0\, \cos \Omega t \tag{11.157}$$

has the Fourier components

$$F(\omega) = \pi F_0 [\delta(\omega - \Omega) + \delta(\omega + \Omega)] \tag{11.158}$$

Hence, the instantaneous rate of work done on a periodically driven oscillator is given by

$$\frac{\partial E}{\partial t} = F_0 \cos \Omega t \int_{-\infty}^{\infty} \frac{d\omega}{2\pi} \exp[i\omega t] \, i\omega \, \tilde{\chi}(\omega) \, F(\omega)$$

$$= \frac{F_0^2}{2} \cos \Omega t \, i\Omega [\exp[i\Omega t] \, \tilde{\chi}(\Omega) - \exp[-i\Omega t] \, \tilde{\chi}(-\Omega)] \tag{11.159}$$

On averaging over a period, T, of oscillation of the driving field given by

$$T = \frac{2\pi}{\Omega} \tag{11.160}$$

one finds that the average rate at which work is done is given by

$$\overline{\frac{\partial E}{\partial t}} = \frac{F_0^2}{2} \frac{1}{T} \int_0^T dt \, \cos \Omega t \, i\Omega [\exp[i\Omega t] \, \tilde{\chi}(\Omega) - \exp[-i\Omega t] \, \tilde{\chi}(-\Omega)]$$

$$= -\left(\frac{F_0^2}{2}\right) \Omega \left[\frac{\tilde{\chi}(\Omega) - \tilde{\chi}(-\Omega)}{2i}\right] \tag{11.161}$$

The dissipation is given by the average intensity of the driving field times the imaginary part of the susceptibility. For the damped Harmonic Oscillator, this reduces to

$$\overline{\frac{\partial E}{\partial t}} = \left(\frac{F_0^2}{2}\right) \Omega \frac{1}{m} \left[\frac{\gamma \Omega}{(\Omega^2 - \omega_0^2)^2 + \gamma^2 \Omega^2}\right] \tag{11.162}$$

Since the driven oscillator is in a steady state, the rate at which work is done on the system is equal to the rate at which energy is dissipated. This interpretation is supported by the results of evaluating of the average rate at which work is done by the dissipative force $-m\gamma \dot{x}$.

11.5   Kramers-Kronig Relations for $\chi_{A,B}$

The "*Kramers-Kronig Relations*" [105, 106] relate the real and imaginary parts of the frequency domain susceptibility $\chi_{A,B}(\omega)$. The Kramers-Kronig relation is found by considering the function

$$\int_{-\infty}^{\infty} dz \left(\frac{\chi_{A,B}(z)}{\omega - z + i\eta}\right) \tag{11.163}$$

where ω is real and $\eta \to 0+$. Thus, the integrand has a pole just above the real z-axis. This integral can be complemented with an integral around the semicircle at infinity in the lower-half complex z plane. Causality ensures that $\chi_{A,B}(z)$ is analytic in the lower-half complex z plane. Since $\chi_{A,B}(z)$ is analytic in the lower-half plane and since $\lim_{z \to \infty} \chi_{A,B}(z) \to 0$, the integral around the contour $\mathcal{C}$ enclosing the lower-half complex plane is zero. Therefore, the integral along the real z-axis vanishes

$$\int_{-\infty}^{\infty} dz \left(\frac{\chi_{A,B}(z)}{\omega - z + i\eta} \right) = 0 \qquad (11.164)$$

The Kramers-Kronig relation is found by expressing $\chi_{A,B}(z)$ in terms of its real and imaginary parts

$$\chi_{A,B}(z) = \Re e \, \chi_{A,B}(z) + i \, \Im m \, \chi_{A,B}(z) \qquad (11.165)$$

and using the identity

$$\frac{1}{\omega - z + i\eta} = \frac{\mathrm{Pr}}{\omega - z} - i\pi \, \delta(\omega - z) \qquad (11.166)$$

and then substituting these into the integral over the real axis. This leads to the relation

$$\int_{-\infty}^{\infty} dz \frac{\mathrm{Pr}}{\omega - z} \chi_{A,B}(z) - i \int_{-\infty}^{\infty} dz \, \pi \, \delta(z - \omega) \chi_{A,B}(z) = 0$$

$$\int_{-\infty}^{\infty} dz \, \frac{\mathrm{Pr}}{\omega - z} \chi_{A,B}(z) - i\pi \, \chi_{A,B}(\omega) = 0 \qquad (11.167)$$

Separating the integral into its real and imaginary parts, produces the two equalities

$$\int_{-\infty}^{\infty} dz \, \frac{\mathrm{Pr}}{\omega - z} \Re e \, \chi_{A,B}(z) = -\pi \, \Im m \, \chi_{A,B}(\omega)$$

$$\int_{-\infty}^{\infty} dz \, \frac{\mathrm{Pr}}{\omega - z} \Im m \, \chi_{A,B}(z) = \pi \, \Re e \, \chi_{A,B}(\omega) \qquad (11.168)$$

that constitutes the Kramers-Kronig relations. The Kramers-Kronig relations are a consequence of causality. Usually frequency-dependent experiments measure combinations of the real and imaginary parts of the susceptibility. In such cases, two sets of measurements have to be made in order to extract the frequency-dependence of the real and imaginary parts of the response function. It is often more convenient to only perform one set of measurements and use the Kramers-Kronig relations to analyze the results and extract the real and imaginary parts of $\chi_{A,B}(\omega)$.

The Kramers-Kronig relations can be put in an alternate form by using the relations

$$\chi_{A,B}^{*}(\omega) = \chi_{A,B}(-\omega) \qquad (11.169)$$

For example, since the imaginary part satisfies the relation

$$\Im m\, \chi_{A,B}(z) = -\Im m\, \chi_{A,B}(-z) \tag{11.170}$$

it can be combined with the equation

$$\int_{-\infty}^{\infty} dz\, \frac{\mathrm{Pr}}{\omega - z}\, \Im m\, \chi_{A,B}(z) = \pi\, \Re e\, \chi_{A,B}(\omega) \tag{11.171}$$

to obtain an alternate form of the relation. The integral can be broken up as integrals over the positive and negative z-axis.

$$\begin{aligned}
\pi\, \Re e\, \chi_{A,B}(\omega) &= \int_{0}^{\infty} dz\, \frac{\mathrm{Pr}}{\omega - z}\, \Im m\, \chi_{A,B}(z) + \int_{-\infty}^{0} dz\, \frac{\mathrm{Pr}}{\omega - z}\, \Im m\, \chi_{A,B}(z) \\
&= \int_{0}^{\infty} dz\, \frac{\mathrm{Pr}}{\omega - z}\, \Im m\, \chi_{A,B}(z) + \int_{0}^{\infty} dz\, \frac{\mathrm{Pr}}{\omega + z}\, \Im m\, \chi_{A,B}(-z) \\
&= \int_{0}^{\infty} dz\, \frac{\mathrm{Pr}}{\omega - z}\, \Im m\, \chi_{A,B}(z) - \int_{0}^{\infty} dz\, \frac{\mathrm{Pr}}{\omega + z}\, \Im m\, \chi_{A,B}(z)
\end{aligned} \tag{11.172}$$

where the equality has been used to obtain the last line. Combining the two integrals leads to the alternate form of the Kramers-Kronig relation

$$\pi\, \Re e\, \chi_{A,B}(\omega) = 2 \int_{0}^{\infty} dz\, \frac{\mathrm{Pr}\, z}{\omega^2 - z^2}\, \Im m\, \chi_{A,B}(z) \tag{11.173}$$

Likewise, the other partner equation of the Kramers-Kronig relation can be combined with the symmetry condition

$$\Re e\, \chi_{A,B}(z) = \Re e\, \chi_{A,B}(-z) \tag{11.174}$$

to yield an alternative form of the partner Kramers-Kronig relation. This is achieved by breaking up the integration

$$\int_{-\infty}^{\infty} dz\, \frac{\mathrm{Pr}}{\omega - z}\, \Re e\, \chi_{A,B}(z) = -\pi\, \Im m\, \chi_{A,B}(\omega) \tag{11.175}$$

into an integral over the positive and negative real z-axis

$$\int_{0}^{\infty} dz \frac{\mathrm{Pr}}{\omega - z} \Re e\, \chi_{A,B}(z) + \int_{-\infty}^{0} dz \frac{\mathrm{Pr}}{\omega - z} \Re e\, \chi_{A,B}(z) = -\pi\, \Im m\, \chi_{A,B}(\omega)$$

$$\int_{0}^{\infty} dz\, \frac{\mathrm{Pr}}{\omega - z} \Re e\, \chi_{A,B}(z) + \int_{0}^{\infty} dz \frac{\mathrm{Pr}}{\omega + z} \Re e\, \chi_{A,B}(-z) = -\pi\, \Im m\, \chi_{A,B}(\omega)$$

$$2\, \omega \int_{0}^{\infty} dz \frac{\mathrm{Pr}}{\omega^2 - z^2} \Re e\, \chi_{A,B}(z) = -\pi\, \Im m\, \chi_{A,B}(\omega)$$

$$\tag{11.176}$$

Use of these equations in analyzing experiments does, however, require that experimentally known quantities are known over the entire range of positive frequencies. This requirement is usually overcome by extrapolating the experimental data to high frequencies.

Theoretical Interlude: The Response on the Imaginary Frequency Axis

The response function $\chi_{A,B}(\omega)$ does not take on real values at any finite point in the lower-half complex ω plane, except on the imaginary axis $\omega = ix$ where it decreases monotonically from a positive value at $\omega = 0$ to zero at $\omega \to -i\infty$. Thus, on the imaginary axis the susceptibility satisfies

$$\Im m \; \chi_{A,B}(ix) = 0$$
$$\Re e \; \chi_{A,B}(ix) > 0 \qquad (11.177)$$

Hence, $\chi_{A,B}(x)$ has no zeros in the lower half-complex plane.

To prove the above statement, consider the integral around a contour $\mathcal{C}$ composed of the real axis and the semi-circle at infinity located in the lower-half complex plane

$$\frac{1}{2\pi i} \int_{\mathcal{C}} d\omega \; \frac{\left(\frac{d\chi_{A,B}(\omega)}{d\omega} \right)}{\chi_{A,B}(\omega) - \chi_0} \qquad (11.178)$$

If the function $\chi_{A,A}(\omega) - \chi_0$ has a zero of multiplicity n_0 at ω_0, the integrand has a simple pole at ω_0 with residue n_0. Likewise, if $\chi_{A,B}(\omega)$ has a pole ω_p of order n_p, the integrand has a simple pole at ω_p with residue

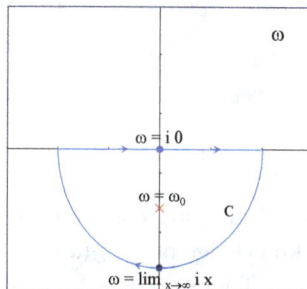

Fig. 11.7 The contour composed of the real axis which is closed by the semi-circle at infinity in the lower-half complex plane, $\mathcal{C}$, for the evaluation of the integral in Eq. (11.178). Since the poles of $\chi_{A,B}(\omega)$ are located in the upper half-complex plane, the integral only counts the zeros of $\chi_{A,B}(\omega) - \chi_0$ enclosed by the contour.

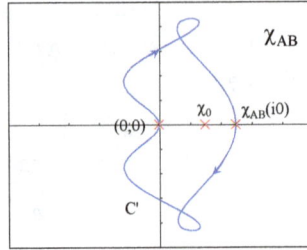

Fig. 11.8   A sketch of the contour of integration $\mathcal{C}'$ in the complex $\chi_{A,B}$ plane. The range of integration over the semi-circular contour at infinity in the ω plane is mapped onto the origin $(0,0)$ since $\lim_{\omega \to \infty} \chi_{A,B}(\omega) = 0$. The point $\chi_{A,B}(i0)$ is on the real axis. For any real value of χ_0, there is at most one solution of the equation $\chi_{A,B} = \chi_0$ and the solution must be bounded by $\chi_{A,B}(i0)$ and $(0,0)$.

$-n_p$. Thus, the integral around the contour $\mathcal{C}$ is equal to the difference in the number of zeros and poles of the function $\chi_{A,B}(\omega) - \chi_0$, including their multiplicities, enclosed by the contour.

We shall choose χ_0 to be a real number then, due to the analyticity of $\chi_{A,B}(\omega)$, their are no poles of $\chi_{A,B}(\omega) - \chi_0$ enclosed by the contour. Then the integrand simply gives the number of solutions of

$$\chi_{A,B}(\omega) = \chi_0 \tag{11.179}$$

which is just the number of points at which $\chi_{A,B}(\omega)$ has the real value χ_0. The integral is evaluated as

$$\frac{1}{2\pi i} \int_{\mathcal{C}'} \frac{d\chi_{A,B}}{\chi_{A,B} - \chi_0} \tag{11.180}$$

The integral is equal to the phase change of $\chi_{A,B} - \chi_0$ on traversing the contour $\mathcal{C}'$, since

$$\frac{1}{2\pi i} \ln(\chi_{A,B} - \chi_0)|_{\mathcal{C}'} \tag{11.181}$$

The contour $\mathcal{C}'$ traced out in $\chi_{A,B}$ space is found from the contour $\mathcal{C}$ traced out in ω space. Detailed knowledge of $\chi_{A,B}(\omega)$ for complex ω is needed to determine the details of $\mathcal{C}'$. There are five facts of the mapping that can be surmised.

(i) The infinite semi-circle in the complex ω plane is mapped onto the point $\chi_{A,B} = 0$ in the complex $\chi_{A,B}$ plane.

(ii) The origin $w = 0$ of the complex w plane is mapped onto a real value $\chi(i0)$, since

$$\Im m \; \chi_{A,B}(\omega) < 0 \qquad \text{for } \omega > 0$$
$$\Im m \; \chi_{A,B}(\omega) > 0 \qquad \text{for } \omega < 0 \tag{11.182}$$

(iii) The positive real ω axis, $\omega > 0$, is mapped onto a complicated curve in the lower-half complex χ plane.

(iv) The negative real ω axis, $\omega < 0$, is mapped onto a complicated curve in the upper-half complex χ plane.

(v) The two curves in the upper and lower halves of complex χ-space only meet at the two point $\chi = 0$ and $\chi(i0)$. This can be discerned since, for ω on the real axis, $\chi(\omega)$ does not take on a real value except at $\omega = 0$.

If $\chi_{A,B}(i0) > \chi_0 > 0$, then the total change of phase of $\chi_{A,B} - \chi_0$ around $\mathcal{C}$ is simply 2π. Therefore,

$$\frac{1}{2\pi i} \int_{\mathcal{C}} d\omega \; \frac{\left(\frac{d\chi_{A,B}(\omega)}{d\omega}\right)}{\chi_{A,B}(\omega) - \chi_0} = 1 \tag{11.183}$$

so $n_0 = 1$ if $\chi_{A,B}(i0) > \chi_0 > 0$.

If $|\chi_0| > \chi_{A,B}(i0)$, then the total change of phase of $\chi_{A,B} - \chi_0$ is simply 0. This leads to

$$\frac{1}{2\pi i} \int_{\mathcal{C}} d\omega \; \frac{\left(\frac{d\chi_{A,B}(\omega)}{d\omega}\right)}{\chi_{A,B}(\omega) - \chi_0} = 0 \tag{11.184}$$

so $n_0 = 0$ if $|\chi_0| < \chi_{A,B}(i0)$. Since both $\chi_{A,B}(i0)$ and χ_0 are real, these are the only two possibilities.

Therefore, the function $\chi_{A,B}(\omega)$ only takes on the value χ_0 just once, if χ_0 is in the range

$$\chi_{A,B}(i0) \geq \chi_0 \geq 0 \qquad \text{for} \quad \Im m \; \omega < 0 \tag{11.185}$$

The function $\chi_{A,B}(\omega)$ does not take on any other values.

Thus $\chi(\omega)$ is real on the imaginary axis $\chi(ix)$ and does not have any maximum or minimum. If it had a maximum or minimum, $\chi(ix)$ would take on the same value at least twice. Consequently $\chi(\omega)$ varies monotonically on the negative imaginary axis, decreasing from its maximum value at the origin to zero as $\omega \to -\infty$.

11.6   Sum Rules for $\chi_{A,B}$

The response function $\chi_{A,B}(\omega)$ must satisfy certain sum rules. The sum
rules can be deduced from the expression for the response function in the
time-domain

$$\chi_{A,B}(t - t') = \frac{i}{\hbar} \, \Theta(t - t') \langle [\hat{A}(t), \hat{B}(t')] \rangle$$

$$= \int_{-\infty}^{\infty} \frac{d\omega}{2\pi} \, \exp[i\omega(t - t')] \, \chi_{A,B}(\omega) \qquad (11.186)$$

On taking the limit $t - t' \to 0$, one finds the sum rule

$$\int_{-\infty}^{\infty} \frac{d\omega}{2\pi} \, \chi_{A,B}(\omega) = \frac{i}{\hbar} \langle [\hat{A}, \hat{B}] \rangle \qquad (11.187)$$

which is given by the expectation value of the equal-time commutator of $\hat{A}$
and $\hat{B}$.

Higher-order sum rules can be obtained from multiple derivatives wrt t
of the expression for the real-time response function. The first derivative
yields the equation of motion

$$i\hbar \, \frac{\partial}{\partial t} \, \chi_{A,B}(t - t') = -\delta(t - t') \langle [\hat{A}(t), \hat{B}(t')] \rangle$$

$$+ \frac{i}{\hbar} \, \Theta(t - t') \langle [[\hat{A}(t), \hat{H}_0(t)], \hat{B}(t')] \rangle \qquad (11.188)$$

The Fourier transform of $\chi_{A,B}(t)$ is defined as

$$\chi_{A,B}(\omega) = \int_{-\infty}^{\infty} dt \, \exp[-i\omega t] \, \chi_{A,B}(t) \qquad (11.189)$$

and the inverse Fourier transform is given by

$$\chi_{A,B}(t) = \int_{-\infty}^{\infty} \frac{d\omega}{2\pi} \, \exp[+i\omega t] \, \chi_{A,B}(\omega) \qquad (11.190)$$

Hence, the equation of motion can be written as

$$\int_{-\infty}^{\infty} \frac{d\omega}{2\pi} \, \hbar\omega \, \exp[+i\omega t] \, \chi_{A,B}(\omega) = \delta(t) \langle [\hat{A}(t), \hat{B}(0)] \rangle$$

$$- \frac{i}{\hbar} \, \Theta(t) \langle [[\hat{A}(t), \hat{H}_0(t)], \hat{B}(0)] \rangle \qquad (11.191)$$

Taking the limit $t \to 0$, yields

$$\int_{-\infty}^{\infty} \frac{d\omega}{2\pi} \, \hbar\omega \chi_{A,B}(\omega) = -\frac{i}{\hbar} \, \langle [[\hat{A}, \hat{H}_0], \hat{B}] \rangle \qquad (11.192)$$

The equal time commutators can be easily evaluated leading to a condition
on the first moment of the susceptibility. By taking further derivatives, one

can obtain sum rules for the higher-order moments of $\chi_{A,B}(\omega)$ in terms of higher-order nested equal-time commutators of $\hat{H}_0$. However, the expectation values of the various commutators are not guaranteed to converge.

11.7   The Kubo Conductivity Formula

The electrical conductivity can be expressed in term of the (causal) paramagnetic current-current response function $R^{i,j}(\underline{q},\omega)$. The non-interacting Hamiltonian in the presence of a vector potential is written as

$$\hat{\mathcal{H}}_0(\underline{r},t) = \hat{\psi}^\dagger(\underline{r},t)\left[\frac{1}{2m}\left(\hat{\underline{p}} - \frac{q}{c}\underline{A}(\underline{r},t)\right)^2 + q\phi(\underline{r},t)\right]\hat{\psi}(\underline{r},t) \qquad (11.193)$$

On expanding in powers of the vector potential, one has

$$\hat{\mathcal{H}}_0(\underline{r},t) = \hat{\psi}^\dagger(\underline{r},t)\left[\frac{1}{2m}\left[\hat{\underline{p}}^2 - \frac{q}{c}\hat{\underline{p}}\cdot\underline{A}(\underline{r},t) - \frac{q}{c}\underline{A}(\underline{r},t)\cdot\hat{\underline{p}} + \left(\frac{q}{c}\right)^2\underline{A}^2(\underline{r},t)\right]\right.$$
$$\left. + q\phi(\underline{r},t)\right]\hat{\psi}(\underline{r},t) \qquad (11.194)$$

In the first term of the interaction with the vector potential, the momentum operator acts on the field operator $\hat{\psi}(\underline{r},t)$ and the vector potential. An equivalent Hamiltonian density operator can be found from the Hamiltonian by integrating this term by parts, so that the term changes sign and the momentum operator only acts on the Hermitean conjugate of the field operator $\hat{\psi}^\dagger(\underline{r},t)$. Since the interaction of a charged particle with the electromagnetic field can be expressed as

$$\mathcal{H}_{int}(\underline{r},t) = -\frac{1}{c}\,\hat{\underline{j}}(\underline{r},t)\cdot\underline{A}(\underline{r},t) \qquad (11.195)$$

the current density operator can be defined as the functional derivative

$$\hat{\underline{j}}(r,t) = -c\left(\frac{\delta H_0}{\delta\underline{A}(r,t)}\right) \qquad (11.196)$$

The current density can be decomposed as

$$\hat{\underline{j}}(\underline{r},t) = \hat{\underline{j}}_p(\underline{r},t) + \hat{\underline{j}}_d(\underline{r},t) \qquad (11.197)$$

where the paramagnetic current density is defined by

$$\hat{\underline{j}}_p(r,t) = \frac{q}{2m}\,[\hat{\psi}^\dagger(\underline{r},t)\,\hat{\underline{p}}\,\hat{\psi}(\underline{r},t) - (\hat{\underline{p}}\,\hat{\psi}^\dagger(\underline{r},t))\,\hat{\psi}(\underline{r},t)] \qquad (11.198)$$

and the diamagnetic current density is defined as

$$\hat{\underline{j}}_d(\underline{r},t) = -\left(\frac{q^2}{mc}\right)\hat{\psi}^\dagger(\underline{r},t)\,\underline{A}(\underline{r},t)\,\hat{\psi}(\underline{r},t) \qquad (11.199)$$

The i-th component of the Fourier transformed paramagnetic current density can be expressed as

$$\hat{j}_p^i(\underline{q}) = \frac{q\hbar}{2m} \sum_{\underline{k},\sigma} (2k^i + q^i) \, c_{\underline{k}+\underline{q},\sigma}^\dagger c_{\underline{k},\sigma} \tag{11.200}$$

where

$$q = -|e| \tag{11.201}$$

is the charge of the electron. The paramagnetic current can be determined from linear response theory. The diagonal component of the $T = 0$ paramagnetic current-current correlation function reduces to

$$R^{i,i}(\underline{q},\omega) = -\sum_n \left[\frac{|\langle \Phi_0| \hat{j}_p^i(\underline{q}) |\Phi_n \rangle|^2}{\hbar\omega - E_n + E_0 - i\eta} - \frac{|\langle \Phi_n| \hat{j}_p^i(\underline{q}) |\Phi_0 \rangle|^2}{\hbar\omega + E_n - E_0 - i\eta} \right] \tag{11.202}$$

Since the total current is given by

$$j^i(\underline{q},\omega) = \sum_j \left[\frac{1}{c} R^{i,j}(\underline{q},\omega) - \left(\frac{\rho e^2}{mc} \right) \delta^{i,j} \delta(\underline{q}) \right] A^j(\underline{q},\omega) \tag{11.203}$$

when the vector potential is expressed in term of the electric field, the conductivity tensor is identified as

$$\sigma^{i,j}(\underline{q},\omega) = \frac{\left[\frac{1}{c} R^{i,j}(\underline{q},\omega) - \left(\frac{\rho e^2}{mc} \right) \delta^{i,j} \delta(\underline{q}) \right]}{i\omega/c} \tag{11.204}$$

For a normal metal the two terms in the numerator cancel in the limit $\omega \to 0$ but the conductivity is given by the terms of the numerator which are first-order in ω.

11.7.1 *Example: Superconductivity and the Meissner Effect*

For a superconductor, the total current is given by

$$j^i(\underline{q},\omega) = \sum_j \left[\frac{1}{c} R^{i,j}(\underline{q},\omega) - \left(\frac{\rho e^2}{mc} \right) \delta^{i,j} \delta(\underline{q}) \right] A^j(\underline{q},\omega) \tag{11.205}$$

However, the paramagnetic current is zero, so a diamagnetic current flows in response to an applied static vector potential. The diamagnetic current is given by the London equation

$$\underline{j}(\underline{r}) = -\left(\frac{\rho e^2}{mc} \right) \underline{A}(\underline{r}) \tag{11.206}$$

where ρ is the electron density. This result is gauge dependent, and only holds true in the London gauge $\nabla \cdot \underline{A} = 0$. The gauge condition follows from the conservation of charge as expressed by the continuity equation

$$\frac{\partial \rho}{\partial t} + \nabla \cdot \underline{j} = 0 \tag{11.207}$$

For a static vector potential, the charge density $\rho(\underline{r}, t)$ is expected to be time independent so

$$\frac{\partial \rho}{\partial t} = 0 \tag{11.208}$$

Therefore, the current density must satisfy the condition

$$\nabla \cdot \underline{j}(\underline{r}) = 0 \tag{11.209}$$

The London equation then demands that the vector potential must satisfy the London gauge condition

$$\nabla \cdot \underline{A}(\underline{r}) = 0 \tag{11.210}$$

It is possible for the superconducting gap to be dependent on the vector potential. However, rotational invariance demands that the vector potential be combined in a scalar product. The only vector that is available is $\underline{q}$ but, with the choice of the London gauge, one has

$$\underline{q} \cdot \underline{A}(\underline{q}) = 0 \tag{11.211}$$

Hence, the choice of the London gauge guarantees the validity of our results. The diamagnetic current that flows shields the superconductor from the applied magnetic field. A superconductor is to be contrasted to a normal metal in which the diamagnetic and paramagnetic currents cancel in the dc limit, so no current flows in response to a static vector potential.

The Meissner effect follows from combining Maxwell's equation

$$\nabla \wedge \underline{B} = \frac{4\pi}{c} \underline{j} \tag{11.212}$$

with the London equation

$$\underline{j}(\underline{r}) = -\left(\frac{\rho e^2}{mc}\right) \underline{A}(\underline{r}) \tag{11.213}$$

yielding

$$\nabla \wedge \underline{B} = -\left(\frac{4\pi \rho e^2}{mc^2}\right) \underline{A} \tag{11.214}$$

On taking the curl of the equation, one finds

$$\nabla \wedge (\nabla \wedge \underline{B}) = -\left(\frac{4\pi \rho e^2}{mc^2}\right) \underline{B} \tag{11.215}$$

or

$$\nabla^2 \underline{B} = \left(\frac{4\pi \rho e^2}{mc^2}\right) \underline{B} \tag{11.216}$$

The above equation defines the London penetration depth λ_L which is given by

$$\frac{1}{\lambda_L^2} = \left(\frac{4\pi \rho e^2}{mc^2}\right) \tag{11.217}$$

The solution of the partial differential equation for $\underline{B}$ shows that the penetration depth λ_L is the distance inside the surface of a superconductor over which an external magnetic field is screened. The Meissner effect describes the exclusion of magnetic fields from the bulk of a superconductor.

11.7.2 *Example: The Drude Conductivity*

Conduction in metals is frequently described by the Drude conductivity which can be derived from the Kubo formula. The Kubo formula for the conductivity tensor is expressed as

$$\sigma^{i,j}(\underline{q},\omega) = \frac{\left[\frac{1}{c} R^{i,j}(\underline{q},\omega) - \left(\frac{\rho e^2}{mc}\right)\delta^{i,j}\delta(\underline{q})\right]}{i\omega/c} \tag{11.218}$$

where

$$R^{i,j}(\underline{q},\omega) = -\sum_n \left[\frac{\langle\Phi_0|\hat{j}_p^j(\underline{q})|\Phi_n\rangle\langle\Phi_n|\hat{j}_p^i(\underline{q})|\Phi_0\rangle}{\hbar\omega - E_n + E_0 - i\eta}\right.$$
$$\left. - \frac{\langle\Phi_0|\hat{j}_p^i(\underline{q})|\Phi_n\rangle\langle\Phi_n|\hat{j}_p^j(\underline{q})|\Phi_0\rangle}{\hbar\omega + E_n - E_0 - i\eta}\right] \tag{11.219}$$

The first term of the conductivity tensor can be rewritten by using the identity

$$\frac{1}{\omega}\frac{1}{\hbar\omega - x} = \frac{1}{x}\left(\frac{\hbar}{\hbar\omega - x} - \frac{1}{\omega}\right) \tag{11.220}$$

The term proportional to $\frac{1}{\omega}$ in the conductivity can be combined with the diamagnetic term of the response, yielding an apparently infrared divergent term in the conductivity which is represented by

$$\frac{1}{i\omega}\left[\sum_n \left(\frac{\langle\Phi_0|\hat{j}_p^j(\underline{q})|\Phi_n\rangle\langle\Phi_n|\hat{j}_p^i(\underline{q})|\Phi_0\rangle}{(E_n - E_0)}\right.\right.$$
$$\left.\left. + \frac{\langle\Phi_0|\hat{j}_p^i(\underline{q})|\Phi_n\rangle\langle\Phi_n|\hat{j}_p^j(\underline{q})|\Phi_0\rangle}{(E_n - E_0)}\right) - \left(\frac{\rho e^2}{m}\right)\delta^{i,j}\delta(\underline{q})\right] \tag{11.221}$$

However, since terms proportional to $\frac{1}{\omega}$ cancel in the limit $q \to 0$, the infrared divergence does not exist. The cancellation can be demonstrated by writing the diamagnetic term as

$$
\begin{aligned}
\frac{\rho e^2}{m} \delta^{i,j} &= \frac{e^2}{im\hbar} \langle \Phi_0 | [\hat{r}_i, \hat{p}_j] | \Phi_0 \rangle \\
&= \frac{e^2}{im\hbar} \sum_n \langle \Phi_0 | \hat{r}_i | \Phi_n \rangle \langle \Phi_n | \hat{p}_j | \Phi_0 \rangle \\
&\quad - \frac{e^2}{im\hbar} \sum_n \langle \Phi_0 | \hat{p}_j | \Phi_n \rangle \langle \Phi_n | \hat{r}_i | \Phi_0 \rangle
\end{aligned}
\tag{11.222}
$$

Since the unperturbed Hamiltonian and $\hat{r}_j$ satisfy the commutation relation

$$
[\hat{r}_j, \hat{H}_0] = \frac{i\hbar}{m} \hat{p}_j
\tag{11.223}
$$

one may evaluate the matrix elements of the commutation relation between the eigenstates of $\hat{H}_0$

$$
(E_n - E_0) \langle \Phi_0 | \hat{r}_j | \Phi_n \rangle = \frac{i\hbar}{m} \langle \Phi_0 | \hat{p}_j | \Phi_n \rangle
\tag{11.224}
$$

which yields an expression for $\langle \Phi_0 | \hat{r}_j | \Phi_n \rangle$. On substituting the expression in the diamagnetic term, one finds that it has exactly the same form as the $\omega \to 0$ limit of $R^{i,j}(q, \omega)$

$$
\begin{aligned}
\frac{\rho e^2}{m} \delta^{i,j} &= \frac{e^2}{m^2} \sum_n \frac{\langle \Phi_0 | \hat{p}_i | \Phi_n \rangle \langle \Phi_n | \hat{p}_j | \Phi_0 \rangle}{(E_n - E_0)} \\
&\quad + \frac{e^2}{m^2} \sum_n \frac{\langle \Phi_0 | \hat{p}_j | \Phi_n \rangle \langle \Phi_n | \hat{p}_i | \Phi_0 \rangle}{(E_n - E_0)}
\end{aligned}
\tag{11.225}
$$

and, therefore, the pair of terms cancel. A potential infrared divergence in the conductivity has been eliminated by the use of a sum rule. Hence, the conductivity can be expressed as

$$
\begin{aligned}
\sigma^{i,j}(q, \omega) = i\hbar \frac{e^2}{m^2} \sum_n &\left[\frac{\langle \Phi_0 | \hat{p}^j(q) | \Phi_n \rangle \langle \Phi_n | \hat{p}^i(q) | \Phi_0 \rangle}{(E_n - E_0)(\hbar\omega - E_n + E_0 - i\eta)} \right. \\
&\left. + \frac{\langle \Phi_0 | \hat{p}^i(q) | \Phi_n \rangle \langle \Phi_n | \hat{p}^j(q) | \Phi_0 \rangle}{(E_n - E_0)(\hbar\omega + E_n - E_0 - i\eta)} \right]
\end{aligned}
\tag{11.226}
$$

When the current is expressed in second quantized form in a basis of single-particle states ϕ_n with energies ϵ_n

$$\hat{p}^i(\underline{q}) = \sum_{n,m} \langle \phi_m | p^i(\underline{q}) | \phi_n \rangle \hat{c}_m^\dagger \hat{c}_n \tag{11.227}$$

one finds that the conductivity reduces to

$$\sigma^{i,j}(\underline{q}, \omega) = -i\hbar \frac{e^2}{m^2} \sum_{n,m} \left(\frac{f_n - f_m}{\epsilon_m - \epsilon_n} \right) \tag{11.228}$$

$$\times \left[\frac{\langle \phi_n | \hat{p}^j(\underline{q}) | \phi_m \rangle \langle \phi_m | \hat{p}^i(\underline{q}) | \phi_n \rangle}{(\hbar\omega - \epsilon_m + \epsilon_n - i\eta)} \right] \tag{11.229}$$

where f_n are Fermi-functions. The small imaginary parts in the denominators are responsible for producing a real part of the conductivity. This is expected since the conductivity describes the rate of dissipation of EM energy in a metal. The first factor in the expression for the conductivity is only non-zero if the pair of energies ϵ_n and ϵ_m are either on opposite sides of the Fermi-energy or are close to the Fermi-energy. The current matrix elements only connect states with momenta $\underline{k}$ and $\underline{k} + \underline{q}$ which in the limit $q \to 0$ have similar energies. Thus, the effect of the matrix elements together with the restrictions imposed by the first factor, the conductivity only involves states close to the Fermi-energy and, thus, in a metal one may use the approximation

$$\left(\frac{f_n - f_m}{\epsilon_m - \epsilon_n} \right) \approx -\left(\frac{\partial f_n}{\partial \epsilon_n} \right) \delta_{m,n} \tag{11.230}$$

Therefore, the conductivity can be written as

$$\sigma^{i,j}(0, \omega) = \frac{\hbar(\eta - i\hbar\omega)}{(\hbar^2\omega^2 + \eta^2)} \frac{e^2}{m^2} \sum_n \left(-\frac{\partial f(\epsilon_n)}{\partial \epsilon_n} \right)$$

$$\times \langle \phi_n | \hat{p}^j | \phi_n \rangle \langle \phi_n | \hat{p}^i | \phi_n \rangle \tag{11.231}$$

The conductivity is diagonal for an isotropic system. We shall evaluate the conductivity using an isotropic dispersion relation of the form

$$\epsilon(p) = \frac{1}{2m} \sum_{i=1}^d p_i^2 \tag{11.232}$$

Furthermore, we shall replace η by $\frac{\hbar}{\tau}$ which corresponds to the sum of the imaginary parts of ϵ_n and ϵ_m that represents their scattering rates. With these assumptions, the conductivity reduces to the Drude formula

$$\sigma^{i,j}(0,\omega) = \left(\frac{1 - i\omega\tau}{1 + \omega^2\tau^2}\right)\frac{\rho e^2\tau}{m}\,\delta^{i,j} \tag{11.233}$$

which has the d.c. limit

$$\sigma^{i,j}(0,0) = \frac{\rho e^2\tau}{m}\,\delta^{i,j} \tag{11.234}$$

which provides a simple and adequate description for most metals.

Example: Quantized Hall Response of Topological Insulators

We shall examine the Hall conductivity response for a simple model of the surface states of a three-dimensional Topological Insulator. A model of topological surface states can be obtained by generalization of the results of Jackiw and Rebbi [107]. A one-dimensional Dirac equation with spatially-dependent mass can be written in a two by two block diagonal form

$$i\hbar\frac{\partial}{\partial t}\begin{pmatrix}\phi_A \\ \phi_B\end{pmatrix} = \left[-i\hbar c\begin{pmatrix}0 & \sigma^z \\ \sigma^z & 0\end{pmatrix}\frac{\partial}{\partial z} + m(z)\,c^2\begin{pmatrix}0 & \sigma^z \\ \sigma^z & 0\end{pmatrix}\right]\begin{pmatrix}\phi_A \\ \phi_B\end{pmatrix}$$

where ϕ_A and ϕ_B are two-component spinors. Since this model is fully-relativistic, it includes spin-orbit coupling. As shown by Jackiw and Rebbi, if one imposes a requirement that the spinors are not independent but are related by

$$\phi_A = \pm i\sigma^z\,\phi_B \tag{11.235}$$

where the sign has yet to be chosen, the coupled equations reduce to

$$i\hbar\frac{\partial}{\partial t}\phi_A = \left[\pm\hbar c\frac{\partial}{\partial z} + m(z)\,c^2\right]\phi_A$$

$$i\hbar\frac{\partial}{\partial t}\phi_B = -\left[\pm\hbar c\frac{\partial}{\partial z} + m(z)\,c^2\right]\phi_B \tag{11.236}$$

Since the spinors ϕ_A are not independent but are related by a rotation, these equations have a stationary solution with $E = 0$ if

$$\left[\pm\hbar\frac{\partial}{\partial z} + m(z)\,c\right]\phi_A = 0 \tag{11.237}$$

The first-order differential equation has a solution given by

$$\phi_A(z) = A\,\exp\left[\mp\int_0^z dz'\,\frac{m(z')\,c}{\hbar}\right]\chi \tag{11.238}$$

where χ is an arbitrary constant two-component spinor. The solution is normalizable and describes a localized state if $m(z)$ changes sign so that exponential decays on both sides of the point $z = 0$. That is

$$\lim_{z \to \infty} \pm \int_0^z dz' \, \frac{m(z') \, c}{\hbar} \to \infty$$

$$\lim_{z \to -\infty} \pm \int_0^z dz' \, \frac{m(z') \, c}{\hbar} \to \infty \qquad (11.239)$$

The normalizability of the bound state wave function fixes the choice of the sign $\pm$. To be sure, if for $z > 0$, $m(z) > 0$ and $m(z) < 0$ for $z < 0$, one must chose the upper sign, while if the inequalities are reversed one must chose the lower sign. To summarize the findings of Jackiw and Rebbi, in a modified Dirac equation, the point $z = 0$ at which the mass switches sign is associated with localized states with zero energy. The change the sign of the mass in the Dirac equation results in interchanging the parity of the eigenstates at the time-reversal invariant point $k = 0$ of the insulator.

A simple extension to three-dimensions leads to the in-gap $E = 0$ state of the one-dimensional problem to become dispersive due to the surface states being extended parallel to the $z = 0$ surface. The appropriate three-dimensional Dirac equation, including the constraint is

$$i\hbar \frac{\partial}{\partial t} \begin{pmatrix} \phi_A \\ \phi_B \end{pmatrix} = \left[-i\hbar c \begin{pmatrix} 0 & \underline{\sigma} \cdot \underline{\nabla} \\ \underline{\sigma} \cdot \underline{\nabla} & 0 \end{pmatrix} + m(z) \, c^2 \begin{pmatrix} 0 & \sigma^z \\ \sigma^z & 0 \end{pmatrix} \right] \begin{pmatrix} \phi_A \\ \mp i\sigma^z \phi_A \end{pmatrix}$$

which with the ansatz

$$\phi_A(z) = A \, \exp \left[\mp \int_0^z dz' \, \frac{m(z') \, c}{\hbar} \right] \, \exp[i\underline{k}_\parallel \cdot \underline{r}] \, \chi_A \qquad (11.240)$$

leads to a Rashba energy eigenvalue equation

$$E \, \chi_A = \mp \, \hbar c (\underline{k}_\parallel \wedge \underline{\sigma}) \cdot \hat{e}_z \, \chi_A \qquad (11.241)$$

Since ϕ_B is completely determined by ϕ_A, all physical properties can be obtained directly from ϕ_A. The dispersion relation describes a gapless Weyl cone

$$E = \tau \hbar c \, |\underline{k}_\parallel| \qquad (11.242)$$

where $\tau = \pm 1$. The gapless surface states are described by the spinors

$$\chi_A = \frac{1}{\sqrt{2}} \begin{pmatrix} \tau \, e^{-i(\frac{\varphi_k}{2} + \frac{\pi}{4})} \\ e^{+i(\frac{\varphi_k}{2} + \frac{\pi}{4})} \end{pmatrix} \qquad (11.243)$$

where φ_k signifies the direction of the in-plane momentum $\underline{k}_\parallel = |\underline{k}_\parallel| \, (\cos \varphi_k, \sin \varphi_k, 0)$. Therefore, the states are non-degenerate and exhibit spin-momentum locking. The spins point in the $x - y$ plane and have

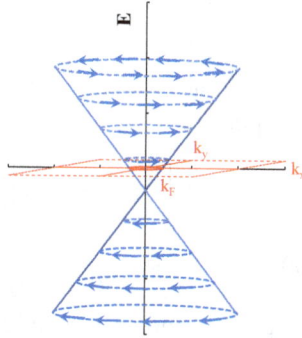

Fig. 11.9 The dispersion relation in two-dimensional momentum space (k_x, k_y) for a semi-infinite slab forms a Weyl cone. The directions of the states spins are indicated by the blue arrows. The Fermi-energy, E_F, for the surface states is denoted by the red plane.

the azimuthal angle $\varphi_k + \tau \frac{\pi}{2}$, so that they are perpendicular to the direction of the momentum (see Fig. 11.9). The exception to the rule is found at the singularity of the coordinate system at the point $k_\parallel = 0$, i.e. at the vertex of the cone, where the states become degenerate. It is this singularity that is the source of a non-trivial Berry phase [108].

The Berry phase of the surface states is manifested in the Landau level quantization by applied magnetic fields and a quantized Hall conductance. In the presence of a magnetic field, the momenta are replaced according to the minimum coupling scheme

$$\hbar k^\mu \rightarrow \hbar k^\mu + \frac{e}{c} A^\mu \tag{11.244}$$

For a uniform magnetic field B applied in the z-direction, the vector potential can be chosen to be

$$\underline{A} = xB\hat{e}_y \tag{11.245}$$

The effective Hamiltonian for the surface states takes the form

$$\hat{H} = \hbar c \left[\sigma^x \left(\hat{k}^y + \frac{e}{\hbar c} Bx \right) - \sigma^y \hat{k}^x \right] \tag{11.246}$$

and the energy-eigenstates can be expressed as

$$\Psi = \frac{1}{\sqrt{L_y}} \exp[ik^y y] \, \Phi(x) \tag{11.247}$$

where the spinor $\Phi(x)$ satisfies

$$E\Phi(x) = \hbar c \left[\sigma^x \left(k^y + \frac{e}{\hbar c} Bx \right) - \sigma^y \hat{k}^x \right] \Phi(x) \tag{11.248}$$

On introducing a dimensionless shifted Harmonic Oscillator variable s via

$$s = \sqrt{\frac{eB}{\hbar c}} \left(x + \frac{\hbar k^y c}{eB} \right) \tag{11.249}$$

the eigenvalue equations reduce to

$$E^2 \phi^+(x) = \hbar ceB \left[s^2 - \frac{\partial^2}{\partial s^2} + 1 \right] \phi^+(x)$$

$$E^2 \phi^-(x) = \hbar ceB \left[s^2 - \frac{\partial^2}{\partial s^2} - 1 \right] \phi^-(x) \tag{11.250}$$

which are recognized as the Harmonic Oscillator eigenvalue equations. Hence, the energy eigenfunctions are given by

$$\Phi_{\tau,n}(x) = \frac{1}{\sqrt{2}} \begin{pmatrix} \tau \phi_{n-1}(s) \\ \phi_n(s) \end{pmatrix} \tag{11.251}$$

where $\tau = \pm 1$ and $\phi_n(z)$ are the normalized wave functions of the shifted Harmonic Oscillator. The eigenvalues are

$$E_{\tau,n} = \tau \sqrt{2\hbar ceBn} \tag{11.252}$$

which does not include the usual factor of $n + \frac{1}{2}$ since the Berry phase of π produces a cancellation. The density of states consists of discrete levels as seen in Fig. 11.10. A direct observation of the Landau levels in the surface density of states is expected to be challenging for large n, since the energy separation between successive Landau levels is proportional to $n^{-\frac{1}{2}}$, which may be smaller than the broadening of the individual levels. The solutions are localized but are highly degenerate. The degeneracy of the state is

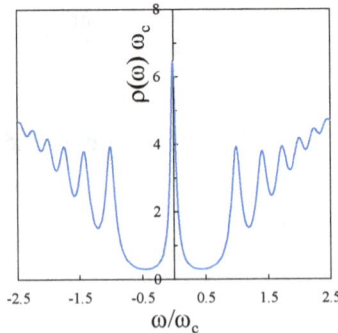

Fig. 11.10   A plot of the density of states $\rho(\omega)$ of the spin-orbit coupled Landau levels. The delta functions have been given a finite width. Note that for large energies, the density of states recovers the linear $|\omega|$ dependence expected for zero field.

found by counting the number of k^y states for which the eigenfunctions are centered within the an interval $L_x > x > 0$, such that the states are confined to the surface. The degeneracy D of the Landau levels for a surface of area $L_x L_y$ is found as

$$D = L_x L_y \left(\frac{eB}{2\pi\hbar c} \right)$$

$$= \frac{\Phi}{\Phi_0} \tag{11.253}$$

where $\Phi = B L_x L_y$ is the magnetic flux threading the surface and Φ_0 is the value of the fundamental flux quantum

$$\Phi_0 = \left(\frac{2\pi\hbar c}{e} \right) \tag{11.254}$$

To calculate the Hall conductivity, one needs to know the expression for the current density. The continuity equation for the probability density of the two-dimensional Rashba-Weyl equation is expressed in terms of the density, ρ, and current density, j.

$$\frac{\partial \rho}{\partial t} + \underline{\nabla} \cdot \underline{j} = 0 \tag{11.255}$$

where the current density ρ and probability current j are given by

$$\rho = \Psi^\dagger \Psi$$

$$\underline{j} = c\Psi^\dagger \left(\hat{e}_y \sigma^x - \hat{e}_x \sigma^y \right) \Psi \tag{11.256}$$

The thermally averaged current density j is given by

$$\langle \underline{j} \rangle = \text{Trace } \hat{\rho}\hat{\underline{j}} \tag{11.257}$$

where $\hat{\rho}(t)$ is the density operator. The current density, evaluated with the equilibrium density operator, is zero due to the temporal homogeneity of the equilibrium state. The system is perturbed by the introduction of a static electric field aligned along the x-axis, which is represented by the electrostatic interaction

$$\hat{H}_{int} = eEx \tag{11.258}$$

which has the effect of perturbing the density operator by $\delta\hat{\rho}$. In the presence of the perturbation k_y remains a good quantum number. To linear-order in the perturbing interaction, the current density is found to be

$$\langle \underline{j} \rangle = \frac{D}{L_x L_y} \sum_{n,\tau \neq m,\tau'} \left(\frac{f_{\tau,n} - f_{\tau',m}}{E_{\tau,n} - E_{\tau',m}} \right)$$

$$\times \langle \Phi_{\tau,n} | \underline{\hat{j}} | \Phi_{\tau',m} \rangle \langle \Phi_{\tau',m} | \hat{H}_{int} | \Phi_{\tau,n} \rangle \tag{11.259}$$

where $f_{n,\tau}$ are the Fermi-Dirac distribution for the occupation of the energy eigenstate $\Phi_{n,\tau}$. The above result corresponds to that found with linear response theory for a static perturbation.

The Hall conductivity is given by the y, x components of the tensor $\sigma^{\mu,\nu}$

$$\sigma^{y,x} = -e^2 c \frac{D}{L_x L_y} \sum_{\tau, n \neq \tau', m} \left(\frac{f_{\tau,n} - f_{\tau',m}}{E_{\tau,n} - E_{\tau',m}} \right)$$

$$\times \langle \Phi_{\tau,n} | \sigma^x | \Phi_{\tau',m} \rangle \langle \Phi_{\tau',m} | x | \Phi_{\tau,n} \rangle \qquad (11.260)$$

The matrix elements are evaluated as

$$\langle \Phi_{\tau,n} | \sigma^x | \Phi_{\tau',m} \rangle = \frac{1}{2} \left[\delta_{m,n-1} \, sign(\tau) + \delta_{m,n+1} \, sign(\tau') \right] \qquad (11.261)$$

which gives rise to the selection rule $n = m \pm 1$. The matrix elements allows transitions between segments of the Weyl cone with the same signs of the energy as well as between segments with opposite signs of the energy. The matrix elements of x are evaluated as

$$\langle \Phi_{\tau',m} | x | \Phi_{\tau,n} \rangle = \frac{1}{2} \sqrt{\frac{\hbar c}{2eB}} \left[\delta_{m,n+1} \sqrt{n+1} + \delta_{m,n-1} \sqrt{n} \right.$$

$$\left. + sign(\tau, \tau')(\delta_{m,n+1} \sqrt{n} + \delta_{m,n-1} \sqrt{n-1}) \right]$$

$$(11.262)$$

The Hall conductivity is evaluated as the sum of two terms, a term where the indices τ, τ' have the same signs and a term where the signs are different

$$\sigma^{y,x} = \left(\frac{e^2}{8\pi\hbar} \right) \sum_{\tau,n} (f_{\tau,n} - f_{\tau,n+1}) \left[\frac{\sqrt{n+1} + \sqrt{n}}{\sqrt{n+1} - \sqrt{n}} \right]$$

$$+ \left(\frac{e^2}{8\pi\hbar} \right) \sum_{\tau,n} (f_{\tau,n} - f_{-\tau,n+1}) \left[\frac{\sqrt{n+1} - \sqrt{n}}{\sqrt{n+1} + \sqrt{n}} \right] \qquad (11.263)$$

The first term represents excitation between an initial and final state that are either both in the upper part of the Weyl cone or both are in the lower part of the Weyl cone. The second term represents transitions between the upper and lower parts of the Weyl cone. The chemical potential will be considered to be constant. This is justified by the assumption that in-gap impurity states in the bulk acts as an electron reservoir. We shall assume that the Fermi-energy E_F is located in the upper-part of the Weyl cone, as seen in Fig. 11.9. The Fermi-energy defines the Fermi-wave vector k_F via $E_F = \hbar c k_F$. At $T = 0$, the Fermi-functions reduce to

$$f_{+,n} = \Theta\left(\frac{\pi k_F^2 \Phi_0}{4\pi^2 B} - n \right) \qquad (11.264)$$

where $\Theta(x)$ is the Heaviside step function and

$$f_{-,n} = 1 \qquad (11.265)$$

The two terms can be combined to yield

$$\sigma^{y,x} = \left(\frac{e^2}{8\pi\hbar}\right) \sum_n (f_{+,n} - f_{+,n+1})$$

$$\times \left[\frac{\sqrt{n+1} + \sqrt{n}}{\sqrt{n+1} - \sqrt{n}} + \frac{\sqrt{n+1} - \sqrt{n}}{\sqrt{n+1} + \sqrt{n}}\right]$$

$$= \left(\frac{e^2}{2\pi\hbar}\right) \sum_n (f_{+,n} - f_{+,n+1}) \left(n + \frac{1}{2}\right) \qquad (11.266)$$

The zero temperature limit of the Hall conductivity of the surface state is quantized in multiples of the conductance quantum $\frac{e^2}{2\pi\hbar}$. The Hall conductivity as a function of B^{-1} exhibits a regular series of steps of height $\frac{e^2}{2\pi\hbar}$. Although $\sigma^{x,y}$ diverges in the limit $B \to 0$, it vanishes at $B = 0$ since $\sigma^{x,y}$ is antisymmetric under the transformation $B \to -B$. The vanishing of $\sigma^{x,y}$ at $B = 0$ is a consequence of time-reversal invariance. The high-field intercept of the Hall conductivity is $\frac{1}{2}$ in contrast to the result of Thouless *et al.* [109]. The factor of $\frac{1}{2}$ can be traced to the relativistic nature of the zero-field dispersion relation.

11.8 Time-Dependent Correlation Functions

There are two main types of time-dependent fluctuations found in a system. These are, (i) fluctuations induced by external perturbations and (ii) spontaneous fluctuations occurring in the system. The relaxation of fluctuations that are caused by the switching off of an external perturbation must be identical to the relaxation of a spontaneous fluctuation in the system which temporarily produces a non-equilibrium state. This is Onsager's "*Regression Hypothesis*".

The fluctuations of the system can be quantified by their correlation functions. The fluctuation of an observable $\Delta\hat{A}$ measures the deviation of $\hat{A}(t)$ from its equilibrium expectation value

$$\langle A \rangle = \text{Trace } \hat{\rho}\hat{A}(t) \qquad (11.267)$$

The equilibrium value A is time-independent since $\hat{U}_0$ commutes with $\hat{\rho}$. This is a manifestation of the invariance of the thermal equilibrium state under time translations. Specifically the deviation operator is given

$$\Delta\hat{A}(t) = \hat{A}(t) - \langle A \rangle \qquad (11.268)$$

The mean-squared fluctuation gives a measurement of the fluctuation at time t

$$\langle \Delta \hat{A}(t)^2 \rangle = \langle \hat{A}(t)^2 \rangle - \langle A \rangle^2 \tag{11.269}$$

The mean-squared fluctuation correlates ΔA to itself at the same time t. When evaluated in an equilibrium state, the mean-squared fluctuation is also found to be independent of time.

The time-dependence of the fluctuations can be observed by examining the correlation between $\Delta \hat{A}$ at time t with its value at a time τ later.

$$S_{A,A}(\tau) = \frac{1}{2} \langle [\Delta \hat{A}(t+\tau) \, \Delta \hat{A}(t) + \Delta \hat{A}(t) \, \Delta \hat{A}(t+\tau)] \rangle \tag{11.270}$$

which can be generalized to describe correlations between the fluctuation of A at time $t + \tau$ to the fluctuations of B at an earlier time t.

$$S_{A,B}(\tau) = \frac{1}{2} \langle [\Delta \hat{A}(t+\tau) \, \Delta \hat{B}(t) + \Delta \hat{B}(t) \, \Delta \hat{A}(t+\tau)] \rangle \tag{11.271}$$

Note that the definition of the correlation function introduces a symmetry under the interchange of A and B

$$S_{A,B}(\tau) = S_{B,A}(-\tau) \tag{11.272}$$

If A and B are independent variables, then

$$S_{A,B}(\tau) = \langle \Delta \hat{A}(t+\tau) \rangle \langle \Delta \hat{B}(t) \rangle$$
$$= 0 \tag{11.273}$$

The Fourier transform of $S_{A,B}(\tau)$ is defined as

$$S_{A,B}(\omega) = \int_{-\infty}^{\infty} d\tau \, \exp[-i\omega\tau] \, S_{A,B}(\tau) \tag{11.274}$$

and the inverse Fourier transform is

$$S_{A,B}(\tau) = \int_{-\infty}^{\infty} \frac{d\omega}{2\pi} \, \exp[+i\omega\tau] \, S_{A,B}(\omega) \tag{11.275}$$

Properties of $S_{A,B}$

On setting $\tau = 0$ in the inverse Fourier transform, one finds

$$S_{A,B}(\tau = 0) = \int_{-\infty}^{\infty} \frac{d\omega}{2\pi} \, S_{A,B}(\omega)$$
$$= \frac{1}{2} \langle [\Delta \hat{A}(t) \, \Delta \hat{B}(t) + \Delta \hat{B}(t) \, \Delta \hat{A}(t)] \rangle \tag{11.276}$$

which is the expectation value of an anti-commutator.

From the definition of the correlation function, one has

$$S_{A,B}(\tau) = S_{B,A}(-\tau) \tag{11.277}$$

On substituting the above relation into the Fourier transform, one finds that

$$S_{A,B}(\omega) = S_{B,A}(-\omega) \tag{11.278}$$

On taking the complex conjugate of $S_{A,B}(\tau)$ the order of the Hermitean operators are reversed, so one finds

$$S_{A,B}(\tau) = S_{A,B}^*(\tau) \tag{11.279}$$

since the correlation function is symmetrized. On taking the Fourier transform of $S_{A,B}(\tau)$, taking the complex conjugate and then noting the reality of $S_{A,B}(\tau)$, one finds

$$S_{A,B}(\omega) = S_{A,B}^*(-\omega) \tag{11.280}$$

The frequency domain correlation function $S_{A,A}(\omega)$ is real, as can be seen by combining the two relations for the Fourier transforms

$$S_{A,A}(\omega) = S_{A,A}(-\omega) = S_{A,A}^*(\omega) \tag{11.281}$$

Therefore, $S_{A,A}(\omega)$ is a real and symmetric function of ω.

One may define a more fundamental correlation by

$$\tilde{S}_{A,B}(\tau) = \frac{1}{2}\langle[\hat{A}(t+\tau)\hat{B}(t) + \hat{B}(t)\ \hat{A}(t+\tau)]\rangle \tag{11.282}$$

which is related to $S_{A,B}$ via

$$S_{A,B}(\tau) = \tilde{S}_{A,B}(\tau) - \langle\hat{A}\rangle\langle\hat{B}\rangle \tag{11.283}$$

The two correlation functions only differ by a time-independent term.

Inelastic Scattering

The correlation function can be measured via inelastic scattering experiments, when the probe beam couples weakly to the system [110]. In these experiments the probe beam falls incident on the sample and is inelastically scattered by the fluctuations of the system and the scattered particles are detected. If a momentum $\hbar q$ and an energy $\hbar\omega$ is transferred between the particle and the sample, generally, the double-differential scattering cross-section is given by

$$\left(\frac{d^2\sigma}{d\Omega\,d\omega}\right) = \left(\frac{d\sigma}{d\Omega}\right)\frac{S_{A,A}(q,\omega)}{\pi} \tag{11.284}$$

where $\frac{d\sigma}{d\Omega}$ is the differential scattering for a single-particle and $S_{A,A}(\underline{q},\omega)$ is the scattering function.

An Explicit Expression for $\tilde{S}_{A,B}$

The correlation function $\tilde{S}_{A,B}$ is defined as

$$\tilde{S}_{A,B}(\tau) = \frac{1}{2} \langle [\hat{A}(t+\tau)\,\hat{B}(t) + \hat{B}(t)\,\hat{A}(t+\tau)] \rangle \qquad (11.285)$$

It can be evaluated in terms of the eigenstates of $\hat{H}_0$

$$\hat{H}_0|n\rangle = E_n|n\rangle \qquad (11.286)$$

and by noting that the density operator $\hat{\rho}$ is a function of $\hat{H}$ so it is diagonal in the basis of eigenstates. The diagonal matrix is denoted by p_n

$$p_n = \langle n|\hat{\rho}|n\rangle \qquad (11.287)$$

The correlation function can be expressed as

$$\begin{aligned}
\tilde{S}_{A,B}(\tau) &= \frac{1}{2}\,\text{Trace}\,\hat{\rho}[\hat{A}(\tau)\hat{B}(0) + \hat{B}(0)\hat{A}(\tau)] \\
&= \frac{1}{2}\sum_{n,m} p_n \langle n|\hat{A}(\tau)|m\rangle\langle m|\hat{B}(0)|n\rangle \\
&\quad + \frac{1}{2}\sum_{n,m} p_n \langle n|\hat{B}(0)|m\rangle\langle m|\hat{A}(\tau)|n\rangle \qquad (11.288)
\end{aligned}$$

after introducing a complete set of states

$$\hat{I} = \sum_m |m\rangle\langle m| \qquad (11.289)$$

between the pair of operators $\hat{A}$ and $\hat{B}$. Interchanging the summation indices n and m in the second term leads to

$$\begin{aligned}
\tilde{S}_{A,B}(\tau) &= \frac{1}{2}\sum_{n,m}(p_n + p_m)\langle n|\hat{A}(\tau)|m\rangle\langle m|\hat{B}(0)|n\rangle \\
&= \frac{1}{2}\sum_{n,m}(p_n + p_m)\,\exp\left[+i\,\frac{(E_n - E_m)\,\tau}{\hbar}\right] \\
&\quad \times \langle n|\hat{A}|m\rangle\langle m|\hat{B}|n\rangle \qquad (11.290)
\end{aligned}$$

On Fourier transforming the correlation function, one finds that the spectral density $S_{A,B}(\omega)$ is given by

$$\tilde{S}_{A,B}(\omega) = \pi\hbar \sum_{n,m} (p_n + p_m)\, \delta(\hbar\omega - E_n + E_m)$$

$$\times \langle n|\hat{A}|m\rangle\langle m|\hat{B}|n\rangle \tag{11.291}$$

Illustration: Scattering From a Harmonic Oscillator

Consider a beam of particles that scatter from the particle density via a point contact interaction. The Fourier component of the particle density is given by

$$\hat{\rho}_q = \exp[i\underline{q} \cdot \underline{r}] \tag{11.292}$$

where $\underline{q}$ is the momentum transfer and $\underline{r}$ describes the position of a particle. Consider the scattering of the particle from a mass m which is described by a one-dimensional quantum mechanical Harmonic Oscillator

$$\hat{H} = \frac{\hat{p}^2}{2m} + \frac{m\Omega^2}{2}\,\hat{x}^2 \tag{11.293}$$

The position operator of the oscillator can be expressed in terms of creation and annihilation operators

$$\hat{x} = \sqrt{\frac{\hbar}{2m\Omega}}\,(\hat{a}^\dagger + \hat{a}) \tag{11.294}$$

When expressed in terms of the creation and annihilation operators, the Hamiltonian becomes

$$\hat{H} = \hbar\Omega\left(\hat{a}^\dagger \hat{a} + \frac{1}{2}\right) \tag{11.295}$$

The scattering experiment probes the density-density correlation function which can be expressed as

$$S(t) = \langle \exp[-iq\hat{x}(t)]\, \exp[iqx(0)]\rangle \tag{11.296}$$

Note that the correlation function satisfies the sum rule $S(0) = 1$. The exponentials factors can be combined by using the Baker-Campbell-Hausdorf expansion

$$\exp[\hat{A}]\, \exp[\hat{B}] = \exp\left[\hat{A} + \hat{B} + \frac{1}{2}\,[\hat{A}, \hat{B}] + \dots\right] \tag{11.297}$$

The commutator is evaluated as

$$[\hat{x}(t), \hat{x}(0)] = \frac{\hbar}{2m\Omega} [\hat{a}^\dagger \exp[i\Omega t] + \hat{a} \exp[-i\Omega t], \hat{a}^\dagger + \hat{a}]$$

$$= -\frac{\hbar}{2m\Omega} (\exp[i\Omega t] - \exp[-i\Omega t]) \qquad (11.298)$$

which shows that the expansion truncates. Hence, the correlation function reduces to

$$S(t) = \langle \exp[-iq(\hat{x}(t) - x(0))] \rangle \exp\left[-i \frac{\hbar q^2}{2m\Omega} \sin \Omega t \right] \qquad (11.299)$$

However, the Bloch identity states that

$$\langle \exp[-iq(\hat{x}(t) - x(0))] \rangle = \exp\left[-\frac{q^2}{2} \langle (\hat{x}(t) - x(0))^2 \rangle \right] \qquad (11.300)$$

Hence, the expectation value now appears in the exponent.

We shall now prove Bloch's identity. Bloch's identity holds for any operator $\hat{B}$ that is linear in the creation and annihilation operators

$$\hat{B} = c_1 \hat{a}^\dagger + c_2 \hat{a} \qquad (11.301)$$

It can be proved by using the Baker-Campbell-Hausdorf identity to re-write the exponential of $\hat{B}$ as

$$\exp[\hat{B}] = \exp[c_1 \hat{a}^\dagger] \exp[c_2 \hat{a}] \exp\left[\frac{c_1 c_2}{2} [\hat{a}, \hat{a}^\dagger] \right]$$

$$= \exp[c_1 \hat{a}^\dagger] \exp[c_2 \hat{a}] \exp\left[\frac{c_1 c_2}{2} \right] \qquad (11.302)$$

The expectation value of the exponential of $\hat{B}$ is then given by the expansion

$$\langle \exp[\hat{B}] \rangle = \langle \exp[c_1 \hat{a}^\dagger] \exp[c_2 \hat{a}] \rangle \exp\left[\frac{c_1 c_2}{2} \right]$$

$$= \sum_{m,n} \frac{c_1^n}{n!} \frac{c_2^m}{m!} \langle (\hat{a}^\dagger)^n (\hat{a})^m \rangle \exp\left[\frac{c_1 c_2}{2} \right]$$

$$= \sum_{n=0}^{\infty} \frac{c_1^n c_2^n}{n! \, n!} \langle (\hat{a}^\dagger)^n (\hat{a})^n \rangle \exp\left[\frac{c_1 c_2}{2} \right] \qquad (11.303)$$

The sum over m is trivially evaluated since the expectation value vanishes unless $n = m$. On setting $x = \exp[-\beta\hbar\omega]$ the thermal average can be expressed as

$$\langle\exp[\hat{B}]\rangle = (1-x)\sum_{n=0}^{\infty}\frac{c_1^n c_2^n}{n!\,n!}\,x^n\,\frac{\partial^n}{\partial x^n}\sum_{m=0}^{\infty}x^m$$

$$= (1-x)\sum_{n=0}^{\infty}\frac{c_1^n c_2^n}{n!\,n!}\,x^n\,\frac{\partial^n}{\partial x^n}\left(\frac{1}{1-x}\right)$$

$$= \sum_{n=0}^{\infty}\frac{c_1^n c_2^n}{n!}\left(\frac{x}{1-x}\right)^n \tag{11.304}$$

On noting that the average occupation number is given by

$$\bar{n} = \frac{x}{1-x} \tag{11.305}$$

the correlation function reduces to

$$\langle\exp[\hat{B}]\rangle = \sum_{n=0}^{\infty}\frac{(c_1 c_2 \bar{n})^n}{n!}\,\exp\left[\frac{c_1 c_2}{2}\right]$$

$$= \exp\left[\frac{c_1 c_2}{2}\,(2\bar{n}+1)\right]$$

$$= \exp\left[\frac{1}{2}\langle\hat{B}^2\rangle\right] \tag{11.306}$$

which proves "*Bloch's Identity*".

The expectation value in the exponent of the correlation function can then be evaluated as

$$\frac{q^2}{2}\langle(\hat{x}(t)-x(0))^2\rangle = \frac{\hbar q^2}{4m\Omega}$$

$$\times \langle(\hat{a}^\dagger(e^{i\Omega t}-1)+\hat{a}(e^{-i\Omega t}-1))^2\rangle$$

$$= \frac{\hbar q^2}{2m\Omega}\,(2\bar{n}+1)(1-\cos\Omega t)$$

$$= \frac{\hbar q^2}{2m\Omega}\,\coth\frac{\beta\hbar\Omega}{2}\,(1-\cos\Omega t) \tag{11.307}$$

The term in the exponent

$$W = \frac{\hbar q^2}{4m\Omega}\,(2\bar{n}+1)$$

$$= \frac{\hbar q^2}{4m\Omega}\,\coth\frac{\beta\hbar\Omega}{2} \tag{11.308}$$

is known as the Debye-Waller factor. On introducing the Debye-Waller factor, the correlation function takes the form

$$S(t) = \exp[-2W] \ \exp\left[\frac{\hbar q^2}{2m\Omega} \left(\coth\frac{\beta\hbar\Omega}{2} \cos\Omega t - i \sin\Omega t\right)\right]$$

$$= \exp[-2W] \ \exp\left[\frac{\hbar q^2}{2m\Omega} \left(\bar{n}e^{i\Omega t} + (\bar{n}+1) e^{-i\Omega t}\right)\right]$$

$$= \exp[-2W] \ \exp\left[\frac{\hbar q^2}{4m\Omega} \operatorname{cosech}\frac{\beta\hbar\Omega}{2} \left(e^{\frac{\beta\hbar\Omega}{2}} e^{-i\Omega t} + e^{-\frac{\beta\hbar\Omega}{2}} e^{i\Omega t}\right)\right]$$

$$(11.309)$$

One may use the generating function expansion for the modified Bessel Functions of the first kind,

$$\exp\left[\frac{z}{2}\left(s + s^{-1}\right)\right] = \sum_{n=-\infty}^{\infty} s^n I_n(z) \qquad (11.310)$$

to expand the correlation function in terms of its Fourier components

$$S(t) = \exp[-2W] \sum_{n=-\infty}^{\infty} I_n\left(\frac{\hbar q^2}{4m\Omega} \operatorname{cosech}\frac{\beta\hbar\Omega}{2}\right)$$

$$\times \exp\left[-n\frac{\beta\hbar\Omega}{2}\right] \exp[in\Omega t] \qquad (11.311)$$

On Fourier Transforming with respect to time, the result takes the form of a sum of delta functions

$$S(\omega) = 2\pi \ \exp[-2W] \sum_{n=-\infty}^{\infty} I_n\left(\frac{\hbar q^2}{4m\Omega} \operatorname{cosech}\frac{\beta\hbar\Omega}{2}\right)$$

$$\times \exp\left[-\frac{\beta\hbar\omega}{2}\right] \delta(\omega - n\Omega) \qquad (11.312)$$

The inelastic scattering cross-section consists of a set of well separated peaks involving energy transfers of $\omega = \pm n\hbar\Omega$. The $\omega = 0$ peak is the elastic scattering peak. It is diminished by the Debye-Waller factor that represents dephasing due to the thermally activated and zero-point oscillations of the mass. The intensity of the elastic peak decreases exponentially with increasing momentum transfer q.

The $n = -1$ term, represents a process in which the incoming beam transfers an energy of Ω to excite the oscillator by a single quantum. This is known as the Stokes peak. The Stokes process can occur even when the oscillator is in the ground state. The process that corresponds to the $n = 1$ term represents the transfer of one quantum from the excited oscillator to the beam, which can only occur if the oscillator is in a thermally excited state. This peak is known as the anti-Stokes peak. Since the modified Bessel Functions satisfy $I_n(z) = I_{-n}(z)$, the intensities of the Stokes and anti-Stokes processes only differ by a factor of $\exp[\beta\hbar\Omega]$. This factor represents the relative states of occupation of the oscillator $(N(\Omega)+1)/N(\Omega)$. The peaks with other values of $|n|$ involve excitations involving the excitation or de-excitation of the oscillator by n quanta.

Example: The Dynamical Structure Factor

The density is given by the expression

$$\rho(\underline{r}) = \sum_j \delta^3(\underline{r} - \underline{r}_j) \tag{11.313}$$

and its spatial Fourier Transform is given by

$$\rho_{\underline{q}} = \sum_j \exp[-i\underline{q} \cdot \underline{r}_j] \tag{11.314}$$

The "*Dynamical Structure Factor*", $S(\underline{q}, \omega)$ is the Fourier Transform of the density-density correlation function

$$S(\underline{q}, \omega) = 2\pi \sum_{n,m} p_n |\langle n|\hat{\rho}_{\underline{q}}|m\rangle|^2 \delta(\omega - \omega_m + \omega_n)$$

The above expression is not symmetrized.

The "*Static Structure Factor*" can be obtained from the relation

$$\int_{-\infty}^{\infty} \frac{d\omega}{2\pi} S(\underline{q}, \omega) = S(\underline{q}) \tag{11.315}$$

is evaluated as

$$\int_{-\infty}^{\infty} \frac{d\omega}{2\pi} S(\underline{q}, \omega) = \sum_{n,m} p_n |\langle n|\hat{\rho}_{\underline{q}}|m\rangle|^2$$

$$= \sum_n p_n \langle n|\hat{\rho}_{\underline{q}}\hat{\rho}_{\underline{q}}^\dagger|n\rangle$$

$$= \sum_{i,j} \exp[i\underline{q} \cdot (\underline{r}_i - \underline{r}_j)]$$

$$= N \sum_i \exp[i\underline{q} \cdot \underline{r}_i]$$

$$= NS(\underline{q}) \tag{11.316}$$

which is recognized as the structure factor $S(\underline{q})$ that is measured in diffraction experiments.

The dynamic structure factor satisfies the sum rule

$$\int_{-\infty}^{\infty} \frac{d\omega}{2\pi} \, \hbar\omega S(\underline{q}, \omega) = \int_{-\infty}^{\infty} d\omega \sum_{n,m} p_n \hbar\omega \delta(\omega - \omega_m + \omega_n)$$
$$\times |\langle n|\hat{\rho}_{\underline{q}}|m\rangle|^2$$
$$= \int_{-\infty}^{\infty} d\omega \sum_{n,m} p_n \hbar(\omega_m - \omega_n)$$
$$\times \langle n|\hat{\rho}_{\underline{q}}|m\rangle\langle m|\hat{\rho}_{\underline{q}}^{\dagger}|m\rangle$$
$$= \sum_{n,m} p_n \langle n|\hat{\rho}_{\underline{q}}|m\rangle\langle m|[\hat{H}, \hat{\rho}_{\underline{q}}^{\dagger}]|n\rangle$$
$$= \sum_{n} p_n \langle n|\hat{\rho}_{\underline{q}}[\hat{H}, \hat{\rho}_{\underline{q}}^{\dagger}]|n\rangle \tag{11.317}$$

On relating the factor of $\hbar(\omega_m - \omega_n)$ with the factor $\hat{\rho}_q$, one obtains an equivalent result

$$\int_{-\infty}^{\infty} \frac{d\omega}{2\pi} \, \hbar\omega S(\underline{q}, \omega) = -\sum_{n} p_n \langle n|[\hat{H}, \hat{\rho}_{\underline{q}}]\hat{\rho}_{\underline{q}}^{\dagger}|n\rangle \tag{11.318}$$

For inversion invariant systems, the system is invariant under the transformation $\underline{q} \to -\underline{q}$ under which

$$\hat{\rho}_{\underline{q}}^{\dagger} \to \hat{\rho}_{\underline{q}} \tag{11.319}$$

Therefore, on combining the above two results (which is equivalent to symmetrizing the correlation function), one finds the sum rule

$$\int_{-\infty}^{\infty} \frac{d\omega}{2\pi} \, \hbar\omega S(\underline{q}, \omega) = \frac{1}{2} \sum_{n} p_n \langle n|[\hat{\rho}_{\underline{q}}, [\hat{H}, \hat{\rho}_{\underline{q}}^{\dagger}]]|n\rangle \tag{11.320}$$

which involves a double commutator.

Since the Hamiltonian is given by

$$\hat{H} = -\frac{\hbar^2}{2m} \sum_{i} \nabla_i^2 + V(\{\underline{r}_i\}) \tag{11.321}$$

and

$$\rho_{\underline{q}}^{\dagger} = \sum_{j} \exp[i\underline{q} \cdot \underline{r}_j] \tag{11.322}$$

one can evaluate the commutator

$$[\hat{H}, \hat{\rho}_{\underline{q}}^{\dagger}] = -\frac{\hbar^2}{2m} \sum_{i,j} [\underline{\nabla}_i^2, \exp[i\underline{q} \cdot \underline{r}_j]]$$

$$= -\frac{\hbar^2}{2m} \sum_i \exp[i\underline{q} \cdot \underline{r}_i](2i\underline{q} \cdot \underline{\nabla}_i - \underline{q}^2) \qquad (11.323)$$

which is independent of the potential. The double commutator can then be evaluated via

$$[\hat{\rho}_{\underline{q}}, [\hat{H}, \hat{\rho}_{\underline{q}}^{\dagger}]] = -\frac{\hbar^2}{2m} \sum_{i,j} \exp[i\underline{q} \cdot \underline{r}_i]$$

$$\times [\exp[-i\underline{q} \cdot \underline{r}_j], (2i\underline{q} \cdot \underline{\nabla}_i - \underline{q}^2)]$$

$$= N \frac{\hbar^2 \underline{q}^2}{m} \qquad (11.324)$$

which is given by the kinetic energy. Hence, the dynamical structure factor is given by

$$\int_{-\infty}^{\infty} \frac{d\omega}{2\pi} \hbar\omega S(\underline{q}, \omega) = N \frac{\hbar^2 \underline{q}^2}{2m} \qquad (11.325)$$

11.9 The Fluctuation-Dissipation Theorem

The *"Fluctuation-Dissipation Theorem"* [111] is a relationship between $\chi_{A,B}(\omega)$ and $S_{A,B}(\omega)$. Mathematically, this is a relationship between a commutator and an anti-commutator. Physically $S_{A,B}$ is related to the spontaneous fluctuations of the system and $\chi_{A,B}$ is the response to the external perturbation, which leads to the dissipation of energy.

The relation can be derived from the decomposition of

$$\chi_{A,B}(t) = \frac{i}{\hbar} \Theta(t) \, \text{Trace} \, \hat{\rho}[\hat{A}(t), \hat{B}(0)] \qquad (11.326)$$

in terms of the eigenstates of $\hat{H}_0$

$$\hat{H}_0 |n\rangle = E_n |n\rangle \qquad (11.327)$$

As has been shown previously, the response function has the form of a commutator

$$\chi_{A,B}(t) = \frac{i}{\hbar} \Theta(t) \sum_n p_n \langle n| \, [\hat{A}(t), \hat{B}(0)] \, |n\rangle \qquad (11.328)$$

where the probability p_n is given by

$$p_n = \frac{\exp[-\beta E_n]}{Z} \tag{11.329}$$

On inserting the completeness relation

$$\hat{I} = \sum_m |m\rangle\langle m| \tag{11.330}$$

between the pair of operators $\hat{A}$ and $\hat{B}$, the susceptibility reduces to

$$
\begin{aligned}
\chi_{A,B}(t) &= \frac{i}{\hbar}\,\Theta(t) \sum_{n,m} p_n [\langle n|\hat{A}(t)|m\rangle\langle m|\hat{B}(0)|n\rangle \\
&\quad - \langle n|\hat{B}(0)|m\rangle\langle m|\hat{A}(t)|n\rangle] \\
&= \frac{i}{\hbar}\,\Theta(t) \sum_{n,m} p_n \left[\exp\left[+i\,\frac{(E_n - E_m)\,t}{\hbar} \right] \right.\\
&\quad \times \langle n|\hat{A}|m\rangle\langle m|\hat{B}|n\rangle \\
&\quad - \exp\left[-i\,\frac{(E_n - E_m)\,t}{\hbar} \right] \\
&\quad \left. \times \langle n|\hat{B}|m\rangle\langle m|\hat{A}|n\rangle \right]
\end{aligned} \tag{11.331}
$$

On exchanging the summation indices in the second term, one obtains

$$
\begin{aligned}
\chi_{A,B}(t) &= \frac{i}{\hbar}\,\Theta(t) \sum_{n,m} (p_n - p_m)\,\exp\left[+i\frac{(E_n - E_m)\,t}{\hbar} \right] \\
&\quad \times \langle n|\hat{A}|m\rangle\langle m|\hat{B}|n\rangle
\end{aligned} \tag{11.332}
$$

The Fourier transform of the response function is given by

$$
\begin{aligned}
\chi_{A,B}(\omega) &= \sum_{n,m} \frac{(p_n - p_m)}{\hbar\omega - i\hbar\eta - E_n + E_m} \\
&\quad \times \langle n|\hat{A}|m\rangle\langle m|\hat{B}|n\rangle
\end{aligned} \tag{11.333}
$$

In the limit $\eta \to 0$, the term nominally proportional to η reduces to

$$
\begin{aligned}
\Im m\, \chi_{A,B}(\omega) &= \pi \sum_{n,m} (p_n - p_m)\,\delta(\hbar\omega - E_n + E_m) \\
&\quad \times \langle n|\hat{A}|m\rangle\langle m|\hat{B}|n\rangle
\end{aligned} \tag{11.334}
$$

which is denoted as the "imaginary part" of $\Im m\, \chi_{A,B}(\omega)$.

The correlation function $\tilde{S}_{A,B}(\omega)$ has the similar form

$$\tilde{S}_{A,B}(\omega) = \pi\hbar \sum_{n,m} (p_n + p_m)\, \delta(\hbar\omega - E_n + E_m)$$
$$\times \langle n|\hat{A}|m\rangle\langle m|\hat{B}|n\rangle \qquad (11.335)$$

In the Canonical Ensemble, the probability p_n is given by

$$p_n = \frac{\exp[-\beta E_n]}{Z} \qquad (11.336)$$

where Z is the partition function. In the expressions for $\tilde{S}_{A,B}(\omega)$ and $\chi_{A,B}(\omega)$, p_n is multiplied by the Dirac delta function which sets

$$E_m = E_n - \hbar\omega \qquad (11.337)$$

Hence, one may replace p_m by

$$p_m \rightarrow p_n\, \exp[\beta\hbar\omega] \qquad (11.338)$$

in the above expressions. This yields

$$\Im m\, \chi_{A,B}(\omega) = \pi(1 - \exp[\beta\hbar\omega]) \sum_{n,m} p_n\, \delta(\hbar\omega - E_n + E_m)$$
$$\times \langle n|\hat{A}|m\rangle\langle m|\hat{B}|n\rangle \qquad (11.339)$$

and

$$\tilde{S}_{A,B}(\omega) = \pi\hbar(1 + \exp[\beta\hbar\omega]) \sum_{n,m} p_n\, \delta(\hbar\omega - E_n + E_m)$$
$$\times \langle n|\hat{A}|m\rangle\langle m|\hat{B}|n\rangle \qquad (11.340)$$

Comparison of the above two results leads to the Fluctuation-Dissipation Theorem

$$\tilde{S}_{A,B}(\omega) = -\hbar\, \coth\left(\frac{\beta\hbar\omega}{2}\right) \Im m\, \chi_{A,B}(\omega) \qquad (11.341)$$

The temperature and frequency dependent factor can be re-written as

$$\coth\left(\frac{\beta\hbar\omega}{2}\right) = \left(\frac{\exp[\beta\hbar\omega] + 1}{\exp[\beta\hbar\omega] - 1}\right)$$
$$= 1 + 2\, N(\hbar\omega) \qquad (11.342)$$

where $N(\hbar\omega)$ is the Bose-Einstein distribution function. The factor is equal to the energy the harmonic oscillator with frequency ω, $E(\omega)$ in units of the zero-point Energy $\hbar\omega/2$

$$\coth\left(\frac{\beta\hbar\omega}{2}\right) = \frac{2E(\omega)}{\hbar\omega} \qquad (11.343)$$

The content of the Fluctuation-Dissipation Theorem is that the relation between the absorption or emission by a system and the spontaneous fluctuations is a universal function of ω and T.

In the limit $T \to 0$, the relation reduces to

$$\tilde{S}_{A,B}(\omega) = -\hbar \, \Im m \, \chi_{A,B}(\omega) \tag{11.344}$$

which is valid for $\hbar\omega \gg k_B T$.

The high temperature limit, $k_B T \gg \hbar\omega$, the Fluctuation Dissipation Theorem reduces to

$$\tilde{S}_{A,B}(\omega) = -2 \left(\frac{k_B T}{\omega} \right) \Im m \, \chi_{A,B}(\omega) \tag{11.345}$$

The factor $\hbar$ no longer appears in the high temperature relation. This is in accordance with the fact that at high temperatures thermal fluctuations are classical.

11.10   Onsager's Reciprocity Relations

The *"Onsager Reciprocity Relations"* are statements about time-reversal invariance [104]. Specifically they are statements about the symmetry of the response functions $\chi_{A,B}(\omega)$ on interchanging A and B. The reciprocity relations can be deduced from the Fluctuation Dissipation Theorem by using the properties of $\tilde{S}_{A,B}(\omega)$. The homogeneity of time implies that

$$\tilde{S}_{A,B}(t) = \tilde{S}_{B,A}(-t) \tag{11.346}$$

Also, since $S_{A,B}(t)$ is real, one has

$$\tilde{S}^*_{A,B}(t) = \tilde{S}_{A,B}(t) \tag{11.347}$$

The spectral density $\tilde{S}_{A,B}(\omega)$ satisfies

$$\tilde{S}_{A,B}(\omega) = \tilde{S}_{B,A}(-\omega) \tag{11.348}$$

and

$$\tilde{S}^*_{A,B}(\omega) = \tilde{S}_{A,B}(-\omega) \tag{11.349}$$

The Onsager relations can be derived from the Fluctuation-Dissipation Theorem in the form

$$\tilde{S}_{A,B}(\omega) = -\hbar \, \coth \left(\frac{\beta\hbar\omega}{2} \right) \Im m \, \chi_{A,B}(\omega) \tag{11.350}$$

and

$$\tilde{S}_{B,A}(-\omega) = \hbar \, \coth \left(\frac{\beta\hbar\omega}{2} \right) \Im m \, \chi_{B,A}(-\omega) \tag{11.351}$$

which on using Eq. (11.348), leads to

$$\Im m \ \chi_{A,B}(\omega) = -\Im m \ \chi_{B,A}(-\omega) \qquad (11.352)$$

or, equivalently

$$\Im m \ \chi_{A,B}(\omega) = \Im m \ \chi_{B,A}(\omega) \qquad (11.353)$$

Then, on using the Kramers-Kronig relations, one finds the Onsager relations

$$\chi_{A,B}(\omega) = \chi_{B,A}(\omega) \qquad (11.354)$$

or

$$\chi_{A,B}(t) = \chi_{B,A}(t) \qquad (11.355)$$

The proof of this relation depends on time-reversal symmetry. The symmetry is altered if the system is subjected to an external magnetic field or if it is rotating.

Onsager's original formulation of the reciprocity relations was couched in terms of the time-independent kinetic coefficients that describe the rate of change of the system's response to time-independent fields. Consider the time derivative of the response $\Delta A(t)$ produced by the application of an external field $F_B(t)$ which couples to the operator $\hat{B}(t)$

$$\hat{H}_{int}^{B}(t) = -\hat{B}(t) \ F_B(t) \qquad (11.356)$$

then

$$\frac{\partial}{\partial t}\Delta A(t) = \int_{-\infty}^{\infty} dt' \ \frac{\partial}{\partial t}\chi_{A,B}(t - t') \ F_B(t') \qquad (11.357)$$

Likewise, the interaction

$$\hat{H}_{int}^{A}(t) = -\hat{A}(t) \ F_A(t) \qquad (11.358)$$

leads to

$$\frac{\partial}{\partial t}\Delta B(t) = \int_{-\infty}^{\infty} dt' \ \frac{\partial}{\partial t}\chi_{B,A}(t - t') \ F_A(t') \qquad (11.359)$$

The kinetic coefficients may be defined via

$$\gamma_{A,B} = \int_{-\infty}^{\infty} dt' \ \frac{\partial}{\partial t}\chi_{A,B}(t - t') \qquad (11.360)$$

The rate of change of the response to a time-independent applied field F_B can then be expressed in terms of the kinetic coefficients as

$$\frac{\partial}{\partial t}\Delta A(t) = \gamma_{A,B}F_B \qquad (11.361)$$

On transforming $t' \rightarrow -t'$, one has

$$\gamma_{A,B} = \int_{-\infty}^{\infty} dt' \, \frac{\partial}{\partial t} \chi_{A,B}(t + t') \tag{11.362}$$

which ensures that $\gamma_{A,B}$ is time independent and is given by

$$\gamma_{A,B} = \lim_{t \to \infty} \chi_{A,B}(t) \tag{11.363}$$

since $\lim_{t \to -\infty} \chi_{A,B}(t) = 0$ because of causality. Hence, the rate of change $\frac{\partial}{\partial t}\Delta A$ is time independent. Therefore, the kinetic coefficients are time independent and are symmetric

$$\gamma_{A,B} = \gamma_{B,A} \tag{11.364}$$

since

$$\chi_{A,B}(t) = \chi_{B,A}(t) \tag{11.365}$$

Example: Electric Polarization

As an example, if one considers the classical electric polarization

$$\underline{P}(\underline{r}) = q \sum_i \underline{r}_i \, \delta^3(\underline{r} - \underline{r}_i) \tag{11.366}$$

where the sum is over i is the sum over the electron positions, then in the dipole approximation, the coupling to a static electric field $\underline{E}(\underline{r})$ is given by

$$\hat{H} = -\underline{P}(\underline{r}) \cdot \underline{E}(\underline{r}) \tag{11.367}$$

The quantum dipole operator $\underline{\hat{P}}(\underline{r}, t)$ is defined by

$$\underline{\hat{P}}(\underline{r}, t) = q \Psi^\dagger(\underline{r}, t) \, \underline{r} \, \Psi(\underline{r}, t) \tag{11.368}$$

The time-derivative $\frac{\partial}{\partial t}\underline{\hat{P}}(\underline{r}, t)$ is related to the electric current density operator $\hat{\underline{j}}(\underline{r})$ since

$$\frac{\partial \underline{\hat{P}}(\underline{r}, t)}{\partial t} = q \, \frac{i}{\hbar} \, \Psi^\dagger(\underline{r}, t) \, [\underline{r}, \hat{H}] \, \Psi(\underline{r}, t)$$

$$= q \Psi^\dagger(\underline{r}) \, \frac{\partial \hat{\underline{r}}(t)}{\partial t} \, \Psi(\underline{r}) \tag{11.369}$$

where we have transformed from the Schrödinger to the Heisenberg representation. The above expression corresponds to the classical current density $\underline{j}(\underline{r})$ defined by

$$\underline{j}(\underline{r}) = q \sum_i \underline{\dot{r}}_i \, \delta^3(\underline{r} - \underline{r}_i) \tag{11.370}$$

The linear response relation for the α-th Cartesian component of current density is found to take the form

$$j^\alpha(\underline{r}) = \lim_{t \to \infty} \sum_\beta \int d^3\underline{r}' \, \chi_{P,P}^{\alpha,\beta}(\underline{r} - \underline{r}') \, E_\beta(\underline{r}') \qquad (11.371)$$

where $\chi_{P,P}^{\alpha,\beta}(\underline{r} - \underline{r}')$ is the non-local dielectric tensor and in which α and β denote the Cartesian components of $\underline{P}$ and $\underline{E}$. It should be noted that, close to an inhomogeneity, such as a surface, one expects that translational invariance will be lost. In this case, the susceptibility $\chi_{P,P}^{\alpha,\beta}(\underline{r}, \underline{r}')$ should depend separately on $\underline{r}$ and $\underline{r}'$.

11.11   Non-Linear Response

Non-linear Optics uses the generalization of linear response relation by including higher-order terms in the expression for the α-th component of the frequency domain induced polarization $\underline{P}(\omega)$. For convenience of presentation we adopt the long-wavelength approximation, so the linear relation can be expressed as

$$P^\alpha(\omega) = \sum_\beta \chi_{0,0}^{\alpha,\beta}(\omega) \, E_\beta(\omega)$$

$$= \int d\omega_1 \sum_\beta \chi_{0,0}^{\alpha,\beta}(\omega) \, \delta(\omega - \omega_1) \, E_\beta(\omega_1) \qquad (11.372)$$

The lowest-order non-linear response can then be written in terms of an integral over ω_1 and ω_2 of

$$\chi^{\alpha,\beta}(E, B)(\omega : \omega_1, \omega_2)$$

$$= \sum_\gamma \chi_{1,0}^{\alpha,\beta,\gamma}(\omega : \omega_1, \omega_2)\delta(\omega - \omega_1 - \omega_2) \, E_\beta(\omega_1)E_\gamma(\omega_2)$$

$$+ \sum_\gamma \chi_{0,1}^{\alpha,\beta,\gamma}(\omega : \omega_1, \omega_2)\delta(\omega - \omega_1 - \omega_2) \, E_\beta(\omega_1)B_\gamma(\omega_2) \qquad (11.373)$$

where $\chi_{1,0}$ is a real third-rank tensor that is symmetric under the interchange of the last two indices, and ω_1 and ω_2 are the frequencies of the perturbing fields. In the lowest-order non-linear terms, the frequency of the response ω is given by the sum of the frequencies of the perturbing fields $\omega = \omega_1 + \omega_2$. In practice, the above relation must be generalized to include the effects of non-locality.

In the presence of a current density $\underline{j}$ and charge density ρ, Maxwell's equations assume the form

$$\underline{\nabla} \wedge \underline{B} - \frac{1}{c} \frac{\partial \underline{E}}{\partial t} = \frac{4\pi}{c} \underline{j}$$

$$\underline{\nabla} \wedge \underline{E} + \frac{1}{c} \frac{\partial \underline{B}}{\partial t} = 0$$

$$\underline{\nabla} \cdot \underline{E} = 4\pi\rho$$

$$\underline{\nabla} \cdot \underline{B} = 0 \tag{11.374}$$

Maxwell's field equations ensure that the sources $\underline{j}$ and ρ satisfy a continuity equation. Under spatial inversion the position vector transforms as

$$\underline{r} \rightarrow -\underline{r} \tag{11.375}$$

Since all polar vectors transform in the same way, if $\underline{j}$ is a polar vector so that $\underline{j} \rightarrow -\underline{j}$, then one finds that inversion invariance of Maxwell's equation requires that $\underline{E} \rightarrow -\underline{E}$ which identifies $\underline{E}$ as a polar vector. However, inversion invariance also requires that $\underline{B}$ doesn't change sign which identifies $\underline{B}$ as an axial vector. Moreover, Maxwell's equations are also invariant under time-reversal in which time is transformed according to

$$t \rightarrow -t \tag{11.376}$$

and where the current $\underline{j}$ reverses its direction of flow. Time reversal-invariance requires that under the transformation $\underline{E}$ doesn't change sign and $\underline{B}$ reverses its sign.

Electro-Optic Non-linearities

The non-linear electro-optic susceptibilities $\chi_{1,0}$ have generic symmetries that follow from the definition and the symmetries of $\underline{E}$ and $\underline{B}$.

1. Permutation Symmetry: Since β and γ are summed over, the susceptibilities are invariant under the transformation

$$(\beta, \omega_1) \leftrightarrow (\gamma, \omega_2) \tag{11.377}$$

This implies

$$\chi_{1,0}^{\alpha,\beta,\gamma}(\omega : \omega_1, \omega_2) = \chi_{1,0}^{\alpha,\gamma,\beta}(\omega : \omega_2, \omega_1) \tag{11.378}$$

2. Reality of χ in the time domain: Since the electric field and polarization are real quantities in the time domain, their Fourier Transforms

satisfy $P^{\alpha*}(\omega) = P^{\alpha}(-\omega)$ and $E^*_{\beta}(\omega) = E_{\beta}(-\omega)$. Therefore, on taking the complex conjugate of the non-linear relation, one obtains

$$\chi_{1,0}^{\alpha,\beta,\gamma}(\omega : \omega_1, \omega_2)^* = \chi_{1,0}^{\alpha,\beta,\gamma}(-\omega : -\omega_1, -\omega_2) \qquad (11.379)$$

3. Causality: In the time domain, the response must be zero at any time before the perturbing fields are applied. If the response occurs at time t and the fields are applied at times τ_1 and τ_2, the non-linear susceptibility must satisfy

$$\chi_{1,0}^{\alpha,\beta,\gamma}(t - \tau_1, t - \tau_2) = 0 \qquad (11.380)$$

for any $\tau_s > t$. This implies that

$$\chi_{1,0}^{\alpha,\beta,\gamma}(\tau_1, \tau_2) \equiv \Theta(\tau_1) \, \Theta(\tau_2) \, \chi_{1,0}^{\alpha,\beta,\gamma}(\tau_1, \tau_2) \qquad (11.381)$$

which leads to a Kramers-Kronig relation of the form

$$i\pi\chi_{1,0}^{\alpha,\beta,\gamma}(\omega : \omega_1, \omega_2) = \text{Pr} \int d\omega_1' \, \frac{\chi_{1,0}^{\alpha,\beta,\gamma}(\omega' : \omega_1', \omega_2)}{\omega_1 - \omega_1'} \qquad (11.382)$$

where $\omega' = \omega_1' + \omega_2$.

If the crystal structure of the sample has an inversion center, then $\chi_{1,0}$ must change sign under inversion, since P^{α}, E^{β} and E^{γ} all reverse sign under inversion. Hence, invariance under inversion requires that

$$\chi_{1,0} = -\chi_{1,0} \qquad (11.383)$$

so $\chi_{1,0}$ must vanish. On the other hand if the system does not posses inversion symmetry, then $\chi_{0,1}$ can be non-zero.

Example: Sum and Difference Frequency Generation

Simple use of second-order perturbation in the electromagnetic interaction

$$\hat{H}_{int}(t) = -\frac{1}{c} \, \hat{j}(t) \cdot \underline{A}(t) \qquad (11.384)$$

leads to an expression for the current density $\langle \hat{j}^{\alpha}(t) \rangle$ generated by to two-photon processes as

$$\langle \hat{j}^{\alpha}(t) \rangle = -\frac{1}{c^2 \hbar^2} \int_{-\infty}^{t} d\tau_1 \int_{-\infty}^{\tau_1} d\tau_2 \sum_{\beta,\gamma} A_{\beta}(\tau_1) \, A_{\gamma}(\tau_2)$$

$$\times \langle |[[\hat{j}^{\alpha}(t), \hat{j}^{\beta}(\tau_1)], \, \hat{j}^{\gamma}(\tau_2)]|\rangle \qquad (11.385)$$

which, due to the homogeneity of time, can be expressed as

$$\langle j^\alpha(t)\rangle = \int_{-\infty}^{\infty} d\tau_1 \int_{-\infty}^{\infty} d\tau_2 \sum_{\beta,\gamma} \mathcal{R}^{\alpha;\beta,\gamma}(t-\tau_1, t-\tau_2)\, A_\beta(\tau_1)\, A_\gamma(\tau_2)$$

(11.386)

where we have introduced the second-order conductivity response function as

$$\mathcal{R}^{\alpha;\beta,\gamma}(-\tau_1, -\tau_2) = -\frac{1}{2c^2\hbar^2}\, \Theta(-\tau_1)\, \Theta(-\tau_2)$$
$$\times \langle |[[\hat{j}^\alpha(0), \hat{j}^\beta(\tau_1)], \hat{j}^\gamma(\tau_2)]]|\rangle$$

(11.387)

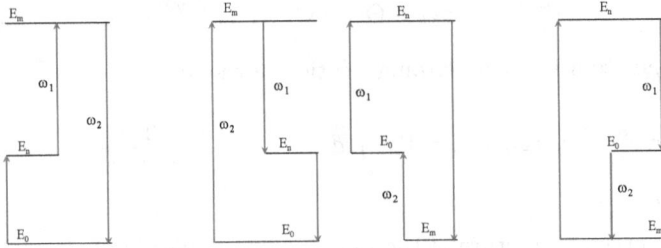

Fig. 11.11   The resonance structure for generation of radiation with the sum and difference frequencies.

The vector potentials are real and can be expressed in terms of their complex Fourier components through

$$A_\sigma(\tau) = \int_{-\infty}^{\infty} \frac{d\omega_s}{2\pi}\, \exp[i\omega_s\tau]\, A_\sigma(\omega_s)$$

(11.388)

and so has both positive and negative frequency components. Likewise, the Fourier transform of the induced current density is defined by

$$\langle \hat{j}^\alpha(\omega)\rangle = \int_{-\infty}^{\infty} dt\, \exp[-i\omega t]\langle \hat{j}^\alpha(t)\rangle$$

(11.389)

The frequency-dependent induced current density is then given by

$$\langle \hat{j}^\alpha(\omega)\rangle = \int_{-\infty}^{\infty} d\omega_1 \int_{-\infty}^{\infty} d\omega_2\, \delta(\omega - \omega_1 - \omega_2)$$
$$\times \left[\sum_{\beta,\gamma} \frac{1}{2\pi} \mathcal{R}^{\alpha;\beta,\gamma}(\omega_1, \omega_2)\, A_\beta(\omega_1)\, A_\gamma(\omega_2) \right]$$

(11.390)

which leads to the non-linear conductivity tensor for frequency sum and difference processes being written as

$$\sigma^{\alpha;\beta,\gamma}(\omega:\omega_1,\omega_2) = -\frac{c^2}{2\pi}\frac{\mathcal{R}^{\alpha;\beta,\gamma}(\omega_1,\omega_2)}{\omega_1\omega_2} \tag{11.391}$$

where, at $T = 0$, $\mathcal{R}^{\alpha;\beta,\gamma}(\tau_1,\tau_2)$ is expressed in terms of the unperturbed energy eigenstates by

$$\mathcal{R}^{\alpha;\beta,\gamma}(\omega_1,\omega_2) = \sum_{n,m}\left[\frac{\langle 0|\hat{j}^\alpha|n\rangle\langle n|\hat{j}^\beta|m\rangle\langle m|\hat{j}^\gamma|0\rangle}{2c^2\hbar(\omega_1+\omega_{n,m})\hbar(\omega_2+\omega_{m,0})}\right.$$
$$+\frac{\langle 0|\hat{j}^\gamma|m\rangle\langle m|\hat{j}^\beta|n\rangle\langle n|\hat{j}^\alpha|0\rangle}{2c^2\hbar(\omega_1-\omega_{n,m})\hbar(\omega_2-\omega_{m,0})}$$
$$-\frac{\langle 0|\hat{j}^\beta|n\rangle\langle n|\hat{j}^\alpha|m\rangle\langle m|\hat{j}^\gamma|0\rangle}{2c^2\hbar(\omega_1+\omega_{0,n})\hbar(\omega_2+\omega_{m,0})}$$
$$\left.-\frac{\langle 0|\hat{j}^\gamma|m\rangle\langle m|\hat{j}^\alpha|n\rangle\langle n|\hat{j}^\beta|0\rangle}{2c^2\hbar(\omega_1-\omega_{0,n})\hbar(\omega_2-\omega_{m,0})}\right] \tag{11.392}$$

The above expression should be symmetrized w.r.t. the interchange of the pairs of indices (β,ω_1) and (γ,ω_2), since these are dummy indices which are summed and integrated over. For the case of resonant transitions, the lifetimes of the states should be included by replacing the frequencies $\omega_{n,m}$ with $\omega_{n,m}+\frac{i}{2\tau_n}+\frac{i}{2\tau_m}$.

The Kramers-Kronig relations for the second-order response come in various forms, but can be written compactly as

$$\sigma^{\alpha;\beta,\gamma}(\omega_T;\omega_1+p_1\omega,\omega_2+p_2\omega) = \frac{i}{\pi}\,\mathrm{Pr}\int_{-\infty}^{\infty}d\omega'$$
$$\times\left[\frac{\sigma^{\alpha;\beta,\gamma}(\omega_T';\omega_1+p_1\omega',\omega_2+p_2\omega')}{\omega'-\omega}\right] \tag{11.393}$$

where

$$\omega_T = \omega_1+\omega_2+(p_1+p_2)\,\omega \tag{11.394}$$

etc., in which $p_i \geq 0$ and at least one p_i must be non-zero. The above relations can be obtained from the response function

$$\mathcal{R}^{\alpha;\beta,\gamma}(\tau_1,\tau_2) = -\frac{1}{2c^2\hbar^2}\,\Theta(\tau_1)\,\Theta(\tau_2)$$
$$\times\langle|[[\hat{j}^\alpha(0),\hat{j}^\beta(-\tau_1)],\hat{j}^\gamma(-\tau_2)]|\rangle \tag{11.395}$$

and its Fourier Transform

$$\mathcal{R}^{\alpha;\beta,\gamma}(\omega_1,\omega_2) = \int_{-\infty}^{\infty} d\tau_1 \int_{-\infty}^{\infty} d\tau_2 \ \mathcal{R}^{\alpha;\beta,\gamma}(\tau_1,\tau_2) \ \exp[-i(\omega_1\tau_1 + \omega_2\tau_2)]$$

(11.396)

The Kramers-Kronig relations can then be obtained by noting that the frequency-domain response is regular and analytic in the lower-halves of the ω_1 and ω_2 complex planes. Consider the integral

$$0 = \int_C d\omega' \ \frac{\mathcal{R}^{\alpha;\beta,\gamma}(\omega_1 + p_1\omega',\omega_2 + p_2\omega')}{\omega' - \omega}$$

(11.397)

which runs over the real axis and excludes the pole at $\omega' = \omega$ and is completed by a semi-circular contour at infinity in the lower-half complex plane. Since the integrand is analytic and enclosed no poles, the contour integral is zero. Expressing the frequency dependent response in terms of the time dependent response, results in the equation

$$0 = \int_{-\infty}^{\infty} d\tau_1 \int_{-\infty}^{\infty} d\tau_2 \ \mathcal{R}^{\alpha;\beta,\gamma}(\tau_1,\tau_2) \ \exp[-i(\omega_1\tau_1 + \omega_2\tau_2)]$$
$$\times \int_C d\omega' \ \frac{\exp[-i\omega'(p_1\tau_1 + p_2\tau_2)]}{\omega' - \omega}$$

(11.398)

where the integration over ω' is solely contained in the last factor. For the contour at infinity to vanish, one requires that $(p_1\tau_1 + p_2\tau_2) \geq 0$, and causality requires that $\tau_1 \geq 0$ and $\tau_2 \geq 0$. Thus, requiring $p_1\tau_1 + p_2\tau_2$ to be positive, imposes the restriction that the p_i cannot be negative and that at least one p_i must be positive. With this condition, the factor involving the integral over ω' reduces to the principal part of the integral along the real axis and the half circular contour that excludes the pole at $\omega = \omega'$

$$\mathrm{Pr} \int_{-\infty}^{\infty} d\omega' \ \frac{\exp[-i\omega'(p_1\tau_1 + p_2\tau_2)]}{\omega' - \omega} + i\pi\delta(\omega - \omega')$$

(11.399)

On using the above expression for the factor and transforming back from the time to the frequency domain, one finds the desired Kramers-Kronig relation

$$\frac{i}{\pi} \mathrm{Pr} \int_{-\infty}^{\infty} d\omega' \ \frac{\mathcal{R}^{\alpha;\beta,\gamma}(\omega_1 + p_1\omega',\omega_2 + p_2\omega')}{\omega' - \omega} = \mathcal{R}^{\alpha;\beta,\gamma}(\omega_1 + p_1\omega,\omega_2 + p_2\omega)$$

(11.400)

For a quasi-static electric field E_γ with $\omega_2 \sim 0$, the non-vanishing of $\chi_{1,0}$ leads to the linear electro-optic or Pockels effect. However, if the perturbing field $\underline{E}$ is monochromatic with frequency $\omega_1 = \omega_2 = \omega$, the frequency-dependent response will occur at twice the frequency, 2ω, which results in the generation of light with frequency 2ω. This phenomenon is known as second-harmonic generation. It should be noted that the imaginary part of $\chi_{1,0}^{\alpha,\beta,\gamma}(\omega : \omega_1, \omega_2)$ does not represents a dissipative process, unless $\omega_2 = 0$, but simply gives rise to a phase difference. Since the surface of a solid breaks inversion symmetry, second-harmonic generation should occur at surfaces even if the bulk crystal is centro-symmetric. Furthermore, as the dielectric constant is in general frequency-dependent, second harmonic generation produces light with wavevector q which differs from twice the wavevector of the incident light. Since the electric field with frequency 2ω builds up from a zero value at the surface to a finite value within the solid, the difference in wavelengths results in a phase mismatch. Due to the phase mismatch, the effect of the finite wavelengths of light are observable.

Magneto-Optic Non-linearities

When static magnetic and electric fields are applied, they may be considered as having zero frequencies, so that the response can be derived by using the non-linear formalism. The susceptibility must be Hermitean so

$$\chi^{\alpha,\beta}(\underline{E}, \underline{B}) = \chi^{\beta,\alpha}(\underline{E}, \underline{B})^* \qquad (11.401)$$

The susceptibility must also be symmetric under time-reversal. Since time-reversal is achieved by complex conjugation and reversing $\underline{B}$, the susceptibility must also satisfy

$$\chi^{\alpha,\beta}(\underline{E}, \underline{B}) = \chi^{\alpha,\beta}(\underline{E}, -\underline{B})^* \qquad (11.402)$$

as follows from Onsager's reciprocity relation. Therefore, the non-linear term in the generalized susceptibility $\chi_{0,1}$ must be imaginary and antisymmetric under the interchange of α and β. Hence

$$\chi_{0,1}^{\alpha,\beta,\gamma} B_\gamma = i\kappa \sum_\gamma \epsilon^{\alpha,\beta,\gamma} B_\gamma \qquad (11.403)$$

For an electromagnetic field propagating along the direction of a static magnetic field B ($\omega_2 = 0$), one has

$$- \underline{k} \wedge (\underline{k} \wedge \underline{E}) = \left(\frac{\omega}{c}\right)^2 \epsilon \underline{E} \qquad (11.404)$$

where

$$\epsilon = 1 + 4\pi\chi \qquad (11.405)$$

and, in Cartesian coordinates where both $\underline{k}$ and $\underline{B}$ are along the z-direction, χ has the form

$$\chi = \begin{pmatrix} \chi_0 & i\kappa B & 0 \\ -i\kappa B & \chi_0 & 0 \\ 0 & 0 & \chi_0 \end{pmatrix} \qquad (11.406)$$

If the sample is ferromagnetic or ferrimagnetic, the static magnetic field $\underline{B}$ is usually replaced by the static magnetization, $\underline{M}$. For circularly polarized light, the contribution to the non-linear term in the dielectric tensor will lead to wave-vectors that depend on the sense of the circular polarization. The electric field $\underline{E}$ of a linearly polarized beam can be decomposed into a left and right circular polarized components

$$\underline{E} = \underline{E}_L + \underline{E}_R \qquad (11.407)$$

where

$$\underline{E}_L = \frac{E}{\sqrt{2}} \exp[+i(\omega t - \underline{k} \cdot \underline{r})] \begin{pmatrix} 1 \\ i \\ 0 \end{pmatrix} \qquad (11.408)$$

and

$$\underline{E}_L = \frac{E}{\sqrt{2}} \exp[+i(\omega t - \underline{k} \cdot \underline{r})] \begin{pmatrix} 1 \\ -i \\ 0 \end{pmatrix} \qquad (11.409)$$

Both circularly polarized components of E separately satisfy the wave equation, but the off-diagonal (non-linear) terms in the susceptibility have the effect of introducing different dielectric constants for the left and right circularly polarized components

$$\epsilon_L = 1 + 4\pi(\chi_0 - \kappa B)$$
$$\epsilon_R = 1 + 4\pi(\chi_0 + \kappa B) \qquad (11.410)$$

The different dielectric constants results in the different circularly polarized components of the beam having different wave vectors and so the components will accumulate a phase difference after traversing the solid. However, the amplitudes of the components remain unchanged and so the amplitudes remain equal. Hence, when the circularly polarized components are recombined, the resulting field will be linearly polarized, but the phase difference between the components produces a polarization with a rotated direction. The angle of the rotation depends on κ, the magnetic field B, the frequency ω and the path length. This is the linear magneto-optic or Faraday effect. If the optical path length in the solid is denoted by L, the polarization is rotated by an amount ΦL, where Φ is the Faraday rotation constant.

11.12 Problems

Problem 11.1

Consider a toy system comprised of one fermion in a two-level system which is coupled to an indefinite number of bosons with frequency ω_0. The system is described by the Hamiltonian

$$\hat{H} = \hat{H}_0 + \hat{H}_{int}$$

where

$$\hat{H}_0 = \sum_{\alpha=1}^{2} \epsilon_\alpha c_\alpha^\dagger c_\alpha + \hbar \omega_0 a^\dagger a$$

$$\hat{H}_{int} = g \sum_{\alpha > \beta} (c_\alpha^\dagger c_\beta + c_\beta^\dagger c_\alpha)(a^\dagger + a)$$

where the c operators obey anti-commutation relations and the a operators obey commutation relations, and the fermionic indices α have values of either one or two.

(i) Is the Hamiltonian Hermitean? Are the total numbers of fermions and bosons conserved?

(ii) Assume that at $t \to -\infty$ the system is described by a density operator

$$\hat{\rho} = \frac{1}{Z} \exp[-\beta \hat{H}_0]$$

Using time-dependent perturbation theory, assuming that the interaction g is turned on adiabatically at $t \to -\infty$, i.e. $g(t) = g \exp[\eta t]$ where η is a small positive constant, determine how the combined fermion and boson number states $c_\alpha^\dagger |0\rangle |n\rangle$ evolve with time. Hence, find the lowest-order corrections to the density matrix. By taking the expectation value of the time-evolved density operator and then taking a time-derivative, determine an expression for the rates at which the fermion makes a transition from state 1 to state 2 and vice versa, if the boson is initially in an eigenstate of the number operator $a^\dagger a$ with eigenvalue n and if the transition is energetically allowed.

(iii) Take a partial Trace over the boson number eigenstates and show that the "Principle of Detailed Balance" holds. Compare your results for the rates of emission and absorption of a boson with Einstein's A and B coefficients.

(iv) Assume that the bosons are initially in a coherent state. How do the fermionic transition rates change?

Problem 11.2

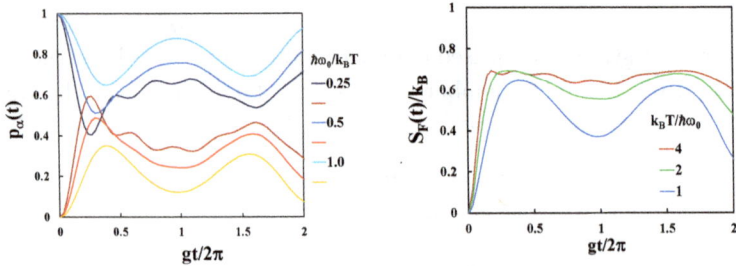

Fig. 11.12   The time-dependent probabilities, $p_1(t)$ and $p_2(t)$, of the reduced density operator for a two-state system coupled to a bosonic bath after a measurement at $t = 0$. The measurement uniquely specifies a state of the two-state subsystem. However, since the total system undergoes unitary evolution, the subsystem's probabilities and the von Neumann entropy $S_F(t)$ evolve due to the coupling. Thermal excitation processes are seen to reduce the coherence of the dynamics.

Consider a toy system comprised of one fermion in a two-level system which is coupled to an indefinite number of bosons of frequency ω_0. The system is described by the Hamiltonian

$$\hat{H} = \hat{H}_0 + \hat{H}_{int}$$

where

$$\hat{H}_0 = \sum_{\alpha=1}^{2} \epsilon_\alpha c_\alpha^\dagger c_\alpha + \hbar\omega_0 a^\dagger a$$

$$\hat{H}_{int} = g \sum_{\alpha>\beta} (c_\alpha^\dagger c_\beta + c_\beta^\dagger c_\alpha)(a^\dagger + a)$$

where α and β have values of 1 or 2 and the c operators obey anti-commutation relations and the a operators obey commutation relations. We shall assume that

$$\epsilon_2 = \epsilon_1 + \hbar\omega_0$$

so that the eigenstates of $\hat{H}_0$ are doubly-degenerate except for the ground state of $\hat{H}_0$ which is non-degenerate.

(i) Using the interaction representation and by considering low-order time-dependent perturbation theory and neglecting non-resonant terms, show that, to order $g^2/\hbar\omega_0$, the excited states and excitation energies coincide with those of set of a quasi-two-level systems

$$|\pm, n\rangle \approx \frac{1}{\sqrt{2}} \left(c_1^\dagger |0\rangle |n\rangle \pm c_2^\dagger |0\rangle |n-1\rangle \right)$$

$$E^\pm(n) \approx \epsilon_\alpha + n\hbar\omega_0 \pm g\sqrt{n}$$

where $|0\rangle$ is the fermionic vacuum state and $|n\rangle$ are eigenstates of the boson number operator. The approximate excited states are entangled. The above approximation is only exact for $g = 0$. However, for non-zero g, the approximation is assumed to be reasonable for large t only for pedagogical purposes!

(ii) Determine an expression for the density operator $\hat{\rho}$ for the total system in equilibrium at temperature T. If necessary, you may evaluate your result as the first few terms of a power series in g, e.g.

$$Z = \exp[-\beta\epsilon_\alpha] \left[1 + 2 \sum_{n=1}^{\infty} \cosh \beta g\sqrt{n} \, \exp[-\beta\hbar\omega_0 n] \right]$$

$$\approx \exp[-\beta\epsilon_\alpha] \left[1 + \frac{2}{\exp[\beta\hbar\omega_0] - 1} \left(1 + \frac{(\beta g)^2/2!}{1 - \exp[-\beta\hbar\omega_0]} + \cdots \right) \right]$$

Find the entropy S.

A measurement is made on the fermionic system at time $t = 0$ and it is found that the fermion is in state 1.

(iii) Determine the form of the density operator $\hat{\rho}(t)$ appropriate to the system for times $t > 0$. Evaluate the entropy of the total system.

(iv) Show that the reduced density operator for the fermionic system, $\hat{\rho}_F(t)$, can be written as

$$\hat{\rho}_F(t) = [p_1(t)|1\rangle\langle 1| + p_2(t)|2\rangle\langle 2|]$$

where

$$p_1(t) = \frac{\exp[-\beta\epsilon_\alpha]}{Z_F} \sum_{n=0}^{\infty} \exp[-\beta\hbar\omega_0 n] \, \cosh \beta g\sqrt{n} \, \cos^2\left(\frac{gt\sqrt{n}}{\hbar}\right)$$

$$p_2(t) = \frac{\exp[-\beta\epsilon_\alpha]}{Z_F} \sum_{n=1}^{\infty} \exp[-\beta\hbar\omega_0 n] \, \cosh \beta g\sqrt{n} \, \sin^2\left(\frac{gt\sqrt{n}}{\hbar}\right)$$

Note that the time-dependence originates from the coupling to the bosons. Furthermore, if the above expressions are considered as being valid for large times, the frequencies are incommensurate so that the probabilities would not recur.

(v) Evaluate the von-Neumann entropy of the fermionic subsystem, S_F. Plot your result.

Problem 11.3

(i) Determine, quantum mechanically, an expression for a system described by a Hamiltonian $\hat{H}_0$, with energy eigenvalues E_n and energy eigenstates $|n\rangle$, the rate of transitions from the state $|n\rangle$ to the state $|m\rangle$ due to the perturbation

$$\hat{H}_{int}(t) = -\hat{A}(t) \ F(t)$$

where $F(t)$ has the form

$$F(t) = F_0 \ \cos(\omega t + \varphi)$$

(ii) Using this result show that the net power absorbed by the system is given by

$$-\frac{1}{2} \ \omega \ \Im m \ \chi_{A,A}(\omega) \ F_0^2$$

Problem 11.4

Consider the simple harmonic oscillator with the Hamiltonian

$$\hat{H} = \frac{\hat{p}^2}{2m} + \frac{m}{2} \ \omega_0^2 \hat{x}^2$$

(i) Use the harmonic oscillator raising and lowering operators to solve the Heisenberg equations of motion for $\hat{x}(t)$ and show that

$$\hat{x}(t) = \hat{x} \ \cos \omega_0 t + \frac{\hat{p}}{m\omega_0} \ \sin \omega_0 t$$

(ii) Hence, determine $\chi_{x,x}(t)$ in the time domain and show that in the frequency domain

$$\Im m \ \chi_{x,x}(\omega - i\eta) = -\frac{\pi}{2m\omega_0} \ [\delta(\omega - \omega_0) - \delta(\omega + \omega_0)]$$

so $\omega \ \Im m \ \chi_{x,x}(\omega) < 0$.

Problem 11.5

Consider a spin one-half particle described by the Hamiltonian

$$\hat{H}_0 = -g\mu_B B_0 S^z$$

which is subject to the time-dependent perturbation

$$\hat{H}_{int} = -g\mu_B \underline{B}(t) \cdot \underline{S}$$

(i) Find a general expression for the $T = 0$ susceptibility tensor $\chi^{\alpha,\beta}(\omega)$.

(ii) Determine the only non-zero components of the tensor.

Problem 11.6

Consider a classical system which initially has a Canonical Distribution function $\rho(\{q_i, p_j\})$ which is perturbed by an external field $F(t)$. The Hamiltonian is described by

$$H = H_0 - BF(t)$$

The change in the distribution function induced by the field is denoted by $\delta\rho$

$$\rho = \rho_0 + \delta\rho$$

(i) Show that the total derivative of the distribution function is determined by the equation

$$\frac{d\delta\rho}{dt} = \beta\rho_0 \frac{dB}{dt} F(t)$$

(ii) Show that the response of the system ΔA induced by a static field $F(t) = F_s$, is given by

$$\delta A(t) = \beta \int d\Gamma \rho_0 \, ABF_s$$

(iii) Show that, for an applied field in the form of a delta function pulse

$$F(t) = F_0 \delta(t)$$

the system's subsequent relaxation is given by

$$\delta A(t) = \beta \int d\Gamma \, \rho_0 \, A(t) \frac{dB}{dt} F_0$$

Problem 11.7

Derive the Kramers-Kronig Relations directly from the definition of the principle part of an integral

$$\mathrm{Pr} \int_{-\infty}^{\infty} dz \, \frac{\chi_{A,B}(z)}{\omega - z} \, dz = \int_{-\infty}^{\omega-\eta} dz \, \frac{\chi_{A,B}(z)}{\omega - z} + \int_{\omega+\eta}^{\infty} dz \, \frac{\chi_{A,B}(z)}{\omega - z}$$

by using Cauchy's Theorem and by completing the contour by segments in the lower-half complex plane.

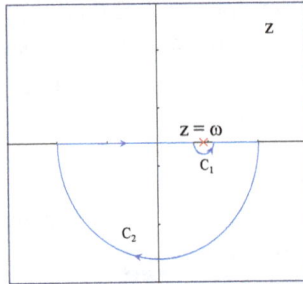

Fig. 11.13   An alternate method of determining the Kramers-Kronig relation starts from the definition of the principle part of the integral. The principle part of the integral is evaluated by contour integration in which the contour is completed by the segments C_1 and C_2.

Problem 11.8

Demonstrate the validity of the Kramers-Kronig relations by direct calculation of the real and imaginary parts of the susceptibility $\chi_{A,B}(\omega)$.

Problem 11.9

Consider the function

$$\Im m\, \chi(\omega) = -\chi_0 \frac{\omega\Gamma}{\omega^2 + \Gamma^2}$$

where $\Gamma > 0$. This expression is usually invoked to describe relaxational dynamics in metals.

(i) Using the Kramers-Kronig relation, find the real part of $\chi(\omega)$ on the real axis.

(ii) Show that if an analytic function $\chi(z)$ is determined from Cauchy's Theorem for z in the lower-half complex plane, one has

$$\chi(z) = -\chi_0 \frac{i\Gamma}{z - i\Gamma} \qquad \text{if} \quad \Im m\, z < 0$$

The relaxational approximation to the equation of motion for the current density is given by

$$\frac{\partial \underline{j}}{\partial t} + \frac{1}{\tau} \underline{j} = \frac{\rho e^2}{m} \underline{E}(t)$$

(iii) Find the response function.

Problem 11.10

Calculate $\chi(z)$ using the Kramers-Kronig relations for the three spectral functions $\Im m \, \chi(z)$ given below

$$-\frac{1}{1+z^2}$$

$$-\frac{\pi}{2} \, \Theta(1-z)\Theta(1+z)$$

$$-2\sqrt{1-z^2} \, \Theta(1-z)\Theta(1+z)$$

where $\Theta(z)$ is the Heaviside step function.

Problem 11.11

Prove the sum rule for the mass

$$\frac{1}{m} = -\frac{1}{\pi} \int_{-\infty}^{\infty} d\omega \, \Im m \, \chi_{x,x}(\omega) \, \omega$$

Verify, by direct integration, that

$$\frac{1}{m} = \frac{1}{\pi} \int_{-\infty}^{\infty} d\omega \left[\frac{\gamma\omega^2}{m[(\omega^2 - \omega_0^2)^2 + \gamma^2\omega^2]} \right]$$

Problem 11.12

Consider the response function $\chi_{A,A}(\omega)$ and its n-th order moment

$$\int_{-\infty}^{\infty} \frac{d\omega}{2\pi} \, \omega^n \chi_{A,A}(\omega)$$

(i) Find a sum rule for the n-th order moment in terms of the real and imaginary part of susceptibility.
(ii) Using the symmetry properties of $\chi_{A,A}(\omega)$, express the moments in terms of an integration over positive values of ω. Simplify the resulting expression for odd and even values of n.

Problem 11.13

The dielectric constant $\epsilon(\underline{q},\omega)$ is related to the electron density-density susceptibility, $\chi(\underline{q},\omega)$ by

$$\epsilon(\underline{q},\omega)^{-1} = 1 + \frac{4\pi e^2}{q^2} \chi(\underline{q},\omega)$$

where, at $T = 0$, the susceptibility for non-interacting electrons, $\chi_0(q, \omega)$, is given by

$$\chi_0(q, \omega) = \sum_n |\langle 0|\hat{\rho}_q|n\rangle|^2 \left[\frac{1}{\hbar\omega - E_n + E_0} - \frac{1}{\hbar\omega - E_0 + E_n} \right]$$

(i) Simplify the large frequency limit of the expression for $\chi(\omega)$, by using a sum rule derivable from considerations of the dynamic structure factor

$$\sum_n (E_n - E_0)|\langle 0|\hat{\rho}_q^\dagger|n\rangle|^2 = \rho_0 \frac{\hbar^2 q^2}{2m}$$

where ρ_0 is the electron density. Frequently, the effect of electron-electron interactions is approximately taken into account by replacing the susceptibility of non-interacting electrons $\chi_0(q, \omega)$ by $\chi_0(q, \omega)/\epsilon(q, \omega)$ which assumes that the induced charge density is produced by the external and induced potential. That is the induced charge density is assumed to be the result of the screened potential and not the external potential.

(ii) Find a high frequency approximation for $\epsilon(q, \omega)$ and identify the plasmon frequency ω_p.

(iii) If the Coulomb interaction were short-ranged, would electromagnetic radiation be reflected by a metal?

Problem 11.14

Consider a system has been perturbed by a static external field λ where the Hamiltonian is given by

$$\hat{H}(\lambda) = \hat{H}_0 - \lambda\hat{A}$$

The full thermodynamic response of the system can be expressed as

$$-\left(\frac{\partial F}{\partial \lambda} \right) = \frac{k_B T}{Z} \frac{\partial}{\partial \lambda} [\text{Trace } \exp[-\beta\hat{H}(\lambda)]]$$

where F is the Helmholtz Free-Energy.

(i) Use the Baker-Campbell-Hausdorf relation in the form

$$\frac{\partial}{\partial \lambda} \exp[\hat{\kappa}(\lambda)] = \left[\frac{\partial\hat{\kappa}}{\partial \lambda} + \frac{1}{2!} \left[\hat{\kappa}(\lambda), \frac{\partial\hat{\kappa}}{\partial \lambda} \right] + \frac{1}{3!} \left[\hat{\kappa}, \left[\hat{\kappa}, \frac{\partial\hat{\kappa}}{\partial \lambda} \right] \right] + \dots \right] \exp[\hat{\kappa}(\lambda)]$$

to show that, even if $\hat{A}$ and $\hat{H}(\lambda)$ do not commute, the full response is given by

$$-\left(\frac{\partial F}{\partial \lambda}\right) = \langle \hat{A} \rangle$$

where the expectation value is evaluated with the Hamiltonian $\hat{H}(\lambda)$.

Problem 11.15

Consider a system that has been perturbed by an external field. The Hamiltonian is given by

$$\hat{H}(\lambda) = \hat{H}_0 - \lambda \hat{A}$$

The operator $\hat{A}$ does not commute with $\hat{H}$. The isothermal susceptibility, χ_T, is defined by

$$\chi_T = -\left(\frac{\partial^2 F}{\partial \lambda^2}\right)$$

Use the identity

$$\frac{\partial}{\partial \lambda} \exp[-\beta \hat{H}] = \int_0^\beta d\tau \; \exp[-\tau \hat{H}] \; \hat{A} \; \exp[+\tau \hat{H}] \; \exp[-\beta \hat{H}]$$

to obtain an exact expression for the isothermal susceptibility.

Problem 11.16

(i) Examine the linear response relation for the case, in which either the operator $\hat{B}$ of the perturbing interaction $-F(t)\hat{B}$ or the observed quantity $\hat{A}$ commutes with the Hamiltonian. Show that the method does not lead to a finite response since, when the interaction is turned on adiabatically, the system remains in the ground state.

(ii) In this case and for a static field F, one should use the isothermal susceptibility in which the interaction is included in the density operator $\hat{\rho}$. By using the interaction representation, show that to first order in the applied field

$$\Delta A = \int_0^\beta d\tau [\langle \hat{B}(\tau) \; \hat{A}(0) \rangle - \langle \hat{B}(\tau) \rangle \langle \hat{A}(0) \rangle] \, F$$

where τ is an imaginary time and where

$$\hat{B}(\tau) = \exp[-\tau \hat{H}_0] \; \hat{B} \; \exp[+\tau \hat{H}_0]$$

Hence, the susceptibility measures the correlation between the fluctuations of $\hat{A}$ and $\hat{B}$.

Problem 11.17

Consider an isolated spin of magnitude S governed by the Zeeman Hamiltonian

$$\hat{H} = -\mu_B \hat{S}_z B^z$$

(i) Show that S^z is a constant of motion.

(ii) By using the interaction representation, show that the isothermal susceptibility is given by

$$\chi^{z,z} = \mu_B^2 \, \beta[\langle S_z^2 \rangle - \langle S_z \rangle^2]$$

(iii) Hence, derive the expression for the Curie susceptibility

$$\chi^{z,z} = \mu_B^2 \, \frac{S(S+1)}{3k_B T}$$

This is to be contrasted with the adiabatic linear response result for a conserved quantity. The adiabaticity leads to the conserved magnetization retaining initial value, before the magnetic field is turned on, so $S^z(t) = S^z(-\infty) = 0$.

Problem 11.18

The a.c. current flowing in an electrical circuit as a response to a frequency dependent voltage is described by

$$V(\omega) = Z(\omega) \, I(\omega)$$

where $Z(\omega)$ is the impedance. The resistive part of the impedance is given by

$$\Re\left[\frac{1}{Z(\omega)}\right] = \frac{R(\omega)}{|Z(\omega)|^2}$$

where $R(\omega)$ is the resistive part of the impedance. Since the voltage couples to the charge Q, not $I = \dot{Q}$, the response function is related to the impedance by

$$\chi_{I,Q}(\omega) = \frac{1}{Z(\omega)}$$

The current-current response function is related to the current-charge response function via

$$\chi_{I,I}(\tau) = -\frac{\partial}{\partial \tau} \, \chi_{I,Q}(\tau)$$

(i) Use the Fluctuation-Dissipation Theorem [112] to determine an expression for the high-temperature limit of the current fluctuations.

Problem 11.19

Derive the high-temperature Curie Law for a spin system which interacts with a static external field through the Zeeman interaction

$$\hat{H}_{int} = -\underline{\mu} \cdot \underline{B}$$

by noting that in the limit $B \to 0$

$$S_{\underline{\mu},\underline{\mu}}(t) = \langle \underline{\mu} \cdot \underline{\mu} \rangle$$

and then using the Fluctuation-Dissipation Theorem.

Bibliography

[1] W. Nernst, "*Über die Beziehung zwischen Wärmeentwicklung und maximaler Arbeit bei kondensierten Systemen*", Ber. Kgl. Pr. Akad. Wiss. **52**, 933-940, (1906).

[2] F. Simon and F. Lange, "*Zur Frage die Entropie amorpher Substanzen*", Zeit. für Physik, **38**, 227-236 (1926).

[3] J.D. Bekenstein, "*Black Holes and Entropy*", Physical Review D, **7**, 2333-2346 (1973).

[4] E. Hopf, "*Complete Transitivity and the Ergodic Principle*", Proc. Nat. Acad. Sci. **18**, 204-209 (1932).

[5] L. Boltzmann, "*Einige allgemeninen Sätze über das Wärmegleichgewicht*", Wien Ber. **63**, 670-711 (1871).

[6] G.D. Birkhoff, "*Proof of the Ergodic Theorem*", Proc. Natl. Acad. Sci. **17**, 656-660 (1930).

[7] J. von Neumann, "*Proof of the Quasi-ergodic Hypothesis*", Proc. Natl. Acad. Sci. USA, **18**, 70–82, (1932).

[8] J. von Neumann, "*Physical Applications of the Ergodic Hypothesis*", Proc. Natl. Acad. Sci. **18**, 263-266, (1932).

[9] H. Poincaré, "*Sur les courbes définies par une équation différentielle*", Oeuvres, 1, Paris, (1892).

[10] A.N. Kolmogorov, "*On Conservation of Conditionally Periodic Motions for a Small Change in Hamilton's Function*", Dokl. Akad. Nauk SSSR **98**, 527-530, (1954).

[11] V.I. Arnol'd, "*Proof of a Theorem of A.N. Kolmogorov on the Preservation of Conditionally Periodic Motions under a Small Perturbation of the Hamiltonian*", Uspehki Mat. Nauk **18**, 13-40, (1963).

[12] J. Moser, "*On Invariant Curves of Area-Preserving Mappings of an Annulus*", Nachr. Akad. Wiss. Göttingen Math.-Phys. Kl. II, 1-20, (1962).

[13] M. Falcioni, U.M.B. Marconi and A. Vulpiani, "*Ergodic Properties of High-Dimensional Symplectic Maps*", Physical Review A, **44**, 2263-2270 (1991).

[14] E. Schrödinger, "*Statistical Thermodynamics*", Courier Corporation (1989).

[15] C.E. Shannon, "*A Mathematical Theory of Communication*", Bell System Tech. J., **27**, 379-423, 623-656, (1948).

[16] G-L. L. de Comte de Buffon, de l'Acad. Roy. des. Sciences (1733), 43-45; naturelle, generale et particuliere Supplement 4, p. 46. (1777).

[17] H. Weyl, *"Uber die Gibbs'sche Erscheinung und verwandte Konvergenzphanomene"*, Rendiconti del Circolo Matematico di Palermo, **330**, 377-407, (1910).

[18] A.-T. Petit and P.-L. Dulong, *"Recherches sur quelques points importants de la Théorie de la Chaleur"*, Annales de Chimie et de Physique **10**, 395-413 (1819).

[19] A. Einstein, *"The Planck Theory of Radiation and the Specific Heat"*, Ann. Physik **22**, 180-190 (1906).

[20] E. Ising, *"Beitrag zur Theorie des Ferromagnetismus"*, Z. Phys. **31**, 253-258, (1925).

[21] P. Debye, *"Theory of Specific Warmth"*, Annalen der Physik, **39**, 789-839 (1912).

[22] H.B.G. Casimir, *"On the Attraction between Two Perfectly Conducting Plates"*, Proc. Kon. Nederland. Akad. Wetensch. B **51**, 793 (1948).

[23] J. Stefan, *"Über die Beziehung zwischen der Wärmestrahlung und der Temperatur"*, Sitzungsberichte der mathematisch-naturwissenschaftlichen Classe der kaiserlichen Akademie der Wissenschaften, Wien, Bd. **79**, 391-428, (1879).

[24] L. Boltzmann, *"Ableitung des Stefan'schen Gesetzes, betreffend die Abhängigkeit der Wärmestrahlung von der Temperatur aus der electromagnetischen Lichttheori"*, Ann. Physik, **22**, 291-294 (1884).

[25] M. Planck, *"Über das Gesetz der Energieverteilung im Normalspectrum"* Annalen der Physik, **3**, 553-563 (1901).

[26] Lord Rayleigh, *"Remarks on the Complete Theory of Radiation"*, Phil. Mag. **49**, 539-540 (1900).

[27] J.H. Jeans, *"On the Partition of Energy between Matter and Æther"*, Phil. Mag. **10**, 91-98, (1905).

[28] D.J. Fixen, E.S. Cheng, J.M. Gales, J.C. Mather, R.A. Shafer and E.L. Wright, *"The Cosmic Ray Background from the full COBE-FIRAS data set"*, The Astrophysical Journal, **473**, 576-587 (1996).

[29] P.A.M. Dirac, *"The Quantum Theory of the Emission and Absorption of Radiation"*, Proc. Roy. Soc. A, **114**, 243-265, (1927).

[30] A. Einstein, *"Strahlungs-Emission und Absorption nach der Quantentheorie"*. Verhandlungen der Deutschen Physikalischen Gesellschaft, **18**, 318-323, (1916).

[31] E. Schrödinger, *"Der stetige Ubergang von der Mikro-zur Makromechanik"*, Naturwissenschaften **14**, 664-666, (1926).

[32] R.J. Glauber, *"Coherent and Incoherent States of Radiation Field"*, Phys. Rev. **131**, 2766-2788, (1963).

[33] A.M. Gleason, *"Measures on the Closed Subspaces of a Hilbert Space"*, Journal of Mathematics and Mechanics **6**, 885-893, (1957).

[34] H. Araki and Elliott H. Lieb, *"Entropy Inequalities"*, Communications in Mathematical Physics, **18**, 160-170 (1970).

[35] Y. Aharonov, P.G. Bergmann and J.L. Lebowitz, *"Time Symmetry in Quantum Process of Measurement"*, Phys. Rev. B. **134**, 1410-1416 (1960).

[36] J. von Neumann, *"Proof of the Ergodic Theorem and the H-Theorem in the new Mechanics"*, Zeit. fur Physik, **57**, 30-70 (1929).

[37] C. Radin, *"Approach to Equilibrium in a Simple Model"*, J. Math. Phys. **11**, 2945-2955 (1971).

[38] G. Emsch and C. Radin, *"Relaxation of Local Thermal Deviations from Equilibrium"*, J. Math. Phys. **12**, 2043-2046 (1971).

[39] J.M. Deutsch, *"Quantum Statistical Mechanics in a Closed System"*, Phys. Rev. A, **43**, 2046-2049, (1991).

[40] M. Srednicki, *"Chaos and Quantum Thermalization"*, Phys. Rev. E, **50**, 888-901 (1994).

[41] A.M.L. Messiah and O.W. Greenberg, *"Symmetrization Postulate and its Experimental Foundations"*, Phys. Rev. **136**, B 248-267 (1964).

[42] J.M. Leinaas and J. Myrheim, *"On the Theory of Identical Particles"*, Il Nuovo Cimento B **37**, 1-23 (1977).

[43] M. Fierz, *"Über die relativistische Theorie kräftefreier Teilchen mit beliebigem Spin"*, Helvetica Physica Acta, **12**, 3-37, (1939).

[44] W. Pauli, *"The Connection Between Spin and Statistics"*, Phys. Rev. **58**, 716-722 (1940).

[45] R. Hanbury Brown, and R.Q. Twiss, *"A Test of a New Type of Stellar Interferometer on Sirius"*, Nature, **178**, 1046-1048 (1956).

[46] P.W. Anderson, B.I. Halperin and C.M. Varma, *"Anomalous Low-Temperature Properties of Glasses and Spin Glasses"*, Phil. Mag. **25**, 1-9 (1972).

[47] A. Sommerfeld, *"Zur Elektronentheorie der Metalle auf Grund der Fermischen Statistik"*, Zeitschrift für Physik **47**, 1-3, (1928).

[48] J.C. Slater, *"A Simplification of the Hartree-Fock Method"*, Phys. Rev. **81**, 385-390 (1951).

[49] T. Koopmans, *"Uber die Zuordnung von Wellenfunktionen und Eigenwerten zu den einzelnen Elektronen eines Atoms"*, Physica, **1**, 104-113 (1934).

[50] P. Hohenberg and W. Kohn, *"Inhomogeneous Electron Gas"*, Phys. Rev. **136**, B 864-871 (1964).

[51] W. Kohn and L.J. Sham *"Self-Consistent Equations including Exchange and Correlation Effects"*, Phys. Rev. **140**, A 113, 561-567, (1966).

[52] E.C. Stoner, *"The Limiting Density of White Dwarf Star"*, Phil. Mag. **41**, 63-70 (1929): E.C. Stoner and F. Tyler, *"A Note on Condensed Stars"*, Phil. Mag. **11**, 986-995 (1931): E.C. Stoner, *"The Minimum Pressure of a Degenerate Electron Gas "*Monthly Notices of the Royal Astronomical Society, **92**, 651-661, E.C. Stoner, *"Upper Limits for Densities and Temperatures in Stars"*, Monthly Notices of the Royal Astronomical Society, **92**, 662-667 (1931).

[53] S. Chandrasekhar, *"The Highly Collapsed Configurations of a Stellar Mass"*, Monthly Notices of the Royal Astronomical Society, **91**, 456-466 (1930), *"The Density of White Dwarf Stars"* Phil. Mag. **11**, 592-596 (1931),

"*The Maximum Mass of Ideal White Dwarfs*", Astrophysical Journal, **74**, 81-82 (1931).

[54] S. Bose, "*Plancks Gesetz und Lichtquantenhypothese*", Zeit. für Physik, **26**, 178–181 (1924).

[55] A. Einstein, "*Quantentheorie des einatomigen idealen Gases*", Sitzungs-berichte der Preussischen Akademie der Wissenschaften, **1**, 3-14, (1925).

[56] J.E. Robinson, "*Note on Bose-Einstein Integral Functions*", Phys. Rev. **83**, 678-679 (1951).

[57] O. Penrose, "*On the Quantum Mechanics of He II*", Phil. Mag. **42**, 1373-1377 (1951).

[58] O. Penrose and L. Onsager, "*Bose-Einstein Condensation and Liquid Helium*", Phys. Rev. **104**, 576-584 (1956).

[59] C.N. Yang, "*Concept of Off-Diagonal Long-Ranged Order and Quantum Phases of Liquid Helium and Superconductors*", Rev. Mod. Phys. **34**, 694-704 (1962).

[60] P.W. Anderson, "*Considerations of Flow in Superfluid Helium*", Rev. Mod. Phys. **38**, 298-310 (1966).

[61] J.C. Slater and J.G. Kirkwood, "*The van der Waals Forces in Gases*", Phys. Rev. **37**, 682-697 (1931).

[62] P. Kapitza, "*Viscosity of Liquid Helium below the λ-point*", Nature, **141**, 74-74, (1938).

[63] J.F. Allen and A.D. Misener, "*Flow of Liquid Helium II*", Nature, **141**, 75-75, (1938).

[64] E.I. Andronikashvili, "*Direct Observation of Two Kinds of Motion for Helium II*", Zh. Eksp. Theor. Fiz., **10**, 201 (1946).

[65] F. London, "*The Lambda Phenomenon of Liquid Helium and the Bose-Einstein Degeneracy*", Nature, **141**, 643-644 (1938).

[66] N.N. Bogoliubov, "*On the Theory of Superfluidity*", J. Phys. USSR, **11**, 23 (1947).

[67] M. Girardeau and R. Arnowitt, "*Theory of Many-Boson Systems: Pair Theory*", Phys. Rev. **113**, 755-761 (1959).

[68] E.F. Sears and E.C. Svensson, "*Pair Correlations and the Condenstate Fraction in Superfluid He-4*", Phys. Rev. Lett. **43**, 2009-2011, (1979).

[69] L.D. Landau, "*The Theory of Superfluidity of He II*", J. Phys. USSR, **5**, 71, (1941).

[70] R.P. Feynman and M. Cohen, "*Energy Spectrum of the Excitations in Liquid Helium*", Physical Review, **102**, 1189-1204, (1956).

[71] L. Onsager, "*Statistical Hydrodynamics*", Nuovo Cim. **6** Suppl. 2 pp. 279-287 (1949).

[72] H.E. Hall and W.F. Vinen, "*The Rotation of Liquid He II*", Proc. Roy. Soc. London, Series A, **238**, 204-215 (1956): **238**, 215-234 (1956).

[73] M.H. Anderson, J.R. Ensher, M.R. Matthews, C.E. Wieman and E.A. Cornell, "*Observation of Bose-Einstein Condensation in a Dilute Atomic Vapor*", Science, **269**, 198-201, (1995).

[74] L. Onsager, "*Crystal Statistics: A Two-Dimensional Model with an Order-Disorder Transition*", Phys. Rev. **65**, 117-149 (1944).

[75] C.N. Yang, "*The Spontaneous Magnetization of a Two-Dimensional Ising Model*", Phys. Rev. **85**, 808 (1952).

[76] C.N. Yang and T.D. Lee, "*Statistical Theory of Equations of State and Phase Transitions: 1. Theory of Condensation*", Phys. Rev. **87**, 404-409, (1952).

[77] T.D. Lee and C.N. Yang, "*Statistical Theory of Equations of State and Phase Transitions: 2. Lattice Gas and Ising Model*", Phys. Rev. **87**, 410-419, (1952).

[78] H.N.V. Temperley, "*Statistical Mechanics of the Two-Dimensional Assembly*", Proc. Roy. Soc. London, Ser. A **202**, 202-207 (1950).

[79] G.H. Wannier, "*Antiferromagnetism: The Triangular Ising Net*", Phys. Rev. B **79**, 357-364 (1950).

[80] L. Onsager, unpublished: Nuovo Cimento 6, Suppl. p. 261 (1949).

[81] P. Ehrenfest, "*Phase Conversions in a General and Enhanced Sense, Classified According to the Specific Singularities of the Thermodynamic Potential*", Proc. of the Koniklijke Akademie Van Wetenschappen Te Amsterdam, **36**, 153-157 (1933).

[82] B. Widom, "*Equations of State in Neighborhood of Critical Point*", J. Chem. Phys. **143**, 3898-3905, (1965).

[83] L.P. Kadanoff, "*Scaling Laws for Ising Models Near T_c.*", Physics, **2**, 263-272 (1966).

[84] J. Goldstone, "*Field Theories with Superconductor Solutions*", Nuovo Cimento, **19**, 154-164, (1961).

[85] P.W. Anderson, "*An Approximate Quantum Theory of the Antiferromagnetic Ground State*", Phys. Rev. **86**, 694-701, (1952).

[86] P.W. Anderson, "*Random-Phase Approximation in the Theory of Superconductivity*", Phys. Rev. **112**, 1900-1916, (1958).

[87] P.W. Anderson, "*Plasmons, Gauge Invariance, and Mass*", Phys. Rev. **130**, 439-442 (1963).

[88] P. Higgs, "*Broken Symmetries and the Masses of Gauge Bosons*", Phys. Rev. Lett. **13**, 508-509, (1964).

[89] G. Guralnik, C.R. Hagen and T.W.B. Kibble, "*Global Conservation Laws and Massless Particles*", Phys. Rev. Lett. **13**, 585-587 (1964).

[90] F. Englert and R. Brout, "*Broken Symmetry and the Mass of Gauge Vector Mesons*", Phys. Rev. Lett. **13**, 321-323 (1964).

[91] H.A. Kramers and G.H. Wannier, "*Statistics of the Two-Dimensional Ferromagnet*", Phys. Rev. **60**, 252-262, (1941), Phys. Rev. **60**, 263-276 (1941).

[92] E.K. Riedel and F.J. Wegner, "*Effective Critical and Tricritical Exponents*", Phys. Rev. B, **9**, 294-315 (1974).

[93] K.G. Wilson, "*Renormalization Group and Critical Phenomena. I. Renormalization Group and the Kadanoff Scaling Picture*", Physical Review B, **4**, 3174-3183 (1971).

[94] K.G. Wilson and M.E. Fisher, "*Critical Exponents in 3.99 Dimensions*", Phys. Rev. Lett. **28**, 240-243 (1972).

[95] Th. Niemeijer and J.H.J. van Leeuwen, "*Wilson Theory For Two-Dimensional Ising Spin Systems*", Physica, **71**, 17-40 (1974).

[96] D. Mermin and H. Wagner, *"Absence of Ferromagnetism and Antiferromagnetism in One or Two-Dimensional Isotropic Heisenberg Models"*, Phys. Rev. Lett. **17**, 1133-1136 (1966).

[97] A.M. Polyakov, *"Interaction of Goldstone Particles in Two-Dimensions: Application to Ferromagnets and Massive Yang-Mills Fields"*, Physics Letters B, **59**, 79-81, (1975).

[98] J.M. Kosterlitz and D.J. Thouless, *"Long Range Order and Metastability in Two-Dimensional Solids and Superfluids"*, J. Phys. C: Solid State Phys. **5**, L124-L126 (1972).

[99] J.M. Kosterlitz and D.J. Thouless, *"Ordering, Metastability and Phase Transitions in 2 Dimensional Systems"*, J. Phys. C: Solid State Phys. **6**, 1181-1203 (1973).

[100] E.T. Jaynes and F.W. Cummings, *"Comparison of Quantum and Semi-classical Radiation Theories with Application to the Beam Maser"*, Proc. IEEE. **51** 89-109 (1963).

[101] J.W. Strutt, *"Some General Theorems Relating to Vibrations"*, Proc. London Math. Soc. **4**, 357-368 (1873).

[102] F.W. Cummings, *"Stimulated Emission of Radiation in a Single-Mode"*, Phys. Rev. **140**, A 1051-1056 (1965).

[103] R. Kubo, *"Statistical Mechanical Theory of Irreversible Processes I: General Theory and Applications to Magnetic and Conduction Problem"*, J. Phys. Soc. Jpn. **12**, 570-586 (1957).

[104] L. Onsager, *"Reciprocal Relations in Irreversible Processes. I"*, Phys. Rev. **37**, 405-426 (1931).

[105] R. de L. Kronig, *"On the Theory of Dispersion of X-rays"*, J. Opt. Soc. Am, **12**, 547-557 (1926).

[106] H.A. Kramers, *"La Diffusion de la Lumiere par les Atomes"*, Atti Cong. Intern. Fisica, Como, **2**, 545-557 (1927).

[107] R. Jackiw and C. Rebbi, *"Solitons with Fermion Number 1/2"*, Phys. Rev. D, **13**, 3398-3409 (1976).

[108] M.V. Berry, *"Quantal Phase-Factors Accompanying Adiabatic Changes"*, Proc. Roy. Soc. London, A **392**, 45-57 (1984).

[109] D.J. Thouless, M. Kohmoto, M.P. Nightingale and M. den Nijs, *"Quantized Hall Conductance in a Two-Dimensional Periodic Potential"*, Phys. Rev. Lett. **49**, 405-408 (1982).

[110] L. Van Hove, *"Correlations in Space and Time and Born Approximation Scattering in Systems of Interacting Particles"*, Phys. Rev. **95**, 249-262 (1954).

[111] H.B. Callen and T.A. Welton, *"Irreversibility and Generalized Noise"*, Phys. Rev. **83**, 34-40 (1951).

[112] H. Nyquist, *"Thermal Agitation and Electric Charge in Conductors"*, Phys. Rev. **32**, 110-113 (1928).

Index